Biofuels, Solar and Wind as Renewable Energy Systems

David Pimentel
Editor

Biofuels, Solar and Wind as Renewable Energy Systems

Benefits and Risks

 Springer

Editor
Dr. David Pimentel
Cornell University
College of Agriculture and Life Sciences
5126 Comstock Hall
Ithaca, NY 15850
USA
dp18@cornell.edu

ISBN: 978-1-4020-8653-3   e-ISBN: 978-1-4020-8654-0

Library of Congress Control Number: 2008931413

Cover Images
Dutch windmill (Courtesy of Schoen Photography, www.schoenphotography.com)
© Schoen Photography, Colorado, USA
Wind turbine © 2008 JupiterImages Corporation

Printed on acid-free paper

9 8 7 6 5 4 3 2 1

springer.com

Preface

The petroleum age began about 150 years ago. Easily available energy has supported major advances in agriculture, industry, transportation, and indeed many diverse activities valued by humans. Now world petroleum and natural gas supplies have peaked and their supplies will slowly decline over the next 40–50 years until depleted. Although small amounts of petroleum and natural gas will remain underground, it will be energetically and economically impossible to extract. In the United States, coal supplies could be available for as long as 40–50 years, depending on how rapidly coal is utilized as a replacement for petroleum and natural gas.

Having been comfortable with the security provided by fossil energy, especially petroleum and natural gas, we appear to be slow to recognize the energy crisis in the U.S. and world. Serious energy conservation and research on viable renewable energy technologies are needed. Several renewable energy technologies already exist, but sound research is needed to improve their effectiveness and economics. Most of the renewable energy technologies are influenced by geographic location and face problems of intermittent energy supply and storage. Most renewable technologies require extensive land; a few researchers have even suggested that one-half of all land biomass could be harvested in order to supply the U.S. with 30% of its liquid fuel!

Some optimistic investigations of renewable energy have failed to recognize that only 0.1% of the solar energy is captured annually in the U.S. by all the green plants, including agriculture, forestry, and grasslands. Photovoltaics can collect about 200 times more solar energy per year than green plants. The green plants took more than 700 million years to collect and then be stored as the concentrated energy found in petroleum, natural gas, and coal supplies.

This book examines various renewable energy technologies and reports on their potential to supply the United States and other nations with needed energy. Some chapters examine several renewable energy technologies and their potential to replace fossil fuel, while others focus on one specific technology and its potential, as well as its limitations. In this volume, the aim of the contributors is to share their analyses as a basis for more research in renewable energy technologies. Basic to all the renewable energy technologies is that they attempt to minimize damage to the environment that supports all life.

Several of the chapters reflect the current lack of agreement in the field, as pressure mounts to explore and develop potential energy alternatives. The reader will notice considerable variability in the energy inputs and potential energy outputs in some of the studies. This is evidence of the complexity of assessing the large number of energy inputs that go into production of a biofuel crop and the extraction of its useful energy. As research continues, we will discover if current analyses of renewable energy technologies have adequately estimated energy requirements, outputs and environmental consequences. Hopefully, this research will help guide energy policy makers toward the most viable choices and away from energy costly missteps, as we collectively encounter energy descent.

The authors of each of these chapters have done a superb job in presenting the most up to date perspective of various renewable energy technologies in a highly readable fashion.

NY, USA   D. Pimentel

Acknowledgements

I wish to express my sincere gratitude to the Cornell Association of Professors Emeriti for the partial support of our research through the Albert Podell Grant Program. In addition, I wish to thank Anne Wilson for her valuable assistance in the preparation of our book.

Contents

About our Authors

Bruno J. R. Alves graduated in Agronomy from UFRRJ (Federal University of Rio de Janeiro) in 1987. He concluded the Master's Degree (1992) and PhD (1996) in Agronomy at the same University, specializing in techniques for the study of the dynamics of N in the soil and for the quantification of biological N2 fixation in legume and non-legume species. He is a researcher at the Brazilian Corporation of Agricultural Research (Embrapa) and a teacher-advisor in the post-graduation program in Agronomy at UFRRJ. His research covers the quantification of soil C sequestration, greenhouse gas emissions, and energy balance for biomass production.

Robert Boddey graduated in 1975 from Leeds University, UK, with a BSc in Agricultural Chemistry. He earned a PhD at the University of the West Indies (Trinidad) in 1980, with a thesis on biological nitrogen fixation (BNF) associated with wetland rice. He then moved to the Soil Microbiology Centre of the Brazilian Corporation for Agricultural Research (Embrapa Agrobiologia) in Seropédica, Rio de Janeiro. There he developed various techniques, including those using the stable isotope 15N, to quantify inputs of BNF to grasses and cereals. His team also works on the impact of BNF on N dynamics in various agroecosystems. Boddey has published almost 100 papers in international journals, and over 60 chapters in books and conference proceedings.

Marcelo E. Dias de Oliveira graduated in 1997 as an Agronomic Engineer at University of São Paulo-Brazil, working as an undergrad student with GIS and Remote Sensing. In 2001 he started his Master's Degree at Washington State University – Richland - USA, concluding his work in 2004. During this time he did research on hazardous materials at the Hanford site, and developed his thesis on energy balance, carbon dioxide emissions and environmental impacts of ethanol production. Currently he works as consultant for an environmental company in Brazil and is about to start his PhD studies.

Jeroen Dooper holds a Bachelor degree in Ecological Material Technology and is currently completing a Master's degree in Sustainable Development, Energy and

Resources at Utrecht University in the Netherlands. His research is focused on energy conversion technologies, energy policies, greenhouse gas mitigation and life cycle assessments. In 2007, he began collaboration with REAP-Canada to pursue research on various bioenergy conversion technologies and their efficiencies. His previous work experience includes environmental education at Econsultancy, and environmental consulting, environmental product development, and optimizing processing efficiency with the Avans University of Professional Education.

Andrew Ferguson, after National Service flying training in Canada, joined BEA (later British Airways). In the 1960s, he tried to see if it was possible to persuade his flying colleagues that there was an environmental crisis ahead due to growing population. Finding that it was impossible to locate even one person to acquiesce in this proposition, he waited for more propitious times to engage in wider efforts. In the 1990s, he became a member of the Optimum Population Trust (UK), started by the late David Willey, and since 2002 has been editor of the biannual OPT Journal.

Thomas Edgar Gangwer has a B.S. in Chemistry from Lebanon Valley College and a PhD in Physical Chemistry from the University of Notre Dame. His career spans basic research, applied research, regulatory compliance, and technology implementation in the chemistry, engineering, licensing, and environmental arenas. His materials processing experience includes chemical, radioactive, hazardous, sanitary, and byproduct feed stocks and wastes. For both commercial and government (NRC, DOE, DOD) clients, he has performed methodology development, process modeling, process evaluation, and project/program management covering diverse treatment, transport, pollution prevention, and disposal activities. In addition to client reports, he has over 40 scientific/technical literature publications.

Mario Giampietro is an ICREA Research Professor at ICTA – Institute of Science and Technology for the Environment - Universitat Autonoma Barcelona, SPAIN. He has been visiting scholar at: Cornell University; Wageningen University; European Commission Joint Research Center, Ispra; Wisconsin University Madison; Penn State University, Arizona State University. His research addresses technical issues associated with "Science for Governance" such as Multi-Scale Integrated Analysis of Societal and Ecosystem Metabolism, Participatory Integrated Assessment of Scenarios and Technological Changes. He has published more than 150 papers and chapters of books and is the author of: "Multi-Scale Integrated Analysis of Agro-ecosystems" 2003 (CRC press), and co-author of "The Jevons Paradox" 2008 (Earthscan).

Tiziano Gomiero holds a degree in Nature Science from Padua University and a PhD in Environmental Science from the Universitat Autonoma de Barcelona, Spain. His work concerns integrated analysis of farming systems (which takes into consideration the environmental, social and economic domains) and rural development, including theoretical and epistemological issues, modeling, practical applications (he worked in Italian and South-East Asia contexts). He is currently a Professor of Ecology and Agroecology at Padua University.

Nathan John Hagens is currently at the Gund Institute of Ecological Economics at the University of Vermont studying the impacts that a decline in liquid fuels will have on planetary ecosystems and society. On the supply side, he is exploring net-energy comparisons of the primary alternate fuel sources to oil: coal, wind, nuclear and biomass. Prior to coming to the Gund Institute, Nate developed trading algorithms for commodity systems and was President of Sanctuary Asset Management, Managing Director of Pension Research Institute, and Vice President at the investment firms Salomon Brothers and Lehman Brothers. He holds an undergraduate degree from the University of Wisconsin and an MBA with honors from the University of Chicago.

Charles A. Hall is a Systems Ecologist who received his PhD from Howard T. Odum. Dr. Hall is the author of seven books and more than 200 scholarly articles. He is best known for his development of the concept of EROI, or energy return on investment, which is an examination of how organisms, including humans, invest energy into obtaining additional energy to increase biotic or social fitness. He has applied these approaches to fish migrations, carbon balance, tropical land use change and petroleum extraction, in both natural and human-dominated ecosystems. He is developing a new field, biophysical economics, as a supplement or alternative to conventional neoclassical economics.

Edwin Kessler graduated from Columbia College in 1950 and received the Sc.D. in Meteorology from MIT in 1957. From 1954-1961 he specialized in radar meteorology with the Air Force Cambridge Research Laboratories in Massachusetts, and from 1961-1964 he was Director of the Atmospheric Physics Division, Travelers Research Center in Hartford, Connecticut. From 1964 until retirement in 1986, he was Director of the National Oceanic and Atmospheric Administration's National Severe Storms Laboratory in Norman, Oklahoma. In 1989, he received the Cleveland Abbe award of the American Meteorological Society. He has been Chair of Common Cause Oklahoma and is now Vice-Chair. He manages 350 acres of pastures with woodlands and stream in central Oklahoma.

Doug Koplow is the founder of Earth Track in Cambridge, MA (www.earthtrack. net), an organization focused on making the scope and cost of environmentally harmful subsidies more visible. The author of Biofuels - At What Cost? Government support for ethanol and biodiesel in the United States (Global Subsidies Initiative, Geneva: 2006 and 2007), Doug has worked on natural resource subsidy issues for nearly twenty years. He holds an MBA from the Harvard Graduate School of Business Administration and a BA in economics from Wesleyan University.

Claudia Ho Lem is currently a Project Manager for REAP-Canada's International Development and Bioenergy Programs. A rural development specialist with over 10 years of experience in environmental project management, Ms. Ho Lem holds a B.Sc. in Environmental Science specializing in Biology and Chemistry from the University of Calgary. She has worked on bioenergy, climate change and agro-ecological development in China, the Philippines, Cuba, West Africa and Canada, supporting farming communities in increasing their self-sufficiency through participatory assessments, training and research. Her experience has given her an integrated understanding of the social, economic, biological, ecosystem and health impacts of agricultural development.

Kozo Mayumi, a former student of Georgescu-Roegen, has been working in the field of energy analysis, ecological economics and complex hierarchy theory. He is a professor at the University of Tokushima and an editorial board member of Ecological Economics, Organization and Environment, and International Journal of Transdisciplinary Research. He is the author of The Origins of Ecological Economics: The Bioeconomics of Georgescu-Roegen, published by Routledge in 2001, and The Jevons Paradox and The Myth of Resource Efficiency Improvements from Earthscan in 2008. Together with Dr. Mario Giampietro and three other researchers, Mayumi started a biennial international workshop, ("Advances in Energy Studies,") in 1998.

Kenneth Mulder obtained his PhD in Ecological Economics from the Gund Institute for Ecological Economics at the University of Vermont. His research is multidisciplinary, applying systems modeling and analysis to problems in ecology, economics and agriculture. He is particularly interested in the development of meaningful indicators for alternative energy technologies. Dr. Mulder currently manages an integrated student farm at Green Mountain College and teaches in the Environmental Studies Department there.

Maurizio G. Paoletti is a Professor of Ecology at Padova University, Padova, Italy. With a background in biology and human ecology, he is an internationally recognized researcher in biodiversity, agroecology, entomology and ethnobiology. He has held visiting professorships in a number of countries (Finland, China, USA and Australia). He has organised more than ten international conferences on agroecology, sustainable agriculture, biodiversity, and is very active in public conferences to inform citizens on sustainability issues. Overall, he has completed 260 scientific papers and 18 edited books.

Tad Patzek is a professor of Geoengineering at U.C. Berkeley. Prior to joining Berkeley in 1990, he was a scientist at Shell Development, a research company managed for 20 years by M. King Hubbert. Patzek's current research involves mathematical modeling of earth systems with emphasis on fluid flow in soils and rocks. He also works on the thermodynamics and ecology of human survival and energy supply for humanity. Currently, he teaches courses in hydrology, ecology and energy supply, computer science, and mathematical modeling of earth systems. Patzek is a coauthor of over 200 papers and reports, and is writing five books.

David Pimentel is a professor of ecology and agricultural sciences at Cornell University, Ithaca, NY. His PhD is from Cornell University. His research spans the fields of energy, biotechnology, sustainable agriculture, and environmental policy. Pimentel has published more than 600 scientific papers and 25 books. He has served on many national and government committees including the National Academy of Sciences; President's Science Advisory Council; U.S Department of Agriculture and U.S. Department of Energy; Office of Technology Assessment of the U.S. Congress; and the U.S. State Department. In 2008 he received an Honorary Doctorate from the University of Massachusetts for his work in recognizing and publicizing critical trends in interactions between humans and the environment.

Robert Powers is finishing a BS in Environmental Science at the State University of New York College of Environmental Science & Forestry under Dr. Charles Hall. He is interested in the intersection of energy and economic issues, specifically in modeling problems to find innovative solutions. He has also started a Masters in System Dynamics at the University of Bergen (Norway) to further develop his modeling skills.

Marco Raugei obtained a Master's degree in Chemistry and a PhD in Chemical Sciences at the University of Siena (Italy), with a thesis on Life Cycle Assessment. He is currently working as a researcher and consultant in Life Cycle Assessment and Environmental Management, with active collaborations with Ambiente Italia Research Institute (Rome, Italy), University Parthenope (Naples, Italy), Brookhaven National Laboratory (NY, USA), Columbia University (NY, USA), and Escola Superior de Commerç Internacional - Universitat Pompeu Fabra (Barcelona, Spain). He has published over 35 peer-reviewed papers in various international journals, books and conference proceedings.

Robert Rapier has Bachelor's Degrees in Chemistry and Mathematics, and a Master's Degree in Chemical Engineering from Texas A&M University. Passionate about energy and sustainability issues, his R-Squared Energy Blog is devoted to debate and discussion of those topics. Robert has over 15 years of experience in the petrochemicals industry, including experience with cellulosic ethanol, gas-to-liquids (GTL), refining, and butanol production. He holds several U.S. and international patents, and works for a major oil company. Robert is currently based in Scotland where he lives with his wife and three children.

Daniela Russi earned a Master's Degree in Environmental Economics at the University of Siena (Italy). She did an internship at the Wuppertal Institute for Climate, Environment and Energy, in Wuppertal (Germany). She obtained a PhD in

Environmental Sciences at the Autonomous University of Barcelona (Spain) with a thesis on Social Multi-Criteria Evaluation (SMCE) applied to a conflict concerning rural electrification and large-scale biodiesel use in Italy. She has published peer-reviewed papers in international journals, and contributed to various books and conference proceedings on these topics. She is presently working for the environmental consultancy Amphos21.

Roger Samson is the Executive Director of Resource Efficient Agricultural Production (REAP)-Canada, a charitable organization working to develop and commercialize ecological solutions to energy, fibre and food production. Mr. Samson is a leading world expert in biomass energy development. He has authored over 60 publications on bioenergy, ecological farming, and climate change mitigation and has been working on bioenergy projects in North America, Europe, China, the Philippines, and West Africa since 1991. His work has pioneered ecological approaches for bioenergy production and thermodynamically efficient bioenergy conversion systems. Mr. Samson holds a B.Sc. (Crop Science) from Guelph University and a M.Sc. (Plant Science) from McGill University in Montreal.

William Schoenberg graduated from the State University of New York College of Environmental Science & Forestry with a Bachelors Degree in Environmental Studies. He is very interested in energy issues, especially peak oil and its ramifications for society. He is continuing his studies at the University of Bergen, Norway in the System Dynamics program, where he will be able to more fully explore dynamic modeling and its ability to help society prepare for the backside of the peak oil curve.

Luis Henrique de B. Soares is an Agronomist who graduated from Federal University of Rio Grande do Sul State (UFRGS, Brazil). He received a Master's degree in Environmental Microbiology, and his PhD in Molecular and Cellular Biology (Biotechnology Centre, Federal University fo Rio Grande do Sul, 2003), working on microbial enzymes for industrial applications. Dr. Soares is currently a Research Scientist at Embrapa Agrobiologia, Rio de Janeiro, studying principally agroenergy. The areas of his research include biofuels production and processing, enzymology, and energy balances for the assesmenet of agroecosystem sustainability.

Ms. Bailey Stamler is the Climate Change Project Manager with REAP-Canada. She has been working with REAP developing business plans for international carbon trading projects using small scale biomass energy technologies in the Philippines, Nigeria and Ethiopia since 2005. Ms. Bailey Stamler is experienced in bioenergy and bioheat pellet potential in Canada, focusing on the use of energy crops, agriculture and crop milling residues for heating applications. She also has experience quantifying GHG emissions, mitigation potential and relative efficiencies of biofuels. Ms. Bailey Stamler holds B.Sc. (Environmental Science) from Laurentian University and an M.Sc. from McGill University (Natural Resource Sciences).

Ronald Steenblik at the time of writing was Director of Research for the Global Subsidies Initiative (GSI) of the International Institute for Sustainable Development (IISD). Ronald's career spans three decades, in industry, academia, the U.S. federal government, and intergovernmental organizations, working on policy issues related to natural resources, the environment, or trade. Prior to joining the IISD, he was a senior trade policy analyst in the Trade Directorate of the Organisation for Economic Co-operation and Development (OECD), where his analyses supported the WTO negotiations on environmental goods and services. Ronald holds degrees from Cornell University and the University of Pennsylvania.

David Swenson is an associate scientist in economics at Iowa State University and a lecturer there in community and regional planning as well as in the graduate program in urban and regional planning at The University of Iowa. His primary area of research focuses on regional economic changes and their fiscal and demographic implications for communities. He has completed scores of economic impact studies, and written and presented extensively on the uses of impact models for decision making. Of late, he has scrutinized the potential community economic outcomes associated with biofuels development in the Midwest and the Plains.

Sergio Ulgiati received an education in Physics and Environmental Chemistry. He is a Professor of Life Cycle Assessment and General Systems Theory at Parthenope University in Napoli, Italy. He has expertise in Energy Analysis, LCA, Environmental Accounting and Emergy Synthesis. He has published over 200 papers in national and international journals and books. His research in LCA covers renewable and nonrenewable energy systems (wind, geothermal, hydro, bioenergy; solar thermal and photovoltaic, hydrogen and fuel cells; thermal fossil-powered power plant, including cogeneration and NGCC plants), as well as zero emission technologies and strategies (ZETS). He is the organizer and Chair of the Biennial International Workshop "Advances in Energy Studies."

Segundo Urquiaga graduated in Agronomy in 1973 from the Agrarian University "La Molina", Lima, Perú, with BSc, and defended his PhD thesis in 1982 in the Agricultural college "Luiz de Queiroz" of the São Paulo State University, Piracicaba, São Paulo. In 1984 he moved to the Brazilian Corporation for Agricultural Research (Embrapa Agrobiologia) in Seropédica, Rio de Janeiro. At present he is studying the influence of biological nitrogen fixation (BNF) on the energy balance of several renewable energy sources such as sugar cane, soybean and elephant grass. Urquiaga has published over 120 papers in national and international journals, and over 50 chapters in books and conference proceedings.

Contributors

Bruno J.R. Alves
Embrapa-Agrobiologia, BR-465, Km 07, Caixa Postal 75.505, Seropédica, 23890-000, Rio de Janeiro, Brazil

Robert M. Boddey
Embrapa-Agrobiologia, BR-465, Km 07, Caixa Postal 75.505, Seropédica, 23890-000, Rio de Janeiro, Brazil, e-mail: bob@cnpab.embrapa.br

Jeroen Dooper
Resource Efficient Agricultural Production (REAP) – Canada, Box 125 Centennial Centre CCB13, Ste. Anne de Bellevue, Quebec, Canada H9X 3V9

Andrew R.B. Ferguson
11 Harcourt Close, Henley-on-Thames, RG9 1UZ, England, e-mail: andrewrbferguson@hotmail.com

Tom Gangwer
739 Battlefront Trail, Knoxville, TN 37934, USA, e-mail: fcha-tom@charter.net

Mario Giampietro
ICREA Research Professor, Institute of Environmental Science and Technology (ICTA), Autonomous University of Barcelona , Building Q – ETSE - (ICTA), Campus of Bellaterra 08193 Cerdanyola del Vallès (Barcelona), Spain, e-mail: mario.giampietro@uab.cat

Tiziano Gomiero
Department of Biology, Padua University, Italy, Laboratory of Agroecology and Ethnobiology, via U. Bassi, 58/b, 35121-Padova, Italy, e-mail: tiziano.gomiero@libero.it

Nathan John Hagens
Gund Institute for Ecological Economics, University of Vermont, 617 Main Street, Burlington, VT 05405, USA, e-mail: Nathan.Hagens@uvm.edu

Charles A. S. Hall
State University of New York, College of Environmental Science and Forestry, Syracuse, New York, NY 13210, USA, e-mail: chall@esf.edu

Edwin Kessler
1510 Rosemont Drive, Norman, OK 73072, e-mail: kess3@swbell.net

Doug Koplow
Earth Track, Inc., 2067 Massachusetts Avenue, 4th Floor, Cambridge, MA 02140,
USA, e-mail: dkoplow@earthtrack.net

Claudia Ho Lem
Resource Efficient Agricultural Production (REAP) – Canada, Box 125 Centennial
Centre CCB13, Ste. Anne de Bellevue, Quebec, Canada H9X 3V9

Kozo Mayumi
Faculty of IAS, The University of Tokushima, Minami-Josanjima 1-1, Tokushima
City 770-8502, Japan, e-mail: mayumi@ias.tokushima-u.ac.jp

Kenneth Mulder
Green Mountain College, Poultney VT, USA

Marcelo Dias De Oliveira
Avenida 10, 1260, Rio Claro - SP – Brazil, CEP 13500-450,
email: dias_oliveira@msn.com

Maurizio G. Paoletti
Dept. of Biology, Padua University, Italy, Lab. of Agroecology and Ethnobiology,
via U. Bassi, 58/b, 35121-Padova, Italy, e-mail: paoletti@bio.unipd.it

Tad W. Patzek
Department of Civil and Environmental Engineering, University of California, 425
David Hall, MC1716, Berkeley, CA 94720, USA,
e-mail: patzek@patzek.ce.berkeley.edu

David Pimentel
College of Agriculture and Life Sciences, Cornell University, 5126 Comstock Hall,
Ithaca, NY 15850, USA, e-mail: Dp18@cornell.edu

Robert Powers
State University of New York, College of Environmental Science and Forestry,
Syracuse, New York, NY 13210, USA

Robert Rapier
Accsys Technologies PLC, 5000 Quorum Drive, Suite 310, Dallas, TX 75254,
USA, e-mail: rrapier1@yahoo.com

Marco Raugei
Department of Sciences for the Environment, Parthenope University of Napoli,
Centro Direzionale – Isola C4, 80143 Napoli, Italy

Daniela Russi
Autonomous University of Barcelona, Department of Economics and Economic
History, Edifici B, Campus de la UAB, 08193 Bellaterra (Cerdanyola del V.),
Barcelona, Spain

Roger Samson
Resource Efficient Agricultural Production (REAP) – Canada, Box 125 Centennial
Centre CCB13, Ste. Anne de Bellevue, Quebec, Canada H9X 3V9,
e-mail: rsamson@reap-canada.com

William Schoenberg
State University of New York, College of Environmental Science and Forestry,
Syracuse, New York, NY 13210, USA

Luis Henrique de B. Soares
Embrapa-Agrobiologia, BR-465, Km 07, Caixa Postal 75.505, Seropédica,
23890-000, Rio de Janeiro, Brazil

Stephanie Bailey Stamler
Resource Efficient Agricultural Production (REAP) – Canada, Box 125 Centennial
Centre CCB13, Ste. Anne de Bellevue, Quebec, Canada H9X 3V9

Ronald Steenblik
Global Subsidies Initiative of the International Institute for Sustainable
Development, Maison Internationalle de l'Environment 2, 9, chemin de Balexert,
1219 Châtelaine Genève, Switzerland, e-mail: ronald.steenblik@gmail.com

David Swenson
Department of Economics, 177 Heady Hall, Iowa State University, Ames IA 50011,
e-mail: dswenson@iastate.edu

Sergio Ulgiati
Department of Sciences for the Environment, Parthenope University of Napoli,
Centro Direzionale – Isola C4, 80143 Napoli, Italy,
e-mail: Sergio.ulgiati@uniparthenope.it

Segundo Urquiaga
Embrapa-Agrobiologia, BR-465, Km 07, Caixa Postal 75.505, Seropédica,
23890-000, Rio de Janeiro, Brazil

Chapter 1
Renewable and Solar Energy Technologies: Energy and Environmental Issues

David Pimentel

Abstract A critical need exists to investigate various renewable and solar energy technologies and examine the energy and environmental issues associated with these various technologies. The various renewable energy technologies will not be able to replace all current 102 quads (quad = 10^{15} BTU) of U.S. energy consumption (USCB 2007). A gross estimate of land and water resources is needed, as these resources will be required to implement the various renewable energy technologies.

Keywords Biomass energy · conversion systems · ethanol · geothermal systems · hydroelectric power · photovoltaic systems · renewable energy · solar · wind power

1.1 Introduction

The world, and the United States in particular, face serious energy shortages and associated high energy prices during the coming decades. Oil, natural gas, coal, and nuclear power provide more than 88% of world energy needs; the other 12% is provided by various renewable energy sources (Table 1.1). Oil, natural gas, coal, and nuclear provide more than 93% of U.S. energy needs; the other 9% consists of various renewable and non-renewable energy sources (Table 1.1).

The U.S., with slightly more than 45% of the world's population, accounts for nearly 25% of the world's energy consumption (Table 1.1). On average, each American uses nearly 8,000 L of oil equivalents per year for all purposes, including transportation, industry, heating and cooling.

The United States now imports more than 63% of its oil at an annual cost of approximately $200 billion (USCB 2007). Projections are that within 20 years the U.S. will be importing more than 90% of its oil. The United States has consumed more than 90% of its proved oil reserves (Pimentel et al. 2004a). Because the U.S.

✉ D. Pimentel
College of Agriculture and Life Sciences, Cornell University, 5126 Comstock Hall, Ithaca, NY 15850
e-mail: Dp18@cornell.edu

D. Pimentel (ed.), *Biofuels, Solar and Wind as Renewable Energy Systems*,
© Springer Science+Business Media B.V. 2008

Table 1.1 Fossil and solar energy use in the U.S. and world (quads = 10^{15} BTU) (USCB 2007)

Fuel	U.S.	World
Petroleum	40.1	168
Natural Gas	23.0	103
Coal	22.3	115
Nuclear	8.2	28
Biomass	3.0	30
Hydroelectric power	3.4	27
Geothermal and windpower	0.4	0.8
Biofuels	0.5	0.9
Total	100.9	472.7

population is growing nearly twice as fast as that of China per capita, and is adding 3.3 million to the population each year, energy resources are becoming scarce (PRB 2006). These shortages are now contributing to greater interest in renewable energy resources.

Diverse renewable energy sources currently provide 6.8% of U.S. needs and about 12% of world needs (Table 1.1). In addition to energy conservation, the development and use of renewable energy is expected to increase as fossil fuel supplies decline and become highly expensive. Eight different renewable technologies are projected to provide the U.S. with most of its energy in the future: hydropower, biomass, wind power, solar thermal, photovoltaics, passive energy systems, geothermal, and biogas. In this chapter, I assess the potential of these 8 renewable energy technologies, including their environmental benefits and risks, and their energetic and economic costs.

1.2 Hydroelectric Power

Hydropower contributes significantly to world energy, providing 6% of the supply (Table 1.1). In the United States, hydroelectric plants produce approximately 3% or 3.4 quads of total U.S. energy (340 billion kWh) (1 kWh = 860 kilocalories [kcal] = 3,440 BT = 3.6 megajoules), or 11% of the nation's electricity, each year at a cost of \$0.02 per kWh (Table 1.2; USCB 2007). Development and rehabilitation of existing dams in the United States could produce an additional 5 quads per year (Table 1.3).

Hydroelectric plants, however, require considerable land for their water storage reservoirs. An average of 75,000 hectares (ha) of reservoir land area and 14 trillion L of water are required per 1 billion kWh per year produced (Table 1.2, Gleick and Adams 2000). Based on regional estimates of US land use and average annual energy generation, reservoirs currently cover approximately 26 million ha of the total 917 million ha of land area in the United States (Pimentel 2001). To develop the remaining best candidate sites, assuming land requirements similar to those in past developments, an additional 7 million ha of land would be required for water storage (Table 1.3).

Table 1.2 Land resource requirements and total energy inputs for construction of renewable and other facilities that produce 1 billion kWh/yr of electricity. Energy return on investment is listed for each technology. (See text for explanations)

Electrical energy Technology	Land required (ha)	Energy input:output	Cost per kWh ($)	Life in years
Hydroelectric power	75,000[a]	1:24	$0.02[a]	30
Biomass	200,000[b]	1:7[b]	0.058[c]	30
Parabolic troughs	1,100[d]	1:5[b]	0.07–09[e]	30
Solar ponds	5,200[f]	1:4[b]	0.15[b]	30
Wind power	9,500[g]	1:4[h]	0.07[i]	30
Photovoltaics	2,800[j]	1:7[j]	0.25[b]	30
Biogas	____[k]	1:1.7–3.3[l]	0.02[l]	30
Geothermal	30[b]	1:48[b]	0.064[b]	20
Coal (non-renewable)	166[b]	1:8[b]	0.03[b]	30
Nuclear (non-renewable)	30[b]	1:5[b]	0.05[b]	30

[a] Based on a random sample of 50 hydropower reservoirs in the United States, ranging in area from 482 ha to 763,00 ha (Pimentel, unpublished).
[b] Pimentel, unpublished.
[c] Production costs based on 70% capacity factor (J. Irving, Burlington Electric, Vermont, personal communication 2001).
[d] Calculated (DOE/EREN 2000).
[e] (DOE/EREN 2000).
[f] Based on 4,000 ha solar ponds plus an additional 1,200 ha for evaporation ponds.
[g] (Andrew Ferguson, Optimum Population Trust (UK), personal communication, June 16, 2007).
[h] (Tyner 2002).
[i] (Peace Energy 2003).
[j] Calculated from DOE 2000.
[k] No data available.
[l] (B. Jewell, Cornell University, Ithaca, NY, personal communication 2001).

Table 1.3 Current and projected US gross annual energy supply from various renewable energy technologies, based on the thermal equivalent and required land area

Energy technology	Current (2005)		Projected (2050)	
	Quads	Million hectares	Quads	Million hectares
Biomass	4.5[a]	75[b]	5	102[b]
Ethanol	0.16	4	0.2	5
Hydroelectric power	3.9[a]	26[c]	5	33
Geothermal energy	1.7[a]	0.5	1.2	1
Solar thermal energy	<0.06	<0.01	10	11
Photovoltaics	<0.06	<0.01	11	3
Wind power	0.11[a]	1.00	7	8
Biogas	<0.001	<0.001	0.5	0.01
Passive solar power	0.3[d]	0	6	1
Total	10.8	107	45.9	164

[a] USCB (2004–2005).
[b] This is the equivalent land area required to produce 3 metric tons per hectare.
[c] Total area based on an average of 75,000 hectares per reservoir area per 1 billion kilowatt-hours per year produced.
[d] Pimentel et al. (2002).

Despite the benefits of hydroelectric power, the plants cause major environmental problems. The impounded water frequently covers valuable, agriculturally productive, alluvial bottomland. Sediments build up behind the dams, reducing their effectiveness and creating another major environmental problem. Further, dams alter the existing plants, animals, and microbes in the ecosystem (Nilsson and Berggren 2000). Fish species may significantly decline in river systems because of these numerous ecological changes.

1.3 Biomass Energy

Most biomass is burned for cooking and heating, however, it can also be converted into electricity and liquid fuel. Under sustainable forest conditions in both temperate and tropical ecosystems, approximately 3 dry metric tons (t/ha) per year of woody biomass can be harvested sustainably (Birdsey 1992, Repetto 1992, Trainer 1995, Ferguson 2003). Although this amount of woody biomass has a gross energy yield of 13.5 million kcal/ha, it requires an energy expense of approximately 33 L of diesel fuel per ha, plus the embodied energy for cutting and collecting wood for transport to an electric power plant. Thus, the energy input per output ratio for a woody biomass system is calculated to be 1:22.

The cost of producing 1 kWh of electricity from woody biomass is about $0.06, which is competitive with other electricity production systems that average $0.07 in the U.S. (Table 1.2) (USCB 2007). Approximately 3 kWh of thermal energy is expended to produce 1 kWh of electricity, an energy input/output ratio of 1:7 (Table 1.2). Per capita consumption of woody biomass for heat in the United States amounts to 625 kilograms (kg) per year. In developing nations, use of diverse biomass resources (wood, crop residues, and dung) average about 630 kg per capita (Kitani 1999). Developing countries use only about 500 L of oil equivalents of fossil energy per capita compared with nearly 8,000 L of oil equivalents of fossil energy used per capita in the United States (Table 1.1).

Woody biomass could supply the United States with about 5 quads (1.5 × 10^{12} kWh thermal) of its total gross energy supply by the year 2050, provided there was adequate forest land available (Table 1.3). A city of 100,000 people using the biomass from a sustainable forest (3 t/ha per year) for electricity would require approximately 200,000 ha of forest area, based on an average electrical demand of slightly more than 1 billion kWh (electrical energy [e]) (860 kcal = 1 kWh) (Table 1.2).

Environmental impacts of burning biomass are less harmful than those associated with coal, but more harmful than those associated with natural gas (Pimentel 2001). Biomass combustion releases more than 200 different chemical pollutants, including 14 carcinogens and 4 co-carcinogens, into the atmosphere (Burning Issues 2003). Globally, but especially in developing nations where people cook with fuelwood over open fires, approximately 4 billion people suffer from continuous exposure to smoke (Kids for Change 2006). In the United States, wood smoke kills 30,000 people each year (EPA, 2002). However, the pollutants from electric plants that use wood and other biomass can be controlled.

1.4 Wind Power

For many centuries, wind power has provided energy to pump water and to run mills and other machines. Today, turbines with a capacity of at least 500 kW produce most of the commercially wind-generated electricity. Operating at an ideal location, one of these turbines running at 30% efficiency can yield an energy output of 1.3 million kWh (e) per year (AWEA 2000a). An initial investment of approximately $500,000 for a 500 kW capacity turbine operating at 30% efficiency, will yield an input/output ratio of 1:4 over 30 years of operation (Table 1.2). During the 30-year life of the system, the annual operating costs amount to about $50,000. The estimated cost of electricity generated is $0.07 per kWh (e) (Table 1.2). Some report costs ranging from $0.03 to $0.05 per kWh (Sawin 2004). These values are probably located in favorable wind sites.

In the United States, 2502 megawatts (MW) of installed wind generators produce about 6.6 billion kWh of electrical energy per year (Chambers 2000). The American Wind Energy Association (AWEA 2000b) estimates that the United States could support a capacity of 30,000 MW by the year 2010, producing 75 billion kWh (e) per year at a capacity of 30%, or approximately 2% of the annual US electrical consumption. If all economically feasible land sites are developed, the full potential of wind power is estimated to be about 675 billion kWh (e) (AWEA 2000b). Off-shore sites could provide an additional 102 billion kWh (e) (Gaudiosi 1996), making the total estimated potential of wind power 777 billion kWh (e), or 23% of current electrical use.

Widespread development of wind power is limited by the availability of sites with sufficient wind (at least 20 kilometers per hour [km/h]) and the number of wind machines that the site can accommodate. An average area for one 50 kW turbine is 1.3 ha to allow sufficient spacing to produce maximum power (Table 1.2). Based on this figure, approximately 9,500 ha of land are needed to supply 1 billion kWh per year (Table 1.2). Because the turbines themselves only occupy approximately 2% of the area, most of the land can be used for vegetables, nursery stock, and cattle (Natural Resources Canada 2002). However, it may be impractical to produce corn or other grains because of the heavy equipment used in this type of farming.

An investigation of the environmental impacts of wind energy production reveals a few hazards. Locating the wind turbines in or near the flyways of migrating birds and wildlife refuges may result in birds flying into the supporting towers and rotating blades. For this reason, it is suggested that wind farms be located at least 300 meters (m) from nature reserves to reduce their risk to birds. The estimated 13,000 wind turbines installed in the United States kill an estimated 2,600 birds per year (Sinclair 2003). Choosing a proper site and improving repellant technology with strobe lights or paint patterns might further reduce the number of birds killed.

Bat fatalities are another serious concern. It is projected that by 2020 annual bat fatalities caused by wind turbines will range from 33,000 to 62,000 individuals annually (Kunz et al. 2007). Most bat fatalities are from species that migrate long distances and are tree roosting. Eastern U.S wind turbines installed along forested ridgetops have the highest rate of bat kills, ranging from 15.3 to 41.1 bats per MW of installed capacity per year (Kunz et al. 2007). Monitoring for bat and bird fatalities and research for the reduction of these should be included in all wind energy planning.

The rotating magnets in the turbine electrical generator produce a low level of electromagnetic interference that can affect television and radio signals within 2–3 km of large installations (Sagrillo 2006). Fortunately, with the widespread use of cable networks or line-of-sight microwave satellite transmission, both television and radio are unaffected by this interference.

The noise caused by rotating blades is another unavoidable side effect of wind turbine operation. Beyond 2.1 km, however, the largest turbines are inaudible even downwind. At a distance of 400 m, the noise level is estimated to be about 60 decibel, corresponding roughly to the noise level of a home air-conditioning unit.

1.5 Solar Thermal Conversion Systems

Solar thermal energy systems collect the sun's radiant energy and convert it into heat. This heat can be used directly for household and industrial purposes or produce steam to drive turbines that produce electricity. The complexity of these systems ranges from solar ponds to electricity-generating parabolic troughs. In the following analyses, I convert thermal energy into electricity to facilitate comparison to the other solar energy technologies.

1.5.1 Solar Ponds

Solar ponds are used to capture radiation and store the energy at temperatures of nearly 100 degrees Celsius ($^\circ$C). Constructed ponds can be made into solar ponds by creating a layered salt concentration gradient. The layers prevent natural convection, trapping the heat collected from solar radiation in the bottom layer of brine. The hot brine from the bottom of the pond is piped out to use for heat, for generating electricity, or both.

For successful operation of a solar pond, the salt concentration gradient and the water level must be maintained. A solar pond covering 4,000 ha loses approximately 3 billion L of water per year (750,000 L/ha per year) under arid conditions (Tabor and Doran 1990). Recently, solar ponds in Israel have been closed because of such difficulties. To counteract the water loss and the upward diffusion of salt in the ponds, the dilute salt water at the surface of the ponds has to be replaced with fresh water and salt added to the lower layer (Solar Pond 2007).

The efficiency of solar ponds in converting solar radiation into heat is estimated to be approximately 1:4, assuming a 30-year life for the solar pond (Table 1.2). A 100 ha (1 km^2) solar pond can produce electricity at a rate of approximately $0.30 per kWh (Australian Government 2007).

Some hazards are associated with solar ponds, but most can be avoided with careful management. It is essential to use plastic liners to make the ponds leakproof and prevent contamination of the adjacent soil and groundwater with salt.

1.5.2 Parabolic Troughs

Another solar thermal technology that concentrates solar radiation for large-scale energy production is the parabolic trough. A parabolic trough, shaped like the bottom half of a large drainpipe, reflects sunlight to a central receiver tube that runs above it. Pressurized water and other fluids are heated in the pipe and used to generate steam that drives turbogenerators for electricity production or provides heat energy for industry.

Parabolic troughs that have entered the commercial market have the potential for efficient electricity production because they can achieve high turbine inlet temperature. Assuming peak efficiency and favorable sunlight conditions, the land requirements for the central receiver technology are approximately 1,100 ha per 1 billion kWh per year (Table 1.2). The energy input:output ratio is calculated to be 1:5 (Table 1.2). Solar thermal receivers are estimated to produce electricity at approximately $0.07–$0.09 per kWh (DOE/EREN 2001).

The potential environmental impacts of solar thermal receivers include the accidental or emergency release of toxic chemicals used in the heat transfer system. Water availability can also be a problem in arid regions.

1.6 Photovoltaic Systems

Photovoltaic cells have the potential to provide a significant portion of future U.S. and world electrical energy (Energy Economics 2007). Photovoltaic cells produce electricity when sunlight excites electrons in the cells. The most promising photovoltaic cells in terms of cost, mass production, and relatively high efficiency are those manufactured using silicon. Because the size of the unit is flexible and adaptable, photovoltaic cells can be used in homes, industries, and utilities.

However, photovoltaic cells need improvements to make them economically competitive before their use can become widespread. Test cells have reached efficiencies of about 25% (American Energy 2007), but the durability of photovoltaic cells must be lengthened and current production costs reduced several times to make their use economically feasible.

Production of electricity from photovoltaic cells currently costs about $0.25 per kWh (DOE 2000). Using mass-produced photovoltaic cells with about 18%

efficiency, 1 billion kWh per year of electricity could be produced on approximately 2,800 ha of land, and this is sufficient electrical energy to supply 100,000 people (Table 1.2, DOE 2001). Locating the photovoltaic cells on the roofs of homes, industries, and other buildings would reduce the need for additional land by an estimated 20% and reduce transmission costs. However, because storage systems such as batteries cannot store energy for extended periods, photovoltaics require conventional backup systems.

The energy input for making the structural materials of a photovoltaic system capable of delivering 1 billion kWh during a life of 30 years is calculated to be approximately 143 million kWh. Thus, the energy input per output ratio for the modules is about 1:7 (Table 1.2, Knapp and Jester 2000).

The major environmental problem associated with photovoltaic systems is the use of toxic chemicals, such as cadmium sulfide and gallium arsenide, in their manufacture. Because these chemicals are highly toxic and persist in the environment for centuries, disposal and recycling of the materials in inoperative cells could become a major problem.

1.7 Geothermal Systems

Geothermal energy uses natural heat present in Earth's interior. Examples are geysers and hot springs, like those at Yellowstone National Park in the United States. Geothermal energy sources are divided into three categories: hydrothermal, geopressured-geothermal, and hot dry rock. The hydrothermal system is the simplest and most commonly used for electricity generation. The boiling liquid underground is produced using wells, high internal pressure drives, or pumps. In the United States, nearly 3,000 MW of installed electric generation comes from hydrothermal resources, and this is projected to increase by 4,500 MW.

Most of the geothermal sites for electrical generation are located in California, Nevada, and Utah. Electrical generation costs for geothermal plants in the West range from $0.06 to $0.30/kWh (Gawlik and Kutscher 2000), suggesting that this technology offers potential to produce electricity economically. The US Department of Energy and the Energy Information Administration (DOE/EIA 2001) project that geothermal electric generation may grow three- to fourfold during the next 20–40 years. However, other investigations are not as optimistic and, in fact, suggest that geothermal energy systems are not renewable because the sources tend to decline over 40–100 years (Bradley 1997, Youngquist 1997, Cassedy 2000). Existing drilling opportunities for geothermal resources are limited to a few sites in the United States and world (Youngquist 1997).

Potential environmental problems of geothermal energy include water shortages, air pollution, waste effluent disposal, subsidence, and noise. The wastes produced in the sludge include toxic metals such as arsenic, boron, lead, mercury, radon, and vanadium. Water shortages are an important limitation in some regions. Geothermal systems produce hydrogen sulfide, a potential air pollutant; however, this could be processed and removed for use in industry. Overall, these environmental costs of geothermal energy appear to be minimal relative to those of fossil fuel systems.

1.8 Biogas

Wet biomass materials can be converted effectively into usable energy using anaerobic microbes. In the United States, livestock dung is normally gravity fed or intermittently pumped through a plug-flow digester, which is a long, lined, insulated pit in the earth. Bacteria break down volatile solids in the manure and convert them into methane gas (65%) and CO_2 (35%) (Pimentel 2001). A flexible liner stretches over the pit and collects the biogas, inflating like a balloon. The biogas may be used to heat the digester, to heat farm buildings, or to produce electricity. A large facility capable of processing the dung from 500 cows costs nearly \$300,000 (EPA 2000). The Environmental Protection Agency (EPA 2000) estimates that more than 2000 digesters could be economically installed in the United States.

The amount of biogas produced is determined by the temperature of the system, the microbes present, the volatile solids content of the feedstock, and the retention time. A plug-flow digester with an average manure retention time of about 16 days under winter conditions ($17.4°C$) produced 452,000 kcal/day and used 262,000 kcal/day to heat the digester to $35°C$ (Jewell et al. 1980). Using the same digester during summer conditions ($25°C$) but reducing the retention time to 10.4 days, the yield in biogas was 524,000 kcal/day, and it used 157,000 kcal/day for heating the digester (Jewell et al. 1980). The energy input per output ratios for these winter and summer conditions for the digester were 1:1.7 and 1:3.3, respectively. The energy output of biogas digesters is similar today (Hartman et al. 2000).

In developing countries such as India, biogas digesters typically treat the dung from 15 to 30 cattle from a single family or a small village. The resulting energy produced for cooking saves forests and preserves the nutrients in the dung. The capital cost for an Indian biogas unit ranges from \$500 to \$900 (Kishore 1993). The price value of a kWh biogas in India is about \$0.06 (Dutta et al. 1997). The total cost of producing about 10 million kcal of biogas is estimated to be \$321, assuming the cost of labor to be \$7/h; hence, the biogas has a value of \$356. Manure processed for biogas has fewer odors and retains its fertilizer value (Pimentel 2001).

1.9 Ethanol and Energy Inputs

The average costs in terms of energy and dollars for a large modern corn ethanol plant are listed in Table 1.4. In the fermentation/distillation process, the corn is finely ground and approximately 15 L of water are added per 2.69 kg of ground corn. After fermentation, to obtain a liter of 95% pure ethanol from the 8% ethanol and 92% water mixture, the 1 L of ethanol must be extracted from the approximately 13 L of the ethanol/water mixture. To be mixed with gasoline, the 95% ethanol must be

Table 1.4 Inputs per 1,000 L of 99.5% ethanol produced from corn[a]

Inputs	Quantity	kcal × 1000	Dollars $
Corn grain	2,690 kg[b]	2,550[b]	287.36
Corn transport	2,690 kg[b]	322[c]	21.40[d]
Water	15,000 L[e]	90[f]	21.16[f]
Stainless steel	3 kg[g]	165[h]	10.60[d]
Steel	4 kg[g]	92[h]	10.60[d]
Cement	8 kg[g]	384[h]	10.60[d]
Steam	2,546,000 kcal[i]	2,546[i]	21.16[j]
Electricity	392 kWh[i]	1,011[i]	27.44[k]
95% ethanol to 99.5%	9 kcal/L[l]	9[l]	0.60
Sewage effluent	20 kg BOD[m]	69[n]	6.00
Distribution	331 kcal/L[o]	331	20.00[o]
Total		7,569	$436.92

[a] Output: 1 L of ethanol = 5,130 kcal.
[b] Pimentel (2003).
[c] Calculated for 144 km roundtrip.
[d] Pimentel (2003).
[e] 15 L of water mixed with each 2.69 kg of grain.
[f] Pimentel et al. (2004b).
[g] Estimated.
[h] Newton (2001).
[i] Illinois Corn (2004).
[j] Calculated based on the price of natural gas.
[k] $.07 per kWh (USCB 2004–2005).
[l] 95% ethanol converted to 99.5% ethanol for addition to gasoline (T. Patzek, personal communication, University of California, Berkeley 2004).
[m] 20 kg of BOD per 1,000 L of ethanol produced (Kuby et al. 1984).
[n] 4 kWh of energy required to process 1 kg of BOD (Blais et al. 1995).
[o] DOE (2002).

further processed and more water removed, requiring additional fossil energy inputs to achieve 99.5% pure ethanol (Table 1.4). Thus, a total of about 12 L of wastewater must be removed per liter of ethanol produced, and this relatively large amount of sewage effluent has to be disposed of at an energy, economic, and environmental cost.

To produce a liter of 99.5% ethanol uses 43% more fossil energy than the energy produced as ethanol and costs 44¢ per L ($1.66 per gallon or $2.76 per gallon including the subsidy) (Table 1.4). The corn feedstock requires more than 33% of the total energy input. In this analysis the total cost, including the energy inputs for the fermentation/distillation process and the apportioned energy costs of the stainless steel tanks and other industrial materials, is $436.92 per 1,000 L of ethanol produced (Table 1.4).

The largest energy inputs in corn-ethanol production are for producing the corn feedstock, plus the steam energy, and electricity used in the fermentation/distillation process. The total energy input to produce a liter of ethanol is approximately 7,570 kcal (Table 1.4). However, a liter of ethanol has an energy value of only 5,130 kcal. Based on a net energy loss of 2,440 kcal of ethanol produced, 43% more fossil energy is expended than is produced as ethanol.

1.10 Grasslands and Celulosic Ethanol

Tilman's research (Tillman et al. 2006) has merit in the explanation of field experiments with various combinations of species of natural vegetation, and the productivity of diverse experimental systems. The outstanding, 30-year effort by the Land Institute in Kansas (Jackson 1980) to develop multi-species perennial ecosystems that deliver high productivity for long periods has been *de facto* endorsed by Tillman et al., albeit without acknowledgement.

However, there are concerns about two items. First, the statement by Tillman et al. that crop residues, like corn stover, can be harvested and utilized as a fuel source. This would be a disaster for agricultural ecosystems. Without the protection of crop residues, soil loss may increase as much as 100-fold (Fryrear and Bilbro 1994). Already the U.S. crop system is losing soil 10 times faster than sustainability (NAS 2003). Soil formation rates are extremely slow or less than 1 t/ha/yr (NAS 2003, Troeh et al. 2004). Increased erosion will facilitate soil-C oxidation and contribute to the greenhouse problem (Lal 2003).

Tillman et al. assume about 1,032 L of ethanol can be produced through the conversion of the 4 t/ha/yr of grasses harvested. However, Pimentel and Patzek (2007) reported a negative 50% return in switchgrass conversion. Based on the optimistic data of Tillman et al., and converting all 235 million ha of U.S. grassland into ethanol, only 12% of U.S. petroleum would be provided (USDA 2004, USCB 2004–2005).

In addition, to achieve the production of this much ethanol would mean displacing about 100 million cattle, 7 million sheep, and 4 million horses now grazing on 324 million ha of U.S. grassland and rangeland (USDA 2004, Mitchell 2000). Already overgrazing is a problem on U.S. grasslands and a similar problem exists worldwide (Brown 2001). Thus, the assessment of the quantity of ethanol that can be produced on U.S. and world grasslands by Tillman et al. appears to be unduly optimistic.

1.11 Methanol and Vegetable Oils

Methanol can be produced from a gasifier-pyrolysis reactor using biomass as a feedstock (Hos and Groenveld 1987, Jenkins 1999). The yield from 1 ton of dry wood is about 370 L of methanol (Ellington et al. 1993, Osburn and Osburn 2001). For a plant with economies of scale to operate efficiently, more than 1.5 million ha of sustainable forest would be required to supply it (Pimentel 2001). Biomass is generally not available in such enormous quantities, even from extensive forests, at acceptable prices. Most methanol today is produced from natural gas.

Processed vegetable oils from soybean, sunflower, rapeseed, and other oil plants can be used as fuel in diesel engines. Unfortunately, producing vegetable oils for use in diesel engines is costly in terms of economics and energy (Pimentel and Patzek 2005). A slight net return on energy from soybean oil is possible, if the soybeans are grown without commercial nitrogen fertilizer. The soybean under favorable conditions will produce its own nitrogen. Even assuming a slight net energy return with soy, the total United States would have to be planted to soybeans just to provide soy oil for U.S. trucks!

1.12 Transition to Renewable Energy

Despite its environmental and economic benefits, the transition to large-scale use of renewable energy presents several difficulties. Renewable energy technologies, all of which require land for collection and production, will compete with agriculture, forestry, and urbanization for land in the United States and world. The United States is at maximum use of its prime cropland for food production per capita today, but the world has less than half the cropland per capita that it needs for a diverse diet (0.5 ha) and adequate supply of essential nutrients (USDA 2004). In fact, more than 3.7 billion people are already malnourished in the world (UN/SCN 2004, Bagla 2003). With the world and US populations expected to double in the next 58 and 70 years, respectively, all the available cropland and forestland will be required to provide vital food and forest products (PRB 2006).

As the growing U.S. and world populations demand increased electricity and liquid fuels, constraints like land availability and high investment costs will restrict the potential development of renewable energy technologies. Energy use based on current growth is projected to increase from the current U.S. consumption of 102 quads to approximately 145 quads by 2050. Land availability is also a problem, with the US population adding about 3.3 million people each year (USCB 2007). Each person added requires about 0.4 ha (1 acre) of land for urbanization and highways and about 0.5 ha of cropland (Vesterby and Krupa 2001).

Renewable energy systems require more labor than fossil energy systems. For example, wood-fired steam plants require several times more workers than coal-fired plants (Giampietro et al. 1998).

An additional complication in the transition to renewable energies is the relationship between the location of ideal production sites and large population centers. Ideal locations for renewable energy technologies are often remote, such as deserts of the American Southwest or wind farms located kilometers offshore. Although these sites provide the most efficient generation of energy, delivering this energy to consumers presents a logistical problem. For instance, networks of distribution cables must be installed, costing about $179,000 per km 115-kV lines (DOE/EIA 2002). A percentage of the power delivered is lost as a function of electrical resistance in the distribution cable. There are complex alternating current electrical networks in North America, and 3 of these are tied together by DC lines (Nordel 2001). Based on these networks, it is estimated that electricity can be transmitted up to 1500 km.

A sixfold increase in installed technologies would provide the United States with approximately 46 quads (thermal) of energy, less than half of current US consumption (Table 1.1). This level of energy production would require about 159 million ha of land (17% of US land area). This percentage is an estimate, and could increase or decrease depending on how the technologies evolve and energy conservation is encouraged.

Worldwide, approximately 473 quads of all types of energy are used by the population of more than 6.5 billion people (Table 1.1). Using available renewable energy technologies, an estimated 200 quads of renewable energy could be produced worldwide on about 20% of the world land area. A self-sustaining renewable energy system producing 200 quads of energy per year for about 2 billion people (Ferguson 2001) would provide each person with about 5,000 L of oil equivalents per year, approximately half of America's current consumption per year, but an increase for most people of the world (Pimentel et al. 1999).

The first priority of the US energy program should be for individuals, communities, and industries to conserve fossil fuel resources and reduce consumption. Other developed countries have proved that high productivity and a high standard of living can be achieved with the use of half the energy expenditure of the United States (Pimentel et al. 1999). In the United States, fossil energy subsidies of approximately $40 billion per year should be withdrawn and the savings invested in renewable energy research and education to encourage the development and implementation of renewable technologies. If the United States became a leader in the development of renewable energy technologies, then it would likely capture the world market for this industry (Shute 2001).

The current subsidies for ethanol production total $6 billion per year (Koplow 2006). This means that the subsidies per gallon of ethanol are 60 times greater than the subsidies per gallon of gasoline!

1.13 Conclusion

This assessment of renewable energy technologies confirms that these techniques have the potential to provide the nation with alternatives to meet nearly half of future U.S. energy needs. To develop this potential, the United States would have to commit to the development and implementation of non-fossil fuel technologies and energy conservation. People in the U.S. would have to reduce their current energy consumption by more than 50% and this is entirely possible. Eventually we will be forced to reduce energy consumption. The implementation of renewable energy technologies now would reduce many of the current environmental problems associated with fossil fuel production and use.

The United States' immediate priority should be to speed the transition from the reliance on nonrenewable fossil energy resources to reliance on renewable energy technologies. Various combinations of renewable technologies should be developed consistent with the characteristics of the different geographic regions in the United States. A combination of the renewable technologies listed in Table 1.3 should provide the United States with an estimated 46 quads of renewable energy by 2050.

These technologies should be able to provide this much energy without interfering with required food and forest production.

If the United States does not commit itself to the transition from fossil to renewable energy during the next decade or two, the economy and national security will suffer. It is of critical importance that U.S. residents work together to conserve energy, land, water, and biological resources. To ensure a reasonable standard of living in the future, there must be a fair balance between human population density and use of energy, land, water, and biological resources.

References

American Energy. (2007). America's Solar Energy Potential. Retrieved July 9, 2007, from http://www/americanenergyindependence.com/solarenergy.html

Australian Government. (2007). Solar Thermal. Retrieved July 9, 2007, from http://www/greenhouse.gov.au/markets/mret/pubs/7_thermal.pdf

[AWEA] American Wind Energy Association. (2000a). Wind energy: the fuel of the future is ready today. Retrieved October 20, 2002, from http://www.awea.org/pubs/factsheets/wetoday.pdf

[AWEA] American Wind Energy Association. (2000b). Wind energy and climate change. Retrieved July 28, 2006, from www.awea.org/pubs/factsheets/climate.pdf

Bagla, P. (2003). Dream rice to curb malnutrition. Indian Express. Retrieved July 28, 2006, from http://www.indianexpress.com/res/web/pIe/full_story.php?content_id=17506

Birdsey, R. A. (1992). *Carbon storage and accumulation in United States forest ecosystems.* (Washington, DC: USDA Forest Service)

Blais, J. F., Mamouny, K., Nlombi, K., Sasseville, J. L. & Letourneau, M. (1995). Les mesures deficacite energetique dans le secteur de leau. Sassville JL and Balis JF (eds). *Les Mesures deficacite Energetique pour Lepuration des eaux Usees Municipales.* Scientific Report 405. Vol. 3. INRS-Eau, Quebec

Bradley, R. L. (1997). *Renewable Energy: Not Cheap, Not "Green."* (Washington, DC: Cato Institute)

Brown, L. (2001). Eco-Economy: *Building and Economy for the Earth.* (New York: W.W. Norton & Co.)

Burning Issues. (2003). Burning Issues Wood Fact Sheets. Retrieved May 21, 2004, from http://www/webcom.com/~bifact-sheet.htm

Cassedy, E. S. (2000). *Prospects for sustainable energy.* (New York: Cambridge University Press)

Chambers, A. (2000). Wind power spins into contention. *Power Engineering,* 104, 14–18

[DOE] U.S. Department of Energy. (2000). Consumer energy information: EREC fact sheets. Retrieved October 20, 2002, from www.eren.doe.gov/erec/factsheets/eewindows.html

[DOE] U.S. Department of Energy. (2001). Energy efficiency and renewable energy network. Retrieved October 20, 2002, from www.eren.doe.gov/state_energy/tech_solar.cfm?state=NY

[DOE] U.S. Department of Energy. (2002). Review of transport issues and comparison of infrastructure costs for a renewable fuels standard. Retrieved October 8, 2006, from http://tonto.eia.doe.gov/FTPROOT/service/question3.pdf

[DOE/EIA] U.S. Department of Energy-Energy Information Administration. (2001). Annual Energy Outlook with Projections to 2020. Washington (DC): US Department of Energy, Energy Information Administration

[DOE/EIA] U.S. Department of Energy. Energy Information Administration. (2002). Annual Energy Outlook with Projections to 2020. Washington (DC): US Department of Energy, Energy Information Administration

[DOE/EREN] U.S. Department of Energy. Energy Efficiency and Renewable Energy Network. (2001). Solar parabolic troughs: Concentrating solar power. Retrieved October 20, 2002, from www.eren.doe.gov/csp/faqs.html

Dutta, S., Rehman, I. H., Malhortra, P., & Venkata, R. P. (1997). Biogas: the Indian NGO experience. AFPRO-CHF Network Programme. (Delhi: Tata Energy Research Institute)

Ellington, R. T., Meo, M., & El-Sayed, D. A. (1993). The net greenhouse warming forcing of methanol produced from biomass. *Biomass and Bioenergy*, 4, 405–418

Energy Economics. (2007). Wind energy economics. *Wind Energy Manual*. Iowa Energy Center. Retrieved March 16, 2007, from http://www/energy.iastate.edu/renewable/wind/wem.wem-13_econ.html

[EPA] U.S. Environmental Protection Agency. (2000). *AgSTAR Digest*. (Washington, DC: US. Environmental Protection Agency)

[EPA] U.S. Environmental Protection Agency. (2002). Wood smoke. Retrieved October 20, 2002, from http://216.239.51.100/search?q=cache:bnh0QlyPH20C:www.webcom.com/~bi/brochure.pdf+wood+smoke+pollution&hl=en&ie=UTF-8

Ferguson, A. R. B. (2001). *Biomass and energy*. (Manchester, UK: Optimum Population Trust)

Ferguson, A. R. B. (2003). Wind/biomass energy capture: an update. April 2003. Optimum Population Trust. Retrieved March 16, 2007, from http://www.optimumpopulation. org/opt.af.biomass.journal03apr.PDF

Fryrear, D. W. & Bilbro, J. D. (1994). Wind erosion control with residues and related practices. (In P.W. Unger, (Ed.) *Managing Agricultural Residues* (p. 7–18). Boca Raton, FL: Lewis Publisher)

Gaudiosi, G. (1996). Offshore wind energy in the world context. *Renewable Energy*, 9, 899–904

Gawlik, K. & Kutscher, C. (2000). *Investigation of the opportunity for small-scale geothermal power plants in the Western United States*. (Golden, CO: National Renewable Energy Laboratory)

Giampietro, M., Ulgiati, A., & Pimentel, D. (1998). Feasibility of large-scale biofuel production. *BioScience*, 47, 587–600

Gleick, P. H., & Adams, A. D. (2000). *Water: the potential consequences of climate variability and change*. (Oakland, CA: Pacific Institute for Studies in Development, Environment, and Security)

Hartman, H., Angelidake, I., & Ahring, B. K. (2000). Increase of anaerobic degradation of particulate organic matter in full-scale biogas plants by mechanical maceration. *Water Science and Technology*, 41, 145–153

Hos, J. J. & Groenveld, M. J. (1987). Biomass gasification. (In D. O. Hall and R. P. Overend (Eds.), *Biomass* (pp. 237–255). Chichester (UK): John Wiley & Sons)

Illinois Corn. (2004). Ethanol's energy balance. Retrieved August 10, 2004, from http://www.ilcorn.org/Ethanol/Ethan_Studies/Ethan_Energy_Bal/ethan_energy_bal.html

Jackson, W. (1980). *New root of agriculture*. (Lincoln and London: University of Nebraska Press)

Jenkins, B. M. (1999). Pyrolysis gas. (In O. Kitani, T. Jungbluth, R. M. Peart, & A. Ramdani (Eds.), *CIGAR Handbook of Agricultural Engineering* (pp. 222–248). St. Joseph, MI: American Society of Agricultural Engineering.)

Jewell, W. J., Dell'orto, S., Fanfoli, K. J., Hates, T. D., Leuschner, A. P., & Sherman, D. F. (1980). *Anaerobic Fermentation of Agricultural Residue: Potential for Improvement and Implementation*. (Washington, DC: U.S. Department of Energy)

Kids for Change. (2006). 10 biggest environmental issues. Retrieved May 26, 2006, from http://www.infochangeindia.org/kids/10env_issues.jsp

Kishore, V. V. N. (1993). Economics of solar pond generation. (In V. V. N. Kishore (Ed.), *Renewable energy utilization: Scope, economics, and perspectives* (pp. 53–68). New Delhi: Tata Energy Research Institute)

Kitani, O. (1999). Biomass resources. In O. Kitani, T. Jungbluth, R. M. Peart, & A. Ramdami, (Eds.), *Energy and Biomass Engineering* (pp. 6–11). St. Joseph, MI: American Society of Agricultural Engineers.)

Knapp, K. E., & Jester, T. L. (2000, September). *An empirical perspective of the energy payback time for photovoltaic modules*. (Paper presented at the 28th Institute of Electrical and Electronics Engineers (IEEE) Photovoltaic Specialist Conference, Anchorage, AK

Koplow, D. (2006). *Biofuels—at what cost?*: Government support for ethanol and biodiesel in the United States. The Global Initiative (GSI) of the International Institute for Sustainable

Development (IISD). Retieved August 8, 2007, from http://www.globalsubsidies.org/IMG/pdf/ biofuels_subsidies_us.pdf

Kuby, W.R., Markoja, R., & Nackford, S. (1984). Testing and Evaluation of On-Farm Alcohol Production Facilities. Acures Corporation. Industrial Environmental Research Laboratory. Office of Research and Development. U.S. Environmental Protection Agency: Cincinnati, OH. 100p

Kunz, T. H., Arnett, E. B., Erickson, W. P., Hoar, A. R., Johnson, G. D., Larkin, R. P., Strickland, M.D., Thresher, R. E., and Tuttle, M. D. 2007. Ecological impacts of wind energy development on bats: questions, research needs, and hypotheses. *Frontiers in Ecology and the Environment*, 5(6), 315–324

Lal, R. (2003). Global potential of soil C sequestration to mitigate the greenhouse effect. *Critical Reviews in Plant Sciences*, 22(2), 151–184

Mitchell, J. E. (2000). Rangeland resource trends in the United States: A technical document supporting the 2000 USDA Forest Service RPA Assessment. RMRS-GTR-68

NAS. (2003). *Frontiers in agricultural research: Food, health, environment, and communities.* (Washington, DC: National Academy of Sciences Press)

Natural Resources Canada. (2002). Technologies and applications. Retrieved October 20, 2002, from www.canren.gc.ca/tech_appl/index.asp?Cald=6&Pgld=232

Newton, P.W. (2001). Human settlements theme report. Australian State of the Environment Report 2001. Retieved October 6, 2005, from http://www.environment.gov.au/soe/2001/ settlements/settlements02-5c.html

Nilsson, C., & Berggren, K. (2000). Alterations of riparian ecosystems caused by river regulation. *BioScience*, 50, 783–792

Nordel, L. (2001). The Montana electric transmission grid operation, congestion and issues. Briefing paper for the Montana Environmental Quality Council. Prepared by Larry Nordel, Senior Economist, Department of Environmental Quality Council

Osburn, L. & Osburn, J. (2001). Biomass resources for energy and industry. Retrieved October 20, 2002, from www.ratical.com/renewables/biomass.html

Peace Energy. 2003. Wind Energy Facts. Retrieved June 10, 2007, from http://www.peaceenergy. ca/windpower.html

Pimentel, D. (2001). Biomass utilization, limits of. In *Volume 2, Encyclopedia of Physical Sciences and Technology.* (pp. 159–171). (San Diego: Academic Press)

Pimentel, D. (2003). Ethanol fuels: energy balance, economics, and environmental impacts are negative. *Natural Resources Research*, 12(2), 127–134

Pimentel, D. (2007). Unpublished results

Pimentel, D., & Patzek, T. (2005). Ethanol production using corn, switchgrass, and wood: biodiesel production using soybean and sunflower. *Natural Resources Research*, 14(1), 65–76

Pimentel, D. & Patzek, T. (2007). Ethanol production using corn, switchgrass and wood; biodiesel production using soybean. (In K.V. Peters (Ed.), *Plants for Renewable Energy.* Binghamton: Haworth Press. In press)

Pimentel, D., Bailey, O., Kim, P., Mullaney, E., Calabrese, J., Walman, F., Nelson, F., & Yao, X. (1999). Will the limits of the Earth's resources control human populations? *Environment, Development and Sustainability*, 1(1), 19–39

Pimentel, D., Hertz, M., Glickstein, M., Zimmerman, M., Allen, R., Becker, K., Evans, J., Hussain, B., Sarsfield, R., Grosfeld, A., & Seidel, T. (2002). Renewable energy: current and potential issues. *Bioscience*, 52(12), 1111–1120

Pimentel, D., Pleasant, A., Barron, J., Gaudioso, J., Pollock, N., Chae, E., Kim, Y., Lassiter, A., Schiavoni, C., Jackson, A., Lee, M., & Eaton, A. (2004a). U.S. energy conservation and efficiency: benefits and costs. *Environment Development and Sustainability*, 6, 279–305

Pimentel, D., Berger, B., Filberto, D., Newton, M., Wolfe, B., Karabinakis, E., Clark, S., Poon, E., Abbett, E., & Nandagopal, S. (2004b). Water resources: current and future issues: *BioScience*, 54(10), 909–918

[PRB] Population Reference Bureau. (2006). *World population data sheet.* (Washington, DC: Population Reference Bureau)

Repetto, R. (1992). Accounting for environmental assets. *Scientific American*, June, 266(6) 94–100

Sagrillo, M. (2006). Telecommunication Interference from Home Wind Systems. Retrieved May 27, 2006, from http:www.awea.org/faq/sagrillos/ms_telint_0304.html

Sawin, J. L. (2004). *Mainstreaming renewable energy in the 21st century.* Worldwatch Paper 169. (Washington, DC: Worldwatch Institute)

Shute, N. (2001). The weather: global warming could cause droughts, disease, and political upheaval. *US News & World Report*, February 5, 2001, 44–52

Sinclair, K. (2003). Avian Wind Power Research. National Wind Technology Center. Report. April 29, 2003

Solar Pond. (2007). Salient-Gradient Solar Technology Page. Retrieved July 9, 2007, from http://www/ece/utep.edu/research/Energy/Pond/pond.html

Tabor, H. Z. & Doran, B. (1990). The Beith Ha'arva 5MW (e) solar pond power plant (SPPP): Progress report. *Solar Energy*, 45, 247–253

Tillman, D., Hill, J. & Lehman, C. (2006). Carbon-negative biofuels from low-input high-diversity grassland biomass. *Science*, 314, 1598–1600

Trainer, F. E. (1995). Can renewable energy sources sustain affluent society? *Energy Policy*, 23, 1009–1026

Troeh, F. R., Hobbs, J. A. & Donahue, R. L. (2004). Soil and water conservation for productivity and environmental protection. (Upper Saddle River, NJ: Prentice Hall)

Tyner, G. (2002). New energy from wind power. Retrieved June 10, 2007, from http://www.mnforsustain.org/winndpower_tyner_g_net_energy.htm

UN/SCN. United Nations. Standing Committee on Nutrition. (2004). 5th Report on the world nutrition situation: Nutrition for improved development outcomes. Retrieved July 28, 2006, from http://www.unsystem.org/scn/Publications/AnnualMeeting/SCN31/SCN5Report.pdf

[USCB] U.S. Census Bureau. (2004–2005). *Statistical Abstract of the United States* 2004–2005. (Washington, DC: U.S. Census Bureau)

[USCB] U.S. Census Bureau. (2007). *Statistical Abstract of the United States* 2007. (Washington, DC: U.S. Census Bureau)

[USDA] U.S. Department of Agriculture. (2004). *Agricultural statistics.* (Washington, DC: U.S. Department of Agriculture)

Vesterby, M. & Krupa, S. (2001). Major uses of land in the United States, 1997. Resource Economics Division, Economic Research Service, USDA, Statistical Bulletin No. 973. Retrieved October 20, 2002, from www.ers.usda.gov/publications/sb973/

Youngquist, W. (1997). *GeoDestinies: The inevitable control of earth resources over nations and individuals.* (Portland, OR: National Book Company.)

Chapter 2
Can the Earth Deliver the Biomass-for-Fuel we Demand?

Tad W. Patzek

Abstract In this work I outline the rational, science-based arguments that question current wisdom of *replacing* fossil plant fuels (coal, oil and natural gas) with fresh plant agrofuels. This 1:1 replacement is *absolutely* impossible for more than a few years, because of the ways the planet Earth works and maintains life. After these few years, the denuded Earth will be a different planet, hostile to human life. I argue that with the current set of objective constraints a *continuous stable solution* to human life cannot exist in the near-future, unless we *all* rapidly implement much more limited ways of using the Earth's resources, while reducing the global populations of cars, trucks, livestock and, eventually, also humans.

Keywords Agriculture · agrofuel · biomass · biorefinery · boundary · crop · ecology · energy · ethanol · fuel production · model · mass balance · net energy value · plantation · population · sustainability · thermodynamics · tropics · yield

2.1 Introduction

The purpose of this work is to:

1. Show that the current and proposed "cellulosic" ethanol (a "second generation" agrofuel) refineries are inefficient, low energy-density concentrators of solar light.
2. Prove that even if these refineries were marvels of efficiency, they still would be able to make but a dent in our runaway consumption of transportation fuels, because the Earth simply has little or no biomass to spare in the long run.

The fundamental energy unit I use in this work is

$$1 \text{ exajoule (EJ) or } 10^{18} \text{ joules}$$

✉ T. W. Patzek

Department of Civil and Environmental Engineering, The University of California, Berkeley, CA 94720

e-mail: patzek@patzek.ce.berkeley.edu

D. Pimentel (ed.), *Biofuels, Solar and Wind as Renewable Energy Systems*, 19
© Springer Science+Business Media B.V. 2008

A little over four joules heats one teaspoon of water by 1 degree Celsius. One statistical American develops average continuous power of almost exactly 100 W (Patzek, 2007). One exajoule in the digested food feeds amply 300 million people[1] for one year. The actual food available for consumption in the US is ca. 2 EJ yr^{-1}, and the entire food system uses $\sim$ 20 EJ yr^{-1} (Patzek, 2007). Currently, Americans are using about 105 EJ yr^{-1} (340 GJ (yr-person)$^{-1}$), or 105 times more primary energy than needed as food. The EU countries use 80 EJ yr^{-1} of primary energy or 55% less energy per capita than US.

Current consumption of all transportation fuels in the US is about 33 EJ yr^{-1}, see Fig. 2.1. A barely visible fraction of this energy comes from corn ethanol. According to current government plans, the amount of ethanol produced in the US will reach 35 billion gallons in 2017, see Fig. 2.2, but it is difficult to imagine that a 30 billion gallon per year increase will come from corn ethanol.

Before peaking[2] in 2006, the world production of conventional petroleum grew exponentially at 6.6% per year between 1880 and 1970, see Fig. 2.3. The Hubbert

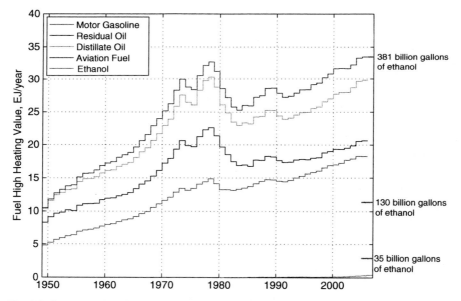

Fig. 2.1 Currently, the US consumes about 33 times more energy in transportation fuels than is necessary to feed its population. This amount of energy is equivalent to 381 billion gallons of ethanol per year. The amount of energy in corn-ethanol is barely visible and it shall always remain so unless we drastically (by a factor of two for starters) lower liquid fuel consumption. Current consumption of ethanol is about 1.2% of the total fuel consumption (without considering energy inputs to the production system)
Source: DOE EIA

---

[1] The US population in 2006.

[2] The short-lived rate peak around 1978 was caused by OPEC limiting its oil production.

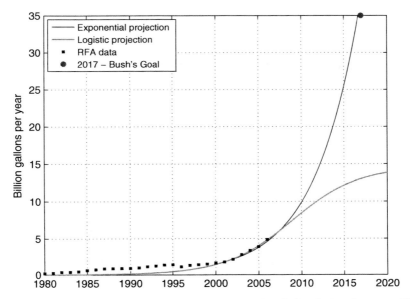

Fig. 2.2 By an exponential extrapolation of ethanol production during the last 7 years at 18.5% per year, one may arrive at 35 billion gallons per year in 2017. The less optimistic logistic fit of the data plateaus at 14 billion gallons per year. Where will the remaining 21 billion gallons of ethanol come from each year?

Sources: DOE EIA, Renewable Fuels Association (RFA)

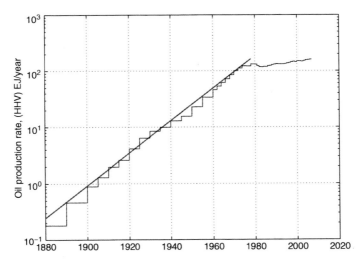

Fig. 2.3 Exponential growth of world crude oil production between 1880 and 1970 proceeded at 6.6% per year

Sources: lib.stat.cmu.edu/DASL/Datafiles/Oilproduction.html, US EIA

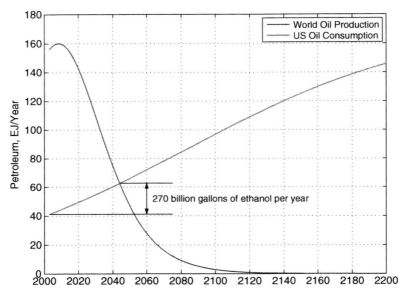

Fig. 2.4 The estimated decline of conventional petroleum production in the world is the red curve. If nothing changes, the current petroleum consumption of petroleum in the US will grow with its estimated population and intercept the global production about 35 years from today
Sources: US EIA, US Census Bureau, (Patzek, 2007)

curves are symmetrical (Patzek, 2007) and predict world production of conventional petroleum to decline exponentially at a similar rate within a decade from now, or so. This decline can be arrested for a while by heroic measures (infill drilling, horizontal wells, enhanced oil recovery methods, etc.), but the longer it is arrested the more precipitous it will become.

If the current per capita use of petroleum in the US is escalated with the expected growth of US population, the US will have to intercept the *entire* estimated production of conventional petroleum[3] in the world by 2042, see Fig. 2.4. In this scenario, the projected *increment* of US petroleum consumption between today and 2042 is equivalent to 270 billion gallons of ethanol per year.

2.2 Background

Humans *are* an integral part of a single system made of all life and all parts of the Earth's near-surface shown in Fig. 2.5. Thus, as President Vaclav Havel said on July 4, 1994: "Our destiny is not dependent merely on what we do to ourselves but also

---

[3] I stress again that I am referring to *conventional*, readily-available petroleum. There will be an offsetting production from *unconventional* sources: tar sands, ultra-heavy oil, and natural gas liquefaction, all at very high energy and environmental costs.

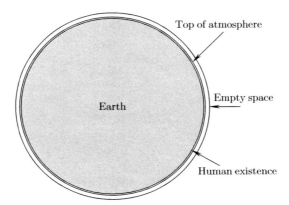

Fig. 2.5 A *system* defined by the mean Earth surface at R_{earth} and the top of the atmosphere at $R_{earth} + 100\,\mathrm{km}$, or outer space at $R_{earth} + 400\,\mathrm{km}$. Almost all of human existence occurs along the surface of the blue sphere (edge of the blue circle). As drawn here, the line thickness actually exaggerates the *thickness* of the life-giving membrane on which we exist. All radii are drawn to scale

on what we do for [the Earth] as a whole. If we endanger her, she will dispense with us in the interest of a higher value - life itself." So how to proceed?

It appears that humanity's survival is subject to these five constraints:

Constraint 1: An almost exponential rate of growth of human population, see Fig. 2.6.
Constraint 2: Too much use of Earth resources; in particular, fossil fuels; and even more specifically, liquid transportation fuels, see Fig. 2.7.
Constraint 3: The Earth that is too small to feed in perpetuity 7 billion people and counting, 1 billion cows, and – now – 1 billion cars, see Fig. 2.8.
Constraint 4: The ossified political structures in which more is better, and more of the same is also safer.
Constraint 5: A global climate change.

Unfortunately, these five constraints prevent existence of a *stable continuous* solution to human life in the near-future. Alternatively, we may choose from the following two *discontinuous* solutions:

Solution 1: Extinguish ourselves and much of the living Earth, or
Solution 2: Fundamentally and abruptly change, while slowly decreasing our numbers.

2.2.1 Problems with Change

The last time humanity ran mostly on living plant carbon was approximately in 1760. There was 1 billion of us, and we certainly knew how to feed ourselves

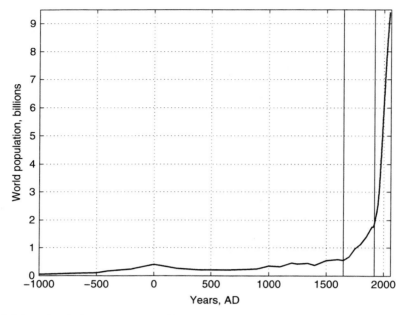

Fig. 2.6 The historical and projected world population. Note the explosive population growth since 1650, the onset of the latest Agricultural Revolution (the *left vertical line*), and its fastest stage since 1920, the start of large-scale production of ammonia fertilizer by the Haber-Bosch process (the *right vertical line*). Imagine yourself standing on the population high in 2050 and looking down
Source: US Census Bureau

due to the latest Agricultural Revolution that started in Europe a century earlier (Osborne, 1970). Our food supply problems then had to do with political madness, inaptitude, and greed – just as they do today (Davis, 2002). Today, however, there is almost 7 times more of us, see Fig. 2.6. We can still feed ourselves, but with huge inputs of fossil carbon in addition to fresh plant carbon, minerals, and soil. These inputs also mine fossil water and pollute surface water, aquifers, the oceans, and the atmosphere.

By extrapolating human population growth between 1650 and 1920 to 2007, one estimates 2.2 billion people who today could live mostly on plant carbon, but use some coal, oil, and natural gas. Therefore it is reasonable to say that today 4.5 billion people[4] owe their existence to the Haber-Bosch ammonia process and the fossil fuel-driven, fundamentally unstable "Green Revolution," as well as to vaccines and antibiotics. Agrofuels are a direct outgrowth of the "Green Revolution," which may be viewed, see Appendix 2, as a short-lived but violent disturbance of terrestrial ecosystems on the Earth.

---

[4] *All* global population increase since 1940.

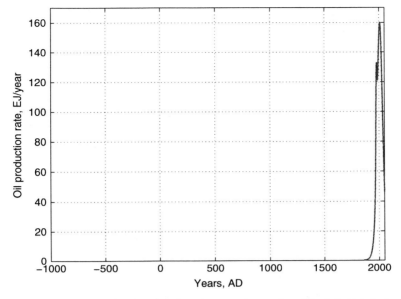

Fig. 2.7 World crude oil production plotted on the same time scale as Fig. 2.6. At today's rate of fossil and nuclear fuel consumption in the US, the global endowment of conventional petroleum would suffice to run the US for 130 years. Of course, by now, one-half of this endowment has been produced, and the US controls little of the remainder

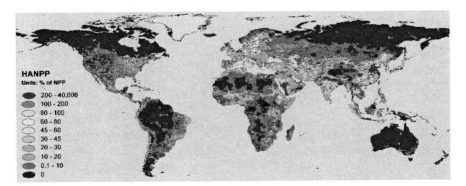

Fig. 2.8 Human-appropriated (HA) Net Primary Production (NPP) of the Earth. Global annual NPP refers to the total amount of plant growth generated each year and quantified as mass of carbon used to build stems, leaves and roots. Note that in the large portions of South and East Asia, Western Europe, Middle East, and eastern US, humans grab up to 1–2 times the net biomass production of local ecosystems. In large cities this ratio increases to 400 times. If this present human commandeering of global NPP is augmented with massive agrofuel production, the Earth ecosystems will collapse
Source: The Visible Earth, NASA images, 06-25-2004, www.nasa.gov/vision/earth/environment/0624_hanpp.html

Since most people have cooked or ridden in a vehicle, many feel empowered to talk about *energy* as though they were experts. It turns out, however, that issues of energy supply, use, environmental impacts, and – especially – of *free* energy are too complicated for the *adlib* homilies we hear every day in the media. Professor Vardaraja Raman, a well-known physicist and humanist, said it best: "A major problem confronting society is the lack of knowledge among the public as to what science is, what constitutes scientific thinking and analysis, and what science's criteria are for determining the correctness of statements about the phonomenological world."

It is a misconception that Constraint 2 can be removed with fresh plant carbon, while forgetting the scale of Constraint 1 and ignoring Constraint 3. Constraint 4 helps us to maintain that more biomass converted to liquid fuels means more of the same lifestyles, and a stable continuation of the current socioeconomic systems – Constraint 3 be damned.[5] Constraint 5 plays the role of a wild card. Its unknown negative impacts may dwarf everything else I have mentioned in this work.

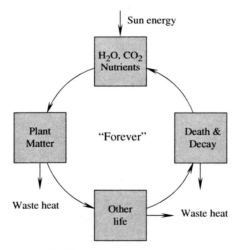

Fig. 2.9 Using sunlight, carbon dioxide, water, and the recycled nutrients, autotrophic plants generate food for heterotrophic fungi, bacteria, and animals. All die in place, and their bodies are decomposed and recycled. Almost *all* mass is conserved, and only low quality heat is exported and radiated back into space. This sustainable *earth household* (ecosystem) may function "forever" compared with the human time scale

---

[5] In his review, Dr. Silin has pointed out to me a beautiful paper by von Engelhardt et al. (1975). This chapter contains several ideas similar or identical to the ideas expressed here. The following statement is particularly salient: "This [*collective human experience of exponential growth*] has fostered the popular notion that growth is synonymous with progress and that further improvements in the quality of human life will be contingent upon steady or increasing rate of growth, even though growth at an increasing rate cannot be sustained indefinitely within the physical limits of a finite earth."

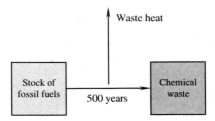

Fig. 2.10 A linear process of converting a stock of fossil fuels into waste matter and heat cannot be sustainable. The waste heat is exported to the universe, but the chemical waste accumulates. To replenish some of the fossil fuel stock, it will take another 50–400 million years of photosynthesis, burial, and entrapment

This leads me to the first major conclusion of this work:

Business as usual will lead to a complete and practically immediate crash of the technically advanced societies and, perhaps, all humanity. This outcome will not be much different from a collapse of an overgrown colony of bacteria on a petri dish when its sugar food runs out and waste products build up. Today, the human "petri dish" is Earth's surface in Fig. 2.5, and "food" is the living matter and water we consume and the ancient plant products and minerals (oil, natural gas, coal, etc.) we mine and burn.

The Earth operates in *endless* cycles as in Fig. 2.9, and modern humans race along *short* line segments, as in Fig. 2.10 and 2.7. At each turn of her cycles, the Earth renews herself, but humans are about to wake up inside a huge toxic waste dump with nowhere to go.

2.3 Plan of Attack

As you are beginning to suspect, it is not sufficient to limit ourselves just to discussing liquid transportation fuels and their future biological sources. These transportation fuels intrude upon every other aspect of life on the Earth: Availability of clean water to drink and clean air to breathe, healthy soil and healthy food supply, destruction of biodiversity and essential planetary services in the tropics, acceleration of global climate change, and so on.

As with many important policy-making decision processes, I start from the end, here the cellulosic ethanol refineries. This is where most public money, attention, and hope are. I show that these refineries are inefficient compared with the existing petroleum- and corn-based refineries, and are difficult to scale up.

Then I return to the beginning and show that even *if* the cellulosic biomass refineries were marvels of efficiency, they still could *not* maintain our current lifestyles by a long stretch, simply because the Earth will not give us the extra

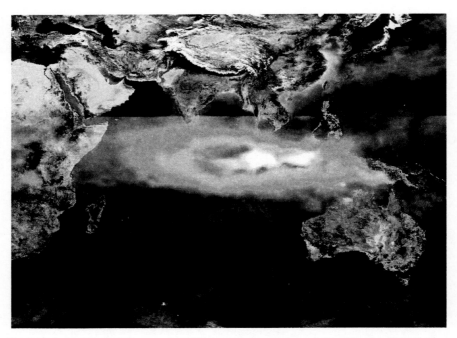

Fig. 2.11 In the fall of 1997, an orgy of 176 fires in Indonesia burned 12 million ha of virgin forest and generated as much greenhouse gases as the US in *one year*. 133 of these illegal fires were started by oil palm plantation/logging companies to steal old-growth trees and burn the rest for new plantations. The smoke and ozone plume had global extent
Sources: NASA's Earth Probe Total Ozone Mapping Spectrometer (TOMS), October 22, 1997; (Schimel and Baker, 2002; Page et al., 2002; Patzek and Patzek, 2007)

biomass needed to keep on existing as we do. For a while we might continue to *rob* this biomass from the poor tropics, but the results are already disastrous for all humanity, see Fig. 2.11.

2.4 Efficiency of Cellulosic Ethanol Refineries

I start from a "reverse-engineering" calculation of energy efficiency of cellulosic ethanol production in an *existing* Iogen pilot plant, Ottawa, Canada. I then discuss the inflated energy efficiency claims of five out-of-six recipients of $385 millions of DOE grants to develop cellulosic ethanol refineries.

2.4.1 Iogen Ottawa Facility

Wheat, oat, and barley straw are first pretreated with sulfuric acid and steam. Iogen's patented enzyme then breaks the cellulose and hemicelluloses down into six- and

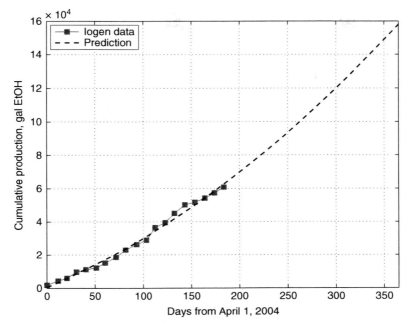

Fig. 2.12 Ethanol production in Iogen's Ottawa plant. Extrapolation to one year yields 158 000 gallons. Note that the data points are evenly spaced as they should be for regularly scheduled batches. Source: Jeff Passmore, Executive Vice President, Iogen Corporation, *Cellulose ethanol is ready to go*, Presentation to Governor's Ethanol Coalition & US EPA Environmental Meeting "Ethanol and the Environment," Feb. 10, 2006

five-carbon sugars, which are later fermented and distilled into ethanol. Normal yeast does not ferment the 5-carbon sugars, so genetically modified, delicate and patented yeast strains are used. Iogen's plant has capacity of 1 million gallons of ethanol per year. The only *published* ethanol production is shown in Fig. 2.12.

From Fig. 2.12 and a presentation[6] by Maurice Hladik, Director of Marketing, Iogen Corp., the following can be deduced:

- 158,000 gallons/year of anhydrous ethanol (EtOH), or 10 bbl EtOH/day = 6.7 bbl of equivalent gasoline/day were actually produced. In press interviews, Iogen claims to be producing 790,000 gallons of ethanol[7] per year.
- There exists $2 \times 52,000 = 104,000$ gallons of fermentation tank volume.
- The actual ethanol production and tank volume give the ratio of 1.5 gallons of ethanol per gallon of fermenter and per year.

---

[6] *Cellulose Ethanol is ready to go*. Renewable Fuels Summit, June 12, 2006.

[7] *It's Happening in Ottawa – Grains become fuel at the world's first cellulosic ethanol demo plant*, Grist, Sharon Boddy, 12 Dec., 2006. It is possible that the notoriously innumerate journalists confused liters with gallons: 790,000 liters is 200,000 gallons, much closer to the published data from Iogen.

- I assume 7-day batches + 2-day cleanups.
- Thus, there is ca. 4% of alcohol in a batch of industrial wheat-straw broth in contrast to 12 to 16% of ethanol in corn-ethanol refinery broths.

Since wheat is the largest grain crop in Canada, I use its straw as a reference (the other two straws are similar). On a dry mass basis (dmb), wheat straw has 33% of cellulose, 23% of hemicelluloses, and 17% total lignin.[8]. Other sources report 38%, 29%, and 15% dmb, respectively, see (Lee et al., 2007) for a data compliation. These differences are not surprising, given experimental uncertainties and variable biomass composition. To calculate ethanol yield, I use the more optimistic, second set of data. The respective conversion efficiencies, assumed after Badger (2002), are listed in Table 2.1.

The calculated ethanol yield, $0.18\,kg$ EtOH (kg straw dmb)$^{-1}$, is somewhat less than a recently reported maximum ethanol yield of 0.24 kg/kg (Saha et al., 2005) achieved in 500 mL vessels, starting from 48.6% of cellulose. Simultaneous saccharification and fermentation yielded 0.17 kg/kg, see Table 2.5 in Saha et al. (2005).

Because enzymatic decomposition of cellulose and hemicelluloses is inefficient, the resulting dilute broth requires 2.4 times more energy to distill than the average $15\,MJL^{-1}$ in an average ethanol refinery (Patzek, 2004; Patzek, 2006a), see Fig. 2.13.

One may argue that Iogen's Ottawa facility is for demonstration purposes only and that the saccharification and fermentation batches were not regularly scheduled. In this case, an alternative calculation yields the same result: At about 0.2 to 0.25 kg of straw/L, the mash is barely pumpable. With Badger's yield of 0.18 kg/kg of EtOH, the highest ethanol yield is $3.5 - 4.4\%$ of ethanol in water.

The higher heating value (HHV) of ethanol is $29.6\,MJ\,kg^{-1}$ (Patzek, 2004). The HHV of wheat straw is $18.1\,MJ\,kg^{-1}$ (Schmidt et al., 1993) and that of lignin $21.2\,MJ\,kg^{-1}$ (Domalski et al., 1987).

With these inputs the first-law (energy) efficiency of Iogen's facility is

$$\eta = \frac{0.28 \times 29.6}{1 \times 18.1 + 0.18 \times 2.4 \times 15/0.787 - 0.15 \times 21.2} \approx 20\% \qquad (2.1)$$

Table 2.1 Yields of ethanol from cellulose and hemicellulose

Step	Cellulose	Hemicellulose
Dry straw	1 kg	1 kg
Mass fraction	×0.38	×0.29
Enzymatic conversion efficiency	×0.76	×0.90
Ethanol stoichiometric yield	×0.51	×0.51
Fermentation efficiency	×0.75	×0.50
EtOH Yield, kg	0.111	0.067

Source: Badger (2002)

---

[8] Biomass feedstock composition and property database. Department of Energy, Biomass Program, www.eere.energy.gov/biomass/progs/searchl.egi, accessed July 25, 2007.

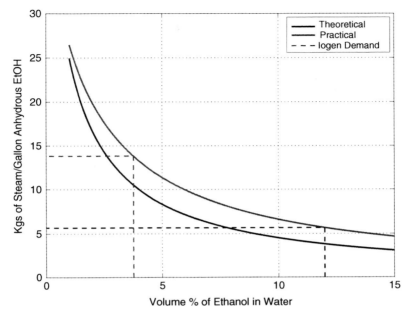

Fig. 2.13 Steam requirement in ethanol broth distillation. The 3.7% broth requires 2.4 times more steam than a 12% broth
Source (Jacques et al., 2003)

where the density of ethanol is $0.787 \, \text{kg} \, \text{L}^{-1}$ and the entire HHV of lignin was used to offset distillation fuel (another optimistic assumption for the wet separated lignin). This calculation disregards the energy costs of high-pressure steam treatments of the straw at 120 or 140°C, and the separated solids at 190°C, sulfuric acid and sodium hydroxide production, etc. Also, the complex enzyme production processes must use plenty of energy.

This analysis leads to the second conclusion:

> The Iogen plant in Ottawa, Canada, has operated well below name plate capacity for three years. Iogen should retain their trade secrets, but in exchange for the significant subsidies from the US and Canadian taxpayers they should tell us what the annual production of alcohols *was*, how much straw was used, and what the fossil fuel and electricity inputs were. The ethanol yield coefficient in kg of ethanol per kg straw dmb is key to public assessments of the new technology. Similar remarks pertain to the Novozymes projects heavily subsidized by the Danes. Until an existing pilot plant provides real, independently verified data on yield coefficients, mash ethanol concentrations, etc., all proposed cellulosic ethanol refinery designs are *speculation*.

2.4.2 *Proposed Cellulosic Ethanol Refineries*

Now I present at *face value* the stated energy efficiencies[9] of the six proposed[10] cellulosic ethanol plants awarded 385 million USD by the US Department of Energy.

Figure 2.14 ranks the rather imaginary claims of 5 out of 6 award recipients. For calibration, after 87 years of development and optimization, the actual energy efficiency of Sasol's Fischer-Tropsch coal-to-liquid fuels plants is about 42% (Steynberg and Nel, 2004). The average energy efficiency of the highly optimized corn ethanol refineries is 37% (not counting grain coproducts as fuels). An average petroleum refinery is about 88% energy-efficient.[11] For details, see (Patzek, 2006a,b,c) The DOE/USDA report by Perlack et al. (2005) has led to the claims by an influential venture capitalist, Mr. Vinod Khosla (2006), of being able to produce 130 billion gallons of ethanol from 1.4 billion tons of biomass (dmb), apparently at a 52% thermodynamic efficiency.

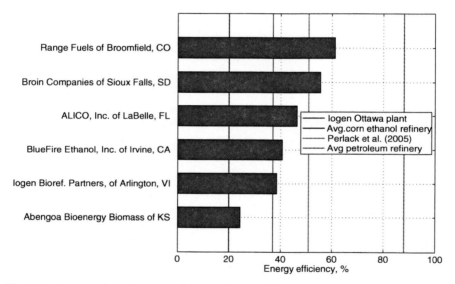

Fig. 2.14 Stated energy efficiencies of the six future cellulosic ethanol refineries awarded $385 millions in DOE grants. The calculated energy efficiency (*left line*) of an existing cellulosic ethanol refinery in Ottawa serves to calibrate the rather inflated efficiency claims of 5/6 grant recipients. Energy efficiencies of an average ethanol refinery and petroleum refinery (Patzek, 2006a) are also shown (second and last line from the *left*)

---

[9] The HHV of ethanol out divided by the HHV of biomass in. No fossil fuels inputs into the plants and the raw materials they use are accounted for.

[10] Environmental and Energy Study Institute, 122 C Street, N. W., Suite 630 Washington, D. C., 2001, www.eesi.org/publications/Press%20Releases/2007/2-28-07_doe_biorefinery_awards.pdf

[11] As pointed out by Drs. John Benemann and John Newman, this comparison may be unfair. No liquid fuel technology will ever match petroleum refining, but petroleum-derived fuels will not last for very long.

To see how very different the new fossil-energy-free world will be, let's compare power from Iogen's plant with that from an oil well in the US. Ever more *power* is what we must have to continue our current way of life (cf. Footnote 5). Iogen's plant delivers the power of 7 barrels of oil per day (68 kW). Average power of petroleum wells in the largely oil-depleted US was 10 bbl (well-day)$^{-1}$ in 2006[12] (98 kW). Therefore, an average US petroleum well delivers more power than a city-block size Iogen facility in Ottawa and its area of straw collection, probably 50 km in radius, which at this time is saturated with fossil fuels outright and their products (ammonia fertilizers, field chemicals, roads, etc.). The petroleum well also uses little input power; unfortunately, soon petroleum will not be a transportation option. Such is the difference between solar energy stocks (depletable fossil fuels) and flows (daily photosynthesis).

One can calculate that an average agricultural worker in the US uses 800 kW of fossil energy inputs and outputs 3,000 kW. An average oil & gas worker in California uses 2,800 kW of fossil energy inputs and outputs 14,500 kW. Due to fossil energy and machines these two workers are supermen, each capable of doing the work of 8,000 and 28,000 ordinary humans, respectively. These two fellows are about to become human again, and we need to get used to this idea.

Now, you may want to go back to Section 2.2.1 and reread it.

2.5 Where will the Agrofuel Biomass Come from?

Collectively, the EU and the US have spent billions of dollars to be able to construct the inefficient behemoth factories, which in the distant future might ingest mega-tonnes or gigatonnes of apparently free biomass "trash" and spit out priceless liquid transportation fuels. It is therefore prudent to ask the following question: Call out using the new paragraph and gray background.

The answer to this question is immediate and unequivocal: Nowhere, close to nothing, and for a very short time indeed. On the average, our planet has **zero** excess biomass at her disposal.

2.5.1 Useful Terminology

Several different ecosystem[13] productivities, i.e., measures of biomass accumulation per unit area and unit time have been used in the ecological literature, e.g., (Reichle et al., 1975; Randerson et al., 2001) and many others. Usually this biomass is expressed as grams of carbon (C) per square meter and per year, or as grams of water-free biomass (dmb) per square meter and year.[14] The conversion

---

[12] See www.cia.doe.gov/emeu/aer/txt/ptb0502.html, accessed July 25, 2007.

[13] An ecosystem is defined in more detail in Appendix 1.

[14] Or as kilograms (dmb) of biomass per hectare and per year.

factor between these two estimates is the carbon mass fraction in the fundamental building blocks of biomass, CH_xO_y, where x and y are real numbers, e.g., 1.6 and 0.6, that express the overall mass ratios of hydrogen and oxygen to carbon. The following definitions are common in ecology:

1. *Gross* Primary Productivity, GPP = mass of CO_2 fixed by plants as glucose.
2. Ecosystem respiration, R_e = mass of CO_2 released by metabolic activity of autotrophs, R_a, and heterotrophs (consumers and decomposers), R_h:

$$R_e = R_a + R_h \tag{2.2}$$

where decomposers are defined as worms, bacteria, fungi, etc. Plants respire about 1/2 of the carbon available from photosynthesis after photorespiration, with the remainder available for growth, propagation, and litter production, see (Ryan, 1991). Heterotrophs respire most, 82–95%, of the biomass left after plant respiration (Randerson et al., 2001).

3. *Net* Primary Productivity, NPP = GPP $- R_a$.
4. Net *Ecosystem* Productivity

$$\text{NEP} = \text{GPP} - R_e - \text{Non} - R \text{ sinks and flows} \tag{2.3}$$

The older NEP definitions would usually neglect the non respiratory losses, e.g., (Reichle et al., 1975). All ecological definitions of NEP I have seen, lump incorrectly mass flows and mass sources and sinks, calling them "fluxes," see, e.g., (Randerson et al., 2001; Lugo and Brown, 1986). For more details, see Appendix 2.

The typical net primary productivities of different ecosystems are listed in Appendix 3.

2.5.2 Plant Biomass Production

The reason for the Earth recycling all of her material parts can be explained by looking again at Fig. 2.5. The Earth is powered by the sun's radiation that crosses the outer boundary of her atmosphere and reaches her surface. The Earth can export into outer space long-wave infrared radiation.[15] But, because of her size, the Earth holds on to all mass of all chemical elements, except perhaps for hydrogen. By maintaining an oxygen-rich atmosphere, *life* has managed to prevent the airborne hydrogen from escaping Earth's gravity by reacting it back to water (and destroying ozone).

---

[15] Therefore, the Earth is an *open* system with respect to electromagnetic radiation. Life could emerge on her and be sustained for 3.5 eons because of this openness.

If all mass must stay on the Earth, all her households must recycle every-thing; otherwise internal chemical waste would build up and gradually kill them. Mother Nature does not usually do toxic waste landfills and spills.

In a mature ecosystem, one species' waste must be another species' food and no *net* waste is ever created, see Fig. 2.9. The little imperfections in the Earth's *surface* recycling programs have resulted in the burial of a remarkably *tiny* fraction of plant carbon in swamps, lakes, and shallow coastal waters[16], see Fig. 2.15. Very rarely the violent anoxic events would kill most of life in the oceanic waters and cause faster carbon burial. Over the last 460,000,000 years (and going back all the way

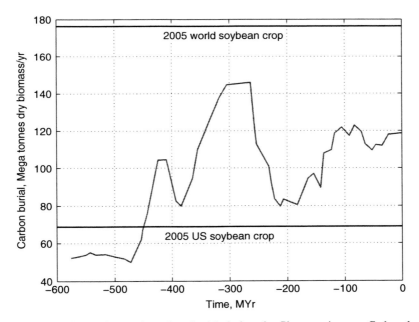

Fig. 2.15 Plot of global organic carbon burial during the Phanerozoic eon. Carbon burial rate modified from Berner (2001, 2003). The units of carbon burial have been changed from 10^{18} mol C Myr^{-1} to Mt biomass yr^{-1}. The very high carbon burial values centered around 300 Myr ago are due predominantly to terrestrial carbon burial and coal formation. Most plants have been buried in swamps, shallow lakes, estuaries, and shallow coastal waters. Note that historically the average rate of carbon burial on the Earth has been *tiny*, half-way between the US- and world crops of soybeans in 2005. This burial rate amounts to $120 \times 10^6 / 110 \times 10^9 \times 100\% = 0.1\%$ of global NPP of biomass

---

[16] Much of this burial has been eliminated by humans. We have paved over most of the swamps and destroyed much of the coastal mangrove forests, the highest-rate local sources of terrestrial biomass transfer into seawater.

to 2,500,000,000 years ago), the Earth has gathered and transformed *some* of the buried ancient plant mass into the fossil fuels we love and loath so much.

The proper mass balance of carbon fluxes in terrestrial ecosystems, see Appendix 2, confirms the compelling, thermodynamic argument that sustainability of any ecosystem requires *all* mass to be conserved on the average. The larger the spatial scale of an ecosystem and the longer the time-averaging scale are, the stricter adherence to this rule must be. Such are the laws of nature.

Physics, chemistry and biology say clearly that there can be *no* sustained net mass *output* from any ecosystem for more than a few years. A young forest in a temperate climate grows fast in a clear-cut area, see Fig. 2.16, and transfers nutrients from soil to the young trees. The young trees grow very fast (there is a positive NPP), but the amount of mass accumulated in the forest is small. When a tree burns or dies some or most of its nutrients go back to the soil. When this tree is logged and hauled away, almost no nutrients are returned. After logging young trees a couple of times the forest soil becomes depleted, while the populations of insects and pathogens are well-established, and the forest productivity rapidly declines (Patzek and Pimentel, 2006). When the forest is allowed to grow long enough, its net ecosystem productivity becomes **zero** on the average.

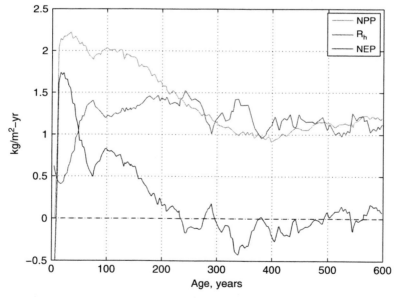

Fig. 2.16 Forest ecosystem biomass fluxes simulated for a typical stand in the H. J. Andrews Experimental Forest. The Net Primary Productivity (NPP), the heterotrophic respiration (R_h), and the Net Ecosystem Productivity (NEP) are all strongly dependent on stand age. This particular stand builds more plant mass than heterotrophs consume for 200 years. After that, for any particular year, an old-growth stand is in steady state and its average net ecosystem productivity is zero. Adapted from Songa and Woodcock (2003)

Therefore, in order to export biomass (mostly water, but also carbon, oxygen, hydrogen and a plethora of nutrients) an ecosystem must import equivalent quantities of the chemical elements it lost, or decline irreversibly. Carbon comes from the atmospheric CO_2 and water flows in as rain, rivers and irrigation from mined aquifers and lakes. The other nutrients, however, must be rapidly *produced* from ancient plant matter transformed into methane, coal, petroleum, phosphates,[17] etc., as well as from earth minerals (muriate of potash, dolomites, etc.), – all *irreversibly* mined by humans. Therefore, to the extent that humans are no longer integrated with the ecosystems in which they live, they are doomed to extinction by exhausting all planetary stocks of minerals, soil and clean water. The question is not *if*, but *how fast*?

It seems that with the exponentially accelerating *mining* of global ecosystems for biomass, the time scale of our extinction is shrinking with each crop harvest. Compare this statement with the feverish proclamations of sustainable biomass and agrofuel production that flood us from the confused media outlets, peer-reviewed journals, and politicians.

2.5.3 Is There any Other Proof of NEP = 0?

I just gave you an abstract proof of no trash production in Earth's Kingdom, except for its dirty human slums.

Are there any other, more direct proofs, perhaps based on measurements? It turns out that there are two approaches that complement each other and lead to the same conclusions. The first approach is based on a top-down view of the Earth from a satellite and a mapping of the reflected infrared spectra into biomass growth. I will summarize this proof here. The second approach involves a direct counting of all crops, grass, and trees, and translating the weighed or otherwise measured biomass into net primary productivity of ecosystems. Both approaches yield very similar results.

2.5.4 Satellite Sensor-Based Estimates

Global ecosystem productivity can be estimated by combining remote sensing with a carbon cycle analysis. The US National Aeronautics and Space Administration

---

[17] Over millions of years, the annual cycles of life and death in ocean upwelling zones have propelled sedimentation of organic matter. Critters expire or are eaten, and their shredded carcasses accumulate in sediments as fecal pellets and as gelatinous flocs termed marine snow. Decay of some of this deposited organic matter consumes virtually all of the dissolved oxygen near the seafloor, a natural process that permits formation of finely-layered, organic-rich muds. These muds are a biogeochemical "strange brew," where calcium – derived directly from seawater or from the shells of calcareous plankton – and phosphorus – generally derived from bacterial decay of organic matter and dissolution of fish bones and scales – combine *over geological time* to form pencil-thin laminae and discrete sand to pebble-sized grains of phosphate minerals. Source: Grimm (1998).

(NASA) Earth Observing System (EOS) currently "produces a regular global estimate of gross primary productivity (GPP) and annual net primary productivity (NPP) of the entire terrestrial earth surface at 1-km spatial resolution, 150 million cells, each having GPP and NPP computed individually" (Running et al., 2000). The MOD17A2/A3 User's Guide (Heinsch et al., 2003) provides a description of the Gross and Net Primary Productivity estimation algorithms (MOD17A2/A3) designed for the MODIS[18] sensor.

The sample calculation results based on the MOD17A2/A3 algorithm are listed in Table 2.2. The NPPs for Asia Pacific, South America, and Europe, relative to North America, are shown in Fig. 2.17. The phenomenal net ecosystem productivity of Asia Pacific is 4.2 larger than that of North America. The South American ecosystems deliver 2.7 times more than their North American counterparts, and Europe just 0.85. It is no surprise then that the World Bank[19], as well as agribusiness and logging companies – Archer Daniel Midlands (ADM), Bunge, Cargill, Monsanto, CFBC, Safbois, Sodefor, ITB, Trans-M, and many others – all have moved in force to plunder the most productive tropical regions of the world, see Fig. 2.18.

Table 2.2 Version 4.8 NPP/GPP global sums (posted: 01 Feb 2007)[a]

Year[b]	GPP (Pg C/yr[c])	NPP[d] (Pg C/yr)
2000	111	53
2001	111	53
2002	107	51
2003	108	51
2004	109	52
2005	108	51

[a] Numerical Terradynamic Simulation Group, The University of Montana, Missoula, MT 59812, images.ntsg.umt.edu/index.php.

[b] 2000 and 2001 were La Niña years, and 2002 and 2003 were weak El Niño years.

[c] 1 Pg C = 1 peta gram of carbon = 10^{15} grams = 1 billion tonnes = 1 Gt of carbon. 50 Gt of carbon per year is equivalent to 1800 EJ yr^{-1}.

[d] This represents all above-ground production of living plants and their roots. Humans cannot dig up all the roots on the Earth, so effectively ~1/2 NPP might be available to humans *if* all other heterotrophs living on the Earth stopped eating.

---

[18] MODIS (or Moderate Resolution Imaging Spectroradiometer) is a key instrument aboard the Aqua and Terra satellites. The MODIS instrument provides high radiometric sensitivity (12 bit) in 36 spectral bands ranging in wavelength from 0.4 to 14.4 μm. MODIS provides global maps of several land surface characteristics, including surface reflectance, albedo (the percent of total solar energy that is reflected back from the surface), land surface temperature, and vegetation indices. Vegetation indices tell scientists how densely or sparsely vegetated a region is and help them to determine how much of the sunlight that could be used for photosynthesis is being absorbed by the vegetation. Source: modis.gsfc.nasa.gov/about/media/modis_brochure.pdf.

[19] Source: (Anonymous, 2007). The World Bank through its huge loans is behind the largest-ever destruction of tropical forest in the equatorial Africa.

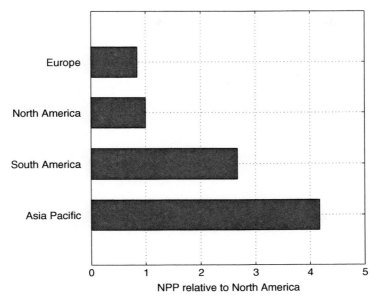

Fig. 2.17 NPP's of Asia-Pacific, South America, and Europe – relative to North America
Source: MOD17A2/A3 model

According to a MODIS-based calculation (Roberts and Wooster, 2007) of biomass burned in Africa in February and August 2004, prior to the fires shown here, the resulting carbon dioxide emissions were 120 and 160 million tonnes per month, respectively.

The final result of this global "end-game" of ecological destruction will be an unmitigated and lightening-fast collapse of ecosystems protecting a large portion of humanity.[20]

2.5.5 NPP in the US

The overall median values of net primary productivity may be converted to the higher heating value (HHV) of NPP in the US, see Fig. 2.19. In 2003, thus estimated net annual biomass production in the US was 5.3 Gt and its HHV was 90 EJ. One must be careful, however, because the underlying distributions of ecosystem productivity are different for each ecosystem and highly asymmetric. Therefore, lumping them together and using just one median value can lead to a substantial systematic error. For example, the lumped value of US NPP of 90 EJ, underestimates the overall

---

[20] For example, in the next 20 years, Australia may gain another 100 million refugees from the depleted Indonesia, look at Haiti for the clues.

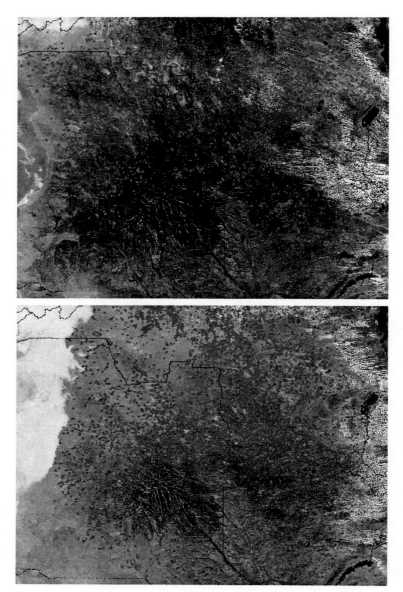

Fig. 2.18 Hundreds of fires were burning in the Democratic Republic of Congo and Angola on Dec 16, 2005 (*top*), and Aug 11, 2006 (*bottom*). Most of the fires are set by humans to clear land for farming, rangelands, and industrial biomass plantations. In this way, vast areas of the continent are being irreversibly transformed
Source: Satellite Aqua, 2 km pixels size. Images courtesy MODIS Land Rapid Response Team at NASA

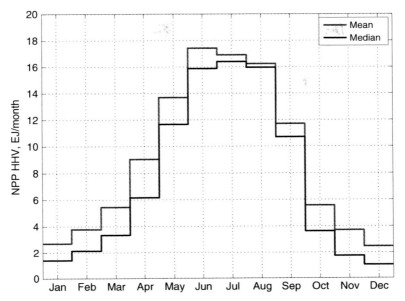

Fig. 2.19 A MOD17A2/A3-based calculation of US NPP in the year 2003. Monthly data for the mean and median GPP were acquired from images.ntsg.umt.edu/browse.php. The land area of the 48 contiguous states plus the District of Columbia $= 7444068\,\text{km}^2$. Conversion to higher heating values (HHV) was performed assuming $17\,\text{MJ kg}^{-1}$ dmb biomass. Conversion from kg C to kg biomass was 2.2, see Footnote b in Table 2.6 in Appendix 3. NPP $= 0.47\times$ GPP for 2003. The robust *median* productivity estimate of the 2003 US NPP is $90\,\text{EJ yr}^{-1}$

2003 estimate[21] of $0.408 \times 7444068 \times 10^6 \times 17 \times 10^6 \times 2.2 \times 10^{-18} = 113\,\text{EJ}$ by some 20%.

To limit this error, one can perform a more detailed calculation based on the 16 classes of land cover listed in Table 2.2 in (Hurtt et al., 2001). The MODIS-derived median NPPs are reported for most of these classes. The calculation inputs are shown in Table 2.3. Since the spatial set of land-cover classes cannot be easily mapped onto the administrative set of USDA classes of cropland, woodland, pastureland/rangeland, and forests, Hurtt et al. (2001) provide an approximate linear mapping between these two sets, in the form of a 16×4 matrix of coefficients between 0 and 1. I have lumped the land-cover classes somewhat differently (to be closer to USDA's classes), and the results are shown in Table 2.4 and Fig. 2.20.

The Cropland + Mosaic class here comprises the USDA's cropland, woodland, and some of the pasture classes. The Remote Vegetation class comprises some of the USDA's rangeland and pastureland classes. The USDA forest class is somewhat larger than here, as some of the smaller patches of forest, such as parks, etc., are in the Mosaic class. Thus calculated 2003 US NPP is $118\,\text{EJ yr}^{-1}$, $74\,\text{EJ yr}^{-1}$ of

---

[21] The median 2003 US NPP of $0.408\,\text{kg Cm}^{-2}\,\text{yr}^{-1}$ was posted at images.ntsg.umt.edu/browse.php.

Table 2.3 The 2003 US NPP by ground cover class

	Class[a]	Area[a] 10^6 ha	2003 US NPP[b] 10^6 t ha^{-1} yr^{-1}	Root:shoot[c]
1	Cropland + Mosaic[d]	219	893	0.318
2	Grassland	123	603	4.224
3	Mixed forest	38	1159	0.456
4	Woody savannah[e]	33	1694	0.642
5	Open shrubland[f]	124	620	1.063
6	Closed shrubland[g]	3	966	1.063
7	Deciduous broadleaf forest	95	1153	0.456
8	Evergreen needleleaf forest	118	1153	0.403

[a] Table 2.2 in (Hurtt et al., 2001).
[b] Numerical Terradynamic Simulation Group, The University of Montana, Missoula, MT 59812, images.ntsg.umt.edu/index.php.
[c] Table 2.2 in (Mokany et al., 2006).
[d] Lands with a mosaic of croplands, forests, shrublands and grasslands in which no one component covers more than 60% of the landscape.
[e] Herbaceous and other understory systems with forest canopy cover over 30 and 60%.
[f] Woody vegetation with less than 2 m tall and with shrub cover 10 to 60%.
[g] Woody vegetation with less than 2 m tall and with shrub cover >60%.

above-ground (AG) plant construction and 44 EJ yr^{-1} in root construction. In addition $12/74 = 17\%$ of AG vegetation is in remote areas, not counting the remote forested areas. Note that my use of land-cover classes and their typical root-to-shoot ratios yields an overall result (118 EJ yr^{-1}) which is very similar to that derived by the Numerical Terradynamic Simulation Group (113 EJ yr^{-1}).

Therefore, the DOE/USDA proposal to produce 130 billion gallons of ethanol from 1400 million tonnes of biomass (Perlack et al., 2005) each year – and year-after-year –, would consume 32% of the remaining above-ground NPP in the

Table 2.4 The 2003 US NPP by lumped ground cover classes

	Class[a]	Area[a] 10^6 ha	2003 US NPP[b] 10^6 t ha^{-1} yr^{-1}	HHV[c] EJ yr^{-1}
1	Cropland + Mosaic	219	1484.8	25.2
2	Pastures	123	142.3	2.4
3	Remote vegetation[d]	160	724.1	12.3
4	Forest[e]	252	2030.0	34.5
5	Roots[f]	754	2575.0	43.8

[a] Derived from Table 2.2 in (Hurtt et al., 2001) and USDA classes
[b] In classes 1 − 4, only above-ground biomass is reported. Class 5 lumps all the roots. The calculations here are based on Table 2.3 with the multiplier of 2.2 to convert from carbon to biomass.
[c] The higher heating value with 17 MJ kg^{-1} on the average.
[d] Classes 4 + 5 + 6 in Table 2.3.
[e] Classes 3 + 7 + 8 in Table 2.3.
[f] Note that roots comprise $44/74 = 59\%$ of NPP. Also the land cover classes here account for 97% of US land area.

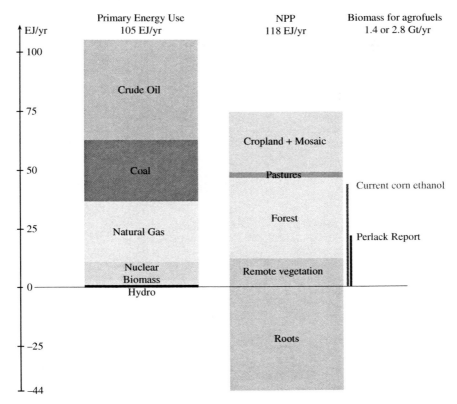

Fig. 2.20 Primary energy consumption and net primary productivity (NPP) in the US in 2003. The annual growth of all biomass in the 48 contiguous states plus the District of Columbia has been translated from gigatonnes per year to the higher heating value of this biomass growth in exajoules per year. The USDA/DOE proposal (Perlack et al., 2005) to produce 130 billion gallons of ethanol per year from 1.4 billion tonnes of biomass would consume 32% of above-ground NPP in the US at a 52% conversion efficiency, or 64% at the current efficiency of the corn-ethanol cycle (Patzek, 2006a)
Sources: EIA, Numerical Terradynamic Simulation Group, and (Patzek, 2007)

US, see Fig. 2.20, if one assumes a 52% energy-efficiency of the conversion.[22] At the current 26% overall efficiency of the corn-ethanol cycle (Patzek, 2006a), roughly 64% of all AG NPP in the US would have to be consumed to achieve this goal with zero harvest losses.[23] To use more than half of all accessible above-ground plant growth in all forests, rangeland, pastureland and agriculture in the US to produce

---

[22] As I mentioned before, this efficiency is close to the *theoretical* thermodynamic efficiency of the Fischer-Tropsch process *never* practically achieved with coal, let alone biomass. After 87 years of research and production experience current F. T coal plants achieve a 42% efficiency, see, e.g., (Steynberg and Nel, 2004).

[23] In forestry, roughly 1/2 of AG biomass is exported as tree logs; the rest is lost and burned.

agrofuels would be a continental-scale ecologic and economic disaster of biblical proportions.[24]

2.6 Conclusions

I have shown that the Earth simply cannot produce the vast quantities of biomass we want to use to prolong our unsustainable lifestyles, while slowly committing suicide as a global human civilization.

In passing, I have noted that the "cellulosic biomass" refineries are very inefficient, currently impossible to scale, and incapable of ever catching up with the runaway need to feed one billion gasoline- and diesel-powered cars and trucks.

Acknowledgments This work was carefully reviewed and critiqued by Drs. John Benemann, Ignacio Chapela, John Newman, Ron Steenblik, Ron Swenson, and Dmitriy Silin, as well as my Ph.D. graduate student, Mr. Greg Croft, and my son Lucas Patzek. I am very grateful to the reviewers for their valuable suggestions, thoroughness, directness, and dry sense of humor.

The opnions expressed in this work are those of the author, who is solely responsible for its content and any errors or omissions.

References

Anonymous 2007, *Carving up the Congo*, Report, Parts I – III, Greenpeace, Washington, DC, www.greenpeace.org/usa/news/rainforest-destruction-in-afri

Badger, P. C. 2002, *Trends in new crops and new uses*, Chapter. Ethanol from Cellulose: A General Review, pp 17–21, ASHS Press, Alexandria, VA.

Berner, R. A. 2001, Modeling atmospheric O_2 over Phanerozoic time, *Geochim. Cosmochim. Acta* **65**: 685–694.

Berner, R. A. 2003, The long-term carbon cycle, fossil fuels and atmospheric composition, *Nature* **426**: 323–326.

Bird, R. B., Stewart, W. E., and Lightfoot, E. N. 1960, *Transport phenomena*, John Wiley & Sons, New York.

Capra, F. 1996, *The Web of Life*, Anchor Books, A Division of Random House, Inc., New York.

Cramer, W. et al. 1995, *Net Primary Productivity – Model Intercomparison Activity*, Report 5, IGBP/GAIM, Washington, DC, gaim.unh.edu/Products/Reports/Report_5/-report5.pdf

Davis, M. 2002, *Late Victorian Holocausts: El Niño Famines and the Making of the Third World*, Verso, London.

Domalski, E. S., Jobe Jr., T. L., and Milne, T. A. (eds.) 1987, *Thermodynamic Data for Biomass Materials and Waste Components*, The American Society of Mechanical Engineers, United Engineering Center, 345 East 47th Street, New York, 10017.

---

[24] We are moving swiftly down this merry path: "Green Energy Resources traveled to Florida and Georgia this week to procure upwards of a million tons of forest fire timber from the region at no cost to the company. The timber is valued at approximately \$15–20 million. Green Energy Resources plans to use the wood to supply biomass power plants in the United States as well as for exports."
Source: Green Energy Resources, May 23, 2007, Press Release. Accessed on June 21, 2007.

Grimm, K. A. 1998, *Phosphorites feed people: Finite fertlizer ores impact Canadian and global food security*, The Monitor, www.eos.ubc.ca/personal/grimm/phosphorites.html

Heinsch, F. A. et al. 2003, *User's Guide GPP and NPP (MOD17A2/A3) Products NASA MODIS Land Algorithm*, Report, NASA, Washington, DC, www.ntsg.ntsg.umt.edu/modis/-MOD17UsersGuide.pdf

Hurtt. G. C., Rosentrater, L., Erolking, S., and Moore, B. 2001, Linking remote-sensing estimates of land cover and census statistics on land use to produce maps of land use of the conterminous united states, *Global Biogeochem. Cycles* **15(3)**: 673–685.

Jacques, K. A., Lyons, T. P., and Kelsall, D. R. (eds.) 2003, *The Alcohol Textbook*, Nottingham University Press, Nottingham, CB, 4 edition.

Khosla, V. 2006, *Biofuels: Think outside the Barrel*, www.khoslaventures.com/presentations/-Biofuels.Apr2006.ppt, Also see the video version at video.google.com/videoplay? docid=-570288889128950913

Lee, D.-K., Owens, V. N., Boe, A., and Jeranyama, P. 2007, *Composition of Herbaceous Biomass Feedstocks*, Report SGINCI-07, Plant Science Department, North Central Sun Grant Center, South Dakota State University, Box 2140C, Brookings, SD 57007.

Lovelock, J. 1979, *Gaia – A new look at life on the Earth*, Oxford University Press, Oxford, GB.

Lovelock, J. 1988, *The Ages of Gaia, A Biography of Our Living Earth*, W. W. Norton & Co. Inc., New York.

Lugo, A. E. and Brown, S. 1986, Steady state terrestrial ecosystems and the global carbon cycle, *Vegetatio* **68**: 83–90.

Mokany, K., Raison, R. J., and Prokushkin, A. S. 2006, Critical analysis of root: shoot ratios in terrestrial biomes, *Glob. Chang. Biol.* **12**: 84–96.

Montgomery, D. R. 2007, Soil erosion and agricultural sustainability, *PNAS* **104(33)**: 13268–13272.

Napitupulu, M. and Ramu, K. L. V. 1982, Development of the Segara Anakan area of Central Java, in *Proceedings of the Workshop on Coastal Resources Management in the Cilacap Region*, pp 66–82, Gadjah Mada University, Yogyakarta.

Osborne, J. W. 1970, *The Silent Revolution: The Industrial Revolution in England as a Source of Cultural Change*, Charles Scribner's Sons, New York.

Page, S. E., Siegert, F., Rieley, J. O., V. Boehm, H.-D., Jaya, A., and Limin, S. 2002, The amount of carbon released from peat and forest fires in Indonesia during 1997, *Nature* **420(6911)**: 61–65.

Patzek, L. J. and Patzek, T. W. 2007, The Disastrous Local and Global Impacts of Tropical Biofuel Production, *Energy Tribune* **March**: 19–22.

Patzek, T. W. 2004, Thermodynamics of the corn-ethanol biofuel cycle, *Critical Reviews in Plant Sciences* **23(6)**: 519–567, An updated web version is at http://petroleum.berkeley.edu/papers/patzek/CRPS416-Patzek-Web.pdf.

Patzek, T. W. 2006a, A First-Law Thermodynamic Analysis of the Corn-Ethanol Cycle, *Natural Resources Research* **15(4)**: 255–270.

Patzek, T. W. 2006b, Letter, *Science* **312(5781)**: 1747.

Patzek, T. W. 2006c, *The Real Biofuels Cycles*, Online Supporting Material for Science Letter, Available at petroleum.berkeley.edu/patzek/BiofuelQA/Materials/RealFuelCycles-Web.pdf

Patzek, T. W. 2007, *Earth, Humans and Energy*, CE170 Class Reader, University of Califonia, Berkeley.

Patzek, T. W. and Pimentel, D. 2006, Thermodynamics of energy production from biomass, *Critical Reviews in Plant Sciences* **24(5–6)**: 329–364, Available at http://petroleum.berkeley.edu/papers/patzek/CRPS-BiomassPaper.pdf

Perlack, R. D., Wright, L. L., Turhollow, A. F., L., G.R., Stokes, B. J., and Erbach, D. C. 2005, *Biomass as feedstock for a bioenergy and bioproducts industry: The technical feasibility of a billion-ton annual supply*, Joint Report, Prepared by U.S. Department of Energy, U.S. Department of Agriculture, Environmental Sciences Division, Oak Ridge National Laboratory, P.O. Box 2008, Oak Ridge, Tennessee 37831–6285, Managed by: UT-Battelle, LLC for the U.S. Department of Energy under contract DE-AC05-00OR22725 DOE/GO-102005-2135 ORNL/TM-2005/66

Randerson, J. T., Chapin, F. S., Harden, J. W., Neff, J. C., and Harmone, M. E. 2001, Net ecosystem production: A comprehensive measure of net carbon accumulation by ecosystems, *Ecological Applications* **12(4)**: 2937–947.

Reichle, D. E., O'Neill, R. V., and Harris, W. F. 1975, *Unifying concepts in ecology*, Chapter Principles of energy and material exchange in ecosystems, pp. 27–43, Dr. W. Junk B. V. Publishers, The Hague, The Netherlands.

Ricklefs, R. E. (ed.) 1990, *Ecology*, W. H. Freeman & Company, New York, 3 edition.

Roberts, G. and Wooster, M. J. 2007, New perspectives on African biomass burning dynamics, *EOS* **88(38)**: 369–370.

Running, S. W., Thornton, P. E., et al. 2000, *Methods in Ecosystem Science*, Chapter. Global terrestrial gross and net primary productivity from the Earth Observing System, pp. 44–57, Springer Verlag, New York.

Ryan, M. G. 1991, Effects of climate change on plant respiration, *Ecolo. Soc. Am.* **1(2)**: 157–167.

Saha, B. C., Iten, L. B., Cotta, M. A., and Wu, Y. V. 2005, Dilute acid pretreatment, enzymatic saccharification and fermentation of wheat straw to ethanol, *Process Biochem.* **40**: 3693–3700.

Schimel, D. and Baker, D. 2002, The wildfire factor, *Nature* **420(6911)**: 29–30.

Schmidt, A., Zschetzsche, A., and Hantsch-Linhart, W. 1993, *Analyse von biogenen Brennstoffen*, Report, TU Wien, Institut für Verfahrens-, Brennstoff- und Umwelttechnik, Vienna, Austria, www.vt.tuwien.ac.at/Biobib/fuel98.html

Smil, V. 1985, *Carbon – Nitrogen – Sulfur – Human Interferences in Grand Biospheric Cycles*, Plenum Press, New York and London.

Songa, C. and Woodcock, C. E. 2003, A regional forest ecosystem carbon budget model: Impacts of forest age structure and landuse history, *Ecol. Modell.* **164**: 33–47.

Steynberg, A. P. and Nel, H. G. 2004, Clean coal conversion options using Fischer-Tropsch technology, *Fuel* **83(6)**: 765–770.

Stocking, M. A. 2003, Tropical Soils and Security: The Next 50 years, *Science* **302(5649)**: 1356–1359.

von Englehardt, W., Goguel, J., Hubbert, M. K., Prentice, J. E., Price, R. A., and Trümpy, R. 1975, Earth Resources, Time, and Man - A Geoscience Perspective, *Environ. Geol.* 1: 193–206.

Webster 1993, *Webster's Third New International Dictionary of the English Language – Unabridged*, Encyclopædia Britannica, Inc., Chicago.

Appendix 1: Ecosystem Definition and Properties

As shown in Fig. 2.9, the *autotrophic*[25] plants capture CO_2 from the atmosphere, and water and dissolved nutrients[26] from soil. Using solar light, plants convert all these chemical inputs into biomass through *photosynthesis*, see Fig. 2.21.

Plants are food to the plant-eating *heterotrophs*:[27] animals, fungi, and bacteria. All die in place and their bodies are recycled for nutrients. Heterotrophs consist of consumers and decomposers. Consumers eat mostly living tissues. Decomposers consume dead organic matter and mineralize[28] it.

---

[25] From Greek *autotrophos* supplying one's own food (Webster, 1993).

[26] Water-soluble chemical compounds rich in N, P, K, Ca, Mg, S, Fe, etc.

[27] Requiring complex organic compounds of nitrogen, phosphorous, sulfur, etc., and carbon (as that obtained from plant or animal matter) for metabolic synthesis (Webster, 1993).

[28] For example, nitrogen can be transformed into inorganic molecules assimilable by plants, such as the aqueous ammonium or nitrate ions, as well as nitrogen dioxide, by (1) microbial *fixation*

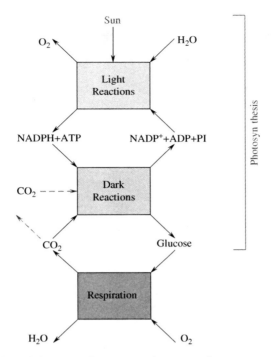

Fig. 2.21 The light reactions use photons to strip protons from water and store energy in NADPH (nicotinamide adenine dinucleotide phosphate) and ATP (adenosine 5′-triphosphate nucleotide). Both these molecules are used to reduce CO_2 and combine carbon with hydrogen and phosphate in the Calvin Cycle or dark reactions: $3CO_2 + 9ATP + 6NADPH \rightarrow$ glyceraldehyde-3-phosphate $+9ADP + 8PI + 6NADP^+$. Here ADP is adenosine diphosphate, PI is inorganic phosphate, and $NADP^+$ is the oxidized form of NADPH. Glyceraldehyde-3-phosphate may be converted to other carbohydrates such as metabolites (fructose-6-phosphate and glucose-1-phosphate), energy stores (sucrose or starch), or cell-wall constituents (cellulose and hemicelluloses). By respiring plants consume O_2 and convert their energy stores back to CO_2 and water

Definition 1. *An ecosystem (an earth household) is a community of living organisms that interact with their non living physical environment (habitat). Most elements of an ecosystem are thoroughly connected (Lovelock, 1979; Lovelock, 1988; Capra, 1996), but over limited spacial scales.[29] In addition to solar energy and inorganic matter, the three basic structural and functional components of an ecosystem are autotrophs, heterotrophs and dead organic matter.*

---

of the atmospheric N_2 and (2) by microbial *mineralization* of organic nitrogen in soil. Conversely, soil nitrogen is returned back to the atmosphere through microbial *denitrification*. The opposite process, oxidation of dissolved ammonia to nitrite and nitrate, is called *nitrification*. For details, see Smil (1985).

[29] In order for an ecosystem to be stable and its emerging properties at a larger scale be independent of the structural details of the smaller scales, the covariances of everything must decline at least exponentially with distance scaled by a yardstick characteristic of the smaller scales.

Inputs to an ecosystem are biotic[30] and abiotic:

1. Abiotic inputs are solar energy, the atmospheric gases (CO_2, O_2, N_2, NO_x and SO_x), mineral nutrients in the soil, rain, surface water, and groundwater.
2. Biotic inputs are organisms that move into the ecosystem, but also organic compounds: proteins, lipids, carbohydrates, humic acid, etc.

Some dead organisms are buried in swamps, lakes, shallow coastal waters, etc., see Fig. 2.15, and some nutrients are imported with floods and rain, while some are exported by rivers and wind. A vast majority of the biomass is, however, recycled within the boundaries of the mother ecosystem[31] in agreement with the Second Law of thermodynamics. This way, a buffalo might eat a wolf, whose bones were incorporated as phosphorous in the prairie grass.

Ecosystems change with time, organisms live and die, and move in and out. Ecosystems are subject to many disturbances: floods, fire, storms, droughts, invasions, and so on.

Appendix 2: Mass Balance of Carbon in an Ecosystem

An eco-*system* is a *system* known to thermodynamics only if a three-dimensional surface[32] fully enveloping the system's contents is imagined for the life-span of the ecosystem. Of course, this surface may itself be time-dependent, but not here.

Once there is a boundary, the carbon mass accumulation *in* the ecosystem is defined through the carbon mass flow *crossing* its boundary, and the *interior* carbon sources and sinks. The general mass-balance equation that describes all physical systems, (see, e.g., Bird et al., 1960), can be written for carbon in the following way:

$$
\underbrace{\frac{dc}{dt}}_{\text{Rate of living carbon accumulation}} = \underbrace{- \oint_{\text{Boundary}} \mathcal{F} \cdot \mathbf{n} \, dA}_{\text{Net rate of flow out}}
$$

$$
\underbrace{+ \sum \text{Sources} - \sum \text{Sinks}}_{\text{Net rate of production inside}} \quad \text{kg} \, \text{C} \, \text{s}^{-1}
$$

(2.4)

Here $\mathcal{F}$ is the overall carbon flux vector, $\mathbf{n}$ is the unit outward normal to the system boundary, and the summation (integral) is over the entire system boundary.

---

[30] Of, relating to, or caused by living organisms (Webster, 1993).

[31] Most ecosystems do not have distinct natural boundaries. Boundaries chosen by us in most cases are arbitrary subdivisions of a continuous gradation of communities.

[32] A 3D curvilinear box extending above the tallest feature of the ecosystem, and below topsoil, river, lake and stream bottoms, etc.

The sources *inside* the system volume are the photosynthesizing autotrophs, and the sinks are the respiring autotrophs and heterotrophs, fires, soil carbon oxidation, volatile hydrocarbon production, etc. The overall carbon flux $\mathcal{F}$ is the vector sum of several different mechanisms of carbon mass exchange, such as convection with air, convection with moving heterotrophs, convection with soil-, river- and flood water, convection with eroded soil, etc. Each of the particular fluxes is nonzero over those parts of the system boundary where it operates and zero elsewhere.

Let's define $\dot{m}_i$, the overall outward carbon mass flow rate due to a specific flux i; Gross Primary Production (GPP), the sum of autotroph photosynthesis sources; R_a, the overall autotroph respiration sink; R_h, the overall heterotroph respiration sink; R_f, the overall fire sink; R_v, the overall volatile hydrocarbon production sink; R_s, the soil carbon oxidation sink; R_b, the carbon burial sink; etc.

$$\dot{m}_i = \oint_{Boundary} F_i \cdot \mathbf{n}\, dA$$

$$GPP = \sum Sources$$

$$R = R_a + R_h + R_f + R_v + R_s + \cdots = \sum Sinks \tag{2.5}$$

Then

$$\frac{dc}{dt} = -\sum_i \dot{m}_i + GPP - \underbrace{(R_a + R_h)}_{\text{Ecosystem Respiration } R_e} - \underbrace{(R_F + R_v + R_s \ldots)}_{\text{Non respiratory sinks of C}} \tag{2.6}$$

In order to correspond to the dominant time scale of observations, the "instantaneous" carbon mass balance equation must be further time-averaged, as denoted by the angular brackets:

$$\frac{1}{\tau_2 - \tau_1} \int_{\tau_1}^{\tau_2} \frac{dC}{dt} dt = -\sum_i \frac{1}{\tau_2 - \tau_1} \int_{\tau_1}^{\tau_2} \dot{m}_i(t) dt +$$

$$+ \frac{1}{\tau_2 - \tau_1} \int_{\tau_1}^{\tau_2} GPP(t) dt - \frac{1}{\tau_2 - \tau_1} \int_{\tau_1}^{\tau_2} R(t) dt$$

$$\frac{C(\tau_2) - C(\tau_1)}{\tau_2 - \tau_1} = \left\langle \frac{dC}{dt} \right\rangle = \underbrace{-\sum_i <\dot{m}_i> + <GPP> - <R>}_{\text{Net Ecosystem Productivity} <NEP>} \tag{2.7}$$

Note that in spirit, the last Eq. (2.7) is similar to Eqs. (2.1) and (2.2) in Randerson et al. (2001), which unfortunately do not distinguish between fluxes and sources and sinks.

NEP is defined here as the net carbon accumulation by an ecosystem, just as in Randerson et al. (2001). It explicitly incorporates all of the carbon fluxes from an ecosystem, and the interior sources and sinks, including lateral transfers among ecosystems, autotrophic respiration, heterotrophic respiration, losses associated with disturbances, dissolved and particulate carbon losses, carbon burial, and volatile organic compound emissions.

Now, if the time of observation is long enough, the average rate of carbon accumulation in a stable ecosystem should tend to zero because of the Second Law of thermodynamics. Global carbon burial has been about 0.1 percent of terrestrial NPP, see Fig. 2.15. Thus, on a time scale of a couple of centuries (Lugo and Brown, 1986; Berner, 2001, 2003), one may postulate that the rate of carbon accumulation is minuscule compared with the fluxes, sources and sinks, and

$$< NEP > \approx 0 \qquad (2.8)$$

Given enough time, stable ecosystems will settle into steady states and recycle almost all carbon (and all other nutrients) in them, see Table 2.5.

Table 2.5 Summary of carbon fluxes in terrestrial ecosystems. Adapted from Tables 2.1 and 2.2 in (Randerson et al., 2001) and NASA MODIS data in Table 2.2

Concept	Acronym symbol	Global flux	Definition
Gross primary production	GPP	110 Gt C/yr	a
Autotrophic respiration	R_a	$\sim 1/2$ of GPP	b
Net primary production	NPP	$\sim 1/2$ of. GPP	GPP R_a
Heterotrophic respiration (on land)	R_h	82 – 95% of NPP	c
Ecosystem respiration	R_e	91 – 97% of GPP	$R_a + R_h$
Non-CO_2 losses	$R_v + R_s$	2.8 – 4.9 Gt C/yr	d
Non-respiratory CO_2 losses (fire)	R_f	1.6 – 4.2 Gt C/yr	e
Net ecosystem production	NEP	0 ± 2.0 Gt C/yr	f

[a] Carbon uptake by plants during photosynthesis, see Table 2.2.
[b] Respiratory (CO_2) loss by plants for construction, maintenance, or ion uptake, see Table 2.2.
[c] Respiratory (CO_2) loss by the heterotrophic community (herbivores, microbes, etc.).
[d] CO, CH_4, isoprene (2-methylbuta-1,3-diene), dissolved inorganic and organic carbon, erosion, etc. These losses are 2.6–4.5% of GPP.
[e] Average combustion flux of CO_2 is 1.5–3.8% of GPP Extreme events, such as the 1997–98 El Niño firestorms in Indonesia are excluded.
[f] Total carbon accumulation within the ecosystem: GPP - $R_e - R_f R_v - R_s - \ldots$. All human crops export about 1.2–1.5 Gt C/yr from agricultural ecosystems, while crop residues contain another 1.3–1.5 Gt C/yr.

Soils, landscapes, and plant communities evolve together through an interdependence on the difference between the rate of soil erosion and soil production (Montgomery, 2007). At steady state this difference must be zero on the average., i.e., the soil erosion rate is equal to the geologic rate of soil production and some equilibrium thickness of soil persists over long time intervals.

Geological erosion rates generally increase from the gently sloping lowland landscape ($<10^{-4}$ to 1 mm/yr), to moderate gradient hillslopes of soil-mantled terrain (0.001–1 mm/yr), and steep tectonically active alpine landscapes (0.1 to >10 mm/yr) (cf. Montgomery (2007) and the references therein).

Rates of soil erosion under conventional agricultural practices almost uniformly exceed 0.029–0.173 mm/yr (the median and mean geological rate of soil production, respectively), according to the data compiled by Montgomery (2007) exhibiting the median and mean values > 1 mm/yr. Erosion rates on the steep mountain slopes in Indonesia easily exceed 30 mm/yr (Napitupulu and Ramu, 1982), and the human-disturbed soil can disappear there within days or months, rather than years.

Rates of erosion reported under native vegetation and conventional agriculture show 1.3- to > 1000-fold increases, with the median and mean ratios of 18- and 124-fold, respectively, for the studies complied by Montgomery (2007). From my work on the tropical plantations (Patzek and Pimentel, 2006) it follows that the respective ratios are even higher in the mechanically-disturbed hilly landscapes.

For this and many other reasons, humanity's experiment with "Green Revolution" is just a large but temporary disturbance of natural ecosystems driven by a gigantic multi-decade subsidy with old plant carbon (fossil fuels, fertilizers, and field chemicals) into the vastly simplified, fasteroding, and – therefore – unstable agricultural systems. As such, these latter systems will never test Eq. (2.8). They will fail much sooner instead.[33]

In addition, a long time-average of the net carbon flow rate out of the system may also be negligible, as most of it is the CO_2 flow rate in for photosynthesis minus the CO_2 flow rate out from respiration. The extreme events,[34] such as fires and floods, will be averaged out and in a stable ecosystem soil erosion should also be low (or the ecosystem would not survive, see Fig. 2.22). The time-averaged rate of

---

[33] "One alternative." Prof. Harvey Blanch notes, "is to bioengineer a low-lignin crop that does not require fertilizer, that doesn't need much water, and that could be grown on land not suitable for food crops. The problem is that lignin is what makes the plant stalks rigid, and without it, a plant would probably be floppy and difficult to harvest. And of course," he adds, "there might be public resistance to huge plantations of a genetically-modified organism." *Global warming - Building a sustainable biofuel production system*, The News Journal, College of Chemistry, University of California, Berkely, 14(1), 2006.

[34] *Disturoances* in the ecology parlance.

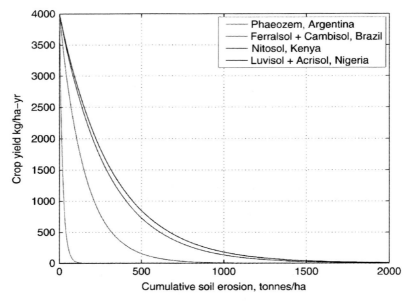

Fig. 2.22 Maize crop yields decay exponentially with eroded soil for a selection of tropical soils: Yields $= 4000 \exp[\text{Cumulative Erosion}/r]$, $r = 20\text{--}300 \, \text{t} \, \text{ha}^{-1}$. The initial yield level is set artificially to 4 tonnes of grain needed by one typical household for 1 year in the subhumid tropics. The cumulative erosion of $10 \, \text{t} \, \text{ha}^{-1} \approx 1 \, \text{mm}$ of soil loss. So a loss of 2 cm of topsoil in the tropics is catastrophic. Adapted from Fig. 2.1 of Stocking (2003)

volatile hydrocarbon emissions must be relatively low too, and, therefore, one may postulate that

$$< \text{GPP} > - < R > \approx 0 \qquad (2.9)$$

When averaged over a sufficiently long time, the gross ecosystem productivity is *roughly* equal to the total rate of carbon consumption inside the ecosystem. The orgin of this postulate is also the Second Law of thermodynamics.

Appendix 3: Environmental Controls on Net Primary Productivity

Net primary productivity is equal to the product of the rate of photosynthesis per unit leaf area and the total surface area of the active leaves per unit area of land, minus the rate of plant respiration per unit area of land. Given sufficient plant nutrients and substrates, temperature and moisture control the rate of photosynthesis.

Extremely cold and hot temperature limit the rate of photosynthesis. Within the range of temperatures that are tolerated, the rate of photosynthesis generally rises with temperature. Most biological metabolic activity takes place between 0 and 50°C. The optimal temperatures for plant productivity coincide with the 15–25°C optimum temperature range of photosynthesis.

A growing season is the period when temperatures are sufficiently warm to support synthesis and a positive net primary production. Warmer temperatures support both higher rates of photosynthesis and a longer growing season, resulting in a higher net primary production – if there are sufficient water and nutrients. The amount of water available to the plant will therefore limit both the rate of photosynthesis and the area of leaves that can be supported.

The influence of temperature and water availability is interrelated. It is the combination of warm temperature and water supply adequate to meet the demands of transpiration that results in the highest values of primary productivity. Net primary production in ecosystems varies widely, cf. Fig. 2.7 in Cramer et al. (1995) and Table 2.6:

1. The most productive terrestrial ecosystem are tropical evergreen rainforests with high rainfall and warm temperatures. Their net primary productivity ranges from 700 to 1400 gCm^{-2} yr^{-1}.
2. Temperate mixed forests produce between 400 and 1000 gCm^{-2} yr^{-1}.
3. Temperate grassland productivity is between 200 and 500 gCm^{-2} yr^{-1}.

Table 2.6 Average net primary productivity of ecosystems

Ecosystem	Value[a] gCm^{-2} yr^{-1}	Value[b] gCm^{-2} yr^{-1}
Swamp and marsh	1130	2500
Algal bed and reef	900	2000
Tropical forest	830	1800
Estuary	810	1800
Temperate forest	560	1250
Boreal forest	360	800
Savanna	320	700
Cultivated land	290	650
Woodland and shrubland	270	600
Grassland	230	500
Lake and stream	230	500
Upwelling zone	230	–
Continental shelf	160	360
Tundra and alpine meadow	65	140
Open ocean	57	125
Desert scrub	32	70
Rock, ice, and sand	15	–

[a] www.vendian.org/envelope/Temporary.URL/draft-npp.html
[b] (Ricklefs, 1990). Note that Column 2 is ~Column 1 × 2.2, corresponding to the mean molecular weight of dry biomass of 26 g/mol per 1 carbon atom, a little less than 27 g/mol in glucose starch, $CH_2O - 1/6H_2O$. A typical molecular composition of dry woody biomass is $CH_{1.4}O_{0.6}$, MW = 23 g/mol.

4. Arctic and alpine tundra have productivities of 0 to $300\,\mathrm{gCm^{-2}\,yr^{-1}}$.
5. Productivity of the open sea is generally low, 10 to $50\,\mathrm{gCm^{-2}\,yr^{-1}}$.
6. Given equal nutrient supplies, productivity in the open waters of the cool temperate oceans tends to be higher than than of the tropical waters.
7. In areas of upwelling, as near the tropical coast of Peru, productivity can exceed $500\,\mathrm{gCm^{-2}\,yr^{-1}}$.
8. Coastal ecosystem and continental shelves have higher productivity than open ocean.
9. Swamps and marshes have a net primary production of $1100\,\mathrm{gCm^{-2}\,yr^{-1}}$ or higher.
10. Estuaries and coral reefs have a net primary productivity of $900\,\mathrm{gCm^{-2}\,yr^{-1}}$. This is caused by the inputs of nutrients from rivers and tides in estuaries, and the changing tides in coral reefs.

High primary productivity results from an energy subsidy to the (generally small) ecosystem. This subsidy results from a warmer temperature, greater rainfall, circulating or moving water that carries in food or additional nutrients. In the case of agriculture, the subsidy comes from fossil fuels for cultivation and irrigation, fertilizers, and the control of pests. Sugarcane has a net productivity of $1700\text{–}2500\,\mathrm{g\ m^{-2}\ yr^{-1}}$ of dry stems, and hybrid corn in the US $800\text{–}1000\,\mathrm{gm^{-2}\ yr^{-1}}$ of dry grain.

Glossary

To be readable, many of the descriptions below are *not* most rigorous:

Ecosystem: A system that consists of living organisms (plants, bacteria, fungi, animals) and inanimate substrates (soil, minerals, water, atmosphere, etc.), on which these organisms live.

Energy: Energy is the ability of a system to lift a weight in a process that involves no heat exchange (is adiabatic). Total energy is the sum of internal, potential and kinetic energies.

Energy, Free That part of internal energy of a system that can be converted into work. You can think of free energy as the amount of electricity that can be generated from something that changes from an initial to a final state (e.g., by burning a chunk of coal in a stove and doing something with the heat of combustion).

Energy, Primary: Here the heat of combustion (HHV) of a fuel (coal, crude oil, natural gas, biomass, etc.), potential energy of water behind a dam, or the amount of heat from uranium necessary to generate electricity in a nuclear power station.

Higher Heating Value (HHV): HHV is determined in a sealed insulated vessel by charging it with a stoichiometric mixture of fuel and air (e.g., two moles of hydrogen and air with one mole of oxygen) at $25°C$. When hydrogen and oxygen are combined, they create hot water vapor. Subsequently, the vessel and its content are cooled down to the original temperature and the HHV of hydrogen is determined by measuring the heat released between identical initial and final temperature of $25°C$.

Petroleum, conventional: Petroleum, excluding lease gases and condensate, as well as tar sands, oil shales, ultra-deep offshore reservoirs, etc.

System: A region of the world *we pick* and separate from the rest of the world (the *environment*) with an imaginary closed *boundary*. We may not describe a system by what happens inside or outside of it, but only by what *crosses* its boundary. An *open* system allows for matter to cross its boundary, otherwise the system is *closed*.

Chapter 3
A Review of the Economic Rewards and Risks of Ethanol Production

David Swenson

Abstract Ethanol production doubled in a very short period of time in the U.S. due to a combination of natural disasters, political tensions, and much more demand globally from petroleum. Responses to this expansion will span many sectors of society and the economy. As the Midwest gears up to rapidly add new ethanol manufacturing plants, the existing regional economy must accommodate the changes. There are issues for decision makers regarding existing agricultural activities, transportation and storage, regional economic impacts, the likelihood of growth in particular areas and decline in others, and the longer term economic, social, and environmental sustainability. Many of these issues will have to be considered and dealt with in a simultaneous fashion in a relatively short period of time. This chapter investigates sets of structural, industrial, and regional consequences associated with ethanol plant development in the Midwest, primarily, and in the nation, secondarily. The first section untangles the rhetoric of local and regional economic impact claims about biofuels. The second section describes the economic gains and offsets that may accrue to farmers, livestock feeding, and other agri-businesses as production of ethanol and byproducts increase. The last section discusses the near and longer term growth prospects for rural areas in the Midwest and the nation as they relate to biofuels production.

Keywords Ethanol · economic impact · biofuels · farmer ownership · scale economies · storage · grain supply · rural development · cellulosic ethanol

3.1 Introduction

The economic, social, political, and environmental impacts of modern ethanol production in the U.S. are highly regionalized. Current ethanol production and most new ethanol plant development in the United States are concentrated in the Corn

✉ D. Swenson
Department of Economics, 177 Heady Hall, Iowa State University, Ames IA 50011
e-mail: dswenson@iastate.edu

D. Pimentel (ed.), *Biofuels, Solar and Wind as Renewable Energy Systems*,
© Springer Science+Business Media B.V. 2008

Belt states of Iowa, Illinois, Indiana, Minnesota, and Nebraska. Those states alone produced nearly 62 percent of the nation's corn in 2006. Not surprisingly, those same states account for about two-thirds of actual or planned ethanol production capacity.

Ethanol production and plant development took on an added urgency in the fall of 2005 after hurricanes Katrina and Rita crippled domestic oil production capacity in the Gulf of Mexico. Those events, coupled with heightened uncertainty about both near-term and long-term oil supplies in light of other international issues, fueled massive amounts of rhetorical, political, and financial resources in support of biofuels production and energy independence.

The growth in U.S. ethanol production has been dramatic: In 2005, 1.6 billion bushels of corn were converted to ethanol, about 12.1 percent of the total corn supply. By the end of 2007 it is estimated that 3.2 billion bushels will be used for that purpose, about a quarter of the nation's corn supply, and an increase of just over 100 percent in only two years (USDA 2007). That much corn will make enough ethanol to account for 3.9 percent of the nation's total demand for motor gasoline that year (EIA 2007). Expansion in ethanol production from corn through the rest of this decade is expected to top out at from 4.0 billion bushels by 2010 (USDA 2007) to 4.3 billion bushels (FAPRI 2007), though some analysts can envision sets of policy and market considerations that might push production higher (Tokgoz et al. 2007).

Responses to this expansion in ethanol production will span many sectors of society and the economy. Already, the expansion in production capacity has driven up corn prices sharply from recent historical levels, which in turn has driven up the number of acres planted in corn: 2007 corn acres nationally are 19 percent higher than 2006. But given a generally fixed supply of arable farmland, there are consequences to this expansion: soybean plantings declined by 15 percent and cotton by 28 percent (USDA June 2007). Over the past two decades, national farm commodity production has been a relatively stable, slowly-adjusting mix of crops and livestock with very distinct regional advantages and production concentrations. The rapid rise in ethanol production from corn, however, likens to dropping a large rock in a calm pond – there are ripples extending in all directions that affect crop production, animal production, food production, and, ultimately, the well-being of households.

As the Midwest gears up to rapidly add new ethanol manufacturing plants, the existing regional economy must accommodate the changes. There are issues for decision makers regarding existing agricultural activities, transportation and storage, regional economic impacts, the likelihood of growth in particular areas and decline in others, and the longer term economic, social, and environmental sustainability. Many of these issues will have to be considered and dealt with in a simultaneous fashion in a relatively short period of time.

This chapter investigates sets of structural, industrial, and regional consequences associated with ethanol plant development in the Midwest, primarily, and in the nation, secondarily. The first section untangles the rhetoric of local and regional economic impact claims about biofuels. The second section describes the economic gains and offsets that may accrue to farmers, livestock feeding, and other agri-businesses as production of ethanol and byproducts increase. The last section discusses the near and longer term growth prospects for rural areas in the Midwest and the nation as they relate to biofuels production.

3.2 Measuring and Mismeasuring Biofuels Economic Impacts

It is important to sort out the rhetoric of claimed economic benefits to be expected from biofuels development in the Midwest and the nation because there are tremendous amounts of public money at stake. In the very early stages of this modern boom in ethanol plant construction, politicians, farm commodity groups, and economic developers hailed the emerging industry as the right and proper evolution of modern agricultural production capacities coupled inexorably with technological breakthroughs and long overdue changes in the nation's energy policies. Amidst this enthusiasm, biofuels trade associations and some agricultural commodity groups reported in various venues that scores of thousands of jobs have been created across the Corn Belt and the nation. Some politicians and government agency representatives parroted those reports uncritically; Midwestern state governments began to specifically and energetically apply government agency services in support of the boom, along with offering lucrative tax credits and incentives to promote even faster growth; land-grant universities promoted their vital scientific contributions in this coming energy revolution; cities and counties scrambled to be the site of a modern ethanol factory, to be on the plus side of economic trends for a change given the historical deterioration of rural Midwestern economies and communities; and some leaders in Midwestern states began to envision a social and economic resurgence in rural areas.

Profound expectations like the aforementioned demand careful scrutiny, especially when massive amounts of national, state, and local government subsidy are at stake. The place to begin is with the measurement of net economic gain attributable to this run-up in ethanol production in the U.S. and the identification of who benefits. Those aggressively promoting private and public investment in more biofuels processing capacities range from farm commodity groups, farm state politicians, some environmental organizations, automobile manufacturers, to both liberal and conservative political orientations.

There are wide ranges of economic activity attributed to biofuels production. The nation's production of ethanol creates jobs at the ethanol plants, boosts the demand for critical mechanical, technical, and service inputs, and helps to improve the prices received by input commodity providers, namely corn producers. Beyond that, few of the conclusions about the economic impacts of biofuels production appear to be based on rigorous, enterprise or industry level research, however (Swenson 2006). Much is of a very rudimentary level using broad assumptions about ethanol industry activity and applying, uncritically and often inappropriately, national economic impact ratios to deduce the size of economic activity attributable to ethanol production. The estimates either at the local level or at the national level are quite diverse and often incredible.

As examples, at the national level, an Urbanchuck (2005) report for the Renewable Fuels Association used US Bureau of Economic Analysis factors to conclude that 114,844 jobs in the national economy depended indirectly on the operation of all ethanol plants and the purchases that are made by workers (and this did not include ethanol plant employment). Earlier in the decade, when the industry was even smaller, Novack (2002) of the Federal Reserve Bank of Kansas City was more upbeat about the job total and reported in a widely read periodical that "... the [ethanol] industry added nearly 200,000 jobs to the U.S. economy." This is a curious claim given that the U.S. Department of Commerce's industrial census for that same year (2002) indicated the ethyl-alcohol industry had just 2,200 jobs. How the author got from 2,200 jobs to 200,000 is not revealed, but the writer went on to predict that "an additional 214,000 jobs [would] be created through the economy over the next decade." Last, as just one example of comments made by many farm state politicians, former South Dakota U.S. Senator Thomas Daschle concluded in a national and widely reprinted publication that the production of 3.1 billion gallons of ethanol in the U.S. created 200,000 jobs (Daschle 2006).

These three examples are emblematic of the rhetoric underscoring ethanol production expansion and public policy development in the U.S. The first was made by a consultancy with long-standing ties with the Renewable Fuels Association, a trade group that aggressively promotes corn ethanol policies and serves as the primary information source for information on renewable fuels opportunities and capacities in the U.S. The second claim came from a writer from the nation's respected public banking regulatory and financial research sector. In this case the Kansas City Federal Reserve Bank also has a specialization in rural development economic studies and affairs; hence, an assumption of rigor and credibility. The third job claim came from a respected and long-time political leader and strong advocate for alternative energy development. Given the implied authority of these three sources it is important to investigate the source of their numerical enthusiasms. A good example for understanding the basis for the robust, yet quite misleading, job claims can be found in recent work sponsored by the Iowa Renewable Fuels Association.

3.2.1 Deconstructing Ethanol Job Impact Claims in the Midwest

An Urbanchuck (2007) report for the Iowa Renewable Fuels Association (IRFA) concluded that Iowa's ethanol industry had created 46,938 jobs and contributed $7.315 billion in state domestic product. Research at Iowa State University (Swenson 2007b) concluded, in contrast, that the state's 28 ethanol producers in processing 600 million bushels of corn into approximately 1.65 billion gallons of ethanol created from 4,100 to 4,700 net new jobs in the Iowa economy through 2005. The public university statistics are a tenth of those produced by the trade group. The following exercise explains most of the differences. Figure 3.1 displays the type and number of jobs the IRFA research credited to Iowa.

First, from the original number of 46,938 jobs are subtracted the 19,733 jobs linked to capital development and construction. There are several good reasons for

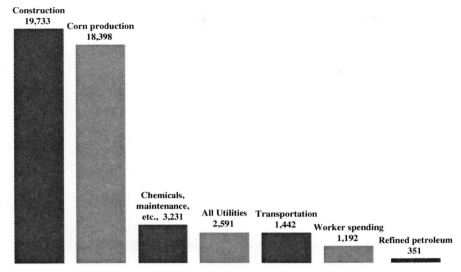

Fig. 3.1 Iowa renewable fuels association estimates of ethanol job impacts in Iowa for 2005

doing this: Those are not net new permanent jobs – the jobs were all ready in the larger regional economy as there is a generally fixed rate of capital formation in the U.S. linked to the availability of investment resources and the overall pace and pattern of capital growth; according to U.S. Bureau of Economic Analysis statistics, the overall national rate of investment in the chemical manufacturing industry where ethanol is located is actually less than the average for all manufacturing for the 2000–2005 period; there is a finite number of plants that can and will be built given this state's current and likely future supply of corn and the rate of national absorption of ethanol; and the capital development that those construction workers are contributing to serves significantly as substitutes for energy-related and other forms of industrial development in Iowa, the greater region, and in the nation. Eliminating the existing and spatially temporary construction jobs leaves us 27,205 jobs to further parse.

Next, a full two-thirds of the purported non-construction ethanol impact jobs were already in the economy whether there was or there was not an ethanol industry. The IRFA study used a set of final demand multipliers to estimate the remaining ethanol job and product impacts (BEA 1997). Final demand means that either the industry is producing for final consumption by households and institutional users within the region or it is producing for consumption by entities external to the region of production. The fundamental assumption in the use of a final demand multiplier and its interpretation, however, is that expansion in ethanol production creates, concomitantly and at fixed rates, expansions in all inter-industrial relations that industry has with all of its inputs suppliers. So the use of a final demand multiplier for a particular industry, like the organic chemical industry where ethanol production is located assumes that as that industry expands production, there are fixed-ratio expansions in all industries that provide its intermediate inputs.

There is a fundamental flaw here because there is no real change in the overall demand for corn in the short run, just a shift in corn deliveries destined for local processing instead of for export. As a consequence, the application of a final demand multiplier to the corn sector is completely spurious. Those jobs already existed and would have existed had there not been an expansion in Iowa ethanol facilities. The ethanol plant did not create the corn production jobs or all of the corn industry's up-stream supply linkages. To claim them as ostensibly having been created by the emerging ethanol industry is misleading. To reiterate: ethanol production is not creating more farmers.

So from the 27,205 total jobs attributed to Iowa's ethanol industry operations in the RFA report we must next subtract the 18,398 jobs linked to its existing corn production sector. That leaves 8,807 jobs to investigate.

Several other items of critical inputs into production into this industry that are listed in the IRFA study after the already discounted corn values must be scrutinized. First, and importantly, the Iowa ethanol industry requires a large amount of natural gas, electricity, and water. The job gains attributable in that study to these three industries combined for 2,591 of the remaining 8,807 potential ethanol economic impact jobs. Those utility suppliers, however, are massive, declining cost industries in which the average costs of delivering their respective commodities up to capacity decline sharply. An industry that is an extremely heavy, and therefore comparatively easy to supply, user of a particular commodity is delivered that commodity at a substantially reduced price due to strong distributional efficiencies. Large users of utilities do not stimulate average job multiplier effects – they stimulate much lower, marginal effects and as a consequence are charged rates that are significantly lower than those charged to smaller users. This is a fundamental flaw in fixed-ratio impact analysis employed by the authors of the study and one of the reasons that experienced analysts conduct additional secondary research before reporting a statistic.

As part of the research conducted at Iowa State University on the potential economic impacts of a biofuels ethanol plant, water, natural gas distributors, and rural electric cooperative professionals were contacted to ascertain the potential new job requirements from a large, single industry increase in demand of their respective commodities in amounts indicative of a modern 50 million gallon per year (MGY) ethanol plant. In all instances, the job requirements reported by those professionals was a tenth or less than the amount assumed in the multiplier-driven modeling systems that are commonly used (Swenson and Eathington 2006). Based on that research and on fundamental scale economy dynamics, it would not be unreasonable to assume that the marginal job gains from all new utility related activities were no greater than 25 percent of the reported values, the much lower estimates of the utility professionals notwithstanding. If that were so, and there is strong economic and practical evidence that it is, the utility job impacts could reasonably be reduced to 648 jobs leaving a total of 6,864 jobs on the operational side of ethanol and other corn processing production in Iowa.

Next to scrutinize is the reasonableness of the transportation assumptions creating 1,442 jobs. Iowa's corn historically was hauled to a mill, to a livestock feeder, or exported out of state. After processing in an ethanol refinery, the amount of weight that must be hauled is roughly the same as it had been when the corn was simply exported, although the nature of the haulage is changed. We can allow for a modicum of new rail capacity, new rail transport needs, and some shifting in local transportation to account for these changes; although, like the corn statistic at the start of this section, we have to conclude that nearly all of the overall transportation had already existed in the region. Consequently, it is not unreasonable to allow for only a 25 percent bump in net new transportation jobs to the region (considering of course a substantial realignment from grain hoppers to ethanol tankers and other hauling substitutes). That would lower the 1,442 transportation jobs to 361 net new transportation jobs, thus leaving 5,782 corn processing jobs in Iowa to consider.

There are several categories of inputs that are not controvertible and would be expected to in fact be new regional indirect industrial demand linked linearly to ethanol plant operations. New ethanol plants will require substantial maintenance and repair services; they will help to stimulate demand for a variety of financial business services, to include banking, accounting, insurance, and other important activities; and they do require a new schedule of industrial chemical inputs into the production process, primarily yeasts, enzymes, and denaturants. For the time being, we can conclude that those inputs and their concomitant output and job multipliers are reasonable.

There is a fundamental question, though, about the likelihood of the bump in petroleum refinery inputs that the IRFA report claims. In all, when one looks at a modern ethanol plant's production recipe, one does not identify a set of refined petroleum product inputs (Tiffany and Eidman 2003). Their energy demands are met overwhelmingly by natural gas and electricity. The organic chemicals industry, the industry that manufactures such diverse commodities as acetone, nail polish, and tear gas along with dozens of others, however, does have strong linkages to refined petroleum products. The assumption that a modern Iowa corn ethanol dry mill operation buys $84.4 million in refined petroleum products from state suppliers as stated in the study is, however, not reasonable. It is especially dubious because Iowa's refineries made just $48.7 million in total sales across the whole state of Iowa and only needed 13 jobs to make those sales. It seems quite appropriate, then to reject the assertion that 351 refinery related jobs were created in Iowa.

After all adjustments, the impact estimate has now been reduced to 5,431 total Iowa jobs that produce ethanol and other processed corn commodities, supplied non-corn inputs, or otherwise produced goods and services for the households that are supported by all of these enterprises.

The Renewable Fuels Association of Iowa report (Urbanchuck 2007) indicated that the operational side of ethanol production in Iowa "...support[ed] 27,200 jobs." After systematically deconstructing the authors' procedures and assumptions, however, it is more likely that somewhere around 5,431 total jobs in Iowa can be attributed to ethanol *and to all other non-fuel, corn processing production that were also counted in that analysis.* That adjusted amount is less than 20 percent of the claimed operational amount and 11.6 percent of the original grand total that included the construction jobs. It is not unreasonable to conclude that the magnitude of misstatement at the national level is often analogous to the Iowa example.

3.2.2 The Policy and Practical Implications of Bloated Economic
Impact Claims

The foregoing assessment assists in understanding the basic job growth potential
of modern ethanol production and the possible magnitude of error common in estimating that potential. The gap between perception and reality is profound and
procedurally troublesome because it has implications for public policy development. Modern industrial development benefits strongly from federal, state, and local government underwriting. New ethanol plants across the U.S. are reaping large
amounts of risk-reducing tax credits, subsidies, and other kinds of public support.
According to one recent study (Koplow 2007), U.S. subsidies in support of ethanol
production ranged from $1.42 to $1.84 per gallon in 2006 considering all capital
development, credits, and other support. Using the same criteria for comparison
that study concluded that subsidies for petroleum averaged just 2.4 percent of those
amounts (Brasher July 2007). In Iowa, newer plants are demanding and receiving up
to 20 year local property tax abatements, along with several other very valuable state
tax breaks under its High Quality Job Creation Program, programs to spur capital
development, and transportation assistance.

    Local, state, and national public policies, incentives, and subsidies are currently
allocated based on an expectation of net gains to regional economies. The IRFA
study and others like it entice conclusions about the economic gains to regions that
are unwarranted, however. Across the nation there is evidence of confusion and a
fusion of the statistics that are used for promotion, which one must necessarily look
at with a grain of salt, and of statistics that are used to justify sound public decision
making, which are supposed to be based on sound scientific, economic, and policy
research. If public resources are allocated on the basis of misleading or exaggerated expectations of economic gain that will not materialize, then public resources
will have been squandered and the competing alternative uses to which those public resources could have been put will have been thwarted. And if so, society
suffers.

3.3 Ethanol Production Economic Opportunities and Offsets

In a mature and relatively stable commodity production and distribution system,
large changes in one segment of that system have consequences for other aspects
of agriculture, non-agriculture industries, the public, and households. Initially it is
important to note that the placement of a modern biofuels plant in a rural economy
will result in an expansion of net regional industrial production. In the short run
there is a positive economic impact to be expected. The rapid run-up in ethanol plant
development in the 2005 through 2007 period, however, has also had consequences
in many other aspects of agriculture, the impacts of which are just starting to be understood. This section works through some of the regional economic opportunities
and offsets that must be considered as this industry matures in the Midwest.

3.3.1 The Incidences and Economic Benefits of Farmer Ownership are Waning

The majority of ethanol plants in Iowa, South Dakota, and Minnesota in the first part of this decade were considered to be "farmer" or otherwise cooperatively or locally owned. The structure of this relationship was such that corn producers as investors linked themselves to a value added production process for their commodity (Gallagher 2005). The reason for this vertical configuration was that transportation costs from some of the nation's best corn production areas ate away at much of the profits to be made from farming. The greater the production costs of shipping corn for export, for example, to the barge terminals on the Mississippi River in Minnesota, Iowa, and Illinois, the lower the price received locally. Areas with a substantial commodity price basis penalty due to transport costs had strong incentives to convert grain to more profitable uses. Livestock feeding is one value added opportunity, and ethanol production is another. A local ethanol plant allowed area farmers to receive a nominally higher price for their corn as it was not sold with the implied shipping penalty.

Most new plants are not in any meaningful sense farmer or even locally owned (Lavigne 2007). Still, there is a strong preference in the Midwest for promoting local ownership of industrial stock (Morris 2007). States like Iowa, the Dakotas, and Minnesota have, to differing degrees enacted programs and policies to promote combinations of local, often-times small or rural investors in emerging enterprises like wind energy and biofuels. The policy and development argument is that local investors will rely on local banks along with financial and legal expertise will be more likely to contract for construction and input services with local suppliers, and most of all will be likely to convert their returns on investment to local consumption and additional local investment.

While local or farmer ownership was the early model for ethanol plant development, as this industry began to rapidly grow, equity investments were sought and received from all kinds of investors from all over the country. Research at Iowa State University (Swenson and Eathington 2006) indicated that, given a 50 MGY ethanol plant, the total added job impacts grew by 29 jobs for every 25 percent that the plant is owned by local residents. In short, local ownership coupled with large returns on investment locally yielded greater main street sales in the plant communities.

Those enhancements to local economic impacts were calculated based on the very robust returns received by investors in 2005 and would not be appropriate in the current market where returns are much more constrained. Importantly, those robust returns were also calculated without measuring the opportunity cost of the locally-supplied investment capital. The opportunity cost would be the normal next best alternative to which this investment money would have been put in that regional investment environment. The net return in excess of the opportunity cost is an unknown as we have no way of knowing exactly how regional investors had hitherto used their savings.

There are, therefore, three considerations that must temper the expectation of localized economic impacts from high levels of regional ownership. The value of alternative uses of that investment capital is not known, but one would assume that the normal investors' returns on all savings would have at least matched the national rate of return. Second, many farmer investors have borrowed against existing assets to invest in biofuels production. That action shifts net gains away from the now mortgaged enterprise, farming, to the new enterprise. That investment option has been widely reported, but the magnitude of it cannot be measured. Last, an increasing number of investors are not farmer-investors, and whether they reside regionally or not, there is no reason to expect those kinds of investors to behave, in the aggregate, any differently than all other investors (Lavigne 2007). Hence, for them, there is no discernible local impact to be assumed.

By the middle of 2007, growth in ethanol production capacity outstripped the national rate of absorption of ethanol and prices moderated considerably leading biofuels researchers to forecast constraints on the profitability in many of the plants, especially the older, smaller, and less efficient operations (Tokgoz et al. 2007). Consequently, one would expect that many plants are not paying substantial dividends as before, and that means the overall benefits of farmer or local ownership are expected to erode.

3.3.2 Higher Returns to Corn Producers and Land Owners Plus Higher Land Rents

Corn producers first promoted ethanol as a mechanism for localized gains in corn prices. The closer a corn farmer was to an ethanol plant, the better the net return on the corn as the comparatively high cost of shipping to alternative buyers was minimized. The farther a farmer was away from a plant, the less of an implied price bump (McNew and Griffith 2005). As the pace of ethanol plant expansion increased through the 2006 production year, however, corn prices nationwide, not just locally, began to climb. Figure 3.2 shows the nominal (not adjusted for inflation) average annual price of corn per bushel over the past several years and as projected through futures. While corn prices demonstrate some strong fluctuations, they averaged near $2.00 for much of the previous decade. In 2006, however, average prices rose sharply as more and more plants began to process ethanol, as demonstrated in Figure 3.2. Accordingly, the average price received nationwide rose by 58 percent over the previous year, though there is the expectation of strong localized volatility in corn prices over time as corn supplies and demand adjust (Hart 2007).

Corn farmers, however, did not see their net receipts increase by 58 percent over those two years, and in fact the U.S. BEA noted that Iowa farm earnings in 2006 were actually 5.3 percent lower than the year previous (BEA 2007) despite the corn price run-up. First, like all producers and consumers in the U.S., higher energy prices have affected farmers' bottom lines. Modern corn farming is energy intensive requiring large amounts of distillates for tractors, fertilizers derived in the main from natural gas, and propane for drying grain. So the same high oil prices boosting ethanol demand, and consequently, the demand and price received by farmers for

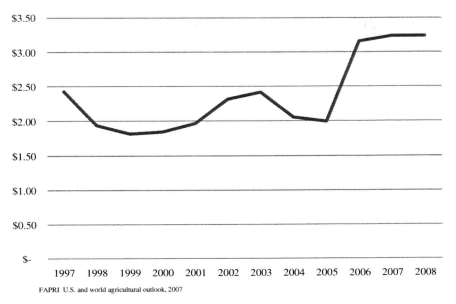

FAPRI U.S. and world agricultural outlook, 2007

Fig. 3.2 U.S. corn prices per Bushel

their corn, is also boosting variable production costs on the farm. Second, as market prices increase, the total amount of government payments to corn farmers decrease, which assuredly is good news for taxpayers but still must be counted when compiling the net change in corn farmer returns and, by extension, the well being of rural economies (Westcott 2007). In all, as price increases the financial position of corn farmers improves, but the exact amount of improvement must be calculated net of subsidy reductions and the changes in all other fixed and variable costs of production changes.

Price driven gains to farmers have two very important outcomes regionally. First, they eventually help bolster the overall profitability of farming as an enterprise, which in turn is realized in higher amounts of on-farm capital and other investment along with boosted farm family spending. Second, sustained higher prices must increase the value of farm land. Over time, farmers who are landowners will realize price-induced capital gains on their land investments. For farmers that must rent their land, however, they will realize higher land use costs, which in turn will limit their net gains on production. In Iowa, according to the 2002 Census of Agriculture, 51 percent of the land in farms was rented. Higher corn prices will therefore result in increased land rent costs for 51 percent of Iowa corn crop production.

3.3.3 Higher Feed and Input Costs for Other Corn Consumers

Most Americans do not eat much corn. They do, on the other hand, eat a tremendous amount of products that are directly or indirectly derived from corn. Nearly all pork, beef, dairy, chicken, turkey, and egg products in the U.S. rely strongly on corn as a feedstock. Also, Americans have increasingly come to rely on high fructose corn syrups (HFCS) as a sugar substitute in many foods, beverages, and confections. It is apparent that there is strong demand for corn as a critical input into food production in the U.S.

Table 3.1 demonstrates the uses of corn historically. In 2000 about 11.3 percent of all corn was made directly into food or high fructose corn syrup. Over 50 percent, however, was a feed to livestock, 16.7 percent was exported, and only 5.4 percent was used for ethanol. By 2005, the amount of feed demanded had increased to 6.1 billion bushels, but ethanol's demand for corn had increased by more than 150 percent. As a consequence of the increased demand for ethanol, the projection for 2010 has the amount of corn available for feed as eight percent lower than in 2005. At that time ethanol is expected to consume 30 percent of the nation's corn supply, up 25 percentage points in just a decade.

The high reliance on corn inputs by the livestock sector is ostensibly offset by the production of distillers' grains at the ethanol plants. Distillers' grains are the high protein residue left after the ethanol fermentation process is completed. Distillers' grains can be fed in varying degrees to livestock, ranging from 30 to 40 percent of diet to feeder cattle down to 10–20 percent for dairy cows, swine, or poultry. No matter the supply and price of distillers' grains and the mix of rations employed, feeders will still have to include some corn input costs in the mix. American cattle producers appear to be cautious about the rapid growth in the ethanol industry and have recently argued against an expansion in federal ethanol production subsidies beyond current levels (NCBA 2007), with increased corn prices as the rationale.

Higher feed prices have several likely expected outcomes that may reduce meat and poultry supply. First, livestock producer net returns will shrink; this is especially the case for those that are located at some distance from ethanol plants and who had historically depended on Midwestern corn supplies. In some cases, less profitable operations will cease production entirely. In other instances, producers will not finish livestock as long – the point at which additional feed yields an optimal return will move towards a smaller animal. Hence, animals will be marketed at a lighter weight.

Table 3.1 Historical and projected uses of corn

	2000	Percent of supply	2005	Percent of supply	2010	Percent of supply
Corn Supply (Millions of Bushels)	11,639.42	100.0	13,237.00	100.0	14,266.60	100.0
Ethanol	627.59	5.4	1,603.00	12.1	4,307.65	30.2
Feed	5,842.09	50.2	6,140.83	46.4	5,657.81	39.7
Food	780.24	6.7	829.90	6.3	861.69	6.0
HFCS	529.75	4.6	528.60	4.0	530.38	3.7
Other	185	1.6	190.20	1.4	196.52	1.4
Seed	19.30	0.2	20.17	0.2	23.33	0.2
Exports	1,941.35	16.7	2,147.34	16.2	1,885.72	13.2

FAPRI U.S. and world agriculture outlook, 2007.

Finally, all consumer prices will increase as consumers absorb the higher costs associated with a lower meat and poultry supply. In all other instances, say for the production of HFCS and other corn to food products, prices will likely be passed on to consumers or otherwise result in lower returns to manufacturing producers (Westcott 2007).

In the longer term, expansion in ethanol production may lead to further concentration and vertical integration in the U.S. meat production sector. The dominant business model for poultry and meat production has a prominent firm like Tyson Foods or Smithfield Foods involved significantly with all aspects of breeding, production, processing, and distribution. As modern ethanol plants produce immense amounts of distillers' grains that are mainly suitable as cattle feed, it is possible that future ethanol plants will include very large integrated cattle feeding operations in order to efficiently feed distillers grains and to capture additional efficiencies by using animal waste as a source of fuel.

Spatial shifts in meat production are another possible outcome. Areas of the Midwest that have the highest concentration of corn production also have some of the nation's greatest concentrations of swine and poultry production because of very strong production efficiencies to be achieved from locating amidst high feed supplies. Iowa, as an example, ranks first nationally in swine and in egg production, and those animal concentrations are centered in the best corn growing areas. Cattle on feed, in large measure, are located much further to the west and southwest. Paradoxically, the animals that are least tolerant of distillers' grains and can only consume it in smaller amounts are found in higher numbers in the areas of the U.S. where there are comparatively high concentrations of ethanol plants, and the animals that are most tolerant are in comparatively lower numbers. It remains to be seen whether production advantages accumulate to the beef industry because it can more readily incorporate distillers' grains as feed and whether those advantages will work at the expense of poultry and swine production.

3.3.4 Grain Storage, Processing, and Distribution Systems Will Change

The nation's grain storage and transportation infrastructure developed over the years in direct response to the historical pace and pattern of crop production in the U.S. As Midwestern states have most of the nation's corn producing capacity, there are extensive systems for storage, marketing, and distributing that bounty. The nation's infrastructure for moving corn includes the inbound systems, the storage systems, grain processing systems, and the outbound systems. The nation's capacity in all aspects of managing its grain supply has developed over a long period of time and, as these are all highly capital intensive systems, that capacity closely matches production. There are several issues affecting this complicated sector of the economy that must be taken into account as the ethanol industry develops (Ginder 2007).

Ethanol plants are able to store anywhere from 10 to 25 days worth of corn. Corn that is delivered directly to the ethanol plant from farm storage, however, is corn that is not conveyed through local grain elevator systems or moved outbound via rail as historically had been the situation. So in the initial stages of ethanol plant development, gains to farmers and the expansion of ethanol production must be assessed in light of a reduction in gross receipts and reduced efficiencies on investments in all grain handling systems. As the industry matures and as competition for corn requires greater grain origination and distribution skills and efficiencies, the nation's elevator systems may come to play an integral role in moving corn into ethanol plants, but the extent and effectiveness of the sector remains to be demonstrated. In the near term, the rapid diversion of grain stocks into ethanol plants has impinged on the profitability of traditional grain handlers.

The rail transportation rolling stock that evolved to move corn is ill-suited to moving either ethanol or the byproducts of ethanol. Ethanol is primarily transported in truck and rail tankers, and cannot be transported by pipeline. Its primary byproduct is distillers' grains, which in either wet or dried form needs special rail stock as well. Furthermore, planned improvements and expansions on the Mississippi River and Illinois River locks and dams have been justified based on controversial expectations of strong growth in corn exports out of the Midwest (WSTB 2004). The expansion of ethanol production interferes with that justification in the long run, and in the short run makes the existing barge and terminal systems in the interior of the country less efficient and, therefore, less profitable.

Corn acre plantings in 2007 are estimated at 19 percent higher than 2006, and soybean plantings are down by 15 percent. Each acre of corn produces from two to three times the bushels per acre as soybeans, the primary crop sacrificed for expanded corn acres. As the nation's grain storage capacity is closely matched to grain production historical development, this rapid rise in corn supply will rapidly exhaust the nation's existing on-farm and elevator storage capacity. Storage capacity is very expensive, and it remains to be seen exactly where the economic incentives will accrue that will induce capital investment in this area. The risk, of course, is that expansion in grain storage will become potentially excess capacity if and when the nation shifts towards cellulosic ethanol production.

3.3.5 Spatial Changes in Crop Production

Which crop can be produced on which acre of land most profitably depends on many factors, but when the price of a commodity rises sharply, as has been the recent experience with corn in the U.S., land that had been primarily suitable for one mix of crops might now be suitable for a different mix.

Corn acreage increased in 45 of the lower 48 states between 2006 and 2007 due primarily to strong futures prices during the crop planning season of post harvest 2006 and planting time in 2007. The states of Indiana, Illinois, Minnesota, California, and North Dakota posted record corn plantings. The amount of greatest gain was in Illinois at 1.9 million more acres. A grain producing state with the strongest shift is North Dakota with nearly a 48 percent rise in corn plantings. Their increase came at the expense of a 7 percent reduction in all wheat planting and a 21 percent reduction in soybean acres. Kansas soybean plantings were down by 24 percent, Nebraska's by 21 percent, Indiana's by 19 percent, and South Dakota's by 16.5 percent. (NASS 2007).

Increased plantings of corn will affect the aforementioned storage issue: corn produces significantly more bushels per acre than either soybeans or wheat. In addition, large shifts in production will have up-stream impacts on normal regional uses of agricultural commodities. Existing processors of oil seeds for food, feed, or other uses will have sharply increased input costs due to the supply reductions. In the longer run, some commodity needs such as soybeans will necessarily be met by increased imports (Westcott 2007).

The large shift in corn acres also places stress on the nation's corn-inputs system. Corn requires fertilizers that derive mainly from natural gas, petroleum distillates for machinery, and large amounts of propane for drying corn. In all, a strong positive shift in corn production in the U.S. increases the demand for a wide array of energy inputs, which in turn drive up the prices charged to other users of those same inputs.

Finally and importantly, there are important environmental issues associated with corn production. The crop's need for high amounts of petroleum based and chemical inputs degrades groundwater and shallow aquifers. Dominant corn tilling practices also result in soil runoff, siltation of streams and rivers, and ultimately the creation of hypoxia zones in the Gulf of Mexico due to, primarily, ag-originated nutrient runoff into that area. These all entail external economic costs that are not borne by the industry or its beneficiaries, but by society at large.

There is pressure to expand the nation's land in production. There are two sources: existing pasture land and land currently enrolled in the Conservation Reserve Program (CRP). In both instances, long term land use preferences and national policy combined to remove vulnerable and marginal land from crop production. The conversion of these acres may exacerbate a wide array of environmental issues, to include increased soil erosion, surface water degradation, and soil nutrient depletion.

3.3.6 The Biofuels Industry will Obtain Scale Economies

Some early ethanol plants produced just 10–20 million gallons yearly (MGY) of ethanol. Over time, ethanol plant sizes increased as investment capital became more available, as public subsidies helped to underwrite and offset risk, and as ethanol prices stabilized and demand demonstrated positive growth. Like many capital intensive industries, there are strong internal economies of scale opportunities. Economies of scale occur as a firm is able to, through more efficient utilization of its capital stock, procurement of inputs, and labor, achieve lower average costs of production per unit of output.

An obvious demonstration of scale economies presents itself readily in the ethanol industry itself. As is demonstrated in Figure 3.3, a 50 million gallon per year (MGY) ethanol plant in Iowa requires 36 jobs. A 100 MGY per year plant only requires 46 jobs. The plant increases its output by 100 percent, but its job needs only go up by 28.5 percent. Similarly, the plants will achieve strong efficiencies in

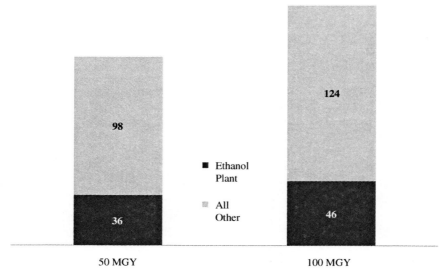

Fig. 3.3 Ethanol plant job impacts by plant capacity in millions of gallons per year (MGY)

the use of storage systems, grain moving and handling infrastructure, its land, much of its technical inputs, and larger bulk purchases of its required inputs.

As the industry shifts, as firms become, on the average, larger and more efficient, larger and better operated firms, usually those that were built most recently will have higher returns per unit of production when compared to smaller and less efficient plants. In consequence there is the expectation that in the very near future several of the nation's smaller, typically locally owned ethanol plants will become less profitable and will likely be forced out of business (Miranowski 2006).

3.4 Bioenergy Promotion and the Overall Sustainability of Rural Economies

In October of 2006 a joint U.S. Department of Energy and U.S. Department of Agriculture conference was held in St. Louis entitled "Advanced Renewable Energy: a Rural Renaissance." New York Senator Hillary Clinton that year noted in a press release that "We can create a rural renaissance and restore the promise of Main Street..." in part by "... investing in renewable energy ...." (Clinton 2006). Along similar lines, U.S. Senators Norm Coleman of Minnesota and Mark Prior of Arkansas jointly proposed a *Rural Renaissance II* program in the U.S. Senate that would provide low-interest loans along with grants to rural areas to develop infrastructure and to entice investment in renewable fuels and energy sources (U.S. Senate 2005). The head of the United Nation's Food and Agriculture Organization,

Alexander Mueller, concluded in 2007 that properly promoting biofuels could be an "important tool for improving the well-being of rural people if governments take into account environmental and food security concerns." (FAO Newsroom 2007).

In each of these instances there is the assumption that the production of renewable energy from wind, corn, and biomass feedstocks will rejuvenate rural areas. Those assumptions are, however, lacking significant substantiating evidence in the near term. For example, wind energy, which is expanding smartly in several places in the Midwest and Plains areas, is disproportionately controlled by existing, regionally dominant investor-owned utility systems. Those companies negotiate land rents for their structures, but otherwise their overall economic impact to regional economies is quite limited – once the machines are up and running, they do not require significant regionally supplied inputs.

The rural economic development potential of cellulosic systems is a complete unknown. Scientists and engineers can agree on many of the technical details and distributional requirements. Technical agreement notwithstanding, economics, however, require that the price of fuel must increase drastically before biomass can be efficiently and competitively processed. The only realistic contemporary laboratory for gauging the revitalization potential of modern biofuels is the current expansion in corn ethanol production in the U.S. and to a lesser extent biodiesel production from oil seeds (Tokgoz et al. 2007). And the market attributes of both of those examples are distorted via the range of subsidies underwriting the current pace of growth.

There are heady expectations for growth, and some recent research (Ugarte et al. 2006) has projected that the attainment of several biofuels production goals in the U.S. will by 2030 create as many as 2.4 million new production related jobs in the U.S. were the nation to produce 60 million gallons of biofuels, many of which could accrue to rural areas. That research is probably much too enthusiastic about the potential: much of it presupposes yet to be proven technical, distributional, investment, and policy developments that would allow for the optimization of production in attaining that optimistic goal. It also projects a future national industrial structure based primarily on the contemporary economy, a dicey prospect in economic modeling. The structure of the national economy in 2030 will be very different from the structure at present.

3.4.1 Putting Biofuels Job Change and Growth into Perspective in the Near Term

The interior economy of the U.S., to include its more rural areas, has not grown at anywhere near the pace as the remainder of the U.S. We also know that manufacturing in the interior of the U.S. has been hard-hit over the past decade. Ethanol production from corn is a form of chemical manufacturing. When we look at the overall value of manufacturing to any economy, two factors are paramount: the number of jobs created and, of course, the associated earnings that workers convert to household consumption. Per unit of output, ethanol requires relatively few jobs as compared to the average manufacturing firm. The jobs produced, however, are good jobs when measured by wage and salary.

There have been very strong declines in manufacturing jobs during the present decade. Nationally, between 2000 and 2005 the nation lost nearly 3 million manufacturing positions, about 18 percent were in non-metropolitan areas of the nation, areas that did not have a central city of 50,000 or more. The chemical manufacturing industry, of which ethanol production is a subset, lost almost 100,000 jobs over the same time period. In 2005 the average earnings of a U.S. manufacturing job considering all wages, salaries, and benefits was $60,100. In the chemical manufacturing sector it was $69,150.

The firm and job growth directly associated with ethanol production in the U.S. can be readily estimated even though current detailed U.S. statistics are not available. In 2005, just over 1.6 billion bushels of corn were converted into ethanol. Assuming that those plants generated at a maximum 2.7 gallons of ethanol per bushel (EEOE 2007), that their average size at that time nationally was 65 million gallons per year (MGY), that they operated at 115 percent of average capacity, and that each plant averaged 38 jobs, then the U.S. ethanol industry directly required 78 plants and 2,910 jobs to process 1.6 billion bushels of corn. Average pay at new U.S. ethanol plants ranged from $45,000 to $55,000 per year – substantially less than either the U.S. manufacturing average or the average for chemical manufacturing, but substantially more than the nonfarm earnings average in most rural areas.

Were the industry to grow to process just over 4.3 billion bushels of corn annually by 2010, and assuming that plants were, on average producing 2.7 gallons of ethanol per bushel of corn, were rated at 85 MGY in average capacity, produced at 120 percent of rated capacity, and had 47 jobs per plant, then the U.S. ethanol industry would require 165 plants and 7,716 jobs in 2010 as shown in Table 3.2. If the rural areas of the U.S. lost some 540,000 manufacturing jobs between 2000 and 2005, it is impossible to conclude that just from corn ethanol the addition of 7,716 jobs will yield a rural renaissance. Figure 6.4, compares just the expected gains in ethanol plant jobs through the end of this decade nationally to the erosions in just chemical manufacturing jobs in the U.S. during the first half of the decade.

Finally, for distributional perspective, if it is assumed that two thirds of the future corn ethanol production capacity were concentrated in Iowa, Indiana, Illinois, Nebraska, and Minnesota, then there would be, on average, one plant per just over four counties, which would work out to slightly fewer than 11.5 new manufacturing jobs per county.

Table 3.2 U.S. ethanol plants and jobs

	Corn bushels in millions	plants	jobs
2005	1,603	78	2,964
2010	4,307	165	7,716

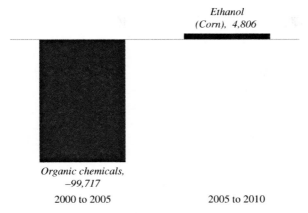

Organic chemicals,
−99,717

2000 to 2005   2005 to 2010

Fig. 3.4 Organic chemical manufacturing job change compared to expected ethanol job growth

3.4.2 The Longer Term Prospects for Rural Areas from Biofuels Development

A hallmark of modern agribusiness and modern manufacturing is the persistent substitution of capital for labor. In 1970 the average farm worker in Iowa tended 200 acres of crop land. In 2005 the average Iowa farm worker tended 300 acres of crop land.

The prospect of increased biofuels production presupposes an extension if not an acceleration in the uses of mechanical and chemical inputs into agricultural production as farmers shift production to accommodate the corn ethanol industry's rapid expansion of late. Simultaneously, the corn ethanol industry itself will expand preferring to develop highly efficient production systems closer to the 100 MGY per year range and larger, which also will require much less labor per gallon of production than is currently the industry average. Both of these assumptions do not portend a rural economic recovery, but rather a continuation if not an acceleration of the fundamental factors undermining most rural areas in the interior of the country: limited and specialized labor demands in only a few dominant industries that are increasingly capital intensive; and production systems that require, over time, fewer and fewer regionally supplied intermediate labor inputs.

The longer term technical and policy outlook contains an expectation of ethanol production deriving significantly from cellulosic stocks, to include ultimately acres of crop land that are dedicated to perennial energy production. If such a situation were to eventuate, then there indeed may be the potential for meaningful expansion in the value of productivity in many places of the U.S. that heretofore had not prospered. Before those unhatched chickens can be counted, however, there are several very important factors that will have to be resolved.

First and foremost, given current technology, cellulosic ethanol production, even under ideal conditions, is not cost effective.

The infrastructure needs for harvesting, converting, separating, transporting, and ultimately processing cellulosic feedstocks currently do not exist and can only be imagined. The production and distributional efficiencies at the plant and spatially are significantly unknown.

The overall labor requirements of processing cellulosic feedstocks is not well understood in light of the current trends in the ratios of labor to all crop acres. Shifts from one form of production, as in the current corn system, to another, such as what might eventuate from energy crop production will require a reallocation of labor and machinery, but not necessarily changes that will indirectly stimulate regional growth, especially in rural households.

The distribution of crop production and processing capacity relative to regional demand will likely favor development closer to built up areas with high demand potential to minimize transport cost and maximize returns.

More remote, yet potentially productive, areas of the U.S. may realize long delays in the timing of biofuels development due to distance, infrastructure, and other constraints.

Global volatility in oil prices may not stimulate the pace and pattern of investment expected to produce expected future levels of biofuels.

The nation's absorption of ethanol as a fuel source will have to increase dramatically.

And finally, an energy policy and a rural development promise that depends on rain has inherent volatility.

There are many important considerations associated with biofuels production and development in the United States that were not dealt with in detail in this chapter. Enterprise-level analysis of the overall costs of operation helps policy makers and decision makers understand the production characteristics of corn and alternative ethanol production and the effects of both external and internal production factors in determining the profitability of ethanol (Tiffany and Eidman 2003). The scope and costs of ethanol subsidies are neither detailed nor assessed here, but it must be recognized that the combined public costs of ethanol production as measured in total or on a per gallon basis is high and promises to grow. Last this analysis does not look at the overall efficacy of this form of energy development vis a vis all others. It is very difficult for many economists to discern net national gains to be derived from the current biofuels policies, and in light of that we see the rationale for ethanol promotion and biofuels development shifting from economics and economic welfare to one of "enhanced national security " (Brown 2007).

There are tangible regional economic and environmental aspects to the current debate on the development of biofuels in the U.S. Some are treating the topic in a race-to-the-moon manner with a promise of technological determinism that will, ultimately, lead to substantial social payoffs and an ultimate rationality to the process. In the meantime, however, public decision makers are charged with maximizing social gains, minimizing the undesirable consequences of public action, and assuring the nation through sound policy research that the economic benefits to be achieved from the nation's biofuels initiatives do indeed outweigh the economic, social, and environmental costs of implementing them and are, on net, better than the alternatives. To date, there is precious little evidence that is so.

References

Brasher, P. (July 15, 2007). The end of the biofuels money train? *The Des Moines Register.* Des Moines, Iowa. Retrieved from http://desmoinesregister.com/apps/pbcs.dll/article?AID=/20070715/BUSINESS01/707150330/-1/biofuels

Brown, R.C. (March 2007). Options for biofuels. Potential to produce liquid fuels from cellulosic feedstock. Alternative crops and alternative policies for bioenergy web program. Iowa State University Cooperative Extension. Retrieved from http://www.extension.iastate.edu/bioeconomy/webcast/3-5-07Webcast.html

Bureau of Economic Analysis (BEA) (1997). Regional multipliers: A users guide for the Regional Input Output Modeling System (RIMS II). U.S. Department of Commerce. Retrieved from http://www.bea.gov/scb/pdf/regional/perinc/meth/rims2.pdf

Bureau of Economic Analysis (BEA) (2007). State income and employment summary, Table SA04. U.S. economic accounts. U.S. Department of Commerce. Retrieved from http://www.bea.gov/regional/spi/default.cfm?satable=SA05N&series=NAICS

Clinton, H.R. (July 21, 2006). Remarks of Senator Hillary Rodham Clinton calling for a rural renaissance to restore the promise and prosperity of main streets and rural communities. Prepared speech delivered in Lockport, NY. Retrieved from http://clinton.senate.gov/news/statements/details.cfm?id=260431&&

Daschle, T. (March 2006). Follow the farmers. *American Prospect.* Retrieved from http://www.prospect.org/cs/articles?article=follow_the_farmers

EERE (Energy Efficiency and Renewable Energy) (March 2007). Useful information about alternative fuels and their feedstocks. U.S. Department of Energy. Retrieved from http://www1.eere.energy.gov/biomass/pdfs/useful_info.pdf

Energy Information Administration. (August 2007). Short term energy outlook, Table 5A. U.S. Department of Energy. Retrieved from http://www.eia.doe.gov/emeu/steo/pub/5atab.html

FAO (Food and Agriculture Organization) (April 23, 2007). Bioenergy could drive rural development. FAO of the United Nations. Retrieved from http://www.fao.org/newsroom/en/news/2007/1000540/index.html

FAPRI 2007 (January 2007). U.S. and world agricultural outlook. Food and Agricultural Policy Research Institute. Iowa State University and the University of Missouri – Columbia. Retrieved from the Center for Agriculture and Rural Development, Iowa State University Web site http://www.fapri.iastate.edu/Outlook2007/text/OutlookPub2007.pdf

Gallagher, P. (2005). Pricing relationships in processors' input market areas: Testing theories for corn prices near ethanol plants. *Canadian Journal of Agricultural Economics.* 53, pp. 117–139.

Ginder, R. (July 2007). Potential infrastructure constraints on current corn-based and future biomass based U.S. ethanol production. Department of Economics Working Paper #7018, Iowa State University. Retrieved from http://www.econ.iastate.edu/research/webpapers/paper_12836_07018.pdf

Hart, C. (Summer 2007). Shifting corn basis patterns. *Iowa Ag Review,* 13:3, pp. 8–10

Koplow, D. (April 2007). Biofuels at what cost? Government support for ethanol and biodiesel in the United States. Global Studies Initiative of the International Institute for Sustainable Development. Retrieved from http://www.globalsubsidies.org/IMG/pdf/biofuels_subsidies_us.pdf

Lavigne, P. (April 29, 2007). Biofuel industry branches out, outside investors flow in. Des Moines Register. Des Moines, IA

McNew, K. and Griffith, D. (2005). Measuring the impact of ethanol plants on local grain prices. *Review of Agricultural Economics* 27:2, pp. 164–180

Miranowski, J. (November 2006). Economic drivers of biofuels expansion. Cooperative Extension report, Iowa State University. Retrieved from http://www.extension.iastate.edu/ag/MIranowskiPresent.indd.pdf

Morris, D. (January 2007). Energizing rural America: Local ownership of renewable energy production is the key. Institute for Local Reliance. Retrieved from http://www.americanprogress.org/issues/2007/01/pdf/rural_energy.pdf

NASS (National Agriculture Statistical Service) (2007). Crop progress and condition reports. U.S. Department of Agriculture. Retrieved from http://www.nass.usda.gov/ Charts_and_Maps/index.asp

NCBA (National Cattlemen's Beef Association) (June 2007). Cattle producers urge equal opportunity energy policy. NCBA News. National Cattlemen's Beef Association. Centennial, CO. Retrieved from http://hill.beef.org/NEWSCattlemenOpposeIncreaseinGrain-BasedEthanolMandate31432.aspx

Novack N. (March 2002). The rise of ethanol in rural America. *The Main Street Economist*. Center for the Study of Rural America, Federal Reserve Bank of Kansas City.

Swenson, D. (June 2006). Input outrageous: The economic impacts of modern biofuels production. (Paper presented at the Mid-Continent Regional Sciences Association and the Biennial IMPLAN Users Conference, Indianapolis, IN). Retrieved from http://www.econ.iastate. edu/research/webpapers/paper_12644.pdf

Swenson, D. (Summer 2007a). Biofueling economic growth in Iowa. *Small Farmer's Journal*. 32:2, 33–34.

Swenson, D. (April 2007b). Understanding biofuels economic impact claims. Department of Economics Staff Report, Iowa State University. Retrieved from http://www.econ.iastate.edu/ research/webpapers/paper_12790.pdf

Swenson, D. and Eathington, L. (September 2006). Determining the regional economic values of ethanol production in Iowa considering different levels of local investment. Department of Economics Staff Report, Iowa State University. Retrieved from http://www.valuechains.org/ bewg/Documents/eth_full0706.pdf

Tiffany, D. and Eidman, V.R. (August 2003). Factors Associated with success of fuel ethanol producers. Staff Paper PO37. Department of Applied Economics. University of Minnesota

Tokgoz, S., Elobeid A., Fabiosa, J.F., Hayes, D.J., Babcock, B.A., Yu, T.S, Dong, F., Hart, C.E., Beghin, J.C. (May 2007). Emerging biofuels: Outlook of effects on U.S. grain, oilseed, and livestock markets. Center for Agriculture and Rural Development, Iowa State University. Retrieved from http://www.card.iastate.edu/publications/DBS/PDFFiles/07sr101.pdf

Ugarte, D., English, B., Jensen, K., Hellwinkel, C., Menard, J., Wilson, B. (2006). Economic and agricultural impacts of ethanol and biodiesel expansion. Agricultural Economics Study Report, University of Tennessee. Retrieved from http://www.ethanol-gec.org/information/ Ethanolagimpacts.pdf

Urbanchuck, J. (January 2005). Contribution of the ethanol industry to the economy of the United States. Renewable Fuels Association. Retrieved from http://www.ethanolrfa.org/objects/ documents/576/economic_contribution_2006.pdf

Urbanchuck, J. (February 2007). Contribution of the biofuels industry to the economy of Iowa. Iowa Renewable Fuels Association. Retrieved from http://www.iowarfa.org/PDF/2006 %20Iowa%20Biofuels%20Economic%20Impact.pdf

USDA. (February 2007). Agricultural Projections to 2016. U.S. Department of Agriculture, Office of the Chief Economist. OCE-2007-1. Retrieved from www.ers.usda.gov/publications/oce071/

USDA. (June 2007). U.S. farmers plant largest corn crop in 63 years. U.S. Department of Agriculture Newsroom. Retrieved from http://www.nass.usda.gov/Newsroom/2007/06_29_2007.asp

S. 1253 (June 15, 2005). Rural renaissance II act of 2005. Senate of the United States, 109th Congress. Retrieved from http://thomas.loc.gov/home/multicongress/multicongress.html

Westcott, P. (May 2007). Ethanol expansion in the U.S.: How will the agriculture sector adjust? Economic Research Service, U.S. Department of Agriculture. Retrieved from http://www.ers.usda.gov/Publications/FDS/2007/05May/FDS07D01/fds07D01.pdf

WSTB (Water and Science Technology Board) (2004). Review of the U.S. Army Corps of Engineers restructured upper Mississippi River-Illinois waterway feasibility study. National Academy of Sciences. Washington, D.S. Retrieved from http://books.nap.edu/catalog. php?record_id=10873#orgs

Chapter 4
Subsidies to Ethanol in the United States

Doug Koplow and Ronald Steenblik

Abstract Ethanol, or ethyl alcohol used for motor fuel, has long been used as a transport fuel. In recent years, however, it has been promoted as a means to pursue a multitude of public policy goals: reduce petroleum imports; improve vehicle emissions and reduce emissions of greenhouse gases; and stimulate rural development. Annual production of ethanol for fuel in the United States has trebled since 1999 and is expected to reach almost 7 billion gallons in 2007. This growth in production has been accompanied by billions of dollars of investment in transport and distribution infrastructure. Market factors, such as rising prices for petroleum products and state bans on methyl tertiary butyl ether (MTBE), a blending agent for which ethanol is one of the few readily available substitutes, drove some of this increase. But the main driving factor has been government support, provided at every point in the supply chain and from the federal to the local level. This chapter reviews the major policy developments affecting the fuel-ethanol industry of the United States since the late 1970s, quantifies their value to the industry, and evaluates the efficacy of ethanol subsidization in achieving greenhouse gas reduction goals. We conclude that not only is total support for ethanol already substantial — $5.8–7.0 billion in 2006 — and set to rise quickly, even under existing policy settings, but its cost effectiveness is low, especially as a means to reduce greenhouse gas emissions.

Keywords Agriculture · biofuel · corn · energy · ethanol · policy · renewable energy · subsidies · support · United States

✉ D. Koplow
Earth Track, Inc., 2067 Massachusetts Avenue, 4th Floor, Cambridge, MA 02140
e-mail: dkoplow@earthtrack.net

R. Steenblik
At the time of article submission, Director of Research for the Global Subsidies Initiative of the International Institute for Sustainable Development, Maison Internationalle de l'Environment 2, 9, chemin de Balexert, 1219 Châtelaine Genève, Switzerland

D. Pimentel (ed.), *Biofuels, Solar and Wind as Renewable Energy Systems*,
© Springer Science+Business Media B.V. 2008

Acronyms & abbreviations

AFV: alternative fuel vehicle
bgpy: billion U.S. gallons per year
mgpy: million U.S. gallons per year
CAFE: corporate average fuel economy
CBERA: Caribbean Basin Economic Recovery Act
CO_2:   carbon dioxide
CRS: Congressional Research Service
E10:   a blended fuel comprised of 10% ethanol and 90% gasoline
E85:   a blended fuel comprised of 85% ethanol and 15% gasoline
EIA: U.S. Energy Information Administration
EPA: U.S. Environmental Protection Agency
EPACT05:   Energy Policy Act of 2005
FFV: flexible-fuel vehicle
GHG: greenhouse gas
GJ: gigajoule (10^9 joules)
GSI: Global Subsidies Initiative
IRS: Internal Revenue Service
JCT: Joint Committee on Taxation (of the U.S. Congress)
MPS: market price support
MTBE: methyl tertiary-butyl ether
NAFTA: North American Free Trade Agreement
OECD: Organisation for Economic Co-operation and Development
OTA: Office of Technology Assessment
RFA: Renewable Fuels Association
RFS: Renewable Fuels Standard
USDA: U.S. Department of Agriculture
VEETC: Volumetric Ethanol Excise Tax Credit

4.1 Introduction

The modern U.S. ethanol industry was born subsidized. The Energy Tax Act of 1978 introduced the first major federal subsidy for ethanol, a 4 cents-per-gallon reduction in the federal excise tax on gasohol, or E10 (a blend of 10% ethanol and 90% gasoline). In that same year, the first commercial ethanol production capacity came online. Between 1980 and 1990, production capacity more than quintupled, ending the decade at around 900 million gallons per year (mgpy). Despite a slower period of growth from the late 1980s through the mid-1990s, production capacity has grown in recent years at a very fast pace over most of the last decade. According to the Renewable Fuels Association (RFA) the main ethanol trade group, production capacity increased from 1.7 billion gallons per year (bgpy) in 1999 to 7.3 bgpy at the end of 2007 (RFA, 2007a). An additional 6.2 bgpy of capacity were under construction, the vast majority of which will rely on corn (RFA, 2007b).[1] Meanwhile, the supply side of the ethanol market is evolving towards ever larger plants, with the largest having annual capacities approaching 300 mgpy (Planet Ark, 2006). This trend will have important effects both on feedstock supply and on the market power of different portions of the supply chain.

Conversion into ethanol serves as an increasingly important outlet for the industry's main feedstock, corn. Estimates of the share of U.S. corn production used for ethanol vary, but most place it above 20% in 2007, and likely to rise above 30% within the next few years.[2] Despite rapid growth in demand and diversion of corn into fuel, ethanol consumption for 2006 (5.4 bgpy) supplied less than 4% of the fuel used by gasoline-powered vehicles in that year (Fig. 4.1).[3]

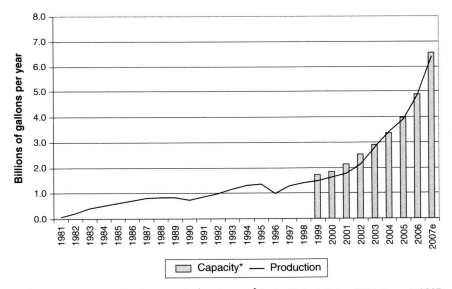

Fig. 4.1 Fuel-ethanol production capacity[1] and output[2] in the United States, 1981 through 2007
[1] Data for 2007 are authors' estimates. Capacity data prior to 1999 are not available.
[2] Capacity represents an estimated mid-year value, obtained by taking the geometric mean of the values reported at the beginning of the year shown and the value at beginning of the following year. Sources: ● **1981–2005:** Energy Information Administration, *Annual Energy Review 2006*, Report No. DOE/EIA-0384(2006), Table 10.3, "Ethanol and Biodiesel Overview, 1981–2006", Retrieved December 7, 2007 from; http://www.eia.doe.gov/emeu/aer/renew.html; ● **2006:** Renewable Fuels Association; "Industry Statistics", Retrieved December 7, 2007, from http://ethanolrfa.org/ industry/statistics/.

---

[1] Sugar from cane or beets, which is an important feedstock in ethanol production in regions such as Brazil and the European Union, has so far played a very small role within the United States. This is largely due to import quotas that make sugar too expensive as a feedstock.
[2] See FAPRI, (2007, February), p. 11; USDA (2007, February), p. 39.
[3] Ethanol consumption data from RFA (2007c); US gasoline consumption data from EIA (2007b).

Industry promotion of expanded purchase mandates and continued protection from imports demonstrate that producers are counting on the government to help keep production viable. Both policies were being considered by Congress in the autumn of 2007. Even more aggressive policy interventions have also been proposed, such as setting a floor price for oil in order to protect the domestic ethanol industry from low oil prices that would render ethanol uncompetitive (see, e.g., Lugar and Khosla, 2006). Clearly, in order to understand the industry, one has to understand the roll of government incentives.

This analysis draws heavily on two in-depth studies conducted for the Global Subsidies Initiative (GSI) of the International Institute for Sustainable Development (Koplow, 2006; 2007) which in turn form part of a multi-country effort by the GSI to more thoroughly characterize and quantify subsidies to biofuels production, distribution and consumption.[4]

This chapter first describes the evolution of government support for ethanol, focusing on the major federal programs. Thereafter follows a more detailed discussion of federal and state support policies, arranged by their point of initial economic incidence. Virtually every production stage of ethanol is subsidized somewhere in the country; in many locations, producers can tap into multiple subsidies at once.

Liquid biofuels have been subsidized largely on the premise that they are domestic substitutes for imported oil; that they reduce greenhouse gas (GHG) emissions; and that they encourage rural development. Critics of subsidization have argued that the production process of these fuels is itself fossil-fuel-intensive, obviating many of the benefits of growing the energy resource; and that there are less expensive options for both GHG mitigation and rural development. Although the most recent work (Farrell et al., 2006a; Hill et al., 2006; U.S. EPA, 2007a) suggests some net fossil fuel displacement when biofuels replace petroleum products, the gains remain moderate, especially for corn-based ethanol. Others strongly contest these conclusions (e.g., Patzek, 2004; Pimentel and Patzek, 2005). Importantly, as additional analysis on modeling life-cycle impacts expands the parameters of assessment to include nitrous oxide emissions from fertilization and associated land-use changes from increased biofuel production, the net benefits of using ethanol produced from dedicated starch crops are looking less positive.

The second part of this chapter provides a variety of quantitative metrics on subsidy magnitude to illustrate how much support is being provided, not only per unit of biofuel produced, but also in terms of greenhouse gas (GHG) reductions. These values are intended to help in evaluating whether other options to diversify transport fuels or mitigate climate change might be more cost-effective.

4.2 Evolution of Federal Policies Supporting Liquid Biofuels

Subsidization of ethanol production at the federal level began with the Energy Tax Act of 1978. That Act granted a 4 cents-per-gallon reduction in the federal motor fuels excise tax for gasohol, a blend of 10% ethanol and 90% gasoline, also called

---

[4] A complete list of the GSI's studies can be found at http://www.globalsubsidies.org.

E10. This rate translates to 40 cents per gallon of pure ethanol at the time, and is equivalent to about $1.00 per gallon in 2007 dollars. The excise tax subsidy rate was adjusted frequently over the ensuing 25 years, until it was replaced by the Volumetric Ethanol Excise Tax Credit (VEETC) in 2004. VEETC is financed by general revenues, rather than through reduced collections for highway funding as occurred with the original exemption.

The US Congress introduced additional measures to support the ethanol industry in 1980. The Energy Security Act of 1980 initiated federally insured loans for ethanol producers, and from 1980–86 alcohol production facilities could access tax-exempt industrial development bonds (Gielecki et al., 2001). Also in 1980, Congress levied a supplemental import tariff of 50 cents per gallon on foreign-produced ethanol (RFA, 2005), which was increased to 60 cents in 1984 (Gielecki et al., 2001) and now stands at 54 cents.

Several states also started to subsidize ethanol around this time. Minnesota introduced a 40 cents per gallon ethanol blenders' credit in 1980 (phased out in 1997), as did North Dakota (Sullivan, 2006). A tally of state measures carried out by the Congressional Research Service two decades ago (CRS, 1986) identified incentives in place in 29 states. By 1986, state excise-tax exemptions alone were costing state treasuries over $450 million per year (in 2007 dollars) in foregone tax receipts.

In 1988, federal legislation began addressing the consumption side of the alternative fuels market. The Alternative Motor Fuels Act passed that year provided credits to automakers in meeting their Corporate Average Fuel Economy (CAFE) standards when they produced cars capable of being fueled by alternative fuels (Duffield and Collins, 2006).[5] Earning these credits did not require that the vehicles actually run on the alternative fuels, and because so few vehicles have (somewhat less than one percent of their mileage, according to a 2002 Report to Congress), the rule has been estimated to have increased domestic oil demand by 80,000 barrels a day (MacKenzie et al., 2005).

Environmental concerns have also helped improve the market position of biofuels. The Clean Air Act Amendments of 1990 mandated changes to the composition of gasoline in an effort to address two specific air-pollution problems. Reformulated gasoline was designed to help reduce ozone-forming hydrocarbons, as well as certain air toxins in motor-vehicle emissions, and was prescribed for areas of the country suffering the most-severe ozone problems. Oxygenated fuels were intended for use in the winter, in certain metropolitan and high-pollution areas, in order to reduce emissions of carbon monoxide. An oxygen-increasing additive, or oxygenate, was required to be added to these types of gasoline reformulations. However, the Amendments did not specify any particular oxygenate (of which there are several) for achieving these goals (Liban, 1997). Mandates to use ethanol for at least 30% of the oxygenates needed to meet these requirements were promulgated by the U.S. Environmental Protection Agency (EPA) in 1994 with the strong support of the

---

[5] The Energy Policy Act of 1992 (EPACT92) formally established E85 as an alternative transportation fuel. In addition, it established alternative-fueled-vehicle mandates for government and state motor fleets, policies that have indirectly encouraged demand for ethanol fuels over time (EIA, 2005a; Schnepf, 2007).

ethanol industry, but they were overturned in a court challenge a year later (Johnson and Libecap, 2001).

MTBE (methyl tertiary butyl ether), a petroleum-derived additive, emerged as the oxygenate of choice, primarily because the oil industry already had more than a decade of experience using it as an octane enhancer. Then, in 2004, concerns over the carcinogenicity of MTBE and contamination of groundwater from leaky storage tanks led several key states, starting with California, New York and Connecticut, to ban the additive (Yacobucci, 2006). By early 2006, nineteen other states had banned or limited the use of MTBE. The demise of MTBE was then accelerated by the Energy Policy Act of 2005 (EPACT05). In addition to not granting MTBE producers liability protection, Congress decided that the oxygenate mandates had yielded mediocre results, and so ended them. Effective 6 May 2006, non-oxygenated reformulated gasoline could be sold in most parts of the country (Yacobucci, 2006). With MTBE effectively no longer an option, ethanol remains as the main surviving competing fuel additive for increasing octane, a position that has helped further boost demand for the fuel.[6]

More significantly, EPACT05 also included the first federal purchase mandates for liquid biofuels. Referred to as the "Renewable Fuels Standard" (RFS), it fixed minimum consumption levels of particular specified fuels for each year, with the mandated level rising over time. Most of the mandated volumes under present law are expected to be fulfilled by ethanol from corn.

4.3 Current Policies Supporting Ethanol

Using a standard economic classification scheme for industry support, we provide an overview of the many types of incentives now in place to support the ethanol industry. As we were able to identify more than 200 support measures benefitting ethanol nationwide in 2006 (some of which also cover biodiesel, which is not discussed here), this section provides illustrations rather than a catalog.

4.3.1 Volume-Linked Support

Volume-linked support takes two main forms. The first, market price support, includes interventions such as import tariffs or purchase mandates that are linked to fuel volumes but operate by raising the price received by commodity producers above what it would be in the absence of such interventions. The second includes direct payments to producers that are linked to their levels of production. In the United States, output-related subsidies for ethanol are generally linked to gallons of fuel produced or blended.

---

[6] Gallagher et al. (2001, p. 3) projected that the MTBE ban alone could double demand for ethanol within 10 years.

4.3.1.1 Market Price Support Associated with Tariffs and Mandates

Market price support (MPS) refers to financial transfers to producers from consumers arising from policy measures that support production by creating a gap between domestic market prices and border prices of the commodity (OECD, 2001). It can be considered the residual support element resulting from the interaction of any number of policies. Three policies play a significant role in supporting market prices for biofuels in the United States: tariffs, blending mandates, and tax credits and exemptions (de Gorter and Just, 2007). Ideally, MPS is measured by comparing actual prices obtained in a market with an appropriate reference price. Because the nature of the information on tax credits is much more concrete than that available on prices, for the purpose of this exercise we treat tax credits separately from the effects of tariffs and blending mandates. These latter two are described briefly below.

Tariffs — Imported fuel ethanol is currently subject to both the normal *ad valorem* tariff and a specific-rate tariff. The applied MFN (most-favored nation) tariff on imports of undenatured ethyl alcohol (80% volume alcohol or higher) is 2.5%, and on denatured ethyl alcohol it is 1.9%. The specific-rate tariff is 54 cents per gallon. Hartley (2006) notes that the supplemental tariff is punitive, since it is applied volumetrically to the full mixture (i.e., including the denaturant), and is actually higher than the domestic subsidy it supposedly offsets.

Not all ethanol imported to the United States is subject to these tariffs, however.[7] Canada and Mexico — the United States' partners in the North American Free Trade Agreement (NAFTA) — for example, can export ethanol to the United States duty-free. Countries that are covered by the Caribbean Basin Economic Recovery Act (CBERA) can export an unlimited amount of ethanol to the United States duty-free if it is made predominantly from local feedstocks, or a volume equivalent of up to seven percent of U.S. fuel-ethanol consumption if it is made mainly from feedstocks grown outside of the region (Etter and Millman, 2007).

Renewable fuels standards — As noted above, federal RFS targets of 4 bgpy in 2006, rising to 7.5 bgpy by 2012, were introduced by EPACT. Post-2012 increases are meant to occur at the same growth rate as for gasoline demand. Higher credits (equal to 2.5 times those for sugar- or starch-based ethanol) are available for cellulosic ethanol until 2012, after which 250 mgpy of cellulosic ethanol usage becomes mandatory (Duffield and Collins, 2006). Biodiesel is included at a higher credit rate as well (1.5 times that of corn ethanol) because of its higher heat rate (EPA, 2006b).

---

[7] Moreover, because of a loophole called the "manufacturer's duty drawback", even the amount of duty actually paid on ethanol imported from countries such as Brazil and China is uncertain. The World Bank (Kojima et al., 2007) points out that an oil marketer can import ethanol as a blending component of gasoline, and obtain a refund ("draw back") on the duty paid if it exports a like-commodity within two years of paying the initial duty. Since jet fuel is considered a like-commodity, and counts as an export when sold for use in aircraft that depart the United States for a foreign country, this has allowed some oil marketers to count such jet-fuel exports against ethanol imports and recover the duty paid on ethanol.

Several states have issued mandates of their own; they are often more stringent than the federal one. Minnesota had already established a renewable fuels mandate prior to the federal RFS; it requires that gasoline sold in the state must contain 20% ethanol by 2013. However, many other states have become active as well. In 2006, Iowa set a target to replace 25% of all petroleum used in the formulation of gasoline with biofuels (biodiesel or ethanol). Hawaii wants 10% of highway fuel use to be provided by alternative fuels by 2010; 15% by 2015; and 20% by 2020. A few other states have set more modest requirements, some of which (as for Montana and Louisiana) are contingent on production of ethanol within these states reaching certain minimum levels.

The combined effects of tariffs in the presence of renewable fuel standards — The main effect of a tariff is to protect domestic markets from competition from lower-priced imports, thus allowing domestic prices to rise higher than they would otherwise. When only a tariff is in place, competition from foreign suppliers of ethanol will be reduced, but domestic manufacturers must still compete with non-ethanol alternatives, notably gasoline.[8] Mandating a minimum market share for a good also normally drives up its price. The size of the impact will depend on a variety of factors, including how large the mandated purchases are relative to what consumption would have been otherwise; the degree to which output of the good increases as prices rise; and whether competition from imports is allowed. With a mandate but no tariff, the amount of ethanol sold domestically would possibly be higher than otherwise, but its price would be constrained by foreign sources. A mandate plus a tariff both raises the threshold price at which foreign-sourced ethanol becomes competitive, and protects domestic suppliers from being undercut by the price of gasoline.

A number of parties have tried to estimate how much the RFS mandates alone, or in combination with import tariffs, increase domestic prices of biofuels. Several (e.g., EPA, 2006b; Urbanchuk, 2003) reach the conclusion that increases in wholesale (also known as "rack") prices would be more than offset by government subsidies, resulting in declines in pump prices. The results of both of these studies are of course sensitive to the degree to which state and federal subsidies to ethanol would be passed on to consumers, rather than absorbed into operating margins and profits of ethanol market participants.[9]

Others have looked mainly at producer prices. Elobeid and Tokgoz (2006) (henceforth "E&T"), analyzed the impact of liberalizing ethanol trade between the United States and Brazil using a multi-market international ethanol model calibrated on 2005 market data and policies, taking the United States' renewable fuel standard

---

[8] The price ceiling for all ethanol would be set by the energy-equivalent price of gasoline, as adjusted by any additional value of ethanol as an additive (e.g., to raise octane levels). Foreign suppliers of ethanol in that case would also be price takers, and the main difference for lower-cost foreign supplies between the situation with and without the tariff would be the market share they could capture from domestic producers, especially in coastal-state markets.

[9] For a more detailed discussion of price formation and the economic incidence of subsidies in the ethanol market see Bullock (2007).

and Brazil's blending mandates as givens.[10] Were trade barriers alone to be removed (retaining the existing renewable fuel mandate of 7.5 billion gallons per year, as well as the VEETC), they estimate the average U.S. ethanol prices from 2006 to 2015 would fall by 13.6%, or $0.27 per gallon. These results provide a rough indication of the degree to which the import tariff, in the presence of the existing (EPACT05-established) renewable fuels standard, increases the cost of meeting that standard. Should the import tariff remain in place while a higher RFS is implemented (as are proposed in pending energy legislation), the MPS would be expected to rise significantly.[11]

Estimating market price support for a commodity ideally involves calculating the gap between the average annual unit value, or price, of the good (usually measured at the factory gate) with a reference price, usually either an average (pre-tariff) unit import price or the export price.[12] Since such data are not readily available for the U.S. market, we have used the E&T results to obtain a rough estimate of market price support exclusive of the effect of the VEETC, the subsidy value of which we treat separately.[13] Applying the E&T's price mark-up to domestically-produced ethanol generates an estimate of the contribution of the tariff to MPS of $1.3 billion in 2006, rising to more than $3 billion per year as domestic production grows.

4.3.1.2 Tax Credits and Exemptions

The federal Volumetric Ethanol Excise Tax Credit (VEETC), enacted in 2004 by the Jumpstart Our Business Strength (JOBS) Act, constitutes the single largest subsidy to ethanol. It provides a credit against income tax of 51 cents per gallon of ethanol blended into motor fuel. It is awarded without limit, and regardless of the price of gasoline, to every gallon of ethanol — domestic or imported — blended in the marketplace. Moreover, it is not subject to corporate income tax, which means its

---

[10] Note that neither Elobeid and Tokgoz, nor any other researchers, have incorporated state-level renewable-fuel mandates into their models. Such state-level mandates, if they are both enforced and more stringent than the federal one, can cause additional price distortions.

[11] More recently, Westhoff (2007) simulated the effects on ethanol production and prices of expanding the mandated level of biofuel use in 2015 from 7.8 bgpy (the baseline) to 15 bgpy under a range of possible future petroleum prices scenarios. Current agricultural policies and the VEETC and ethanol tariff were assumed to remain unchanged. Compared with the baseline, he found that plant (i.e., producer) prices for ethanol in the 2015/16 marketing year would be on average 16 percent ($0.25 per gallon) higher. Considering the results of this study with the E&T results suggests that both the tariff and the RFS raise prices, and that the two effects are mutually supporting rather than additive.

[12] A complicating factor is that ethanol can be both a complement to gasoline when it is used as an additive, and a substitute for it when used as an extender. This makes estimating the appropriate reference price more difficult.

[13] Removal of both the import tariff and ethanol volumetric excise tax credit would generate even larger declines in domestic prices (between $0.29 and $0.36 per gallon, per Elobeid and Tokgoz (2006) and Kruse et al. (2007)). However, the tax credit subsidies are captured directly in our totals, while the MPS from the tariffs and RFS are not.

value to recipients is greater than if it were a simple grant, or a price benefit provided
through an exception from an excise tax (Box 4.1).

Box 4.1 The benefit of tax exemption for the VEETC

Tax breaks allow larger than normal deductions from taxable income or reductions in taxes due. A side-effect of the reduced tax payments is that the
remaining revenues of the enterprise rise. Although the tax burden will remain
lower than before the tax break, a portion of the benefit is lost to the recipient
because there is some tax due on the increase in earnings. For example, under
standard rules if a firm gets a $1 production tax credit (PTC), their taxes paid
go down by $1, but their bottom line — which is taxable — rises by that same
$1 amount. If they pay taxes at a 30% rate, they would see their taxes rise by
30 cents, leaving them with only 70 cents of the original PTC. To generate $1
in *after-tax* value to a firm, a revenue-based subsidy would need to be higher
than $1 — basically $1/(1-marginal tax rate), or $1.43 in this example. This
higher value is referred to as the *outlay equivalent* value of tax breaks. It was
routinely reported in US tax expenditure budgets until a couple of years ago.

The question of whether a tax subsidy is exempt from taxation matters
quite a bit to evaluating the distortions in energy markets from government
programs. Because the VEETC is an excise tax credit rather than a production
tax credit it falls into a gray area of the tax code. This ambiguity illustrates
how tiny changes in the interpretation of the tax code can increase the value
of subsidies to the ethanol industry by billions of dollars per year.

From a technical perspective, Section 87 of the tax code specifically requires that tax credits for biofuels under Section 40 (the income tax credits) be
included in taxable income, rendering their outlay equivalent value identical
to the revenue loss. The language on the VEETC is not clear, however. Section 6426 of the Internal Revenue Code, which describes the VEETC, makes
numerous cross-references to Section 40, mostly for definitional issues. There
is no mention of Section 87.

In January of 2005, the Internal Revenue Service issued a guidance document on implementation issues related to the VEETC (IRS, 2005). Because
this guidance was silent on the tax treatment of the credits, a consortium of industry groups filed comments requesting a clarification on the issue (Herman,
2005). The wording of their request indicates their inclination to treat the
VEETC as not includible in taxable income until clearly instructed otherwise:

> One of the major questions facing our members is whether any part of the new excise
> tax credit for alcohol fuel mixtures is taxable, and whether there are any circum-
> stances in which the excise tax credit or refund (payment) must be reported as part
> of gross income. (Herman, 2005)

Sources within both the Joint Committee on Taxation of the U.S. Congress
(JCT) and the U.S. Department of Treasury have confirmed that, as of

September 2007 at least, there had been no technical corrections in how the excise tax credits are treated by the Internal Revenue Service (IRS), implying that the credits are still excludible from taxable income.

The incremental benefit of this exemption was roughly $1.2 billion for ethanol in 2006 on top of a direct revenue loss of $2.8 billion. The incremental subsidy from this tax loophole, supposedly a policy accident, has become the third-largest subsidy to ethanol. By 2015, even if there is no increase in the RFS, the VEETC will generate subsidies of $6.3 billion per year on a revenue loss basis and $8.9 billion per year on an outlay-equivalent basis.

In addition to the federal VEETC, several states provide reductions or exemptions for ethanol from motor fuel excise or sales taxes. The largest subsidies from these programs appear to be in Hawaii, Illinois, Indiana, and Iowa. With ethanol blends of 10% or less widely used in the country, reduced fuel taxes on E10 are becoming increasingly uncommon. Many still provide reduced rates for E85, however, and these can be fairly large per gallon. Based on the states we quantified, the average exemption for E85 was 11.5 cents per gallon; the median exemption was 7 cents per gallon. For now, the amount of ethanol consumed in E85 is small — less than 15 million gallons in 2006 according to the EIA. This is equivalent to roughly 17.4 million gallons of E85, assuming an 85% blend rate.[14] The largest revenue losses tend to come from states that exempt particular fuel blends from *sales* taxes on fuels. The standard reporting of fuel tax rates provides greater clarity on deviations in excise tax rates than for fuel sales taxes. This may be one explanation for the political preference to subsidize via the sales tax. State motor-fuel tax preferences, along with state-level mandates, seem to exert a big influence on where U.S.-produced ethanol ends up being sold.

4.3.2 Payments Based on Current Output

Production payments or tax credits to producers of ethanol have been on offer by the federal government and many states. These programs are normally structured to provide a pre-specified payment or tax credits for each unit (usually gallon) of output a plant produces. Supplier refunds also exist in a number of places, and operate in a similar manner.

At the federal level, the Small Producer Tax Credit, introduced in 1990, grants ethanol and biodiesel plants that produce less than 60 mgpy a 10-cents-per-gallon income-tax credit on the first 15 million gallons they produce (a maximum of $1.5 million per plant each year). Using industry data on plant nameplate capacity, we

---

[14] The actual blend rate is anyone's guess. States such as Minnesota allow winter blends as low as 60 percent ethanol to count as E85. Lower blend rates would drive up the overall subsidy costs of E85 within a state.

estimate the revenue loss from this provision to be over $100 million per year for ethanol. However, newer plants tend to be larger and we expect that by the end of 2009 less than 60% of the nation's ethanol plants will meet the 60 mgpy cutoff. Subsidies likely will not fall, however. When a similar situation occurred only five years ago (at which point less than 40% of the plants fell under the then 30 mgpy limit), Congress simply increased the limit.

Output-linked payments via the USDA's Bioenergy Program until recently paid an additional bounty per gallon of ethanol or biodiesel produced, with higher bounties for new production. These operated through grants rather than tax credits, but were otherwise fairly similar in structure and impact.

Several states also provide production payments or tax credits for producers. Some of the programs require eligible plants to pre-qualify with the government before they can claim a credit. Some cap the total payouts (or allowable tax credits) per year to all plants. This means that the early plants may absorb the entire available funds, or that the actual per-gallon subsidy received is well below the rate nominally noted in the statute.

4.3.3 Subsidies to Factors of Production

Value-adding factors in biofuel production include capital, labor, land and other natural resources. Surprisingly, even labor related to biofuels production does not escape subsidization. The state of Washington, for example, allows labor employed to build biofuels production capacity, or to make biodiesel or biodiesel feedstock, to pay a reduced rate on the state's business and occupation tax.[15]

4.3.3.1 Support for Capital Used in Manufacturing Biofuels

Scores of incentive programs have been targeted at reducing the capital cost of ethanol plants. Many of these are specific to ethanol (or ethanol and biodiesel), though others are open to a broader variety of alternative fuels. Government subsidies are often directed to encourage capital formation in a specific segment of the supply chain.

Generic Subsidies to Capital

The ethanol sector benefits from a number of important general subsidies to capital formation. Though available to a wide variety of sectors, these policies can nonetheless distort energy markets. All of them subsidize capital-intensive energy production more heavily than less capital-intensive methods. As a result, they tend to diminish the value of energy conservation relative to supply expansions. In addition,

---

[15] Rates on manufacturing of ethanol and biodiesel fuel are the lowest of all categories, and less than one-third the normal rate on manufacturing activities. See WA DOR (2007).

the small print in how they are defined can generate differential subsidies by sector.

Depreciation governs the process by which investments into long-lived equipment can be deducted from taxable income. The theoretical goal of depreciation is to match the cost of an asset with the period over which it will produce income, generating an accurate picture of the economics of an industry. Politically, however, depreciation schedules have become another lever used by Congress to subsidize targeted groups. Federal legislation regularly reclassifies specific industries, or shortens the period over which capital investments can be deducted from taxable income for particular sectors. This generates more rapid tax deductions. Due to the time value of money, rapid tax reductions are more valuable than those occurring slowly over time.

Production equipment for ethanol (and biodiesel) is classified as waste reduction and resource recovery plant (Class 49.5) under the Modified Accelerated Cost Recovery System (MACRS).[16] This grouping includes "assets used in the conversion of refuse or other solid waste or biomass to heat or to a solid, liquid, or gaseous fuel," and allows full deduction of plant equipment in only seven years. An additional benefit comes in the form of the highly accelerated 200% declining balance method that can be used for Class 49.5, and that further front-loads deductions into the first years of plant operation.

With over $18 billion invested in ethanol production capacity since 2000 alone, this can constitute a fairly large subsidy. Note that our estimates incorporate only investments into plant capacity. For simplicity, we have not made similar calculations for investments in distribution infrastructure. These investments include terminals, retail facilities, tank trucks, rail cars and barges. During this same period, the ethanol industry's estimated additional spending on infrastructure assets was roughly $1 billion.[17]

Subsidies for Specific Production-Related Capital

In addition to general subsidies to capital that benefit multiple sectors of the economy, a number of subsidies target biofuel capital directly. Capital grants are used in many states and help finance production facilities, refueling or blending infrastructure, or the purchase of more expensive alternative fueled vehicles. Partial government funding of demonstration projects in the ethanol sector is common. The Energy Policy Act of 2005, for example, provided earmarked funds for a number of large biofuel-demonstration projects.

Credit subsidies, such as loans, guarantees, and access to tax-exempt debt, are common methods to subsidize the development of ethanol production and

---

[16] Choosing the proper grouping is not always easy. This classification reflects input from Mark Laser at Dartmouth University, who noted that based on his reading of the IRS classifications, and "discussions with colleagues from NREL and Princeton," class 49.5 seemed the proper fit (Laser, 2006).

[17] Earth Track estimates based on data in EPA (2006a).

infrastructure. Title XVII of EPACT, for example, will guarantee up to 80% of the cost of selected new plants. Liquid biofuels comprised $2.5 billion of the initial round of requests for federal guarantees (DOE, 2007a), and the largest share (6 of 16) of projects chosen by the DOE to submit final funding proposals (DOE, 2007b). Program structures such as this leave little investment risk borne by investors and increase the chances of both poor project selection and of loan defaults. Many of the ethanol loan guarantees issued in the 1980s defaulted.

Some states (e.g., Delaware's Green Energy Fund) provide direct credit subsidies that are open to ethanol production facilities. Others apply their limited allowances to issue tax-exempt bonds to ethanol projects. Hawaii has authorized $50 million of tax-exempt bonds to fund a bagasse-fed ethanol plant, for example. Nebraska has authorized public power districts to build ethanol plants, and to use tax-exempt municipal bonds to finance their construction.[18] New Jersey is another example, having approved $84 million in tax-exempt financing for a privately-owned ethanol plant.

Special tax exemptions for purchasing biofuels-related equipment are also common. Generally, the tax exemptions are not contingent on production levels. For example, Montana exempts all equipment and tools used to produce ethanol from grain from property taxes for a period of 10 years. In Oregon, ethanol plants pay a reduced rate (50% of statute) on the assessed value of their plant for a period of five years. These policies reduce the private cost to build a biofuels facility.

Subsidy Stacking

Subsidy stacking refers to a practice whereby a single plant will tap into multiple subsidy programs. This is common during the construction of a new plant, but unfortunately is often quite difficult to see when surveying subsidies. One $71-million, 20-million-gallon-per-year ethanol plant being built in Harrison County, Ohio, for example, has been able to line up government-intermediated credit or grants from seven different federal and state sources, covering 60% of the plant's capital.[19]

Regulatory Exemptions

The waiver of regulatory requirements normally applied to similar industrial developments, but from which ethanol has been exempted, also provide a benefit equivalent to a subsidy. These exemptions can sometimes be quite surprising given ethanol's claim to be an environmentally-friendly fuel. For example, Minnesota

---

[18] The subsidies associated with this power may not always be direct. The Nebraska Public Power District, for example, can provide coal and operate coal-fired boilers for ethanol plant operators (Dostal, 2006).

[19] *Project Briefing: Harrison Ethanol On Site/Off Site Rail* (2006, January 10). Retrieved December 8, 2007, www.dot.state.oh.us/OHIORAIL/Project%20Briefings/January%202006/ 06-03%20Harrison%20Ethanol%20-%20briefing.htm. See also www.ethanolproducer.com/article. jsp?article_id=1910.

exempts ethanol plants (though not biodiesel) with a production capacity of less than 125 mgpy from conducting an environmental impact assessment so long as the plant will be located outside of the seven-county metropolitan area.[20]

Less stringent regulation of pollutants from the biofuels sector can also provide a benefit to the industry, by reducing its capital or operating costs. In April 2007, the EPA reclassified ethanol fuel plants from their former grouping as "chemical process plants" into a less-regulated grouping in which firms producing ethanol for human consumption had been operating. The Agency characterized the change as one of providing "equal treatment" for all corn milling facilities (EPA, 2007b). However, the change also increased the allowable air emissions from fuel ethanol facilities substantially — from 100 tons per year to 250 tons. In addition, fugitive emissions (i.e., not from the plant stack) no longer have to be tallied in the emissions total. Finally, the plants have less stringent air permitting requirements in that they no longer have to install the Best Available Control Technology (BACT). Even an industry trade magazine (Ebert, 2007) notes that

> [r]egardless of the legislative tributaries that many producers will have to navigate, barring litigation, most facilities will be able to take advantage of the new rule to expand and ramp up production, to build new plants with greater capacities or to potentially switch to a different power source, such as coal.

The majority of ethanol produced in the country is for fuel purposes, not human consumption.[21]

4.3.3.2 Policies Affecting the Cost of Intermediate Inputs: Subsidies for Feedstocks

Government policies in the United States support the use of key biofuel feedstocks indirectly, through farm subsidies. Because of the United States' dominance in the global markets for corn and soybeans, federal subsidies provided to those crops during the nine years following the passage of the 1996 Farm Bill kept their farm-gate prices artificially low — by an average of, respectively, 23% below and 15% below average farm production costs, according to Starmer and Wise (2007). Market prices were depressed by somewhat less than the unit value of the subsidies, though the specifics varied according to market conditions. Adding to the complexity, corn and soybean markets are linked at several points. For one, the crops are often grown on the same land, in rotation. Second, they both yield competing products, such as vegetable oils and protein feeds (in the case of corn, as a byproduct of producing ethanol). These interactions complicate the way in which subsidies operate across the biodiesel and ethanol sectors.

Corn has historically been one of the most heavily subsidized crops within the United States. The Environmental Working Group (EWG), which tracks farm

---

[20] See MN Statutes 2007, section 116D.04, Subd.2a.
[21] Two inquiries to the EPA's manager for this rule seeking information on cost savings to industry from the change went unanswered.

subsidy payments, estimates that corn subsidies totaled nearly $42 billion between 1995 and 2004 from 12 federal programs,[22] reaching a high of $9.4 billion per year in 2005 (Environmental Working Group, 2006; Campbell, 2006). In 2006, corn did not qualify for first installments on counter-cyclical payments because the effective prices for corn exceeded its respective target price (USDA, 2006). Nonetheless, corn growers continued to receive fixed annual payments on their 2006 harvest.

Pro-rating these values to ethanol, based on the share of supply diverted to fuel production, generates an estimate of expenditure on corn subsidies associated with ethanol production of nearly $500 million for 2006, despite the sharp decline in counter-cyclical support. As ethanol production continues to consume a larger share of the domestic corn crop, its absolute (but not per-gallon) share of corn subsidies will rise accordingly.

The linkages between energy and agricultural policy are also having effects on the environment. Already, rapid growth in demand for biofuel feedstocks, particularly corn and soybeans, is changing cropping patterns in the Midwest, leading to more frequent planting of corn in crop rotations, an increase in corn acreage at the expense of wheat, and the ploughing up of grasslands (GAO, 2007). This trend is worrying, as a growing body of evidence suggests that greater carbon sequestration can be achieved through protecting natural ecosystems than by substituting biofuels for petroleum (Righelato and Spracklen, 2007).

US corn production remains chemical-intensive. Moreover, both corn and soybeans, like all row crops, typically experience higher rates of erosion than crops like wheat. Corn production is often water-intensive as well, a problem that is being exacerbated by current trends in corn-based ethanol plants. These are expanding westward, into areas more dependent on irrigation than corn produced in the Central Midwest. Some of that expansion is into counties served by the heavily overpumped[23] Ogallala Aquifer. In addition to corn production, the ethanol plants themselves also require significant volumes of water (Zeman, 2006; National Research Council, 2007).

4.3.4 Support for R&D on the Production Side

Federal spending on biofuels R&D hovered between $50 and $100 million a year between 1978 and 1998 (Gielecki et al., 2001). The U.S. Office of Technology Assessment reported that direct research on ethanol within the DOE was less than $15 million per year between 1978 and 1980 (OTA, 1979). It is notable that the federal government started the Bioenergy Feedstock Development Program at Oak

---

[22] These included production flexibility; loan deficiency; market loss assistance; direct payments; market gains farm; advance deficiency; deficiency; counter-cyclical payment; market gains warehouse; commodity certificates; farm storage; and warehouse storage. EWG data deduct negative payments or federal recaptured amounts from the total. See http://www.ewg.org/farm for more details.

[23] See USGS (2003).

Ridge National Laboratory nearly 30 years ago to focus on new crops and cropping systems for energy production (Schnepf, 2007). The program continues to operate in a similar form today.[24] Ethanol-related R&D is estimated to reach $400 million per year annually by 2009 (Koplow, 2007), mainly related to cellulosic ethanol.

4.3.5 Subsidies Related to Consumption

Numerous federal and state subsidies support investment in infrastructure used to transport, store, distribute and dispense ethanol. A separate set of policies underwrites the purchase or conversion of vehicles capable of using alternative fuels.

4.3.5.1 Subsidies to Capital Related to Fuel Distribution and Disbursement

Getting ethanol from the refinery to the fuel pump requires considerable infrastructure, separate from that used to distribute gasoline. Pure ethanol attracts moisture, which means that it cannot be transported through pipelines built to carry only petroleum products. High ethanol blends, like E85, also have to be segregated and stored in corrosion-resistant tanks, and pumped through equipment with appropriate seals and gaskets. All such investment is expensive.

Since 2004, the federal government and many states have started to offer financial incentives to help defray some of those costs. Under EPACT, a refueling station can obtain a tax credit that covers 30% of eligible costs of depreciable property (i.e., excluding land) for installing tanks and equipment for E85. This is capped at $30,000 per taxable year per location, and is estimated to cost the U.S. Treasury $15–30 million per year.

At least 15 states also provide assistance to establish new E85 facilities at retail gasoline outlets, as well as to support other ethanol distribution infrastructure. The Illinois E85 Clean Energy Infrastructure Development Program, for example, provides grants worth up to 50% of the total cost for converting an existing facility (up to a maximum of $2,000 per site) to E85 operation, or for the construction of a new refueling facility (maximum grant of up to $40,000 per facility). Florida recently created a credit against the state sales and use tax, available for costs incurred between 1 July 2006 and 30 June 2010, covering 75% of all costs associated with retrofitting gasoline refueling station pumps to handle ethanol; equipment for blends as low as E10 can qualify.

4.3.5.2 Support for Vehicles Capable of Running on Ethanol

The emergence of ethanol FFVs on the market provided a means for federal and state agencies to meet federal requirements for alternative fuel vehicles (AFVs) established in the Energy Policy Act of 1992. These requirements stipulated that

---

[24] http://bioenergy.ornl.gov/

certain government entities purchase AFVs for specified fractions (75% in the case of new light-duty vehicles) of their fleets when purchasing new vehicles. One result of this requirement was that, over time, the federal government acquired significant numbers of ethanol FFVs. Support for privately owned FFVs is also provided by several states in the form of rebates and tax credits for purchasing AFVs, or reductions on license fees and vehicle taxes, some of which apply to ethanol FFVs.

The individual states, and even some municipalities, have also provided regulatory incentives that favor AFVs. These include: the right to drive in high-occupancy vehicle (HOV) lanes, no matter how few the number of occupants in the vehicle (Arizona, California, Georgia, Utah and Virginia); the right to park in areas designated for carpool operators (Arizona); and exemptions from emissions testing (Missouri and Nevada) or certain motor-vehicle inspection programs (Ohio). Because every state develops its own definition of what exact vehicles types may participate in their AFV incentives, it is difficult to evaluate how many of these incentives apply to ethanol-powered vehicles.

4.4 Aggregate Support to Ethanol

To develop a better sense of how all of the individual subsidy programs affect the overall environment for ethanol, we have compiled a number of aggregate measures of support. The aggregate data provide important insights into a variety of policy questions, ranging from the financial cost of the support policies to taxpayers and consumers, to estimates of the costs of achieving particular policy goals. Among arguments put forth in support of biofuel subsidies are that they help the country to diversify from fossil fuels in general, and petroleum in particular; and that they have a better environmental profile than fossil fuels.

Quantification is often difficult either because the subsidy's course of action is indirect (e.g., mandated use of ethanol) or because data on the magnitude of support (especially at the state level, or with tax breaks or credit enhancements) are difficult to locate. As a result, there are inevitable gaps in our subsidy tallies.

Despite not counting everything, however, the subsidy picture is striking. We estimate that total support for ethanol was $5.8 billion to $7.0 billion in 2006 and, assuming no change in the RFS, will rise sharply to $11 billion by 2008 and $14 billion by 2014 (Table 4.1). The VEETC at present is the single largest ethanol subsidy and the difference between the high and low estimate is primarily associated with the incremental benefit blenders receive from the VEETC being excludible from income taxes (Box 4.1). We believe the high estimate is a more accurate representation of government support to ethanol than is the low estimate. Subsidies from the VTEEC were $3–4 billion in 2006, and are projected to total $34 to $48 billion over the 2006–12 period.

Total undiscounted subsidies to ethanol from 2006–2012 are estimated to fall within the range of $68 billion to $82 billion. Implementation of a higher RFS (e.g., 36 bgpy by 2022) would increase total subsidies by tens of billions of dollars per year above these levels.

Table 4.1 Estimated total support for ethanol

Element	2006	2007	2008	Total, 2006–12
Market Price Support	1,390	1,690	2,280	17,450
Output-linked Support[1]				
Volumetric Excise Tax Credit (low)	2,810	3,380	4,380	33,750
Volumetric Excise Tax Credit (high)	4,010	4,820	6,260	48,220
USDA Bioenergy Program	80	Ended in '06	–	80
Reductions in state motor fuel taxes	390	410	440	3,210
State production, blender, retailer incentives	120	NQ	NQ	120
Federal small producer tax credit	110	150	170	1,100
Factors of Production – Capital				
Excess of accelerated over cost depreciation	170	220	680	3,250
Federal grants, demonstration projects, R&D[2]	110	290	350	2,140
Credit subsidies	110	110	110	880
Deferral of gain on sale of farm refineries to coops	10	20	20	130
Factors of Production – Labour	NC	NC	NC	NC
Feedstock Production (biofuels fraction)	510	640	740	5,010
Consumption				
Credits for clean fuel refueling infrastructure	10	30	20	140
State vehicle purchase incentives	NQ	NQ	NQ	NQ
AFV CAFE loophole	NQ	NQ	NQ	NQ
Total support[3]				
Low estimate	5,820	6,940	9,200	67,260
High estimate	7,020	8,390	11,070	81,720

[1] Primary difference between high and low estimates is inclusion of outlay equivalent value for the volumetric excise tax credits. A gap in statutory language allows the credits to be excluded from taxable income, greatly increasing their value to recipients.
[2] Values shown reflect half of authorized spending levels where funds have not be appropriated. This reflects the reality that not all authorized spending is actually disbursed.
[3] Total values reflect gross outlays; they have not been converted to net present values. This follows the general costing approach used by the Joint Committee on Taxation.
[4] Totals may not add due to rounding.
[5] NC = Subsidies were quantified but not counted because provision was generally applicable across the economy. NQ = Subsidies exist that were not quantified.
Source: Koplow (2007).

Market price support, related to the combination of high barriers to imports and domestic purchase mandates, comprises the second largest subsidy to ethanol, at $1.3 billion in 2006, rising to more than $3 billion per year by 2010. Should the RFS be increased to 36 or 60 bgpy as is being considered, market price support would become the largest subsidy element, surpassing even the VEETC. Feedstock support also remains important, despite falling countercyclical payments, as direct payments remain high and ethanol is absorbing an ever-higher share of the total corn crop.

Based on 2004–2005 patterns of fuel consumption we estimate state sales and excise tax exemptions for biofuels to generate a subsidy to ethanol of approximately $400 million in 2006. Fuel taxes change regularly. In any given quarter, at least a few states will change their rates. Similarly, different sources for this information also disagree. While many states provide generous exemptions for E85, sales information are hard to come by, making revenue-loss calculations difficult. We have prorated national E85 sales data (also a few years old) by the state share of E85 refueling stations. This approach enables us to generate a rough estimate, despite the limitation of implicitly assuming that all pumps dispense the same amount of fuel per year. Rising demand; large new incentives, such as a full exemption from state taxes for E85 in New York; larger credits in Iowa; and rapidly growing sales of both ethanol blends and E85 suggest subsidies in 2007 and 2008 will be substantially higher.

State policies beyond reductions in motor-fuel taxation were quantified only for 2006, based on Koplow (2006). Had these many state supports been catalogued and quantified, the magnitude of state and county supports would be much larger than what is shown in the table.

4.4.1 Subsidy per Unit Energy Output and as a Share of Retail Price

Estimates of total support provide only a first-level indication of the potential market distortion that the subsidies may cause. Large subsidies, spread across a very large market, can have less of an effect on market structure than much smaller aggregate subsidies focused on a small market segment. As shown in Table 4.2,

Table 4.2 Subsidy-intensity values for ethanol

	2006	2007	2008	Average 2006–12
Subsidy per gallon of pure ethanol	1.05–1.25	1.05–1.25	1.05–1.30	1.00–1.25
Subsidy per GGE of fuel[1]	1.45–1.75	1.40–1.70	1.45–1.75	1.40–1.70
Subsidy per MMBtu	12.55–15.15	12.45–15.05	12.70–15.30	12.15–14.75
Subsidy per GJ	11.90–14.35	11.80–14.25	12.05–14.50	11.50–13.95
Subsidy as share of retail price[2]	39–47%	46–56%	55–66%	50–66%
Estimated retail price ($/gallon of pure ethanol)	2.70	2.25	1.95	2.05

[1] GGE values adjust the differential heat rates in biofuels so they are comparable to a gallon of pure gasoline. This provides a normalized way to compare the subsidy values to the retail price of gasoline.
[2] Retail price projections are for E100 and B100 as estimated in Westhoff and Brown (2007) for 2006–12; and FAPRI (February 2007) for 2013–16.
Source: Koplow (2007).

subsidies on a volumetric basis are \$1–\$1.30 per gallon of ethanol, and roughly \$1.40–\$1.70 per gallon of gasoline equivalent (GGE). The average subsidy per gigajoule (GJ) of ethanol energy produced is between \$11 and \$14 during the 2006–12 period.

Subsidies per unit energy produced via ethanol subsidies top \$11 per GJ in all years, reaching as high as \$14.50 per GJ in 2008. For the 2006–12 period, subsidies to ethanol will be equal to half or more of its projected retail price. Actual price drops for ethanol during the summer of 2007 have brought prices well below the values shown in our calculations. As of October 2007, ethanol subsidies were equal to as much as 80% of the fuel's then spot-market price of roughly \$1.60 per gallon (Kment, 2007; Shirek, 2007).

4.4.2 Subsidies per Unit Greenhouse Gas Displaced

A common claim by biofuels supporters is that ethanol will play an important role in facilitating the transition to a society with a low carbon footprint. To test how efficient existing policies are in getting us there, we examine the subsidy cost per metric ton of CO_2-equivalent displaced, and then compare this cost with the value of carbon offsets on the world's two major climate exchanges in Chicago (CCX) and Europe (ECX). The results are shown in Table 4.3.

The GHG displacement factors show a large variation across data sources. This is likely due to the complexity of the systems being modeled, but the variation forms a critical policy issue. As Kammen et al. (2007: 4) note:

> the indirect impacts of biofuel production, and in particular the destruction of natural habitats (e.g. rainforests, savannah, or in some cases the exploitation of 'marginal' lands which are in active use, even at reduced productivity, by a range of communities, often poorer households and individuals) to expand agricultural land, may have larger environmental impacts than the direct effects. The indirect GHG emissions of biofuels produced from productive land that could otherwise support food production may be larger than the emissions from an equal amount of fossil fuels.

For corn ethanol, researchers cannot even agree on the direction of impact. Thus, at one end of the displacement factors, GHG emissions rise rather than fall from its production. This would imply very large subsidies per metric ton of *extra* CO_2-equivalent emitted (\$600 per metric ton in the case of corn ethanol).

The best possible case for corn-based ethanol uses the lower bound subsidy estimate and divides it by the most favorable studies showing GHG reductions over the ethanol fuel cycle. Even here, subsidies per metric ton displaced are around \$300.[25] Based on historical prices for carbon offsets, this same investment could

---

[25] This value is lower than in our October 2006 study due to the use of a more favorable upper-end displacement value (a scenario with natural gas-fired plant capacity and avoided drying costs by direct use of wet distillers grain byproducts) based on new work by Wang et al. (2007). This scenario performs well above the average corn-ethanol plant of the future, also modeled in that same paper.

Table 4.3 Subsidy cost per unit of CO_2 equivalent displaced

	2006	2007	2008	Average 2006–12
Subsidy cost ($) per metric tonne CO_2 equivalent displaced				
Low estimate	305	300	310	295
High estimate[1]	(600)	(595)	(605)	(585)
Cellulosic hypothetical case – low	110	110	115	110
Cellulosic hypothetical case – high	200	200	205	195
GHG displacement factors				
Displacement factor – worst[1,2]	(24%)	(24%)	(24%)	(24%)
Displacement factor – best	39%	39%	39%	39%
Displacement factor – cellulosic worst	77%	77%	77%	77%
Displacement factor – cellulosic best[3]	114%	114%	114%	114%
Number of tonnes of carbon offsets subsidies could purchase				
European Climate Exchange[4]	12–24	11–22	11–23	11–21
ECX – cellulosic	5–8	4–7	4–8	4–7
Chicago Climate Exchange[4]	130–256	80–157	81–160	84–167
CCX – cellulosic	48–86	29–53	30–54	31–56
Cost of CO_2-equivalent futures contracts[5]				
ECX – Average prices paid for settlements during year noted	24.9	26.7	26.9	27.3
CCX – Historical average prices paid for settlements during year	2.3	3.8	3.8	3.6

[1] Negative values occur when the specific life cycle modeling scenarios estimate that GHG emissions from the biofuels production chain exceed those of the conventional gasoline or diesel they are replacing. This is fairly common with models that more centrally integrate the land use change impacts of the biofuels production system.

[2] Displacement factors represent the high and low values in the range from a variety of studies: Farrell et al. (2006b); Farrell et al. (2007); Hill et al. (2006); EPA (2007a); Wang et al. (2007) and Zah et al. (2007). The most favorable values included generally represent specific technologies rather than the average expected performance of either the current or future batch of plants.

[3] Values above 100% denote net sequestration benefits from the biofuel scenario (in this case, closed-loop poplar farming). It is not clear that the same high level of displacement would be maintained once the production base scaled up to meet the needs of the transportation sector.

[4] Although the subsidies pay for increased GHG emissions in the ethanol and biodiesel examples, subsidy reform would still free up public money that could be used to purchase low cost carbon offsets on the exchanges. The number of offsets is shown here.

[5] CO_2 futures contract data from European and Chicago exchanges, compiled as of October 2007. Prices represent historical averages of daily transactional data for contracts in the year in question. Markets are not interchangeable; higher prices in Europe reflect tighter constraints.

have purchased 80–130 times as much displacement on the CCX, the most appropriate benchmark for the U.S. carbon market. Even on the more expensive ECX, the subsidies could have purchased 11 metric tons of offsets.

We considered also a hypothetical case assuming the same levels of government support for ethanol, but a closed-loop production system based on short-rotation poplar (*Populus* sp.) as a cellulosic feedstock. Such a production system is believed to generate net sequestration (hence its 114% displacement value). Whether the impacts would really be so low once actual crops are produced on a large scale, move outside of their optimal range, and possibly require irrigation, is an open question. The hypothetical cellulosic-ethanol case provides better tradeoffs than for corn ethanol — \$110–204 per metric ton of CO_2-equivalent displaced — but the subsidies are still high: these funds could have purchased 4–8 times the offsets on the EXC or 30–85 times on the CCX.

4.4.3 Comparisons with Other Countries

The United States is by no means the only country that subsidizes ethanol production and consumption. Ethanol was heavily subsidized early in the development of Brazil's industry (from 1976 through 1998; see Boddey, 1993); although production is no longer directly subsidized, domestic consumption is still favored through

Table 4.4 Total support estimates (TSEs) and energy and CO_2 metrics for ethanol in selected OECD countries in 2006

OECD economy	TSE (10⁹US\$)	US\$ per GJ	US\$ per litre of gasoline equivalent[1]	US\$ per metric ton of avoided CO_2-equivalent[2]
United States[3]	5.8–7.0	12–14	0.38–0.46	305–600
EU[4]	1.6	40	1.40	700–5500
Canada[5]	0.15	20	0.65	250–1700
Australia[6]	0.044	16	0.50	300–630
Switzerland[7]	>0.001	28	0.90	330–380

[1] Per litre of gasoline equivalent (LGE) values adjust the differential heat rates in biofuels so they are comparable to a litre of pure gasoline. This provides a normalized way to compare the subsidy values to the retail price of gasoline.

[2] Displacement factors represent the high and low values in the range from a variety of studies (e.g., Farrell et al. (2006); Farrell and Sperling, et al. (2007); Hill et al. (2006); EPA (2007a); Wang et al. (2007) and Zah et al. (2007) comparing the life-cycle emissions of greenhouse gases with that of unleaded gasoline. The most favorable values included generally represent specific technologies rather than the average expected performance of either the current or future batch of plants. The number in parentheses indicates that subsidies are actually generating extra GHGs.

[3] The primary difference between the high and low estimates in the first three columns relates to whether the volumetric excise-tax credits are counted in revenue-loss or outlay-equivalent terms. A gap in statutory language allows the credits to be excluded from taxable income, greatly increasing their value to recipients.

[4] The range in the final column reflects differences in displacement rates between ethanol produced from sugarbeets and ethanol produced from rye.

[5] The range in the final column reflects differences in displacement rates between ethanol produced from C-molasses and ethanol produced from grains.

[6] The range in the final column reflects differences in displacement rates between ethanol produced from waste wheat starch and ethanol produced from maize.

[7] The range in the final column reflects uncertainty in the displacement rates for ethanol produced as a by-product of cellulose production.

Sources: • **Australia:** Quirke et al. (2008); • **United States:** Koplow (2007); • **Other OECD economies:** Steenblik (2007).

much lower fuel-excise taxes than those applied to gasoline, and by rules preventing private ownership of diesel-powered cars. More recently, several OECD member economies have started offering reduced excise-tax rates on ethanol used as fuel, and in some cases financial assistance for ethanol-manufacturing plants.

Compared with these other countries, the United States still leads in terms of absolute support provided, though per gigajoule or litre of gasoline equivalent its subsidization rate is substantially lower than those of the EU and Switzerland, which apply much higher fuel taxes to gasoline (Table 4.4). Measured in terms of dollars per metric ton of avoided CO_2-equivalent emissions, however, the United States falls within the range of values measured for most other OECD member economies, which in all cases are orders of magnitude higher than the prices of CO_2-equivalent offsets on the major climate exchanges, as well as current estimates of the social cost of a metric ton of CO_2 emitted (see, e.g., IPCC, 2007).

4.5 Pending Legislation

Despite a growing awareness of both the fiscal and environmental concerns about biofuels, legislative support has not abated. As of October 2007, the most "aggressive" proposed reforms (both contained in the tax section of the 2007 Farm Bill) involve reducing the excise tax credit by 5 cents per gallon (less than 10%) once the existing mandate is reached. None of the major bills would phase out the tax credits under high oil prices (when biofuels are more competitive) or remove an existing loophole that allows claimants to exclude the tax credits from their taxable income, further increasing the cost of the provision.

Several major bills under consideration by Congress, including a large proposed Energy Bill and the 2007 Farm Bill, seek to increase levels of support for biofuels, particularly ethanol. By increasing the national mandatory consumption requirement (the Renewable Fuels Standard), for example, lawmakers hope to reduce risks to the industry of a sustained market downturn. The Energy Bill under debate in December 2007 (H.R. 6) would mandate 36 billion gallons per year by 2022. Senate Bill 23 includes a 60 billion gallon per year target by 2030. The costs of these rules are likely to be extremely large. The Energy Information Administration recently estimated that the incremental cost of a 25% renewable fuels mandate (on par with 60 billion gallons per year of biofuels) would $130 billion per year within the fuels sector alone. This translates to a cost per metric ton of CO_2-equivalent reduced of more than $115, or roughly 30 times the current cost of a carbon offset on the Chicago Climate Exchange. Costs of vehicle infrastructure and increased food prices would be extra.

While the specifics of the mandates vary, most do not take into account life-cycle environmental impacts of biofuel production chains. As a result, they may encourage expensive fuels that actually worsen GHG emissions. In addition, none provide a neutral framework within which alternative ways to wean the country from imported oil and reduce greenhouse emissions can compete on a level playing field. Such alternatives include improvements in vehicle efficiency, improved maintenance and tires, and hybrid and plug-in hybrid drive trains.

To further boost ethanol consumption, proposals are also being considered to increase the allowable limits for ethanol blends in gasoline for unmodified engines (currently 10%) and improve distribution infrastructure for E85.

Some proposals seek to diversify the current industry by creating specific incentives for ethanol derived from feedstocks other than corn starch, expanding support for cellulosic ethanol and widening the definition of "advanced biofuels" (a definition that in some bills put before Congress would include fossil-derived fuels, and in many includes fuels derived from sugar and sorghum). As such, the new legislation compounds the current distortions to crop markets with a host of new programs to underwrite production, harvesting, storage, and the transport of cellulosic feedstocks. Some legislation makes compliance with the Renewable Fuels Standard contingent on lowering the greenhouse gas profile of biofuels (difficult to verify given problems with existing life cycle models). However, none would similarly restrict access to the excise tax credits.

4.6 Conclusions

A rapidly-expanding production base, combined with a proliferation of policy incentives, has generated a growing level of public subsidization for the ethanol industry. Many of the existing subsidies scale linearly with production capacity or consumption levels, and the resulting rate of growth in the subsidy payments can be quite large. In addition, the subsidies do not decline as the price of gasoline rises, as is the case for some subsidies benefiting petroleum and natural gas, and for some ethanol-support programs elsewhere, such as Canada (Steenblik, 2007). Although the spiraling costs of the VEETC in particular have led to discussions and proposals for subsidy phase-outs when oil prices are high (Bantz, 2006), there are currently no constraints in place.

At some point, the expiration of existing incentives may temper the growth in subsidization, but that point is still quite a few years off. Strong political support has maintained the key subsidies to ethanol for nearly 30 years, and we anticipate that those forces will remain. In the near term, we expect subsidy levels to rise sharply. Of particular interest are higher renewable fuel mandates and the rate of growth of 85% ethanol blends (E85), for which there are a number of large state subsidies that currently apply to only a small base.

Our analysis illustrates not only that subsidies to ethanol are pervasive and large, but that they are not a particularly efficient means to achieve many of the policy objectives for which they have been justified. These subsidies are the result of many independent decisions at different levels of government, resulting in policies that are often poorly coordinated and targeted. Hundreds of government programs have been created to support virtually every stage of production and consumption relating to ethanol, from the growing of the crops that are used for feedstock to the vehicles that consume the biofuels. In many locations, producers have been able to tap into multiple sources of subsidies.

Because the bulk of subsidies are tied to output and output is increasing at double-digit rates of growth, the cost of these programs will continue to climb. Production is subsidized at the federal level even though consumption of it is mandated through the RFS. Ethanol production is supported on the grounds that it helps wean the United States from imported petroleum, but special loopholes in vehicle efficiency standards for flexible fuel vehicles (including those that run on high ethanol blends) result in higher oil imports (MacKenzie et al., 2005). The maintenance of a high tariff on imported ethanol (2.5% plus 54 cents per gallon), in particular, sits at odds with the professed policy of the U.S. government to encourage the substitution of gasoline by ethanol.

The absolute value of the subsidies is not the only, and perhaps not the main, indicator of the market-distorting potential of a set of support policies. Subsidies as a share of market price were above 40% as of mid-2006, for example, which is high in comparison with other fuels. Such high rates of subsidization might be considered reasonable if the industry was new, and ethanol was being made on a small-scale, experimental basis using advanced technologies. But that is not the case: the vast majority of subsidized production relies on mature technologies that, notwithstanding progressive improvements, have been around for decades.

Ethanol also has some greenhouse gas and local-pollution benefits. But the cost of obtaining a unit of CO_2-equivalent reduction through subsidies to the fuel is extremely high: we calculate that it comes to nearly \$300 per metric ton of CO_2 removed for corn-based ethanol, even when assuming an efficient plant using low-carbon fuels for processing. Yet even under such best-case scenario assumptions for GHG reductions from corn-based ethanol, one could have achieved far more reductions for the same amount of money by simply purchasing the reductions in the marketplace. The cost per metric ton of reductions achieved through public support of corn-based ethanol already programmed over the next several years could purchase more than 10 times the offsets on the European Climate Exchange, or nearly 90 times the offsets on the Chicago Climate Exchange.

Most importantly, the U.S. government has neglected what should be its core role: to adopt a neutral strategy equally accessible to all potential options to reduce the country's reliance on imported oil. Such a strategy would not favor ethanol, but would encourage a range of potential solutions such as more efficient vehicles, better fleet maintenance, and alternative drive-trains such as plug-in hybrids. Similarly, the government has yet to indicate an exit strategy to wean the ethanol industry from protection and subsidies. Indeed, as is often the case with subsidies, current legislative proposals appear to entrench existing arrangements. These will ensure that the biofuel industry remains a significant drain on U.S. taxpayers for decades to come; and that improvements in transport-fuel options will be both slower and more expensive than would occur with a technology-neutral approach.

Acknowledgments The authors gratefully acknowledge the research assistance provided by Tara Laan of the Global Subsidies Initiative, and to the peer reviewers of the two GSI studies on which this article is based.

References

Bantz, S., Union of Concerned Scientists (2006, October 2). E-mail communication with Doug Koplow

Boddey, R.M. (1993). Green energy from sugar cane. *Chemistry and Industry, 17*, 355–358

Bullock, D.S. (2007, November). Ethanol policy and ethanol politics. *Corn-Based Ethanol in Illinois and the U.S* (pp. 147–177) Retrieved December 7, 2007, from University of Illinois (Urbana-Champagne) Department of Agricultural and Consumer Economics Web site: http://www.farmdoc.uiuc.edu/policy/research_reports/ethanol_report/index.html

Campbell, C., Environmental Working Group (2006, September 11). E-mail communication with Doug Koplow

CRS (1986). *Alcohol Fuels and Lead Phasedown.* Report prepared for the Subcommittee on Fossil and Synthetic Fuels of the Committee on Energy and Commerce, US House of Representatives. (Washington, D.C.: Congressional Research Service)

de Gorter, H. & Just, D.R. (2007, October 24). *The Economics of a Biofuel Consumption Mandate and Excise-Tax Exemption: An Empirical Example of U.S. Ethanol Policy.* Cornell University Working Paper No. 2007–20. Retrieved December 8, 2007, from http://ssrn.com/abstract=1024525

DOE (2007a, June 15). *Review of Pre-Applications Requesting Loan Guarantees Under August, 2006: Solicitation Sorted by Category.* (Washington, D.C.: U.S. Department of Energy)

DOE (2007b, October 4). *DOE Announces Final Rule for Loan Guarantee Program.* (Washington, D.C.: U.S. Department of Energy). Retrieved December 8, 007, from http://205.254.148.100/news/5568.htm

Dostal, A.L. (2006, August 10). Presentation for the "An Energy Conversion: Making Renewable Energy America's Energy" conference at the University of Nebraska, Lincoln. Retrieved December 8, 2007, from http://research.unl.edu/biofuel_docs_8-06/NPPD.ppt

Duffield, J. A. & Collins, K. (2006). Evolution of renewable energy policy [Electronic version]. *Choices, 21*(1), 9–14. Retrieved December 8, 2007 from http://www.choicesmagazine.org/2006-1/biofuels/2006-1-02.pdf

Ebert, J. (2007, July). Redefining emissions. *Ethanol Producer Magazine*, 82–87

EIA (2005a, November). *Ethanol Timeline.* (Washington, D.C.: U.S. Energy Information Administration). Retrieved December 8, 2007, from http://www.eia.doe.gov/kids/history/timelines/ethanol.html

EIA (2005b, December). U.S. ethanol fuel consumption in three price cases, 1995–2030. *Annual Energy Outlook 2006.* (Washington, D.C.: U.S. Energy Information Administration)

EIA (2006a, February 22). *Eliminating MTBE in gasoline in 2006.* (Washington, D.C.: U.S. Energy Information Administration). Retrieved December 8, 2007, from http://www.eia.doe.gov/pub/oil_gas/petroleum/feature_articles/2006/mtbe2006/mtbe2006.pdf

EIA (2006b, February). Estimated petroleum consumption: transportation sector, 1949–2005. *Annual Energy Review 2005.* (Washington, D.C.: U.S. Energy Information Administration)

EIA (2007a, July). *Basic Petroleum Statistics.* (Washington, D.C.: U.S. Energy Information Administration). Retrieved December 7, 2007 from http://www.eia.doe.gov/neic/quickfacts/quickoil.html

EIA (2007b, August). *Energy and Economic Impacts of Implementing Both a 25-Percent Renewable Standard and a 25-Percent Renewable Fuel Standard by 2025.* Report No. SR/OIAF/2007-05. (Washington, D.C.: U.S. Energy Information Administration)

Elobeid, A. & Tokgoz, S. (2006). *Removal of U.S. Ethanol Domestic and Trade Distortions: Impact on U.S. and Brazilian Ethanol Markets.* Working Paper No. 06-WP 427. (Ames, Iowa: Iowa State University)

Environmental Working Group. *Farm Subsidy Database: November 2005 update.* Extracts for historic subsidies to corn and soy from 1995–2004, nationally and by-state. Retrieved August 4, 2006 from http://www.ewg.org

EPA (2006a, September). *Renewable Fuel Standard Program: Draft Regulatory Impact Analysis*. Report No. EPA420-D-06-008. (Washington, D.C.: U.S. Environmental Protection Agency)

EPA (2006b, September 7). "Regulation of Fuels and Fuel Additives: Renewable Fuel Standard Program." Preliminary notice of proposed rulemaking. Report No. RIN 2060-AN-76. (Washington, D.C.: U.S. Environmental Protection Agency)

EPA (2007a, April). *Regulatory Impact Analysis: Renewable Fuel Standard Program.* (Washington, D.C.: U.S. Environmental Protection Agency)

EPA (2007b, May 11). *Fact Sheet—Final Changes for Certain Ethanol Production Facilities Under Three Clean Air Act Permitting Programs.* (Washington, D.C.: U.S. Environmental Protection Agency). Retrieved August 28, 2007, from http://www.epa.gov/NSR/fs20070412.html

Etter, L. & Millman, J. (2007, March 9). Ethanol tariff loophole sparks a boom in Caribbean—islands build plants to process Brazil's fuel; farm belt cries foul. *Wall Street Journal*, A1.

FAPRI (2007, February) *U.S. Baseline Briefing Book*, FAPRI-UMC Report #02–07. (Columbia, Missouri: Food and Agricultural Policy Research Institute)

Farrell, A., Plevin, R.J., Turner, B.T., Jones, A.D., O'Hare, M. & Kammen, D.M. (2006a, January 27). Ethanol can contribute to energy and environmental goals. *Science*, 311(5760), 506–508

Farrell, A., Plevin, R.J., Turner, B.T., Jones, A.D., O'Hare, M. & Kammen, D.M. (2006b, July 13). *Supporting Online Material for "Ethanol Can Contribute to Energy and Environmental Goals,"* version 1.1.1, updated July 13, 2006. Retrieved October 8, 2006. from http://rael.berkeley.edu/EBAMM/EBAMM_SOM_1_1.pdf

Farrell, A.E., Sperling, D., Arons, S.M., Brandt, A.R., Delucchi, M.A., Eggert, A., Haya, B.K., Hughes, J., Jenkins, B.M., Jones, A.D., Kammen, D.M., Kaffka, S.R., Knittel, C.R., Lemoine, D.M. Martin, E.W., Melaina, M.W., Ogden, J.M., Plevin, R.J., Turner, B.T., Williams, R.B. & Yang, C. (2007, May 29). *A Low-Carbon Fuel Standard for California Part 1: Technical Analysis.* Paper UCB-ITS-TSRC-RR-2007-2. (Berkeley, California: UC Berkeley Transportation Sustainability Research Center). Retrieved December 8, 2007, from http://repositories.cdlib.org/its/tsrc/UCB-ITS-TSRC-RR-2007-2

Gallagher, P., Otto, D., Shapouri, H., Price, J., Schamel, G., Dikeman, M. & Brubacker, H. (2001). *The effects of expanding ethanol markets on ethanol production, feed markets, and the Iowa economy.* Staff Paper #342. Prepared for the Iowa Department of Agriculture and Land Stewardship

GAO (2007, September). *Prairie Pothole Region: At the Current Pace of Acquisitions, the U.S. Fish and Wildlife Service Is Unlikely to Achieve Its Habitat Protection Goals for Migratory Birds.* Report No. GAO-07-1093. (Washington, D.C.: U.S. General Accountability Office). Retrieved December 8, 2007, from http://www.gao.gov/new.items/d071093.pdf

Gielecki, M., Mayes, F. & Prete, L. (2001, February). *Incentives, Mandates, and Government Programs for Promoting Renewable Energy.* (Washington: Energy Information Administration, US Department of Energy)

Hartley, B., Joint Committee on Taxation, retired (2006, September 19). E-mail communication with Doug Koplow

Herman, M.J. (2005, February 14) "Re: IRS Notice 2005-4, *Fuel Tax Guidance; Request for Public Comments.*" Comments submitted on behalf of the American Coalition for Ethanol, Clean Fuels Development Coalition, National Ethanol Vehicle Coalition, and the Nebraska Ethanol Board. Submitted by Herman & Associates

Hill, J., Nelson, E., Tilman, D., Polasky, S. & Tiffany, D. (2006, July 25). Environmental, economic, and energetic costs and benefits of biodiesel and ethanol biofuels. *Proceedings of the National Academy of Sciences, 103*, 11206–11210

IRS (2005, January 10). *Fuel Tax Guidance; Request for Public Comments.* Notice 2005-4. (Washington, D.C.: U.S. Internal Revenue Service)

IPCC (2007). Summary for Policymakers. In *Climate Change 2007: Impacts, Adaptation and Vulnerability. Contribution of Working Group II to the Fourth Assessment Report of the Intergovernmental Panel on Climate Change*, M.L. Parry, O.F. Canziani, J.P. Palutikof, P.J. van der Linden and C.E. Hanson, eds., Cambridge University Press, Cambridge, UK, 7–22

Johnson, R.N. & Libecap, G.D. (2001). Information distortion and competitive remedies in government transfer programs: the case of ethanol. *Economics of Governance*, 2, 101–134

Kammen, D.M., Farrell, A.E., Plevin, R.J., Jones, A.D., Nemet, G.F. & Delucchi, M.A. (2007, September 7). *Energy and Greenhouse Impacts of Biofuels: A Framework for Analysis*. (Paper presented at the OECD Research Round Table, Biofuels: Linking Support to Performance, Paris). Retrieved September 20, 2007, from http://rael.berkeley.edu/files/2007/Kammen_etal_OECD-biofuels.pdf

Kment, R. (2007, October 5). DTN Daily Ethanol Comments. *DTN Ethanol Center*. Retrieved October 5, 2007, from http://dtnag.com/dtnag/common/link.do?symbolicName=/author/template&authorId=23

Kojima, M., Mitchell, D. & Ward, W. (2007, September). *Considering Trade Policies for Liquid Biofuels*. (Washington, D.C.: Energy Sector Management Assistance Program of the World Bank)

Koplow, D. (2006, October). *Biofuels—At What Cost? Government Support for Ethanol and Biodiesel in the United States*. (Geneva: Global Subsidies Initiative of the International Institute for Sustainable Development)

Koplow, D. (2007, October). *Biofuels—At What Cost? Government Support for Ethanol and Biodiesel in the United States: 2007 Update*. (Geneva: Global Subsidies Initiative of the International Institute for Sustainable Development)

Kruse, J., Westhoff, P., Meyer, S. & Thompson, W. (2007, May). Economic Impacts of Not Extending Biofuels Subsidies. FAPRI-UMC Report #17-07. (Columbia, Missouri: Food and Agricultural Policy Research Institute). Retrieved December 8, 2007, from http://www.fapri.missouri.edu/outreach/publications/2007/FAPRI_UMC_Report_17_07.pdf

Laser, M., Dartmouth University (2006, September 20). E-mail communication with Doug Koplow

Liban, C.B. (1997, Summer). On understanding the MTBE controversy. *National Forum*. Retrieved December 8, 2007 from http://findarticles.com/p/articles/mi_qa3651/is_199707/ai_n8779282

Lugar, R. & Khosla, V. (2006, August 3). We can end oil addiction [Electronic version]. *Washington Times*. Retrieved December 8, 2007, from http://lugar.senate.gov/energy/press/articles/060803washtimes.cfm

MacKenzie, D., Bedsworth, L. & Friedman, D. (2005, August). *Fuel Economy Fraud: Closing the Loopholes That Increase U.S. Oil Dependence*. (Cambridge, MA: Union of Concerned Scientists)

National Research Council (2007, October). *Water Implications of Biofuels Production in the United States*. Prepublication copy. (Washington, D.C.: National Academies Press)

OECD (2001). *Agricultural Policies in OECD Countries: Monitoring and Evaluation 2000*. (Paris: Organisation for Economic Co-operation and Development)

OTA (1979, September). *Gasohol: A Technical Memorandum*. (Washington, D.C.: U.S. Office of Technology Assessment)

Patzek T (2004). Thermodynamics of the corn-ethanol biofuel cycle. *Critical Reviews in Plant Sciences, 23*(6), 519–567

Pimentel, D. & Patzek, T. (2005). Ethanol production using corn, switchgrass, and wood and biodiesel production using soybean and sunflower. *Natural Resources and Research, 14*(1), 65–76

Planet Ark (2006, May 11). ADM to build 275-million gallon ethanol facility. Retrieved December 8, 2007, from http://www.planetark.com/avantgo/dailynewsstory.cfm?newsid=36305

Quirke, D., Steenblik, R. & Warner, B. (2008, April). Biofuels—At What Cost? Government Support for Ethanol and Biodiesel in Australia. (Geneva: Global Subsidies Initiative of the International Institute for Sustainable Development)

RFA (2005, June 30). *The Importance of Preserving the Secondary Tariff on Ethanol*. (Washington, D.C.: Renewable Fuels Association)

RFA (2007a, December). *Ethanol Industry Overview*. (Washington, D.C.: Renewable Fuels Association). Retrieved December 7, 2007, from http://ethanolrfa.org/industry/statistics/#EIO

RFA (2007b, December 3). *Ethanol Biorefinery Locations*. (Washington, D.C.: Renewable Fuels Association). Retrieved December 7, 2007, from http://ethanolrfa.org/industry/locations/

RFA (2007c, December). *2007 Monthly U.S. Fuel Ethanol Production/Demand.* (Washington, D.C.: Renewable Fuels Association). Retrieved December 7, 2007, from http://ethanolrfa. org/industry/statistics/#B

Righelato, R. & Spracklen, D.V. (2007). Carbon mitigation by biofuels or by saving and restoring forests. *Science,* 17, 902

Schnepf, R. (2007, March 07 update). *Agriculture-Based Renewable Energy Production.* Report No. RL32712. (Washington, D.C.: U.S. Congressional Research Service)

Shirek, M. (2007, December). Ethanol experiences growing pains. *Ethanol Producer Magazine.* Retrieved December 7, 2007, from http://www.ethanolproducer.com/article. jsp?article_id=3489

Starmer, E. & Wise, T.A. (2007, December). *Feeding at the Trough: Industrial Livestock Firms Saved $35 billion From Low Feed Prices.* GDAE Policy Brief 07-03. (Medford, Massachusetts: Global Development and Environment Institute, Tufts University). Retrieved December 8, 2007, from http://www.ase.tufts.edu/gdae/Pubs/rp/PB07-03FeedingAtTroughDec07.pdf

Steenblik, R.P. (2007, September). *Biofuels: At What Cost? Goverment Support for Ethanol and Biodiesel in Selected OECD countries.* (Geneva, Switzerland: Global Subsidies Initiative of the International Institute for Sustainable Development)

Sullivan, C. (2006, September) The Ethanol Industry in Minnesota. *Minnesota House of Representatives: House Research.* Retrieved December 8, 2007, from http://www.house.leg.state.mn.us/ hrd/issinfo/ssethnl.htm

USDA (2006, October 13). *USDA announces first partial 2006-crop-year counter-cyclical payments.* (Washington, D.C.: U.S. Department of Agriculture). Retrieved December 7, 2007, from http://www.usda.gov/wps/portal/!ut/p/_s.7_0_A/7_0_1OB?contentidonly=true&contentid=2006/ 10/0413.xml

USDA (2007, February). *USDA Agricultural Projections to 2016.* Office of the Chief Economist. OCE-2007-1. (Washington, D.C.: U.S. Department of Agriculture). Retrieved December 7, 2007, from http://www.ers.usda.gov/Publications/OCE071/

USGS (2003, November). *Ground-Water Depletion Across the Nation.* Fact Sheet No. 103-03. (Washington, D.C.: U.S. Geological Survey)

Urbanchuk, J.M. (2003, May). *Consumer Impacts of the Renewable Fuel Standard* (Wayne, PA: LECG Economics & Finance). Retrieved December 8, 2007, from http://www.ethanolrfa.org/ objects/documents/119/rfs_consumer_impacts.pdf

WA DOR (2007, October). *Business and Occupation Tax.* (Olympia, Washington: Washington State Department of Revenue). Retrieved 13 December 2007, from http://dor.wa.gov/ Docs/Pubs/ExciseTax/BO_PubUtil_LitterTax/BOfs.pdf

Wang, M., Wu, M. & Huo, H. (2007). Life-cycle energy and greenhouse gas emission impacts of different corn ethanol plant types. *Environmental Research Letters, 2.* DOI 10.1088/ 1748-9326/2/2/024001

Westhoff, P. (2007, June). *Impacts of a 15 Billion Gallon Biofuel Use Mandate.* Staff Report No. FAPRI-MU #22-07. (Columbia, Missouri: Food and Agricultural Policy Research Institute). Retrieved December 8, 2007, from http://www.fapri.missouri.edu/outreach/publications/ 2007/FAPRI_MU_Report_22_07.pdf

Westhoff, P. and Brown, S. Baseline Update for U.S. Agricultural Markets, Research Report FAPRI-MU #28-07, Food and Agricultural Policy Research Institute, August 2007

Yacobucci, B. (2006, March 3). *Fuel Ethanol: Background and Public Policy Issues.* Report No. CRS RL33290. (Washington, D.C.: U.S. Congressional Research Service)

Zah, R., Böni, H., Gauch, M., Hischier, R. Lehmann, M. & Wäger, P. (2007, May 22): *Life Cycle Assessment of Energy Products: Environmental Assessment of Biofuels.* Executive Summary and data table. (Berne: Empa)

Zeman, N. (2006, October). Running dry? *Ethanol Producer Magazine*

Chapter 5
Peak Oil, EROI, Investments and the Economy in an Uncertain Future

Charles A. S. Hall, Robert Powers and William Schoenberg

Abstract The issues surrounding energy are far more important, complex and pervasive than normally considered from the perspective of conventional economics, and they will be extremely resistant to market-based, or possibly any other, resolution. We live in an era completely dominated by readily available and cheap petroleum. This cheap petroleum is finite and currently there are no substitutes with the quality and quantity required. Of particular importance to society's past and future is that depletion is overtaking technology in many ways, so that the enormous wealth made possible by cheap petroleum is very unlikely to continue very far into the future. What this means principally is that investments will increasingly have to be made into simply getting the energy that today we take for granted, the net economic effect being the gradual squeezing out of discretionary investments and consumption. While there are certainly partial "supply-side" solutions to these issues, principally through a focus on certain types of solar power, the magnitude of the problem will be enormous because of the scale required, the declining net energy supplies available for investment and the relatively low net energy yields of the alternatives. Given that this issue is likely to be far more immediate, and perhaps more important, than even the serious issue of global warming it is remarkable how little attention we have paid to understanding it or its consequences.

Keywords Energy · oil · energy return on investment · investments · U.S. economy

✉ C.A.S. Hall
State University of New York, College of Environmental Science and Forestry, Syracuse, New York 13210, e-mail: chall@esf.edu

R. Powers
State University of New York, College of Environmental Science and Forestry, Syracuse, New York 13210

W. Schoenberg
State University of New York, College of Environmental Science and Forestry, Syracuse, New York 13210

D. Pimentel (ed.), *Biofuels, Solar and Wind as Renewable Energy Systems*,
© Springer Science+Business Media B.V. 2008

5.1 Introduction

The enormous expansion of the human population and the economies of the United States and many other nations in the past 100 years have been accompanied by, and allowed by, a commensurate expansion in the use of fossil (old) fuels, meaning coal, oil and natural gas. To many energy analysts that expansion of cheap fuel energy has been the principal enabler of economic expansion, far more important than business acumen, economic policy or ideology although they too may be important (e.g. Soddy 1926, Tryon 1927, Cottrell 1955, Boulding 1966, Georgescu Roegan 1971, Odum 1971, Daly 1977, Herendeen and Bullard 1975, Hannon 1981, Kummel 1982, Kummel 1989, Jorgenson 1984 and 1988, Hall 1991, Hall et al. 1986 (and others), Cleveland 1991, Dung 1992, Ayers 1996, Cleveland and Ruth 1997, Hall 2000). While we are used to thinking about the economy in monetary terms, those of us trained in the natural sciences consider it equally valid to think about the economy and economics from the perspective of the energy required to make it run. When one spends a dollar, we do not think just about the dollar bill leaving our wallet and passing to some one else's. Rather, we think that to enable that transaction, that is to generate the good or service being purchased, an average of about 8,000 kilojoules of energy (equal to roughly the amount of oil that would fill a coffee cup) must be extracted from the Earth and turned into roughly a half kilogram of carbon dioxide (U.S. Statistical Review, various years). Take the money out of the economy and it could continue to function through barter, albeit in an extremely awkward, limited and inefficient way. Take the energy out and the economy would immediately contract immensely or stop. Cuba found this out in 1991 when the Soviet Union, facing its own oil production and political problems at that time, cut off Cuba's subsidized oil supply. Both Cuba's energy use and its GDP declined immediately by about one third, all groceries disappeared from market shelves within a week and the average Cuban lost 20 pounds (Quinn 2006). Cuba subsequently learned to live, in some ways well, on about half the oil as previously, but the impacts were enormous. While the United States has become more efficient in using energy in recent decades, most of this is due to using higher quality fuels, exporting heavy industry and switching what we call economic activity (e.g. Kaufmann 2004). Many other countries, including efficiency leader Japan, are becoming substantially less efficient (Hall and Ko, 2007, LeClerc and Hall 2007, Smil 2007, personal communication).

5.2 The Age of Petroleum

The economy of the United States and the world is still based principally on "conventional" petroleum, meaning oil, gas and natural gas liquids (Fig. 5.1). *Conventional* means those fuels derived from geologic deposits, usually found and exploited using drill bit technology, and that move to the surface because of their own pressure or with pumping or additional pressure supplied by injecting natural gas, water or occasionally other substances into the reservoir. *Unconventional* petroleum includes

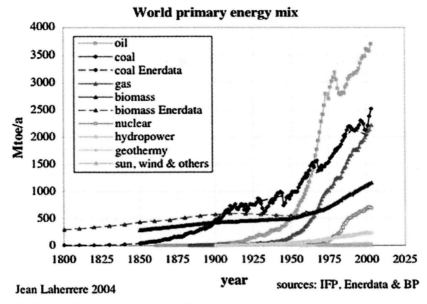

World primary energy mix

Jean Laherrere 2004   year   sources: IFP, Enerdata & BP

Fig. 5.1 Pattern of energy use for the world (Source Jean Laherrere, with permission)

shale oil, tar sands and other bitumens usually mined as solids and also coal bed and certain other methane deposits. For the economies of both the U.S. and the world nearly two thirds of our energy comes from conventional petroleum, about 40 percent from conventional liquid petroleum and another 20–25 percent from gaseous petroleum (EIA 2007; Fig. 5.1). Coal, and natural gas provide most of the rest of the energy that we use. Hydroelectric power and wood together are renewable energies generated from current solar input and provide about five percent of the energy that the US uses. "New renewables" including windmills and photovoltaics, provide less than one percent, and are not growing as rapidly in magnitude globally (although they are as a percent of their own contribution) as petroleum. Thus the annual increase in oil and gas use is much greater than the new quantities coming from the new renewables, at least to date. All of these proportions have not changed very much since the 1970s in the United States or the world. We believe it most accurate to consider the times that we live in as the age of petroleum, for petroleum is the foundation of our economies and our lives. Just look around.

Petroleum is especially important because of its magnitude of current use, because it has important and unique qualitative attributes leading to high economic utility that include very high energy density and transportability (Cleveland 2005), and because its future supply is worrisome. The issue is not the point at which oil actually runs out but rather the relation between supply and potential demand. Barring a massive worldwide recession demand will continue to increase as human populations, petroleum-based agriculture and economies (especially Asian) continue to grow. Petroleum supplies have been growing most years since 1900 at two or three percent per year, a trend that most investigators think cannot continue (e.g. Campbell and Laherrere 1998, Heinberg 2003). Peak oil, that is the time at which an oil field, a nation or the entire world reaches its maximum oil production and then declines, is not some abstract issue debated by theoretical scientists or worried citizens but an actuality that occurred in the United States in 1970 and in some 60 (of 80) other oil-producing nations since (Hubbert 1974, Strahan 2007, Energyfiles 2007). Several prominent geologists have suggested that it may have occurred already for the world, although that is not clear yet (e.g. Deffeyes 2005, see EIA 2007, IEA 2007). With global demand showing no sign of abating at some time it will not be possible to continue to increase petroleum supplies, especially oil globally and natural gas in North America, or even to maintain current levels of supply, regardless of technology or price. At this point we will enter the second half of the age of oil (Campbell 2005). The first half was one of year by year growth, the second half will be of continued importance but year by year decline in supply, with possibly an "undulating plateau" at the top and some help from still-abundant natural gas outside North America separating the two halves and buffering the impact somewhat for a decade or so. We are of the opinion that it will not be possible to fill in the growing gap between supply and demand of conventional oil with e.g. liquid biomass alternatives on the scale required (Hall et al. in review), and even were that possible that the investments and time required to do so would mean that we needed to get started some decades ago (Hirsch et al. 2005). When the decline in global oil production begins we will see the "end of cheap oil" and a very different economic climate.

The very large use of fossil fuels in the United States means that each of us has the equivalent of 60–80 hard working laborers to "hew our wood and haul our water" as well as to grow, transport and cook our food, make, transport and import our consumer goods, provide sophisticated medical and health services, visit our relatives and take vacations in far away or even relatively near by places. Simply to grow our food requires the energy of about a gallon of oil per person per day, and if a North American takes a hot shower in the morning he or she will have already used far more energy than probably two thirds of the Earth's human population use in an entire day. Oil is especially important for the transportation of ourselves and of our goods and services, and gas for heating, cooking, some industries and as a feedstock for fertilizers and plastics.

5.3 How much Oil will we be able to Extract?

So the next important question is how much oil and gas are left in the world? The answer is a lot, although probably not a lot relative to our increasing needs, and maybe not a lot that we can afford to exploit with a large financial and, especially, energy profit. We will probably always have enough oil to oil our bicycle chains. The question is whether we will have anything like the quantity that we use now at the prices that allow the things we are used to having. Usually the issue of how much oil remains is not developed from the perspective of "when will we run out" but rather "when will we reach 'peak oil' globally". World wide we have consumed a little over one trillion barrels of oil. The current debate is fundamentally about whether there are 1, 2 or even 3.5 trillion barrels of economically extractable oil left to consume. Fundamental to this debate, yet mostly ignored, is an understanding of the capital, operating and environmental costs, in terms of money and energy, to find, extract and use whatever new sources of oil remain to be discovered, and to generate whatever alternatives we might choose to develop. Thus the investment issues, in terms of both money and energy, will become ever more important.

There are two distinct camps for this issue. One camp, which we call the "technological cornucopians", led principally by economists such as Michael Lynch (e.g. Lynch 1996, Adelman and Lynch 1997), believes that market forces and technology will continue to supply (at a price) more or less whatever oil we need for the indefinite future. They focus on the fact that we now are able to extract only some 35 percent of the oil from an oil field, that large areas of the world (deep ocean, Greenland, Antarctica) have not been explored and may have substantial supplies of oil, and that substitutes, such as oil shale and tar sands, abound. They are buoyed by the failure of many earlier predictions of the demise or peak of oil, two recent and prestigious analyses by the U.S. Geological Survey and the Cambridge Energy Research Associates that tend to suggest that remaining extractable oil is near the high end given above, the recent discovery of the deepwater Jack 2 well in the Gulf of Mexico and the development of the Alberta Tar Sands, which are said to contain more oil than remains even in Saudi Arabia. They have a strong faith in technology to increase massively the proportion of oil that can be extracted from a given oil field, believe that many additional fields await additional exploration, and believe there are good substitutes for oil.

A second camp, which we can call the "peak oilers", is composed principally of scientists from a diversity of fields inspired by the pioneering work of M. King Hubbert (e.g. 1969; 1974), a few very knowledgeable and articulate politicians such as US Representative Roscoe Bartlett of Maryland, many private citizens from all walks of life and, increasingly, some members of the investment community. All believe that there remains only about one additional trillion barrels of extractable conventional oil and that the global peak – or perhaps a "bumpy plateau", in extraction will occur soon, or, perhaps, has already occurred. The arguments of these people and their organization, the Association for the Study of Peak Oil (ASPO), spearheaded by the analyses and writings of geologists Colin Campbell and Jean Laherrere, are supported by the many other geologists who more or less agree with them, the many peaks that have already occurred for many dozens of oil-producing countries, the recent collapse of production from some of our most important oil fields and the dismal record of oil discovery since the 1960s – so that we now extract and use four or five barrels of oil for each new barrel discovered (Fig. 5.2). They also believe that essentially all regions of the Earth favourable for oil production have been well explored for oil, so that there are few surprises left except perhaps in regions that will be nearly impossible to exploit.

There are several issues that tend to muddy the water around the issue of peak oil. First of all, some people do, and some do not, include natural gas liquids or

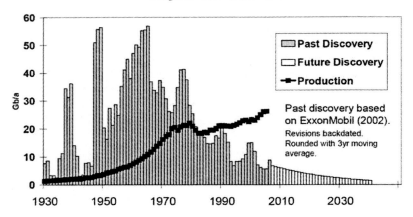

Fig. 5.2 Rate of the finding of oil (where revisions and extensions have been added into the year of initial strike) and of consumption (Source ASPO website)

condensate (liquid hydrocarbons that condense out of natural gas when it is held in surface tanks). These can be refined readily into motor fuel and other uses so that many investigators think they should simply be lumped with oil, which most usually they are. Since a peak in global natural gas production is thought to be one or two decades after a peak in global oil, inclusion of natural gas liquids extends the time or duration of whatever oil peak may occur (Fig. 5.3). Consequently, if indeed peak oil has occurred, a peak in liquid petroleum fuels might still be before us. A second

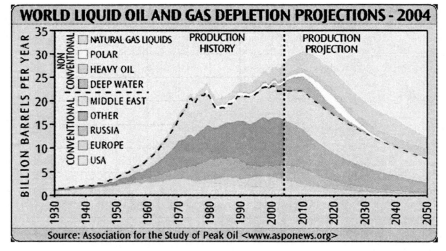

Fig. 5.3 Conventional oil use data and projections with the inclusion of non-conventional liquid fuels (Source ASPO website)

main issue is "how much oil is likely ever to be produced" vs. "when will global production peak, or at least cease growing?" In theory the issues are linked, perhaps tightly, but it is probably far more important to focus on the peak production rate rather than the total quantity that we will ever extract. In terms of ultimate economic impact, and probably prices, the most important issue is almost certainly the ratio between the production rate and its increase or decrease, and the consumption rate and its increase or decrease. Both the production and the consumption of oil and also natural gas have been growing at roughly two percent a year up through at least 2006. The great expansion of the economies of China and India, which at this time show no evidence of a slowdown, have recently more than compensated for some reduced use in other parts of the world. Nevertheless the growth rate of the human population has been even greater so that "per capita peak oil" probably occurred in 1978 (Duncan 2000). What the future holds may have more to do with the consumption rate than the production rate. If and when peak petroleum extraction occurs it is likely to increase prices which should bring an economic slowdown which should decrease oil use which might decrease prices and ... the chickens and eggs can keep going for some time. That is why many peak oilers speak of "a bumpy plateau". However if potential demand keeps growing then the difference between a steady or declining supply and an increasing demand presumably would continue upward pressures on price.

The rates of oil and gas production (more accurately extraction) and the onset of peak oil are dependent upon many interacting factors, including geological, economic and political. The geological restrictions are the most absolute and depend on the number and physical capacity of the world's operating wells. In most fields the oil does not exist in the familiar liquid state but in what is more akin to a complex oil-soaked brick. The rate at which oil can flow through these "aquifers" depends principally upon the physical properties of the oil itself and of the geological substrate, but also upon the pressure behind the oil that is provided initially by the gas in the well. Then, as the field matures, the pressure necessary to force the oil through the substrate to the collecting wells is supplied increasingly by pumping more gas or water into the structure. As with water wells the more rapidly the oil is extracted the more likely the substrate will become compacted, restricting future yields. Detergents, CO_2 and steam can increase yields but too-rapid extraction can cause compaction of the "aquifer" or fragmentation of flows which reduce yields. So our physical capacity to produce oil depends upon our ability to keep finding large oil fields in regions that we can reasonably access, our willingness to invest in exploration and development, and our willingness to not produce too quickly. The usual economic argument is that if supply is reduced relative to demand then the price will increase which will then signal oil companies to drill more, leading to the discovery of more oil and then additional supply. Although that sounds logical the results from the oil industry might not be in accordance to that logic as the empirical record shows that the rate at which oil and gas is found has little to do with the rate of drilling (Fig. 5.4).

It is thought that at this time we are producing oil globally pretty nearly to our present capacity, although future depletion or new fields can change that. Finally,

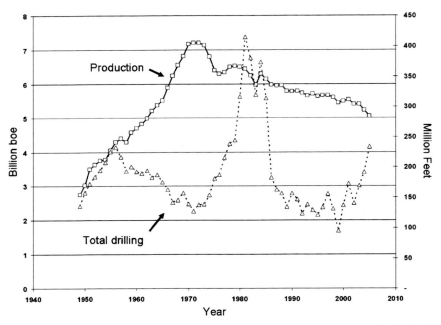

Fig. 5.4 Annual rates of total drilling for and production of oil and gas in the US, 1949–2005 (R^2 of the two = 0.005; source: U.S. EIA and N. D. Gagnon). Since drilling and other exploration activities are energy intensive, other things being equal EROI is lower when drilling rates are high

output can be limited or (at least in the past) enhanced for political reasons – which are even more difficult to predict than the geological restrictions. Empirically there is a fair amount of evidence from post peak countries, such as the U.S., that the physical limitations become important when about half of the ultimately-recoverable oil has been extracted. But why should that be? In the US it certainly was not due to a lack of investment, since most geologists believe that the US had been over drilled. We probably will not know until we have much more data, and much of the data are closely guarded industry or state secrets. According to one analyst if one looks at all of the 60 or so post peak oil-producing countries the peak occurs on average when about 54 percent of the total extractable oil in place has been extracted (Energyfiles.com 2007). Finally oil-producing nations often have high population and economic growth, and are using an increasing proportion of their own production (Hallock et al. 2004).

The United States clearly has experienced "peak oil". In a way this is quite remarkable, because as the price of oil increased by a factor of ten, from 3.50 to 35 dollars a barrel during the 1970s, a huge amount of capital was invested in US oil discovery and production efforts so that the drilling rate increase from 120 million feet per year in 1970 to 400 million feet in 1985. Nevertheless the production of crude oil decreased during the same period from the peak of 3.52 billion barrels a year in 1970 to 3.27 in 1985 and has continued to decline to 1.89 in 2005 even with the addition of Alaskan production. Natural gas production has also peaked and declined, although less regularly (This is included in Fig. 5.4). Thus despite advancement of petroleum discovery and production technology, and despite very significant investment, U.S. production has continued its downward trend since 1970. The technological optimists are correct in saying that advancing technology is important. But there are two fundamental and contradictory forces operating here, technological advances and depletion. In the US oil industry it is clear that depletion is trumping technological progress, as oil production is declining and oil is becoming much more expensive to produce.

5.4 Decreasing Energy Return on Investment

Energy return on investment (EROI or EROEI) is simply the energy that one obtains from an activity compared to the energy it took to generate that energy. The procedures are generally straightforward, although rather too dependent upon assumptions made as to the boundaries, and when the numerator and denominator are derived in the same units, as they should, it does not matter if the units are barrels (of oil) per barrel, Kcals per Kcal or MJoules per Mjoule as the results are in a unitless ratio. The running average EROI for the finding and production of US domestic oil has dropped from greater than 100 kilojoule returned per kilojoule invested in the 1930s to about thirty to one in the 1970s to between 11 and 18 to one today. This is a consequence of the decreasing energy returns as oil reservoirs are increasingly depleted and as there are increases in the energy costs as exploration and development are shifted increasingly deeper and offshore (Cleveland et al. 1984, Hall et al. 1986, Cleveland 2005). Even that ratio reflects mostly pumping out oil fields that are half a century or more old since we are finding few significant new fields. (In other words we can say that new oil is becoming increasingly more costly, in terms of dollars and energy, to find and extract). The increasing energy cost of a marginal barrel of oil or gas is one of the factors behind their increasing dollar cost, although if one corrects for general inflation the price of oil has increased only a moderate amount until 2007.

The same pattern of declining energy return on energy investment appears to be true for global petroleum production. Getting such information is very difficult, but with help from the superb database of the John H. Herold Company, several of their personnel, and graduate student and sometime Herold employee Nate Gagnon we were able to generate an approximate value for global EROI for finding new oil and natural gas (considered together). Our preliminary results indicate that the EROI for global oil and gas (at least for that which was publically traded) was roughly 26:1 in 1992, increased to about 35:1 in 1999, and since has fallen to approximately 19:1 in 2005. The apparent increase in EROI during the late 1990s is during a period when drilling effort was relatively low and may reflect the effects of reduced drilling effort as was seen for oil and gas in the United States (e.g. Fig. 5.4). If the rate of decline continues linearly for several decades then it would take the energy in a barrel of oil to get a new barrel of oil. While we do not know whether that extrapolation is accurate, essentially all EROI studies of our principal fossil fuels do indicate that their EROI is declining over time, and that EROI declines especially rapidly with increased exploitation (e.g. drilling) rates. This decline appears to be reflected in economic results. In November of 2004 The New York Times reported that for the previous three years oil exploration companies worldwide had spent more money in exploration than they had recovered in the dollar value of reserves found. Thus even though the EROI of global oil and gas is still about 18:1 as of 2006, this ratio is for all exploration and production activities. It is possible that the energy break even point has been approached or even reached for finding new oil. Whether we have reached this point or not the concept of EROI declining toward 1:1 makes irrelevant the reports of several oil analysts who believe that we may have substantially more oil left in the world, because it does not make sense to extract oil, at least for a fuel, when it requires more energy for the extraction than is found in the oil extracted.

How well we weather this coming storm will depend in large part on how we manage our investments now. From the perspective of energy there are three general types of investments that we make in society. The first is investments into getting energy itself, the second is investments for maintenance of, and replacing, existing infrastructure, and the third is discretionary expansion. In other words before we can think about expanding the economy we must first make the investments into getting the energy necessary to operate the existing economy, and into maintaining the infrastructure that we have, at least unless we wish to accept the entropy-driven degradation of what we already have. Investors must accept the fact that the required investments into the second and especially the first category are likely to increasingly limit what is available for the third. In other words the dollar and energy investments needed to get the energy needed to allow the rest of the economy to operate and grow have been very small historically, but this is likely to change dramatically. This is true whether we seek to continue our reliance on ever-scarcer petroleum or whether we attempt to develop some alternative. Technological improvements, if indeed they are possible, are extremely unlikely to bring back the low investments in energy that we have grown accustomed to.

The main problem that we face is a consequence of the "best first" principle. This is, quite simply, the characteristic of humans to use the highest quality resources first, be they timber, fish, soil, copper ore or, of relevance here, fossil fuels. This is because economic incentives are to exploit the highest quality, least cost (both in terms of energy and dollars) resources first, as was noted 200 years ago by economist David Ricardo (e.g. 1891). We have been exploiting fossil fuels for a long time. The peak in finding oil was in the 1930s for the United States and in the 1960s for the world, and both have declined enormously since then. An even greater decline has taken place in the efficiency with which we find oil, that is the amount of energy that we find relative to the energy we invest in seeking and exploiting it. As a consequence of the decreasing energy returns as oil depletion increases, and of the increasing energy costs as exploration and development shifted increasingly deeper offshore or into increasingly hostile environments, the energy return on investment (EROI) for US domestic oil has declined to perhaps 15 to one today, even though that contemporary ratio reflects mostly pumping out oil fields that are half a century or older. In other words we can say that new oil is becoming increasingly more costly, in terms of energy (and consequently dollars), to find and extract. The alternatives to oil available to us today are characterized by even lower EROIs, limiting their economic effectiveness. It is critical for CEOs and government officials to understand that the best oil and gas are simply gone, and there is no easy replacement.

That pattern of exploiting and depleting the best resources first also is occurring for natural gas. Natural gas was once considered a dangerous waste product of oil development and was burned or flared at the well head. But during the middle years of the last century large gas pipeline systems were developed in the U.S. and Europe that enabled gas to be sent to myriad users who increasingly discovered its qualities of ease of use and cleanliness, including its relatively low carbon dioxide emissions, at least relative to coal. US natural gas originally came from large fields in Louisiana, Texas and Oklahoma. Its production has moved increasingly to smaller fields distributed throughout Appalachia and, increasingly, the Rockies. As the largest fields that traditionally supplied the country peaked and declined a national peak in production occurred in 1973, and then as "unconventional" fields were developed a second, somewhat smaller peak occurred in 2001. Gas production has fallen by about 6 percent from that peak, and many investigators predict a "natural gas cliff" as traditional fields are exhausted and as it is increasingly difficult to bring smaller unconventional fields on line to replace the depleted giants. There had been an encouragement of electricity production from natural gas because it is relatively clean, but a large loss of petrochemical companies from the US because of the increasing price.

5.5 The Balloon Graph

We pay for imported oil in energy as well as dollars, for it takes energy to grow, manufacture or harvest what we sell abroad to gain the foreign exchange with which we buy fuel, (or we must in the future if we pay with debt today). In 1970 we gained roughly 30 megajoules for each megajoule used to make the crops, jet airplanes and so on that we exported (Hall et al. 1986). But as the price of imported oil increased, the EROI of the imported oil declined. By 1974 that ratio had dropped to nine to one, and by 1980 to three to one. The subsequent decline in the price of oil, aided by the inflation of the export products traded, eventually returned the energy terms of trade to something like it was in 1970, at least until the price of oil started to increase again after 2000. A rough estimate of the quantity and EROI of various major fuels in the U.S., including possible alternatives, is given in Fig. 5.5. An obvious aspect of that graph is that qualitatively and quantitatively alternatives to fossil fuel have a very long way to go to fill the shoes of fossil fuels. This is especially true when one considers the additional qualities of oil and gas, including energy density, ease of transport and ease of use.

The implications of all this is that if we are to supply into the future the amount of petroleum that the US consumed in the first half of this decade it will require

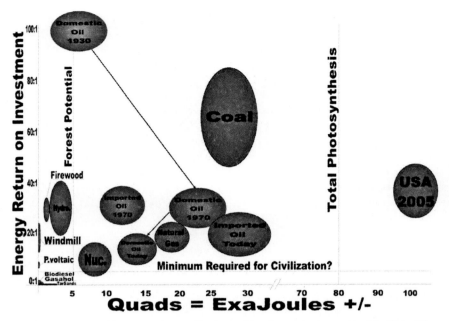

Fig. 5.5 "Balloon graph" representing quality (*y* graph) and quantity (*x* graph) of the United States economy for various fuels at various times. Arrows connect fuels from various times (i.e. domestic oil in 1930, 1970, 2005), and the size of the "balloon" represents part of the uncertainty associated with EROI estimates (Source: US EIA, Cutler Cleveland and C. Hall's own EROI work in preparation)

Source: US EIA, Cleveland et al. 1984, Cleveland 2005, Hall various including 1986 and http://www.theoildrum.com/node/3786.

enormous investments in either additional unconventional sources, in import facilities or as payments to foreign suppliers. That will mean a diversion of investment capital and of money more generally from other uses into getting the same amount of energy just to run the existing economy. In other words investments, from a national perspective, will be needed increasingly just to run what we have, not to generate real new growth. If we do not make these investments our energy supplies will falter or we will be tremendously beholden to foreigners, and if we do the returns may be small to the nation, although of course if the price of energy increases greatly the returns to the individual investor may be large. Another implication is if this issue is as important as we believe it is then we must pay much more attention to the quality of the data we are getting about energy costs of all things we do, including getting energy. Finally the failure of increased drilling to return more fuel (Fig. 5.4) calls into question the basic economic assumption that scarcity-generated higher prices will resolve that scarcity by encouraging more production. Indeed scarcity encourages more exploration and development activity, but that activity does not necessarily generate more resources. It will also encourage the development of alternative liquid fuels, but their EROIs are generally very low (Fig. 5.5).

5.6 Economic Impacts of Peak Oil and Decreasing EROI

Whether global peak oil has occurred already or will not occur for some years or, conceivably, decades, its economic implications will be enormous because we have no possible substitute on the scale required and because alternatives will require enormous investments in money and energy when both are likely to be in short supply. Despite the projected impact on our economic and business life within relatively few years or at most decades, neither government nor the business community is in any way prepared to deal with either the impacts of these changes or the new thinking needed for investment strategies. The reasons are myriad but include: the disinterest of the media, the failure of government to fund good analytic work on the various energy options, the erosion of good energy record keeping at the department of commerce, and the lack of really good options. The second point perhaps is debatable but we stand on that statement because of the top ten or so energy analysts that we are familiar with none are supported by government, or generally any, funding. There are not even targeted programs in NSF or the Department of Energy where one might apply if one wishes to undertake good objective, peer reviewed EROI analyses. Consequently much of what is written about energy is woefully misinformed or simply advocacy by various groups that hope to profit from various perceived alternatives. It is not unlikely that issues pertaining to the end of cheap petroleum will be the most important challenge that Western society has ever faced, especially when considered within the context of our need to deal with climate change and other environmental issues related to energy. Any business leaders who do not understand the inevitability, seriousness and implications of the end of cheap oil, or who make poor decisions in an attempt to alleviate its impact, are likely to be tremendously and negatively impacted as a result. At the same time the investment decisions we will make in the next decades will determine whether civilization is to make it through the transition away from petroleum or not.

What would be the impacts of a large increase in the energy and dollar cost of getting our petroleum, or of any restriction in its availability? While it is extremely difficult to make any hard predictions, we do have the record of the impacts of the large oil price increases of the 1970s as a possible guide. These "oil shocks" had very serious impacts on our economy which we have examined empirically in past publications (e.g. Hall et al. 1986). Many economists then and now did not think that even large increases in the price of energy would affect the economy dramatically because energy costs were but three to six percent of GDP. But by 1980, following the two "oil price shocks" of the 1970s, energy costs had increased dramatically until they were 14 percent of GDP. Actual shortages would have even greater impacts, if for example sufficient petroleum to run our industries or businesses were not available at any price. Other impacts included, and would include, an enhancement of our trade imbalances as more income is diverted overseas, adding to the foreign holdings of our debt and a decrease in discretionary disposable income as more money is diverted to access energy, whether via higher prices, more petroleum exploration or low EROI alternative fuels. This in turn would affect those sectors of the economy that are not essential. Consumer discretionary spending would probably fall dramatically, greatly effecting non-essential businesses such as tourism.

5.7 The "Cheese Slicer" Model

We have attempted to put together a conceptual and computer model to help us understand what might be the most basic implications of changing EROI on the economic activity of the United States. The model was conceptualized when we examined how the U.S. economy responded to the "oil shocks" of the 1970s. The underlying foundation is the reality that the economy as a whole requires energy (and other natural resources derived from nature) to run, and without these most basic components it will cease to function. The other premise of this model is that the economy as a whole is faced with choices in how to allocate its output in order to maintain itself and to do other things. Essentially the economy (and the collective decision makers in that economy) has opportunity costs associated with each decision it makes. Figure 5.6 shows our basic conceptual model parameterized for 1970, before the oil shocks of that decade. The large square represents the structure of the economy as a whole, which we put inside a symbol of the Earth biosphere/geosphere to reflect the fact that the economy must operate within the biosphere (e.g. Hall et al. 2001). In addition, of course, the economy must get energy and raw materials from *outside* the economy, at least as narrowly perceived, that is from nature (i.e. the biosphere/geosphere). The output of the economy, normally considered GDP, is represented by the large arrow coming out of the right side, where the depth of the arrow represents 100 percent of GDP. For the sake of developing our concept we think of the economy, for the moment, as an enormous dairy industry and cheese as the product coming out of the right hand side, moving towards the right. This output (i.e. the entire arrow) could be represented as either money or embodied energy. We use the former in this analysis (as almost all of the relevant data is recorded in monetary, not energy, units), but it is probably not terribly different from using energy outputs. So, our most important question is "how do we slice the cheese", that is how do we, and how will we divide up the output of the economy, or said differently, in what way can the output of the economy be divided up with the least objectionable opportunity cost. Most economists might answer "according to what the market decides," that is according to consumer tastes and buying habits. But we want to think about it a little differently because we think things might be profoundly different in the future.

Most generally the output of the model (and the economy) has two destinations: *investment* or *consumption*. These could be divided further into private vs. government investments and consumptions, or we could add in debt service, exports and imports, and so on. We choose not to do this, at least initially, although we are co-developing a much more complicated model of the economy with the Millennium Institute called T-21 North America. (e.g. Millennium Institute 2007). Our

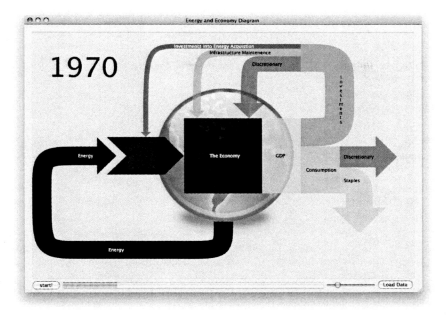

Fig. 5.6 The "Cheese slicer" diagrammatic model, which is a basic representation the fate of the output of the U.S. economy, 1970. The box in the middle represents the U.S. economy, the input arrow from the left represents the energy needed to run the economy, the large arrow on the right of the box represents the output of the model (i.e. GDP) which is then subdivided as represented by the output arrows going to the right—i.e. first into investments (into getting energy, maintenance and then discretionary) and then into consumption (either the basic required for minimal food, shelter and clothing or discretionary). In other words the economic output is "sliced" into different uses according to the requirements and desires of that economy/society. Data principally from the U.S. Department of Commerce. Extrapolations via the Millennium Institute's T-21 model courtesy of Andrea Bassi)

next question is "investment or consumption of/for what?" To do this we ask about what we must spend our money/energy on, that is on the *required* expenditures (without which the economy would cease to function). These include the investments in maintaining societal infrastructure (i.e. repairing and rebuilding bridges, roads, machines, factories, vehicles – represented by the top middle arrow feeding back from output of the economy back to the economy itself), and some kind of minimal food, shelter and clothing for the population (represented by the bottom rightward pointing arrow) required to maintain all individuals in society at the level of the Federal minimum standard of living). Another, and very important, essential need, the one of greatest interest to us here, is the investments into, or payments for, energy (i.e. the amount of diverted economic output that is used to secure and purchase our imported oil) that is required for the energy needed for the economy to operate at any particular level. This energy is absolutely critical for the economy to operate and must be paid for through proper payments and investments – which we consider together as investments to get energy. No investment in energy, no economic output. While there may be other critical inputs from nature (water, minerals etc.) or society (educated workforce etc.) we ignore these for the moment as they are unlikely to be as volatile as energy. This "Energy Investment" feedback is represented by the top-most arrow from the output of the economy back upstream to the "workgate" symbol (Odum 1971). The width of this line represents the investment of energy into getting more energy. Of critical interest here is that as the EROI of our economy's total combined fuel source declines then more and more of the output of the economy *must* be shunted back to getting the energy required to run the economy if the economy is to remain the same size.

Once these necessities are taken care of then what is left is considered the *discretionary* output of the economy. This can be either discretionary consumption (a vacation or a fancier meal, car or house than needed, represented by the upper right pointing arrow in the diagrams) or discretionary investment (i.e. building a new tourist destination in Florida or the Caribbean, represented as the lowest of the top arrows feeding back into the economy. During the last 100 years the enormous wealth generated by the United States economy has meant that we have had an enormous amount of discretionary income. This is in large part because the energy investments represented in Fig. 5.6 have been relatively small.

It turns out that the needed information to construct the above division of the economy is reasonably easy to come by for the U.S. economy, at least if we are willing to make a few major assumptions and accept a fairly large margin of error. Inflation-corrected GDP, i.e. the size of the output of the economy, is published routinely by the U.S. Department of Commerce. The total investments for maintenance in the U.S. economy are available as "Deprecation of Fixed Capital", U.S. Department of Commerce, various years). The minimum needed for food, shelter and clothing is available as "Personal Consumption Expenditures" (or the minimum of that required to be above poverty) which we selected from the U.S. Department of Commerce for various years). The investment into energy acquisition is the sum of all of the capital costs in all of the energy producing sectors of the U.S. plus expenditures for purchased foreign fuel. Empirical values for these components of the economy are plotted in Figs. 5.6–5.10. When these three requirements for maintaining the economy: investments and payments for energy, maintenance of infrastructure and maintenance of people are subtracted from the total GDP then what is left is discretionary income.

We simulated two basic data streams: the U.S. economy from 1949 to 2005 (representing the growth prior to the "oil crises" of 1973 and 1979, the impact of the oil crisis and the recovery from that, which had occurred by the mid-1990s. Then we projected this data stream into the future by extrapolating the data used prior to 2005 along with the assumption that the EROI for society declined from an average of roughly 20:1 in 2005 to 5:1 in 2040. This is an arbitrary scenario but may represent what we have in store for us as we enter the "second half of the age of oil", i.e. a time of declining availability and rising price so that more and more of society's output needs to be diverted into the top arrow of e.g. Fig. 5.6.

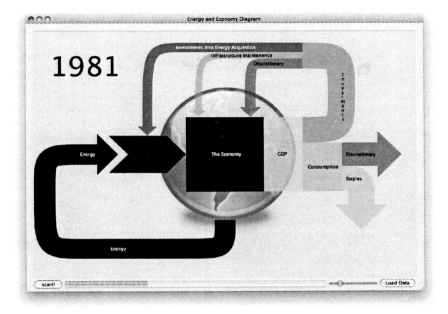

Fig. 5.7 Same as Fig. 5.6 but for 1980, following large increases in the price of oil. Note change in discretionary investments

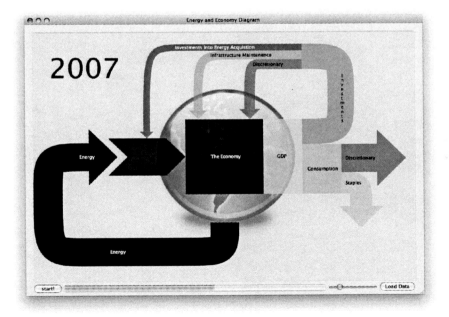

Fig. 5.8 Same as Fig. 5.6 but for 2007, following large decreases then small increases in the price of oil. Note change in discretionary investments

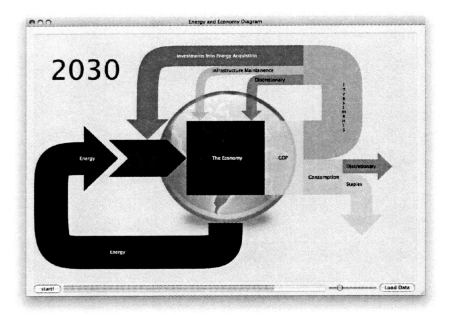

Fig. 5.9 Same as Fig. 5.6 but for 2030, with a projection into the future with the assumption that the EROI declines from 20:1 (on average) to 10:1

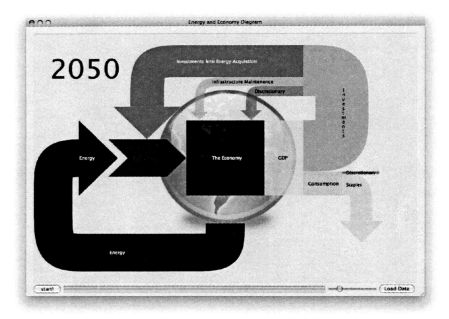

Fig. 5.10 Same as Fig. 5.6 but for 2050, but a projection into the future with the assumption that the EROI declines to 5:1

5.8 Results of Simulation

The results of our simulation suggest that discretionary income, including both discretionary investments and discretionary consumption, will move from the present 50 or so percent in 2005 to about 10 percent by 2050, or whenever (or if) the composite EROI of all of our fuels reaches about 5:1 (Figs. 5.9–5.10).

5.9 Discussion

Individual businesses would be affected by having their fuel costs increase and, for many, a reduction in demand for their products. This simultaneous inflation and recession happened in the 1970s and is projected to happen into the future as EROI for primary fuels declines. The "stagflation" that occurred in the 1970s was not supposed to happen according to an economic theory called the Phillips curve. But an energy-based explanation is easy (e.g. Hall 1992). As more money was diverted to getting the energy necessary to run the rest of the economy disposable income, and hence demand for many non-essential goods and services, declined, leading to economic stagnation. Meanwhile the increased cost for energy led to inflation, as there was no additional production that occurred from this greater expenditure. Although unemployment increased overall during the 1970s it was not as much as demand decreased, as labor at the margin became relatively useful compared to increasingly-expensive energy. Individual sectors might be much more impacted as happened in 2005, for example, with many Louisiana petrochemical companies that were forced to close or move overseas when the price of natural gas increased. On the other hand alternate energy businesses, from forestry operations and woodcutting to solar devices, might do very well.

When the price of oil increases it does not seem to be in the national or in corporate interest to invest in more energy-intensive consumption, as Ford Motor Company seems to be finding out with its large emphasis on large SUVs and pickup trucks. We are likely to have over invested already in the number of remote second homes, cruise ships, and Caribbean semi-luxury hotels, so that it may not a particularly good idea to do more of that now. This is due to the "Cancun effect" – that such hotels require the existence of large amounts of disposable income from the US middle class and cheap energy. That disposable income may have to be shifted into the energy sector with less of an opportunity cost to the economy as a whole. Investors who understand the changing rules of the investment game are likely to do much better in the long run.

So what can the scientist say to the investor? The options are not easy. As noted above worldwide investments in seeking oil have had very low monetary returns in recent years. Investments in many alternatives may not fare much better. Ethanol from corn projects were once financially profitable to the individual investor because they have been highly subsidized by the government, but they are probably a poor investment for the Nation. It is not clear that this fuel makes much of an energy profit, with an EROI of 1.6 at best, and less than one for one at worst, depending upon the study used for analysis (see review in Farrell et al. 2006 and also the many letters on that article in Science Magazine, June 23, 2006). Biodiesel may have an EROI of about three to one. Is that a good investment? Clearly not relative to remaining petroleum, but some day as petroleum EROI declines it may be. However real fuels must have EROIs of 5 or 10 or more returned on one invested to not be subsidized by petroleum or coal in many ways, such as the construction of the vehicles and roads that use them. Other biomass, such as wood, can have good EROIs when used as solid fuel but face real difficulties when converted to liquid fuels, and the technology is barely developed. The scale of the problem can be seen by the fact that we presently use more fossil energy in the US than is fixed by all green plant production, including all of our croplands and all of our forests (Pimentel, D. Personal communication). Biomass fuels may make more sense in nations where biomass is very plentiful and, more importantly, where present use of petroleum is much less than in the US. Alternatively one might argue that if we could bring the use of liquid fuels in the United States down to, say, 20 percent of the present then liquid fuels from biomass could fill in a substantial portion of that demand. Nevertheless we should remember that historically we in the U.S. have used energy to produce food and fibre, not the converse, because we have valued food and fibre more highly. Is this about to change?

Energy return on investment from coal is presently quite favourable compared to alternatives (ranging from perhaps 50:1 to 100:1), but the environmental costs are probably unacceptable as the case for global warming and other pollutants from coal burning becomes increasingly clear. Injecting carbon dioxide into some underground reservoir seems unfeasible for all the coal plants we might build, but it is being pushed hard by many who promote coal. Nuclear has a debateable moderate energy return on investment (5–15:1, some unpublished studies say more), but newer analyses need to be made. Nuclear has a relatively small impact on the atmosphere, but there are large problems with public acceptance and perhaps safety in our increasingly difficult political world.

Windmills have an EROI of 15–20 return on one invested, but this does not include the energy cost of back up or electricity "storage" for periods when the wind is not blowing. They make sense if they can be associated with nearby hydroelectric dams that can store water when the wind is blowing and release water when it is not, but the intermittent release of water can cause environmental problems. Photovoltaics are expensive in dollars and presumably energy relative to their return, but the technology is improving. One should not be confused by all claims for efficiency improvements because many require very expensive "rare-earth" doping materials, and some may become prohibitively expensive if their use expands greatly (e.g. Andersson et al. 1998). According to one savvy contractor the efficiency in energy returned per square foot of collector has been increasing, but the energy returned per dollar invested has been constant as the price of the high end units has increased (Blair May, Waldoboro Maine, personal communication). Additionally while photovoltaics have caught the public's eye the return on dollar investment is about double for hot water installations. Windmills, photovoltaics and some other forms of solar do seem to be a good choice if we are to protect the environment, but the investment costs up front will be enormous compared to fossil fuels.

Energy and money are not the only critical aspects of development of energy alternatives. Recent work by Hirsch et al. (2005) has focused on the investments in time that might be needed to generate some kind of replacement for oil, should that be possible and peak oil occur. They examined what they thought might be the leading alternatives to provide the US with liquid fuel or lower liquid fuel use alternatives, including tar sands, oil shales, deep water petroleum, biodiesel, high MPG automobiles and trucks and so on. They assumed that these technologies would work (a bold assumption) and that an amount of investment capital equal to "many Manhattan projects" (the enormous project that built the first atomic bomb) would be available. They found that the critical resource was time – once we decided that we needed to make up for the decline in oil availability these projects would need to be started one or preferably two decades in advance of the peak for there not to be severe dislocations to the US economy. Given our current petroleum dependence, the rather unattractive aspects of many of the available alternatives, and the long lead time required to change our energy strategy the investment options are not obvious. This, we believe, may be the most important issue facing the United States at this time: where should we invest our remaining high quality petroleum (and coal) with an eye toward insuring that we can meet the energy needs of the future. We do not believe that markets can solve this problem alone or perhaps at all. Research money for good energy analysis unconnected to this or that "solution" simply are not available. Fortunately some private individuals are stepping into the void as per our acknowledgements.

Human history has been about the progressive development and use of ever higher quality fuels, from human muscle power to draft animals to water power to coal to petroleum. Nuclear at one time seemed to be a continuation of that trend, but that is a hard argument to make today. Perhaps our major question is whether petroleum represents but a step in this continuing process of higher quality fuel sources or rather is the highest quality fuel we will ever have on a large scale. There are many other possible candidates for the next main fuel, but few are both quantitatively and qualitatively attractive (Figure 5.5). In our view we cannot leave these decisions up to the market if we are to solve our future climate or peak oil problems. One possible way to look at the problem, probably not a very popular one with investors, is to pass legislation that would limit energy investments to only "carbon-neutral" ones, remove subsidies from low EROI fuels such as corn-based ethanol, and then perhaps allow the market to sort out from those possibilities that remain.

A difficult decision would be whether we should subsidize certain fuels. At the moment alcohol from corn is subsidized three times: in the natural gas for fertilizers, the corn itself through the Department of Agriculture's 100 or so billion dollar general program of farm subsidies, and the additional 50 cents per liter subsidy for the alcohol itself. It seems pretty clear that the corn-based alcohol would not make it economically without the subsidy as it has only a marginal (if that) energy return. Are we simply subsidizing the depletion of oil, natural gas (and soil) to generate an approximately equal amount of energy in the alcohol? Wind energy appears to have about an 18:1 EROI, enough to make it a reasonable candidate, although there are some issues relative to backup technologies for when the wind is not blowing. So should wind be subsidized, or allowed to compete with other zero emission energy sources? A question might be the degree to which the eventual market price would be determined by, or at least be consistent with, the EROI, as all the energy inputs (including that to support labor's paychecks) must be part of the costs. Otherwise that energy is being subsidized by the dominant fuels used by society.

5.10 Conclusion

It seems obvious to us that the U.S. economy is very vulnerable to a decreasing EROI for its principle fuels, whether that comes from an increase in expenditures overseas if and as the price of imported oil increases more rapidly than that of the things that we trade for it, or as domestic oil and gas reserves are exhausted and new reservoirs become increasingly difficult to find, or as we turn to lower EROI alternatives such as biodiesel and or photovoltaics. We do not know exactly what all this means, but our straightforward model suggests that a principal effect will be a decline in discretionary income and a greater investment requirement for getting energy, with all the economic impacts that entails. Since more fuel will be required to run the same amount of economic activity the potential for environmental impacts increasing is very strong. On the other hand protecting the environment, which we support strongly, may mean turning away from some higher EROI fuels to some lower ones. We think all of these issues are very important yet are hardly discussed in our society or even in economic or scientific circles.

Acknowledgments We thank our great teacher, Howard Odum, many students over the years, colleagues and friends including Andrea Bassi, John Gowdy, Andy Groat, Jean Laherrere and Kent Klitgaard, and many others who have helped us to try to understand these issues. Art Smith and Lysle Brinker of John S. Herold Company were generous with their time, insight and data. Nate Gagnon created Fig. 5.4 and Nate Hagens made many useful comments. The Santa Barbara Family Foundation, ASPO-USA, The Interfaith Center on Corporate Responsibility and several individuals who wish not to be named provided much appreciated financial help.

References

Adelman, M. A. & Lynch, M. C. (1997). Fixed view of resource limits creates undue pessimism. *Oil and Gas Journal*, 95, 56–60.
Andersson, B. A., Azar, C., Holmerg, J. & Karlsson, S. (1998). Material constraints for thin-film solar cells. *Energy,* 23, 407–411.
Ayers, R.U. (1996). Limits to the growth paradigm. *Ecological Economics*, 19, 117–134.
Boulding, K. E. (1966). The economics of the coming spaceship earth. (In H. Jarrett (Ed.), *Environmental quality in a growing economy* (pp. 3–14). Baltimore: Johns Hopkins University Press)
Bartlett, R Representative U.S. Congress. http://bartlett.house.gov/

Campbell, C. (2005). *The 2nd half of the age of oil*. Paper presented at the 5th ASPO Conference, Lisbon Portugal

Campbell, C. & Laherrere, J.(1998). The end of cheap oil. *Scientific American* (March), 78–83.

Cleveland, C. J. (1991). Natural resource scarcity and economic growth revisited: economic and biophysical perspectives. (In Costanza R. (Ed.) *Ecological Economics: The Science and Management of Sustainability* (pp. 289–317). New York: Columbia University Press.)

Cleveland, C. J. (2005). Net energy from the extraction of oil and gas in the United States. *Energy: The International Journal*, 30(5), 769–782.

Cleveland C. J. & Ruth, M. (1997). When where, and by how much do biophysical limits constrain the economic process?: A survey of Nicholas Georgescu-Roegen's contribution to ecological economics. *Ecological Economics*, 22, 203–223.

Cleveland C. J., Costanza, R., Hall, C. A. S. & Kaufmann, R. K. (1984). Energy and the US economy: A biophysical perspective. *Science*, 225, 890–897.

Cottrell, F. (1955). *Energy and society*. (Dutton, NY: reprinted by Greenwood Press)

Daly, H. E. (1977). *Steady-state economics*. (San Francisco: W. H. Freeman)

Deffeyes, K. (2005). *Beyond oil: The view from Hubbert's Peak*. (New York: Farrar, Straus and Giroux)

Duncan, R. C. (2000). *Peak oil production and the road to the Olduvai Gorge*. Keynote paper presented at the Pardee Keynote Symposia. Geological Society of America, Summit 2000

Dung, T.H. (1992). Consumption, production and technological progress: A unified entropic approach. *Ecological Economics* Vol. 6, 195–210

EIA (2007). (U.S. Energy Information Agency website, accessed June 2007)

Energyfiles.com Accessed August 2007. www.energyfiles.com

Farrell, A. E., Plevin, R. J., Turner, B. T., Jones, A. D., O'Hare, M. & Kammen, D. M. (2006). Ethanol can contribute to energy and environmental goals. *Science*, 311, 506–508

Georgescu-Roegen, N. (1971). *The Eentropy Law and the economic process*. (Cambridge, MA: Harvard University Press)

Hall, C. A. S. (1991). An idiosyncratic assessment of the role of mathematical models in environmental sciences. *Environment International*, 17, 507–517.

Hall, C. A. S. (1992). Economic development or developing economics? (In M. Wali (Ed.) *Ecosystem rehabilitation in theory and practice, Vol I. Policy Issues* (pp. 101–126) The Hague, Netherlands: SPB Publishing.)

Hall, C. A. S. (Ed.) (2000). Quantifying sustainable development: The future of tropical economies. (San Diego: Academic Press)

Hall, C. A. S. & Ko, J. Y. (2007). The myth of efficiency through market economics: A biophysical analysis of tropical economies, especially with respect to energy, forests and water. (In G. LeClerc & C. A. S. Hall (Eds.) *Making world development work: Scientific alternatives to neoclassical economic theory* (pp. 90–103). Albuquerque: University of New Mexico Press)

Hall, C.A.S., Cleveland, C. J. & Kaufmann R. K. (1986). *Energy and resource quality: The ecology of the economic process*. (New York: Wiley-Interscience. Reprinted 1992. Boulder: University Press of Colorado.)

Hall, C. A. S., Volk, T.A. ,Murphy, D.J., Ofezu, G., Powers R., Quaye A., Serapiglia, M. & Townsend, J. (in review). Energy return on investment of current and alternative liquid fuel sources and their implications for wildlife. *Journal of Wildlife Science*

Hallock, J., Tharkan, P., Hall, C., Jefferson, M. and Wu, W. (2004). Forecasting the limits to the availability and diversity of global conventional oil supplies. *Energy*, 29, 1673–1696.

Hannon B. (1981). Analysis of the energy cost of economic activities: 1963–2000. *Energy Research Group Doc*. No. 316. Urbana: University of Illinois.

Heinberg, R. (2003). *The Party's Over: Oil, War and the Fate of Industrial Societies*. (Gabriella Island, B.C. Canada: New Society Publishers)

Hirsch, R., Bezdec, R. & Wending, W. (2005). Peaking of world oil production: impacts, mitigation and risk management. U.S. Department of Energy. National Energy Technology Laboratory. Unpublished Report.

Hubbert, M. K. (1969). Energy Resources. In *Resources and Man*. National Academy of Sciences. (pp. 157–242). (San Francisco: W.H. Freeman)

Hubbert, M. K. (June 4, 1974). Washington, D.C. *Testimony* before Subcommittee on the Environment of the Committee on Interior and Insular Affairs, House of Representatives, Ninety-Third Congress , Serial no. 93–55 U.S. Government Printing Office, Washington: 1974.

Jorgenson, D. W. (1984). The role of energy in productivity growth. *The American Economic Review,* 74(2), 26–30.

Jorgenson, D. W. (1988). Productivity and economic growth in Japan and the United States. *The American Economic Review*, 78: 217–222.

Herendeen, R. & Bullard, C. (1975). The energy costs of Goods and Services. 1963 and 1967, *Energy Policy*, 268.

IEA. (2007). (European Energy Agency, web page, accessed August 2007).

Kaufmann, R. (2004). The mechanisms for autonomous energy efficiency increases: A cointegration analysis of the US Energy/GDP Ratio. *The Energy Journal*, 25, 63–86.

Kümmel R. (1982). The impact of energy on industrial growth. *Energy - The International Journal*, 7, 189–203.

Kümmel R. (1989). Energy as a factor of production and entropy as a pollution indicator in macroeconomic modelling. *Ecological Economics*, 1, 161–180.

Lynch, M. C. (1996). The analysis and forecasting of petroleum supply: sources of error and bias. (In D. H. E. Mallakh (Ed.) *Energy Watchers VII*. International Research Center for Energy and Economic Development.)

Laherrère. J. *Future Oil Supplies*. Seminar Center of Energy Conversion, Zurich: 2003.

LeClerc, G. & Hall, C. A. S. (2007). Making world development work: Scientific alternatives to neoclassical economic theory. (Albuquerque: University of New Mexico Press)

Odum, H. T. (1972). *Environment, power and society*. (New York: Wiley-Interscience)

Quinn, M. (2005). Peak Oil, Energy, and Local Solutions: Reports from Recent Conferences. Megan Quinn, Global Public Media, 10 June 2005.

Ricardo, D. (1891). *The principles of political economy and taxation*. London: G. Bell and Sons. (Reprint of 3rd edition, originally pub 1821).

Smil, V, 2007. Light behind the fall: Japan's electricity consumption, the environment, and economic growth. Japan Focus. http://japanfocus.org/products/details/2394

Soddy, F. (1926). *Wealth, virtual wealth and debt*. (New York: E.P. Dutton and Co.)

Solow, R. M. (1974). The economics of resources or the resources of economics. *American Economic Review*, 66, 1–14.

Strahan, D. (2007) Open letter to Duncan Clarke. Posted on Wednesday, August 15th, 2007 http://www.davidstrahan.com/blog/?p=35.

Tryon FG. (1927). An index of consumption of fuels and water power. Journal of the American Statistical Association 22: 271–282.

Chapter 6
Wind Power: Benefits and Limitations

Andrew R.B. Ferguson

Abstract Wind turbines have a potential benefit insofar as they have a power density that matches coal, at least according to one measure. Set against this is the uncontrollable nature of their output. This means that without a suitable method of storing output, wind power can satisfy only about 10% of total energy demand. This limit applies to all uncontrollables collectively, with the slight exception that in places using a lot of air conditioning, photovoltaics could be used to help satisfy peak electrical demands.

The basic problem of uncontrollables would resolve if a suitable method of storing electricity could be found. The severe limitations of hydro, hydrogen storage, and vanadium batteries are explored. A storage system that would be both efficient and significant in size, at least in the USA, is Compress Air Energy Storage (CAES), but more experience of this is needed before it can be properly assessed.

Assessment becomes even more difficult when looking ahead to the time when all fossil fuels are scarce, because at present there appears to be no satisfactory solution to the 'liquid' fuel problem, yet the process of manufacturing, installing, and maintaining wind turbines and the associated transmission lines would be very difficult without the help of liquid fossil fuels. In the USA, any likely gain from the use of wind power is likely to be overtaken by the present population growth of at least 1.4% a year.

Keywords Population growth · power density · storage · uncontrollables · wind

6.1 Introduction

The power density[1] that can be achieved using each specific renewable energy source is an important measure of the usefulness of that energy source. To see wind turbines in perspective, it is helpful to look first at a variety of energy sources.

✉ A.R.B. Ferguson
11 Harcourt Close, Henley-on-Thames, RG9 1UZ, England
e-mail: andrewrbferguson@hotmail.com

D. Pimentel (ed.), *Biofuels, Solar and Wind as Renewable Energy Systems*,
© Springer Science+Business Media B.V. 2008

The power density that is likely to be achieved when coal is used to produce electricity has been estimated at 315 kW(e)/ha.[2] Note that the power density is there given in terms of the electrical output. Since the efficiency of producing electricity from coal is about 30%, it can be deduced that, in terms of the coal that produces the electricity, its power density is about 315/0.30 = **1050** kW/ha. The normal route is of course first to calculate the power density of coal itself, but that is incidental.[3]

After establishing the output of electricity from wind turbines, as will be done later, it will be appropriate to discuss whether emphasis should be placed on the power density in terms of just the electrical power produced from the wind turbines or whether, as is often done, that output should be uprated to take account of the fossil fuel required to produce it. For the present, note only that as the output of wind turbines is electricity, the first step will be to measure the power density in terms of the electrical output, i.e. power density measured as kilowatts of electricity, **kW(e)**, rather than **kW** of fossil fuel equivalent.

Before proceeding further into the study of wind power, it will be relevant to look briefly also at the power density of liquid fuels produced from biomass. There are various categories of power density which can be assessed, all of them useful in their own way. The one that is least controversial is to measure the output per hectare of, for example, ethanol, subtracting from it only the amount of energy input that needs to be in liquid form, e.g. as gasoline, diesel or ethanol. That gives the 'useful' ethanol per hectare. In such an assessment, the power density of ethanol from corn (maize) is about 1.9 kW/ha (OPTJ 3/1).[4] Incidentally ethanol from sugarcane, when assessed on this same basis, typically achieves a power density of 2.9 kW/ha, but soil erosion problems are worse with sugarcane than with corn, and the land that is suitable for growing sugarcane is more restricted. Considered against the power density of oil, which is considerably higher than the 1050 kW/ha mentioned for coal, it is clear that these ethanol power densities are very small indeed. For example, in the same paper, OPTJ 3/1, it is calculated that if all the U.S. corn crop were to be used to produce ethanol, it could serve to replace only 6% of the fuel used in the USA for transport.[5]

Another type of power density that can be assessed is by adding to the ethanol output the calorific value of the by-products (e.g. dry distillers' grains that can be fed to cattle), and from that subtracting not only the liquid input but also the non-liquid inputs, e.g. the heat needed for distillation (which constitute about 85% of the inputs). The resultant 'net energy capture' would be a revealing figure if its value could be agreed, but there are huge areas of uncertainty, particularly because we need to know (a) how much of the by-product is actually going to find a use and should therefore be counted as an output; (b) how much of the total crop can be utilized without causing loss of soil quality. For example, in the case of corn total yield is about 15,000 kg/ha (dry), with about half of this being grain and the other half being stover (Pimentel and Pimentel 1996, p. 36). Growing corn is prone to cause soil erosion. All the stover should be either left on or returned to the ground to diminish erosion and return nutrients. Sugarcane is worse than corn at causing soil erosion (Pimentel 1993), so a very significant proportion of the bagasse should be returned to the soil rather than using most of it to produce the heat needed for ethanol distillation (as tends to be done in practice).

All energy balance calculations are crude at best due to such factors, and the 'energy balance' of producing ethanol from corn can be assessed as either positive or negative depending on matters of fine judgement. However, let us be clear about what an approximate zero energy balance means. It means that producing ethanol from biomass is not an energy transformation that produces *useful energy*; it is merely a way of using other forms of available energy to produce energy in a liquid form. The conclusion is twofold: that power density figures need to be hedged about with precise understanding of what is being assessed, and that producing significant quantities of liquid fuels from renewable sources is a difficult problem.

6.2 The Power Density of Electricity from Wind Turbines

In an ideal situation, where the wind always blows from the same direction, and where docile citizens do not mind where the wind turbines are placed, the turbines could be placed fairly close together. But in practice there are few sites where engineers believe that the wind can be trusted to always come from the same direction. Moreover there are often practical restrictions about where the wind turbines can be placed. Due to these factors, the actual placing of wind turbines is such that about 25 ha needs to be 'protected' from interference by other wind turbines for each megawatt (MW) of wind turbine *capacity* (Hayden 2004, pp. 145–149). Note first that this 25 ha/MW is independent of the rated capacity of the wind turbine (e.g. two turbines of 1 MW capacity would require 50 ha and so would one 2 MW turbine), and secondly that the 25 ha/MW refers to the *rated capacity* of the wind turbines not their actual output.

The actual output of a wind turbine, or group of wind turbines, is determined by the capacity factor (also called load factor) that they achieve. In northern Europe (Sweden, Denmark, Germany, the Netherlands) the mean capacity factor achieved over two years was 22% (OPTJ 3/1, p. 4), in the UK for the years 2000–2004 capacity factors achieved were 28%, 26%, 30%, 24%, 27% for an average of 27%,[6] and for the USA for the years 2000–2004 capacity factors were respectively 27%, 20%, 27%, 21%, 27% for an average of 24%.[7] Nevertheless taller wind turbines may produce some improvement, so let us use 30% as a benchmark for the USA. This means that the protected area is 25/0.30 = **83** ha per MW *of output*, which gives a power density of 1000 [kW(e)]/83 = **12** kW(e)/ha. That power density gives an easy way to calculate how much land area would be needed to provide a certain amount of electrical output; e.g., to produce the mean power output of a 1000 MW power station, which delivers over the year say a mean 800 MW, the area needed would be 800,000 [kW]/12 = **66,700** ha, or 667 km^2, or 26 km by 26 km (16 miles by 16 miles). That is a substantial area, the ramifications of which will be considered later, after some other measures of power density have been considered.

Also of considerable relevance is the amount of land that the wind turbines are actually taking up, that is the land taken up by the concrete bases of the turbines and transmission lines, and to provide access roads (obviously this is mainly of concern when the land that is being used is ecologically productive). This has been put at 2–5% of the protected area. Taking a central value of 3.5%, puts the power density of wind turbines — in these terms, when sited on ecologically productive land — at $12/0.035 = \mathbf{343}$ kW(e)/ha. That is to say, it is similar to the power density of electricity from coal. It now becomes obvious why wind turbines are in a different ball park from biomass; that holds true whether the biomass is used to produce ethanol or merely used for its heat value. To touch on the latter briefly, it may well be possible to achieve, at suitable locations, without too many inputs, an annual yield of 10 dry tonnes per hectare using woody short-rotation crop. That would achieve a *gross* power density of about 6 kW/ha.[8] Note that both the wind power density figures being discussed, as well as the 6 kW/ha biomass figure, should really be qualified with the adjective 'gross', because no allowance has been made for inputs. However the difference between 6 kW/ha and 343 kW(e)/ha is so great that it is not necessary to determine to what extent inputs bring the *net* power densities closer together.

6.3 Producing the Output of a Power Station from Wind Power

Returning to the calculation which showed that to replace a 1000 MW power station by wind farms would require 667 km^2 of protected space, a small point to address first is the choice of 800 MW as the mean output. That may be challenged on the basis that power stations generally operate below an 80% load factor. The point though is that many power stations operate below capacity simply because they are *controllable*, which allows their output to be adjusted to suit demand. Clearly wind power cannot be used in that way. Instead it is used in conjunction with a controllable power source. The two operate together, 'in harness', to provide a baseload. Plant operated in that way, that is just to provide a baseload, e.g. nuclear plant, can certainly achieve an 80% load factors. Hayden (2004, p. 246) shows that 7 out of 22 countries operate their nuclear plant at above 80% load factor.

The practical problems of needing such large areas over which to spread the wind turbines is particularly acute in places with high population densities like Europe. But difficulties are encountered in practice in the USA too, due to such things as objections to destroying scenic vistas by putting wind turbines along prominent ridges. Moreover there are other problems in the wide spacing when taking a longer view.

The mean 800 MW of output, with a 30% load factor, would require a capacity of $800/0.3 = \mathbf{2670}$ MW, which might be supplied by 888 wind turbines of 3 MW capacity, for example. The task of installing those, with their access roads, and then connecting them together over an area of 667 km^2, may not seem too daunting to an engineer in the present day, but that is only because fossil fuel oil is available. When liquid fossil fuels become scarce, and a renewable liquid substitute has to be used, most probably one with something like the low power density we considered for ethanol from corn or sugarcane, the challenge would become enormous. *In planning for a fossil free future, it is necessary to continually bear in mind that many things which are easy today because of the availability of suitable fossil fuels, particularly oil, will not be easy in the future.* Whether such tasks as installing and maintaining wind turbines and transmission lines will be possible in the virtual absence of oil must at present be a matter of judgement.

6.4 The Problem of Assessing *Net* Energy with Respect to Wind Turbines

Net energy is simply the energy left over as useful energy once all the inputs have been subtracted. While that is a simple concept, there are practical problems which it is worth dwelling on. The wind industry would most likely respond to the previous paragraph by saying that the 'energy payback' — which is the time it takes to produce enough energy from the wind turbines to produce the same amount of energy as the inputs that are needed for their construction from raw materials and subsequent maintenance — has already been assessed for wind turbines, and it has been put as low as six months, so there must be something misleading in the emphasis being placed on the extent of the inputs needed as per the previous paragraph.

The trouble with such energy payback assessments is that they take only partial account of the different *types* of input and sometimes they do so in a misleading way. For example, in assessing the energy value of the electrical output of wind turbines, that output is valued as the amount of fossil fuel that would be needed to produce it. Since the efficiency of generation of electricity is about 0.33, that means that the electrical output can be valued at $1/0.33 = 3$ times its energy value as electricity. There is some validity in that when electrical energy is so useful to us that society is prepared to suffer the unavoidable loss of energy that occurs in producing it from fossil fuels. However, looking towards a fossil-fuel-free society, the situation is entirely different. We have already noted that the power density of a renewable liquid fuel is below 2 kW/ha, and that of biomass when used merely as heat is around 6 kW/ha, so it would be sound sense to use the high power density of wind turbine output (12 kW(e)/ha or 343 kW(e)/ha depending on the perspective) to replace both heat and *if it is possible* use the electrical output to produce 'liquid' fuels. Thus far from electricity being at a premium value, it is either at no premium, because it is used to replace the heat needed for such industrial processes as glass making, or at a substantial discount in value, because of the large losses that would occur in trying to produce a useful 'liquid' fuel from it, e.g. compressed hydrogen. The extent to which that is viable is a relevant question to be addressed later.

What has become apparent is that wind turbines have a far higher energy density than biomass, on one measure even rivalling that of coal, so the next consideration is to what extent it is advisable to integrate the input from wind turbines into the electrical system just to save fossil fuel now, while we still have the oil to carry out the construction, installation and maintenance processes associated with wind turbines without too much difficulty. That leads on to consideration of the problems of dealing with the *uncontrollable* nature of the output from wind turbines.

6.5 The Implications of the Uncontrollable Nature of the Output from Wind Turbines

To fully understand the problem that uncontrollable inputs of electrical power introduce, perhaps it is best to consider an extreme situation, just to see what effect that would have. Such an extreme is entirely unrealistic, but it will serve to clarify the general principle.

So take, for an imaginary example, a situation in which a widespread group of wind turbines do sometimes produce their full rated power. To be slightly more precise, let us say that the wind turbines are as widely spread as the E.ON Netz network in Germany which covers a distance of 800 km. The assumption of an output of full rated power means, of course, that it is thereby assumed that *at times* the wind blows sufficiently hard to allow every single turbine to produce at its rated power. That is fanciful, but let us now make an even more fanciful assumption that at other times over the course of the year the wind is so desultory that these wind turbines produce only 5% of their rated power. It is immediately obvious that these turbines would be useless for following variations in consumer demand. For that purpose, demand-following plant would have to be used. The only use that could be made of the input from the wind turbines would be to run them 'in harness' with controllable plant which would produce the remaining 95% of the rated power of the wind turbines. Working in harness, the wind turbines and the controllable plant together could produce a baseload equal to the rated power of the wind turbines. In such a clear-cut and extreme situation that is obvious to common sense. Although the actual situation is more complicated, a similar principle applies in reality (covered in greater detail in *The Meaning and Implications of Capacity Factors*, OPTJ 4/1, pp. 18–25).

As already suggested, a suitable benchmark for the capacity factor (also called load factor) of wind turbines is 30%. The 'peak infeed' from wind turbines is defined as the highest output they will reach as a proportion of their rated capacity. Statistics on this parameter are hard to come by except from the distributor E.ON Netz whose network, as mentioned, extends over 800 km. The documentation of their experience from operating wind turbines is superb.[9] From their experience over two years, it seems that peak infeed from their widely spread turbines is about 80% of the rated capacity of the wind turbines. Following the same *principle* as in the previous imaginary example, it can be deduced that in these circumstances wind could provide $30/80 = $ **38%** of the baseload block of electricity, with controllable plant filling in the remaining 62% (using different datums the same point is explained at length on page 20, paragraph 4, of OPTJ 4/1).

A recent modelling study for the UK,[10] based on taller wind turbines located at all the windiest spots spread over the entire UK, showed that during the month of January, in the twelve years studied, the average peak infeed was 98%, and in one year it was 100%. The study's estimate of capacity factor was 35.5%. Note that the all important ratio, in these more windy conditions than Germany, remains much the same, at $35.5/98 = 36\%$.

That is not to say that the wind can satisfy 38% of *total* electrical demand, because, as observed, wind and the plant operating in harness with it can only produce a *baseload*. If there is no nuclear plant operating which needs to be allowed to operate without restrictions to produce a baseload, then wind turbines and the plant operating in harness with them can be set the task of providing a baseload up to the level of low demand. Low demand is about 60% of mean demand. Thus wind output can satisfy 38% of 60% which is 23% of electrical demand, provided that there is no other plant (e.g. current-design, inflexible nuclear plant) that is already fulfilling part of the baseload supply. 23% of electrical demand is only about 10% of total energy demand,[11] but 10% would appear to be worth pursuing provided that it does not too much interfere with the rest of the electrical system. That is what needs to be considered next.

6.6 The Problems of Operating in Harness with Wind Turbines

The effect of introducing wind into an electrical system cannot be judged on the electrical input from wind alone. As we have seen, the task has to be shared: about 38% taken by wind and 62% by a controllable power source. When wind becomes a significant part of the whole, it degrades the efficiency of the rest of the system, not only because of the need to keep plant running to cope with sudden wind changes, but more importantly because of the need to be able to start and stop plant on a frequent basis. Plant designed to do that operates considerably less efficiently than plant optimized to run at constant load. No one knows just how much less efficiently plant actually operates when it has to run in harness with wind turbines, however the effect is not small. In the extreme case of an all-natural-gas system, it can be shown that the loss of efficiency of the plant operating 'in harness' outweighs the benefits of the wind input (OPTJ 5/2, pp. 8–17). In conclusion, while maximum integration of wind turbines may appear capable of saving 10% of fossil fuel use, the actual figure will be lower than this because of:

a) the additional energy needed to construct and maintain the turbines, and
b) the degraded load factor and efficiency of the plant when it operates in harness with the wind turbines.

Also to be borne in mind is that even if the full 10% could be saved, this would rapidly be eaten up by population growth in the USA; a point we will now turn to.
   Electrical production in the USA in 2005 was about 3.8 billion MWh. 23% of that is 0.87 billion MWh, or an annual mean power output 99,000 MW. Thus 99,000/800 = **124** wind turbine farms, each producing a mean 800 MW, would be needed to provide the electricity. They would cover a total area of $124 \times 667 \text{ km}^2 =$ **83,000 km²**. It is hard to imagine such a task being accomplished under a decade. Before the decade was out, the 10% of energy demand saved by introduction of

the wind turbines would be overtaken by the increase in energy demand due to population growth, as can easily be seen.

During the final three decades of the last century, the rate of population growth in the U.S. was 1.06% per year. Even at that growth rate (and it is now higher), by the end of the decade of frantic wind turbine installation, population would have grown by 11%, increasing total energy demand by 11%, and thus outstripping the 10% of energy saved by the newly installed wind turbines. The extent of public opposition can be judged by the fact that so far wind contributes only 0.4% to electrical production in the USA, and that has already caused vociferous complaint. It should be mentioned, too, that the 1.06% per year is an understatement, as it has recently been shown that by the time all the illegal aliens are accounted for, the present rate of population growth in the U.S. is probably in the range of 1.4–1.7% (Abernethy 2006).

6.7 Alternatives to Wind Power

What is often not appreciated is that there is a limit to the contribution from uncontrollable power sources in an electrical supply system. It has been shown that wind turbines can only contribute about 23% of total electricity. A double share could not be achieved by allowing another uncontrollable, say wave power, to also produce 23%. The wave and wind power generators would sometimes produce their maximum output at the same time and thus overwhelm the electrical system. It is therefore necessary to choose only the best form of uncontrollable available at a given time. It should be mentioned that photovoltaics may be an exception, at least in a country that makes heavy use of air conditioning. This is because although peak demand tends to be later than midday, and it is likely to become even later as better insulated houses are built, nevertheless demand at midday will be well above the minimum demand, so to some extent photovoltaics could, cost permitting, reduce fuel use without interfering with other uncontrollables (which are limited to operating below minimum demand). With all other uncontrollables the output correlates poorly with demand; that is true even if the time of output is predictable as it is with tidal flow energy. Thus without storage, it becomes necessary to choose, and go for the best type, provided of course there is sufficient potential output available from that type.

It is clear that wind power has many problems. These stem chiefly from the capacity factor being small in relation to peak infeed, and partly because it is hard to forecast the output from wind to within a few hours, which is desirable for the efficient operation of the plant that has to operate in harness with it. Installation of wind turbines is termed by some as an industrialization of the landscape and, while it is impossible to put a value on the loss of quality of life that would occur for many people thus afflicted, one should not lose sight of that aspect. A further adverse effect of wind turbines is a significant slaughter of birds and bats.[12] Together all these factors suggest that every endeavor should be made to research wave power.

Wave power would certainly be more predictable and less prone to sudden change, and it might offer a better ratio between its capacity factor and peak infeed, thus enabling it to take a larger share of the total demand for electricity than wind ever could. Whether it could be made economically viable is of course another matter.

6.8 The Problems of Storage

The foregoing has not presented a cheerful prospectus for uncontrollables. What everyone hopes is that the problem of uncontrollables will be overcome by finding a way of storing the energy. Storage would solve the problem of not only wind but all uncontrollables, so it deserves detailed consideration.

Hydro. The most useful way to store electricity is in the form of water in a reservoir — using 'pumped storage'. That can be excellent for small amounts of electricity, but calculation soon shows that the capacity available is small compared to the requirements of large populations, especially when it is borne in mind that to produce a steady supply of electricity from wind turbines, only 38% of the block of electricity (according to the above calculation) could be delivered directly, while the remaining 62% would need to be stored first.

Some insight into the problem is gained by looking at the power density of the average reservoir. Based on a random sample of 50 U.S. hydropower reservoirs, ranging in area from 482 ha to 763,000 ha, it has been calculated that the area of reservoir needed to produce 1 billion kWh/yr (a mean 114,155 kW) is 75,000 ha (Pimentel and Pimentel 1996, p. 206). Thus over the course of a year, the power density achieved by these reservoirs is 1.5 kW(e)/ha.

The low power density of water storage arises because to store the energy of 1 kWh, the amount of water which must be raised through 100 m is 3.67 tonnes ($3.67 \, m^3$). And allowing for an overall 75% efficiency in using electrical pumps to elevate the water and then using turbines to regenerate the electricity, 3.67/0.75 = **4.9** tonnes of water must be raised through 100 m in order to store 1 kWh(e). To store one week's output from a 1000 MW plant, running at 80% capacity, would require *660 million tonnes* of water to be raised through 100 m. To put it another way, the area of this substantially elevated reservoir would need to be 66 km^2, or 8 km by 8 km (5 miles by 5 miles), and it would need to tolerate the water level being raised by 10 m. Suitable reservoirs of this kind are hard to come by, quite apart from the extra problem of needing a lower reservoir to hold the water waiting to be pumped back up.

Hydrogen. It is frequently proposed that electrical energy could be stored as hydrogen. There are many problems with that, the first being efficiency of transformation. Hydrogen production by electrolysis is around 70% efficient. About the best efficiency to be expected from fuel cells, including the need to invert their direct current output to AC, is 60%. That makes an overall efficiency of $0.70 \times 0.60 = 42\%$. So to deliver 1 kWh of stored electricity 2.4 kWh would have to generated from the wind turbines, and that is without allowing for further losses in compression which is likely to be necessary for realistic storage of a gas which has an energy density approximately a quarter that of methane (natural gas).[13] For an extended treatment of the problems, see *Hydrogen and Intermittent Energy Sources*, OPTJ 4/1 (pp. 26–29).

Vanadium batteries. Batteries are a possibility, particularly those which store the electrical energy in the form of a liquid in tanks which are separate from the 'engine', for this would appear to offer unlimited expansion using many tanks. A vanadium battery of this kind has been developed, but Trainer (1995, p. 1015) points out various limits, one being that the US Bureau of Mines states that demonstrated world recoverable resources of vanadium total about 69 billion kg.[14] So shortage of vanadium might set an ultimate limit to producing vanadium batteries; but before considering that, let us look at problems concerning the amount of hardware that is needed.

Considerable work has gone into development of vanadium batteries since Trainer's paper. In the 13 January 2007 issue of *New Scientist* there was a three page report on the type of batteries which are being installed by an Australian firm named in the article as Pinnacle VRB. The title of the article, by science journalist Tim Thwaites, was *A Bank for the wind: at last we can store vast amounts of energy and use it when we need it.* While little trust should be placed in the titles of articles in *New Scientist* or other popular science magazines, that does suggest the need for a closer look at the potential of vanadium batteries. After describing how some of the problems of vanadium batteries had been overcome, the article had this to say:

> After more than a decade of development, Skyllas-Kazacos's technology was licensed to a Melbourne-based company called Pinnacle VRB, which installed the vanadium flow battery on King Island. With 70,000 l of vanadium sulphate solution stored in large metal tanks, the battery can deliver 400 kW for 2 h at a stretch.

Those figures indicate that 87 liters of vanadium sulfate are required to store 1 kWh. A source in the firm has confirmed to me that the figure is approximately correct, and that 70 liters per kWh are used at the planning stage. That is a very low power density. As 1 liter of gasoline contains about 9.3 kWh, it would take 650 liters of vanadium sulfate to store the energy contained in a liter of gasoline. Even in station-ary situations, such a low energy density seems likely to engender problems in terms of net energy, because the inputs required to provide and maintain the hardware may become so large as to use most of the output. To consider the overall problem we need to have an idea of how much storage is likely to be required.

Since wind is fairly low for some months, there needs to be storage to cover the low wind months. There are no figures available for the USA, but Windstats provide good month by month data for Denmark, Germany, Netherlands, and Sweden. Dur-ing the months of May thru September in the two years 1998/1999 and 1999/2000, the shortfall in terms of the missing kWh (that is missing on the supposition that delivery needs to be constant each month) through those months, expressed as a fraction of the total year's delivery, was as follows for the two years: Denmark, 14.0%, 9.2%, Germany 13.8%, 14.4%, Netherlands 13.6%, 15.8%, Sweden 13.6%, 15.8%. Considering that just two years of observation are unlikely to have covered the most extreme situation, we may need something more than the worst result of 15.8%, but there is no need for too much accuracy so let us settle for storing 16% of the total annual output to cover the low wind months.[15]

Storage efficiency also needs accounting for. By time the AC output of wind turbines has been changed to DC, and the DC output from the VRBs has been returned to AC, the overall efficiency is probably about 70%, but let us use 75%, resulting in a need to send for storage $16/0.75 = \mathbf{21\%}$ of the annual output of the wind turbines.

Before proceeding with the calculation, there is a possible objection that should be addressed. It may be thought that it is not *really* necessary to be able to store enough energy. Would it matter if for a couple of weeks every two years wind turbine storage was exhausted and thus made peak demands worse by failing to contribute when needed? The answer is that it would matter, because available fossil fuel capacity would have to be kept available just to satisfy those rare occasions when the problem of peak demand were exacerbated by shortfall of wind energy (because it could not maintain its prescribed baseload).

In terms of a plant that delivers a mean 800 MW, the amount to store, 21% of that, amounts to 1470×10^6 kWh. At 70 liters per kWh that would require 103 million cubic meters of electrolyte. Using large storage tanks, say 20 m in height and diameter (about 6300 m^3 capacity), 16,300 such tanks would be needed.

The surface area of one cylindrical tank would be 1885 m^2. The total area would be 30.7 million m^2. Assuming that steel with an average of 10 mm thickness is used, that is $307,000$ m^3 of steel, or about 2.46 Mt or 2640 million kg. The embodied energy in steel is about 21 kWh/kg (Pimentel and Pimentel 1996, p. 206), so the energy embodied in the steel containers alone would be at least 51×10^9 kWh.[16] The annual output of a 1000 MW plant running at 80% capacity would be 7×10^9 kWh, so the steel for delivery of 16% of output after storage alone would cost over seven years of output, without including other construction energy costs associated with storage.

In addition to storage requirements, there would be the 'engine' component. To produce the mean 800 MW from wind turbines, with a 30% capacity factor, $800/0.30 = \mathbf{2667}$ MW of rated capacity would be required. With an 80% peak infeed this would sometimes produce $2667 \times 0.80 = \mathbf{2130}$ MW. However 800 MW of this would be used directly (to maintain the base load of 800 MW, and only the remaining 1330 MW would be an 'overflow' and need to be sent to charge the battery. A 1.5 MW battery system currently being installed requires an 'engine' of about 45 tonnes (50 m^3). On that basis, to provide 1330 MW of battery power would require 40,000 tonnes of material for the 'engine' component. The high dollar cost of the 'engine' component indicates a likely high embodied energy cost.[17]

There are certainly advantages in vanadium batteries. For instance the electrolyte never 'wears out', having a virtually infinite life. But the above figures suggest that until the energy balance calculations have been done, it is idle to claim '*at last we can store vast amounts of energy and use it when we need it*'. The energy inputs need to cover installing and maintaining the wind turbines, transmission lines, plus tanks for electrolyte storage, plus the 'engine' component of the battery and inverters to produce AC current from the DC output. But it is just possible that the outcome on energy balance will look acceptable, so let us turn back to the question of availability of vanadium.

Earlier it was noted that wind turbines might contribute 23% of mean demand, which in relation to the USA could be expressed as an annual mean power output of 99,000 MW. We have also noted the need to store 21% of that output in order to produce a steady baseload through the less windy months. Thus a mean 20,800 MW $=$ 182 billion kWh would need to be stored. At 0.39 kg of vanadium per kWh,[18] that would require **71** billion kg of vanadium. Yet we noted above that the US Bureau of Mines states that demonstrated world recoverable resources of vanadium total about **69** billion kg. Cost would also be a likely barrier.[19]

Clearly even if the energy balance is better than it appears *prima facie*, although vanadium batteries might assist the USA in delivering from store 23% $\times$ 0.16 $=$ **3.2**% of its annual electrical consumption, they cannot provide a worldwide solution, and not much of a solution for the USA, for integration of this storage plant would merely enable the 23% of total electricity which is to be produced from wind to be stabilized at 30% of the rated capacity of the wind turbines (thus avoiding the need to use fossil fuel plant to work in harness). While there is no theoretical bar to installing more wind turbines and vanadium batteries to cover more of U.S. electrical supply than 23%, it is clear that the availability of vanadium means that there is little scope for that, even if the cost were to be bearable.

It should be noted that a storage requirement of 21% of the output of the wind turbines serves only to sustain output through any one year. There is another problem. The U.S. capacity factors in 2001 and 2003, were 20% and 21% respectively. Were the aim to be to provide a reliable output from wind (thus obviating the need to keep fossil fuel back-up for rare occasions), so as to be able to guarantee to produce in every year the 27% capacity factors of 2000, 2002 and 2004, it would be necessary to store $1-(20/27) =$ **26**% of the wind turbine's best annual output, i.e. that achieved with a 27% capacity factor. This would be needed in order to top up the 20% load factor of 2001 to 27%. Moreover to deliver that 26% would require 26/0.75 $=$ **35**% to be sent to storage. This 35% is not instead of the 21% calculated previously but in addition to it. Again it will doubtless be asked whether that is really necessary. Again the answer is that it is not, but to the extent that the storage is not available, a controllable output is needed which can be brought into action during the years in which the wind fails to come up to scratch. The difficulties in making use of an uncontrollable output are very great.

There are other possible batteries, such as nickel-cadmium, sodium-sulfur, and sodium-nickel-chloride, but sufficient data are not available to assess their potential.

The above look at vanadium batteries has been concerned with their effectiveness in solving the overall problem of wind uncontrollability. In that respect, the limitations have been made evident, but perhaps it should be mentioned that there are some limited uses for them provided the cost is tolerable. For instance, Japan has such gusty winds that it is a problem integrating the output from wind turbines. A vanadium battery can be used to damp the wilder excursions. Also it has been suggested that vanadium batteries could take all the output of wind and then sell the output at a much higher price for satisfying peak demands. The principle is sound, but there is insufficient data to determine whether this is is going to prove economically viable.

CAES. Another method of storing electrical energy is compressed air energy storage, CAES, in which air is compressed and stored underground. The compressed air is later used to increase the output of gas turbines by about 200% (by saving the two-thirds of the energy output that would normally go into compression). However the extent of the problem arising from low energy density exceeds even that of hydropower.

There are two operational CAES plants. The plant at Huntorf, located in North Germany, was commissioned in 1978 and has been in operation ever since. It is designed to hold pressures up to 100 bar although 70 bar (1015 psi) is set as the maximum permissible operational pressure. Information available for it[20] suggests that under normal storage, within the $310, 000\,m^3$ space, energy density is about $2\,kWh/m^3$. However there are several ambiguities in the precise meaning of the data, including uncertainty about whether the quoted 300 MW output for 2 h results partly from the natural gas used. Certainly the figure of $2\,kWh/m^3$ energy density appears high in comparison to the McIntosh CAES plant of the Alabama Electric Company, commissioned in 1991.

Moreover the McIntosh plant is said to include 'several improvements over Huntorf, including a waste heat recovery system that reduces the fuel usage by about 25%'. The maximum pressure for storage is reported as being 74 bar (1070 psi), and it is stated that the 5.32 million m^3 cavern can deliver power at 110 MW for 26 h. That indicates an energy density of storage of only $0.54\,kWh/m^3$.

At certain places in the world, the available storage space is vast. I have been assured by an experienced operator in the electricity industry that, in Alabama, 'we are aware that there is tight gas storage of at least 548 billion cubic feet capacity with constant 750 psi pressure from hydro aquifer support'. 548 billion cubic feet equals 15.5 billion m^3. At the aforesaid $0.53\,kWh/m^3$, this would make available from store 8.2 billion kWh. That is equal to the annual output of a 1000 MW power station, operating at 94% capacity. But storage capacity on this scale is not readily available, and even if one is prepared to overlook the need for the turbines to run on natural gas (no commercial solution has yet been demonstrated for running the generators efficiently on compressed air alone), albeit being made more efficient by the infeed of high pressure air, CAES does not appear to offer a worldwide solution to storing electrical energy because of storage space, irrespective of how high the efficiency of the method may be (it has been put as high as 80%).

It has been suggested that with the world emitting about 18 billion tonnes *excess* carbon dioxide each year by burning fossil fuels, there is a need to use most of the available storage space for storing carbon dioxide; but compressed air storage is formed in solution-mined caverns underground, basically very large 'empty' caverns. Carbon dioxide sequestration is best made into old oil deposits for enhanced oil recovery, or into saline aquifers, which can absorb significantly higher amounts of CO_2 than could be obtained from the equivalent amount of open space volume. However it should not be forgotten that the practicality of sequestration into saline aquifers remains to be established.

In summary, while fossil fuels are available, there must be doubts whether a significant amount of net energy could be produced by combining wind turbines with such limited storage capacity as could be made available to assist them. Without fossil fuels, the whole project of producing wind turbines, transmission lines, plus storage capacity and regenerators is likely to be impossible (see problems of 'liquid' fuels below).

6.9 The Problem of 'Liquid' Fuel in a Fossil-Fuel-Free Society

Doubt was previously cast on the possibility of constructing and maintaining large wind farms in the context of a post-fossil-fuel society. The main reason was because of the difficulty of providing fuel in a 'liquid' form. The hope will obviously arise that the relatively high power density of the uncontrollables, including wind turbines, could be used to produce hydrogen by electrolysis. We need to ask whether that idea might be viable.

The essence of producing 'liquid' hydrogen from electricity is to produce the hydrogen from water by electrolysis and then to liquefy it, so that its energy density is sufficient to make it useful for transport. Even as a liquid, it would take 3 liters of liquid hydrogen to move a vehicle over the same distance as 1 liter of gasoline would take a similar car (OPTJ 3/2, pp. 21–27). It would take 9.1 kWh of electricity to produce liquid hydrogen with the same motive energy as 1 liter of gasoline (or 34 kWh(e) per *gallon* of gasoline). The cost of that might seem bearable, except that the output of wind turbines is erratic. It seems unlikely that a production line could be run for producing liquid hydrogen using only the erratic input from wind turbines (which produces some, but often not much, electricity for 95% of the year). Yet the alternative of running the plant continuously would require about two thirds of the electrical energy to come from a controllable power source. Because the efficiency of transformation in producing electricity from fossil fuels is about 33%, if for simplicity we assume for a moment that all the energy needed to produce the equivalent of 1 liter of gasoline were to come from a controllable power source, then that energy needed would amount to 9.1 [kWh(e)]/0.33 = **27** kWh. That would be somewhat alleviated by 38% of the electricity coming directly from the wind turbines, but nevertheless such an inefficient process is unlikely to be attempted while fossil fuels are available; when fossil fuels become scarce, there would be insufficient energy available to contemplate the process. To put it another way, producing liquid hydrogen from renewable sources via a steady production process depends on getting a steady supply by supplementing uncontrollable inputs. Such supplementation could only be achieved if the problem of storage is solved. *The fact is that at present there is no solution in sight to producing the quantities of 'liquid' fuels from renewable sources which would be required to allow present populations to live in even a very frugal version of present lifestyles.*

6.10 Learning from Experience (Denmark)

In the above theoretical analysis, it was noted that the inefficiency introduced into the electrical system by running plant in harness with an uncontrollable power source has not been assessed. For that reason alone it is helpful to try to learn from the experience of a nation which has attempted to make maximum use of wind power, namely Denmark. Inevitably there will be other variables which distort the effect of introducing wind power into the system but some clues can be gained.

Denmark is the nation which should reveal the most about integrating wind power into its electrical system, because in 2004 the electricity produced from its wind turbines amounted to 18.5% of total electricity production. But Denmark can only use a third of this directly, partly because of the very problem of the uncontrollable nature of the output, and partly because the greatest part of the wind turbine electricity is produced in the west of Denmark, and the west Denmark grid is separate from the east Denmark grid.[21,22] This has not inhibited the development of wind power because Denmark has interconnectors to Germany, Norway and Sweden which could carry virtually the whole of west Denmark's wind output. The latter two countries have very substantial hydropower capacity, so they can switch off their hydropower and use Denmark's electricity from wind turbines instead. The Danes can then re-import the electricity as hydropower electricity at a time that suits them (albeit at considerable expense).

Thus although Denmark does not use all its wind turbine electricity directly, wind turbines should serve to reduce its carbon emissions unless the inefficiencies of integrating wind into the system outweigh the advantages of the wind input.

Factors which might distort that assessment are that Denmark has also been trying many other things to reduce its carbon emissions through: (a) greater use of biomass; (b) extensive use of combined heat and power to provide nearly a third of west Denmark's electrical capacity; (c) a high tax on cars together with the provision of excellent public transport, (d) a high standard of insulation for its buildings. If a substantial reduction in carbon emissions had occurred, the picture would be blurred, because any of those items might have been the reason for the reduction, but since there has not been a significant reduction, we can deduce that neither those efforts *nor the input from wind turbines* has had much effect.

To be more precise, carbon emissions per person in Denmark decreased, between 1990 and 2003, by 0.07% compared to an 8.4% decrease in the United Kingdom, which has only a 0.5% wind penetration. Admittedly the decrease in the UK was almost entirely been a result of our dash for gas — replacing coal-fired plant with powered gas generators. In 2003, Denmark's carbon dioxide emissions were 10.9 t/cap compared to the UK's 9.5 t/cap. These figures appear to prove two things. The first is that introducing into an electrical system about 20% of the electricity from wind turbines (the most that countries are likely to be able to introduce) *may* have some effect on reducing carbon emissions, but it is hard to detect. Secondly, it shows that when a nation tries all the things that are often proposed as politically palatable ways of reducing carbon emissions, the actual effect of reducing carbon emissions is also hard to detect. Perhaps it should be noted that it could always be claimed that the carbon emissions in Denmark would have risen considerably more without such efforts. It could also be argued that the savings in energy use have not yet shown up due to the amount of energy being put into constructing and installing wind turbines, but such points probably do not weigh heavily, and it seems a fair conclusion that tackling only what is fairly easy in political terms does not make a significant impact on excessive carbon emissions.

6.11 Making Realistic Assessments of the Cost of Wind Power

The main thrust of this analysis has been at the fundamental level of energy. A brief comment on the potential for misleading statements about wind costs may be useful. The wind industry has for some time been saying that the cost of electricity from wind turbines is about to come down so as to be equal to the cost of electricity derived from fossil fuel. However the cost they are referring to is the total amount of money that the wind turbine operators need to be paid, *for all the kWh that they produce*, in order to bring in a satisfactory profit to the wind turbine operators. In some countries, e.g. Denmark, most wind power is 'prioritized' so that distributors have to use it. In the UK there is effectively a penalty if it is not used.

But what would be satisfactory for the wind turbine operators if all their electricity were to be bought (by whatever forms of compulsion or incentives), is very far from the real cost of wind turbine electricity. Other costs beside those incurred by the wind turbine operator needs to be added: (1) the amortized cost to the distributor of installing, plus the cost of maintaining, the necessary additional transmission lines, and (2) the additional costs incurred when purchasing electricity from controllable sources when the controllable sources are forced to operate at lower capacity in order to make room for wind power when it is available.

The second of these is very significant. It is one thing to make a contract with the operator of a fossil fuel plant to produce a steady output, but quite another to have to make many short term contracts to top up the delivering of electricity only to the extent that wind is not able to deliver it.

6.12 Conclusion

Wind turbines have a potential benefit in that they have a power density that matches coal, at least according to one measure. Set against this is the uncontrollable nature of their output. Looking ahead to when fossil fuels become scarce involves consideration of the low power densities that are likely to be associated with 'liquid' energy sources. At present, it is hard to say whether building wind farms and running a grid will be possible without fossil fuels, especially because no viable renewable fuel in 'liquid' form is evident.

Concerning introducing wind turbines in order to reduce the present use of fossil fuel, while it is probable that wind turbines do save some fossil fuel, there is no evidence of this from Denmark, the country which has taken the experiment further than any other. The maximum penetration that is possible, due to the uncontrollable output of wind turbines, means that they could contribute at best 10% of U.S. energy demand. Even if per capita energy demand remains constant, that 10% would be cancelled out by U.S. population growth in 10 years. In summary, installing wind turbines will not keep up with the present U.S. population growth, let alone give a bulwark of energy security to the present population. However, the whole situation, for wind and other uncontrollables, will need reviewing if compressed air electrical storage, CAES, is shown — even in some countries and the USA is a promising one — to be a practical proposition.

Notes

1. 'Power density' is the flow of energy per unit area, normally given in terms of watts per square meter or kilowatts per hectare (kW/ha). $1\,W/m^2 = 10\,kW/ha$. With biomass, and renewable sources in general, the figure normally refers to the average value over a year. For instance the harvest may be gathered in a few weeks, but what is important is the annual energy capture, which may be expressed in *energy* terms as joules per hectare per year, or worked out as an average *power* density of kW/ha.
2. kW(e) indicates that the kW of energy referred to is in the form of electricity. Often it is so obvious that the reference to kW is electrical that the (e) is omitted. Pimentel and Pimentel (1996, p. 206), quoting Vaclav Smil, give the land requirement for 1 billion kWh of electricity per year from coal as **363** ha. 1 billion kWh(e)/yr $= 114{,}155\,kW(e)$. So in electrical terms the gross power density is $114{,}155/363 = 315\,kW(e)/ha$. The input/output ratio is shown as 1:8. For wind, the ratio shown is 1:5. Such input/output figures are open to much dispute, but they show that there is not such a huge difference in input ratios that comparison of the gross figures is meaningless.
3. Calculating the power density of coal involves taking into account not only the areas at the surface that are being disturbed during the extraction process, but also the areas that are used for transportation.
4. The figure given, 1.9 kW/ha, is calculated from the data on page 12 of OPTJ 3/1, namely an ethanol yield, net of liquid inputs, of 2776 liters/ha $= 2776 \times 21.25 \times 10^6 = 59.0\,GJ/ha/yr = 1.87\,kW/ha$.
5. On page 12 of OPTJ 3/1 it is calculated that 50 million ha of corn could produce sufficient ethanol to satisfy 11% of the oil used in U.S. transport. But since corn is grown on only about 29 Mha, this would yield $11 \times 29/50 = 6.4\%$ of transport fuel.
6. The capacity factors are available for the UK from http://www.dtistats.net/energystats/dukes7_4.xls, accessed 14 Mar. 07.
7. The load factors (capacity factors) can be calculated from Table 11, which gives the installed capacity at mid-year, available at http://www.eia.doe.gov/cneaf/solar.renewables/page/trends/table11.html, and outputs from Table 12 at http://www.eia.doe.gov/cneaf/solar.renewables/page/trends/table12.html, accessed 14 Mar. 07.
8. Dry wood has a slightly higher calorific value than most dry matter – about 20 GJ/t. Thus 10 t/ha/yr would produce 200 GJ/ha/yr $= 200/31.54 = 6.3\,kW/ha$, which at a probably optimistic 30% conversion efficiency would be 1.9 kW(e)/ha.
9. Both of the wind reports from E.ON Netz, *Wind Report 2004* and *Wind Report 2005*, are available as pdf downloads (with text copying permitted) at the E.ON Netz web site at www.eon-netz.com.
10. The title of the report is **25 GW of Distributed Wind on the UK Electricity System**. The full 21 page report is available in pdf format, and is only just over a megabyte in size. It can be printed out or saved to disk without restriction from: http://www.ref.org.uk/images/pdfs/ref.wind.smoothing.08.12.06.pdf

11. In the U.S., 70% of electricity is produced from fossil fuels. So if wind replaces 23% of all electricity, this 23% could be used to replace $0.23/0.70 = \mathbf{33\%}$ of the electricity that is produced by fossil fuels. About 34% of fossil fuels are used for the production of electricity, so the saving would be 33% of $34\% = 0.33 \times 0.34 = \mathbf{11.2\%}$ of fossil fuels. And since fossil fuels supply 86% of all energy used in the U.S., this 11.2% is $0.112\% \times 0.86 = \mathbf{10\%}$ of total energy used.

12. Dr. Smallwood and K. Thelander reported that 2,300 golden eagles, 10,000 other raptors, and 50,000 smaller birds were killed at the Altamont Pass windfarm over 20 years. Sea eagles have been estimated to be killed at the Smola windfarm in Norway at the rate of one per month. Eric Rosenbloom has reported a figure of 350,000 bats, as well as 11,200 birds of prey and 3 million small birds, as having been killed by wind turbines in Spain. A compilation of scientific reports disclosing mortality at wind farms is at: www.iberica2000.org/Es/Articulo.asp?Id=1875.

13. At Standard Temperature and Pressure ($0°C$ and 760 mm mercury), the energy density of natural gas is about $38.5\,MJ/m^3$ and that of hydrogen is $10.8\,MJ/m^3$.

14. The amount of vanadium that is recoverable from the many ores containing vanadium is hard to assess, and supply is another matter, because as Wikipedia tells us, 'Vanadium is usually recovered as a by-product or co-product, and so world resources of the element are not really indicative of available supply'. However the US Bureau of Mines figure of 69 Mt is generous. The Australian assessment of the 'Economic Demonstrated Resources' is only 10 Mt; the reference for this is: http://www.abs.gov.au/Ausstats/abs@.nsf/0/98211B66FB348412CA256DEA000539D8? opendocument

15. Even some people in the industry seem to find this logic hard to follow, so perhaps an analogy will help. The flooding of the river Nile provides one. If there are some years when crop yields are poor and others when crop yields are excellent, then to maintain food availability in the poor years, sufficient grain must be kept in store to balance the shortfall during the lean years. The wind situation is similar, both in terms of months (to tide over the lean summer months) and of years (to tide over the low wind years), unless, in both cases, fossil fuel is used to fill the gap. Both concepts are treated in the main text.

16. I am told that the vanadium sulphate electrolyte is acidic, and steel would need an impermeable lining; or possibly carbon fiber tanks would be used rather than steel. Embodied energy for the latter may be less than for steel, but no precise figures are available.

17. While the VRB company (www.vrbpower.com) is not promulgating costs, sources in the industry suggest a current cost for the power stacks themselves of about US$1500 per kW. The cost of providing the housing structure, tanks, plumbing, pumps, inverters, control system, grid interface is about the same. While some of this could be allocated to storage rather than to providing the 'engine', it is clear that at present the capital cost of the 'engine' exceeds that of a natural gas power station, but then one of the reasons that the company is reticent about costs is because it hopes to greatly reduce those costs as a result of increase in scale.

18. It was hard to get a definitive statement about the vanadium requirement, but sources within the industry told me that 10 kg of vanadium pentoxide (or possibly vanadium pentoxide containing 10 kg of vanadium) are added to 1000 liters of 25% concentration sulphuric acid to produce the vanadium sulfate electrolyte. 70 liters of electrolyte are needed to store 1 kWh. Making the more favorable interpretation that the 10 kg refers to vanadium pentoxide, 70 liters of electrolyte would use $0.7\,kg$ of V_2O_5, and since the atomic weight of vanadium is 51 and that of oxygen is 16, the vanadium content of the 70 liters would be $0.7 \times (102/(102 + 80)) = 0.39\,kg$.

19. Sources within the industry put the cost of the electrolyte at about US$230 per kWh, thus to store 182 billion kWh would cost, in electrolyte alone, US$42 trillion ($42 \times 10^{12}$). One thing that seems likely to mitigate against massive cost reduction in storage costs is that, according to Wikipedia, 'unless known otherwise, all vanadium compounds should be considered highly toxic. Generally, the higher the oxidation state of vanadium, the more toxic the compound is. The most dangerous compound is *vanadium pentoxide*'. However vanadium sulphate is being used rather than vanadium pentoxide.

20. http://www.doc.ic.ac.uk/~matti/ise2grp/energystorage_report/node7.html, (accessed on 18 May 2007), and for further details on the Huntdorf plant, see the 2001 presentation, in Florida, by

Fritz Crotogino, of the long operational experience at this location in Germany, at: http://www.uni-saarland.de/fak7/fze/AKE_Archiv/AKE2003H/AKE2003H_Vortraege/AKE2003H03c_Crotogino_ea_HuntorfCAES_CompressedAirEnergyStorage.pdf

21. Vestergaard, Frede, in *Weekend Avisen* Nr 44, 4, 04 November 2005.
22. Civil engineer Hugh Sharman, who has worked for many years in Denmark, has written a paper on this in Civil Engineering, *Why windpower works for Denmark*, see references.

References

Abernethy, D.V. (2006). Census Bureau Distortions Hide Immigration Crisis: Real Numbers Much Higher. *Population-Environment Balance*, October 2006. (Washington, DC). http://www.Balance.org

Hayden, H. C. (2004). *The Solar Fraud: Why Solar Energy Won't Run the World* (2nd edition). (Vales Lake Publishing LLC. P.O. Box 7595, Pueblo West, CO 81007-0595. 280pp)

OPTJ 3/1. (2003). *Optimum Population Trust Journal*, Vol. 3, No 1, April 2003. Optimum Population Trust. (Manchester, UK). Archived on the web at www.members.aol.com/optjournal2/optj31.doc

OPTJ 3/2. (2003). *Optimum Population Trust Journal*, Vol. 3, No 2, October 2003. *Optimum Population Trust*. (Manchester, UK). Archived on the web at www.members.aol.com/optjournal2/optj32.doc

OPTJ 4/1. (2004). *Optimum Population Trust Journal*, Vol. 4, No 1, April 2004. Optimum Population Trust. (Manchester, UK) Archived on the web at www.members.aol.com/optjournal2/optj41.doc

OPTJ 5/2. (2005). *Optimum Population Trust Journal*, Vol. 5, No 2, October 2005. Optimum Population Trust. (Manchester, UK). Archived on the web at www.members.aol.com/optjournal2/optj52.doc

Pimentel, D. (Ed.). (1993). *World Soil Erosion and Conservation*. (Cambridge, UK: Cambridge University Press)

Pimentel, D., Pimentel, M. (1996). *Food, Energy, and Society*. (Niwot Co.: University Press of Colorado). This is a revised edition; the first one was published by John Wiley and Sons in 1979.

Sharman, H. (2005). Why windpower works for Denmark. *Civil Engineering 158, May 2005, pp. 66–72*

Trainer, F. E. (1995). Can renewable energy sources sustain affluent society? *Energy Policy, Vol 23 No 12 pp. 1009–1026*

Chapter 7
Renewable Diesel

Robert Rapier

Abstract Concerns about the environmental impact of fossil fuels – as well as the possibility that fossil fuel production may soon fall short of demand – have spurred a search for renewable alternative fuels. Distillates, the class of fossil fuels which includes diesel and fuel oil, account for a significant fraction of worldwide fossil fuel demand. Renewable distillates may be produced via several different technologies and from a wide variety of raw materials. Renewable distillates may be categorized as biodiesel, which is a mono-alkyl ester and not a hydrocarbon, or 'green diesel', which is a renewable hydrocarbon diesel produced via either hydrotreating or biomass to liquids (BTL) technology. There are, however, important ecological and economic tradeoffs to consider. While the expansion of renewable diesel production may provide additional sources of income for farmers in tropical regions, it also provides economic incentive for clearing tropical forests and negatively impacting biodiversity. Also, many of the raw materials used to produce renewable diesel are edible, or compete with arable land used to grow food. This creates potential conflicts over the use of biomass for food or for fuel. In contrast to first-generation renewable diesel technologies which utilize primarily edible oils, BTL technology can utilize any type of biomass for diesel production. However, high capital costs have thus far hampered development of BTL technology.

Keywords Biodiesel · biofuels · Fischer-Tropsch · green diesel · renewable diesel

7.1 Introduction

Distillate fuel oils, a category of fuels which includes petroleum diesel and home heating oil, account for almost 30% of worldwide petroleum consumption (EIA 2004). As fossil fuel reserves continue to deplete, sustainable alternatives to petroleum-based products are needed. One potential energy source is renewable distillate fuel oils produced from biomass. Such biofuels have a long history, as

✉ R. Rapier
Accsys Technologies PLC, 5000 Quorum Drive, Suite 310, Dallas, TX 75254
e-mail: rrapier1@yahoo.com

D. Pimentel (ed.), *Biofuels, Solar and Wind as Renewable Energy Systems*,
© Springer Science+Business Media B.V. 2008

peanut oil and whale oil were used as lubricants and energy sources long before they were displaced by petroleum products.

Biomass-derived diesel substitutes can be produced via several different technologies and from a wide variety of starting materials. Renewable diesel may be produced from edible vegetable oils such as soybean oil, cottonseed oil, or rapeseed oil – non-edible oils such as jatropha oil or algal oils – animal fats, and even waste cooking grease.

This chapter will examine the differences between various renewable diesel technologies, the variety of raw materials that can be used to produce renewable diesel, as well as possible trade-offs involved in wide-scale adoption of these alternatives.

7.2 The Diesel Engine

The advantages of using distillates as a fuel source go beyond the fact that distillates and their substitutes are typically more energy dense than gasoline and gasoline substitutes. The diesel, or compression-ignition engine (CIE) is different from a gasoline engine, or spark-ignition engine (SIE) in several respects. Whereas the SIE is normally ignited by a spark plug, the CIE is ignited by compression. The CIE achieves a much higher compression ratio,[1] which allows for a more powerful combustion, thus enabling more useful work to be realized. The result is that the efficiency of the CIE is up to 40% greater than for an SIE. Therefore, on purely the basis of engine efficiency, the CIE and fuels that can run in a CIE are preferred.

A fuel must be resistant to ignition as it is being compressed if it is to be considered as an appropriate fuel for a CIE. Gasoline does not fall into this category, which is why it is not used in CIEs. But diesel fuels do fall into this category. Diesel substitutes produced from biomass are the subject of this chapter.

7.3 Ecological Limits

Before examining potential renewable distillates, consider the question: What is the potential of biofuels with respect to ending the world's petroleum dependence? If biofuels are to make a meaningful dent in present worldwide oil usage of around 85 million barrels per day, then a massive expansion from current production capacity would be required. For example, as of this writing U.S. production of ethanol – seven billion gallons per year – is less than the energy equivalent of 1% of U.S. oil consumption.[2] Yet this is purely on a gross basis, which presumes that there

---

[1] The compression ratio is a measure of the pressure of the fuel at the moment of ignition. A high compression ratio indicates that the fuel was combusted in a small volume, which increases thermal efficiency.

[2] See Calculation 1.

are no petroleum inputs into the production of ethanol. Because fossil fuels are used to grow and harvest corn, and then to operate the ethanol distillery, the net energy added to the U.S. energy supply is much smaller. Yet even this negligible contribution to energy supplies is arguably resulting in a number of undesirable consequences.

But even ignoring the potential negatives, can one presume that biofuels can make a significant contribution to present energy demands? Consider the following thought experiment. There are 148.94 million square kilometers of land area in the world, 13.31% of which are considered to be arable (CIA 2007). Permanent crops occupy 4.71% of the total land area, leaving 12.8 million square kilometers (1.28 billion hectares) of arable land potentially (for the purpose of the thought experiment) available for cultivation of biofuels.[3] There are many different feed stocks from which to make renewable diesel, but most of the world's biodiesel is made from rapeseed oil (Puppan 2002). Rapeseed is an oilseed crop that is widespread and produces relatively high oil production. Unlike ethanol, which has an energy content 1/3rd less than that of gasoline, rapeseed oil has an energy density closer to that of petroleum.

Consider how much petroleum might be displaced if all 1.28 billion hectares of arable land were planted in rapeseed, or an energy crop with an oil productivity similar to rapeseed. While the average worldwide yield is substantially lower, rapeseed growers in Germany have succeeded in pushing oil yields to 2.9 tons/ha (Puppan 2002). If the rest of the world could achieve these high levels, this would result in a hypothetical worldwide oil yield of 3.7 billion tons. The energy content of rapeseed oil is about 10% less than that of petroleum diesel, so the gross petroleum equivalent yield from this exercise is 3.3 billion tons per year.

Because it takes energy to produce the biomass and process into fuel, the net yield will be lower, and in some cases may even be negative (i.e., more energy put into the process than is contained in the final product). Lewis compared several studies that examined the energy inputs required to produce biodiesel from rapeseed (Lewis 1997). Depending on the assumptions made, the energy input estimates ranged from 0.382 to 0.870 joules of input per joule of biodiesel produced and distributed. Assuming the best case value (lowest energy inputs) of 0.382, the net petroleum equivalent yield of rapeseed oil is reduced to 2 billion tons per year.[4]

The world's present usage of petroleum, 85 million barrels per day, is equivalent to 4.25 billion metric tons per year. By making very optimistic assumptions on the amount of land devoted to biofuels, the oil yield per hectare, and the energy inputs to produce the biofuels, the net is still less than half of the world's current demand for petroleum.

---

[3] The present acreage devoted to biofuels is ignored in this analysis as it is minute compared to present petroleum demand. Theoretically, world petroleum demand should have already been reduced by the current acreage planted in energy crops, leaving the rest of the world's arable land as the appropriate metric for displacing current petroleum demand.

[4] See Calculation 2.

Of course this is merely a thought experiment. Positive and negative externalities (e.g., the potential impact on food prices on one hand; the income opportunities for 3rd world farmers on the other) have been ignored. There are many considerations that could influence the result in one direction or another. But the exercise highlights the difficulty the world would face in attempting to replace our petroleum usage with biofuels.

7.4 Straight Vegetable Oil

Unmodified vegetable-derived triglycerides, commonly known as vegetable oil, may be used to fuel a diesel engine. Rudolf Diesel demonstrated the use of peanut oil as fuel for one of his diesel engines at the Paris Exposition in 1900 (Altin et al. 2001). Modern diesel engines are also capable of running on straight (unmodified) vegetable oil (SVO) or waste grease, with some loss of power over petroleum diesel (West 2004). Numerous engine performance and emission tests have been conducted with SVO derived from many different sources, either as a standalone fuel or as a mixture with petroleum diesel (Fort and Blumberg 1982, Schlick et al. 1988, Hemmerlein et al. 1991, Goering et al. 1982).

The advantage of SVO as fuel is that a minimal amount of processing is required, which lowers the production costs of the fuel. The energy return for SVO, defined as energy output over the energy required to produce the fuel, will also be higher due to the avoidance of energy intensive downstream processing steps.

There are several disadvantages of using SVO as fuel. The first is that researchers have found that engine performance suffers, and that hydrocarbon and carbon monoxide emissions increase relative to petroleum diesel. Particulate emissions were also observed to be higher with SVO. However, the same studies found that nitrogen oxide (NOx) emissions were lower for SVO (Altin et al. 2001). On long-term tests, carbon deposits have been found in the combustion chamber, and sticky gum deposits have occurred in the fuel lines (Fort and Blumberg 1982). SVO also has a very high viscosity relative to most diesel fuels. This reduces its ability to flow, especially in cold weather. This characteristic may be compensated for by heating up the SVO, or by blending it with larger volumes of lower viscosity diesel fuels.

7.5 Biodiesel

7.5.1 Definition

Biodiesel is defined as the mono-alkyl ester product derived from lipid[5] feedstock like SVO or animal fats (Knothe 2001). The chemical structure is distinctly different

---

[5] Lipids are oils obtained from recently living biomass. Examples are soybean oil, rapeseed oil, palm oil, and animal fats. Petroleum is obtained from ancient biomass and will be specifically referred to as 'crude oil' or the corresponding product 'petroleum diesel.'

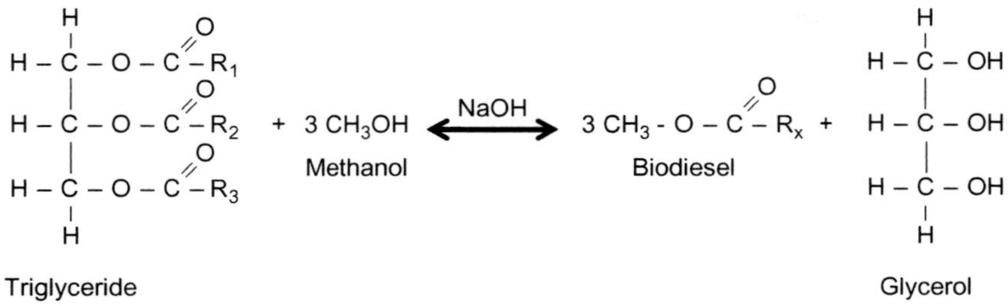

Fig. 7.1 The NaOH-Catalyzed reaction of a triglyceride to biodiesel and glycerol

from petroleum diesel, and biodiesel has somewhat different physical and chemical properties from petroleum diesel.

Biodiesel is normally produced by reacting triglycerides (long-chain fatty acids contained in the lipids) with an alcohol in a base-catalyzed reaction (Sheehan 1998) as shown in Fig. 7.1. Methanol, ethanol, or even longer chain alcohols may be used as the alcohol, although lower-cost and faster-reacting methanol[6] is typically preferred. The primary products of the reaction are the alkyl ester (e.g., methyl ester if methanol is used) and glycerol. The key advantage over SVO is that the viscosity is greatly reduced, albeit at the cost of additional processing and a glycerol byproduct.

7.5.2 Biodiesel Characteristics

Biodiesel is reportedly nontoxic and biodegradable (Sheehan et al. 1998). An EPA study published in 2002 showed that the impact of biodiesel on exhaust emissions was mostly favorable (EPA 2002). Compared to petroleum diesel, a pure blend of biodiesel was estimated to increase the emission of NOx by 10%, but reduce emissions of carbon monoxide and particulate matter by almost 50%. Hydrocarbon emissions from biodiesel were reduced by almost 70% relative to petroleum diesel. However, other researchers have reached different conclusions. While confirming the NOx reduction observed in the EPA studies, Altin et al. determined that both biodiesel and SVO increase CO emissions over petroleum diesel (Altin et al. 2001). They also determined that the energy content of biodiesel and SVO was about 10% lower than for petroleum diesel. This means that a larger volume of biodiesel consumption is required per distance traveled, increasing the total emissions over what a comparison of the exhaust concentrations would imply.

The natural cetane[7] number for biodiesel in the 2002 EPA study was found to be higher than for petroleum diesel (55 vs. 44). Altin et al. again reported a different

---

[6] Methanol is usually produced from natural gas, although some is commercially produced from light petroleum products or from coal. Methanol therefore represents a significant – but often overlooked – fossil fuel input into the biodiesel process.

[7] The cetane number is a measure of the ignition quality of diesel fuel based on ignition delay in a compression ignition engine. The ignition delay is the time between the start of the injection and the ignition. Higher cetane numbers mean shorter ignition delays and better ignition quality.

result, finding that in most cases the natural cetane numbers were lower for biodiesel than for petroleum diesel. These discrepancies in cetane results have been attributed to the differences in the quality of the oil feedstock, and to whether the biodiesel had been distilled (Van Gerpen 1996).

A major attraction of biodiesel is that it is easy to produce. An individual with a minimal amount of equipment or expertise can learn to produce biodiesel. With the exception of SVO, production of renewable diesel by hobbyists is limited to biodiesel because a much larger capital expenditure is required for other renewable diesel technologies.

Biodiesel does have characteristics that make it problematic in cold weather conditions. The cloud and pour points[8] of biodiesel can be 20°C or higher than for petroleum diesel (Kinast 2003). This is a severe disadvantage for the usage of biodiesel in cold climates, and limits the blending percentage with petroleum diesel in cold weather.

7.5.3 Energy Return

The energy return of biodiesel is disputed. Sheehan et al. reported in 1998 that the production of 1 megajoule (MJ) of soy-derived biodiesel required 0.3110 MJ of fossil fuel inputs, for a fossil energy ratio[9] of 3.2 (Sheehan et al. 1998). They further reported that during the production of biodiesel from soybeans, the soybean crushing and soybean conversion steps required the most energy, respectively using 34.25% and 34.55% of the total energy. The remainder of the energy inputs came mostly from agriculture, at approximately 25% of the total energy input.

However, Pimentel and Patzek reported that the energy return for soy biodiesel is slightly less than 1.0, meaning that soy biodiesel is nonrenewable according to their study (Pimentel and Patzek 2005). But there were some differences in the methodology employed. The two studies allocated energy differently between the soy oil product and the soy meal product. This resulted in very different energy input calculations. Sheehan assigned to the soy oil a fossil energy input from the agricultural step equivalent to 0.0656 MJ per MJ of biodiesel produced. Pimentel and Patzek assigned an energy input from the agricultural step equivalent to 0.70 MJ per MJ of biodiesel produced – over 10 times the amount from the Sheehan study.[10] However, the Pimentel and Patzek study found that the energy return from

---

[8] The cloud point is the temperature at which the fuel becomes cloudy due to the precipitation of wax. The pour point is the lowest temperature at which the fuel will still freely flow.

[9] The fossil energy ratio is defined as the energy value of the product divided by the fossil energy inputs. This ratio is also commonly called the energy return, EROI, or EROEI. A fuel having a fossil energy ratio less than 1.0 is considered to be nonrenewable.

[10] Pimentel and Patzek calculated that the production of 1,000 kg of biodiesel with an energy value of 9 million kcal required an agricultural input of 7.8 million kcal. However, an additional credit of 2.2 million kcal from the soy meal was assigned to the biodiesel, for an agricultural input of 7.8 million/11.2 million, or 0.70.

the soybean cultivation step was renewable (considering only energy inputs), with 2.56 MJ of soybeans being returned for an energy input of 1.0 MJ.

7.5.4 Glycerol Byproduct

One of the challenges in the production of biodiesel is disposal of the glycerol[11] byproduct. As shown in Fig. 7.1, production of 3 molecules of biodiesel results in the production of 1 molecule of glycerol. This has created such a glut of glycerol, that some glycerol producers have been forced to shut down plants (Boyd 2007). Excess glycerol is currently disposed of by incineration, prompting the UK's Department for Trade and Industry to fund projects exploring the conversion of glycerol into value-added chemicals (Glycerol Challenge 2007).

7.6 Green Diesel

7.6.1 Definition

Another form of renewable diesel is 'green diesel.' Green diesel is chemically the same as petroleum diesel, but it is made from recently living biomass. Unlike biodiesel, which is an ester and has different chemical properties from petroleum diesel, green diesel is composed of long-chain hydrocarbons, and can be mixed with petroleum diesel in any proportion for use as transportation fuel. Green diesel technology is frequently referred to as second-generation renewable diesel technology.

There are two methods of making green diesel. One is to hydroprocess vegetable oil or animal fats. Hydroprocessing may occur in the same facilities used to process petroleum. The second method of making green diesel involves partially combusting a biomass source to produce carbon monoxide and hydrogen – syngas – and then utilizing the Fischer-Tropsch reaction to produce complex hydrocarbons. This process is commonly called the biomass-to-liquids, or BTL process.

7.6.1.1 Hydroprocessing

Hydroprocessing is the process of reacting a feed stock with hydrogen under elevated temperature and pressure in order to change the chemical properties of the feed stock. The technology has long been used in the petroleum industry to 'crack', or convert very large organic molecules into smaller organic molecules, ranging from those suitable for liquid petroleum gas (LPG) applications through those suitable for use as distillate fuels.

In recent years, hydroprocessing technology has been used to convert lipid feed stocks into distillate fuels. The resulting products are a distillate fuel with properties

---

[11] Glycerol is also commonly referred to as glycerin or glycerine.

very similar to petroleum diesel, and propane (Hodge 2006). The primary advantages over first-generation biodiesel technology are: (1). The cold weather properties are superior; (2). The propane byproduct is preferable over glycerol byproduct; (3). The heating content is greater; (4). The cetane number is greater; and (5). Capital costs and operating costs are lower (Arena et al. 2006).

A number of companies have announced renewable diesel projects based on hydroprocessing technology. In May 2007 Neste Oil Corporation in Finland inaugurated a plant that will produce 170,000 t/a of renewable diesel fuel from a mix of vegetable oil and animal fat (Neste 2007). Italy's Eni has announced plans for a facility in Livorno, Italy that will hydrotreat vegetable oil for supplying European markets. Brazil's Petrobras is currently producing renewable diesel via their patented hydrocracking technology (NREL 2006). And in April 2007 ConocoPhillips, after testing their hydrocracking technology to make renewable diesel from rapeseed oil in Whitegate, Ireland, announced a partnership with Tyson Foods to convert waste animal fat into diesel (ConocoPhillips 2007).

Like biodiesel production, which normally utilizes fossil fuel-derived methanol, hydroprocessing requires fossil fuel-derived hydrogen.[12] No definitive life cycle analyses have been performed for diesel produced via hydroprocessing. Therefore, the energy return and overall environmental impact have yet to be quantified.

7.6.1.2 Biomass-to-Liquids

When an organic material is burned (e.g., natural gas, coal, biomass), it can be completely oxidized (gasified) to carbon dioxide and water, or it can be partially oxidized to carbon monoxide and hydrogen. The latter partial oxidation (POX), or gasification reaction, is accomplished by restricting the amount of oxygen during the combustion. The resulting mixture of carbon monoxide and hydrogen is called synthesis gas (syngas) and can be used as the starting material for a wide variety of organic compounds, including transportation fuels.

Syngas may be used to produce long-chain hydrocarbons via the Fischer-Tropsch (FT) reaction. The FT reaction, invented by German chemists Franz Fischer and Hans Tropsch in the 1920s, was used by Germany during World War II to produce synthetic fuels for their war effort. The FT reaction has received a great deal of interest lately because of the potential for converting natural gas, coal, or biomass into liquid transportation fuels. These processes are respectively referred to as gas-to-liquids (GTL), coal-to-liquids (CTL), and biomass-to-liquids (BTL), and the resulting fuels are 'synthetic fuels' or 'XTL fuels'. Of the XTL processes, BTL produces the only renewable fuel, as it utilizes recently anthropogenic (atmospheric) carbon.

Renewable diesel produced via BTL technology has one substantial advantage over biodiesel and hydrocracking technologies: Any source of biomass may be converted via BTL. Biodiesel and hydrocracking processes are limited to lipids.

---

[12] Hydrogen is produced almost exclusively from natural gas.

This restricts their application to a feedstock that is very small in the context of the world's available biomass. BTL is the only renewable diesel technology with the potential for converting a wide range of waste biomass.

Like GTL and CTL, development of BTL is presently hampered by high capital costs. According to the Energy Information Administration's Annual Energy Outlook 2006, capital costs per daily barrel of production are $15,000–20,000 for a petroleum refinery, $20,000–$30,000 for an ethanol plant, $30,000 for GTL, $60,000 for CTL, and $120,000–$140,000 for BTL (EIA 2006).

While a great deal of research, development, and commercial experience has gone into FT technology in recent years,[13] biomass gasification technology is a relatively young field, which may partially explain the high capital costs. Nevertheless, the technology is progressing. Germany's Choren is building a plant in Freiberg, Germany to produce 15,000 tons/yr of their SunDiesel® product starting in 2008 (Ledford 2006).

7.7 Feed Stocks

While renewable diesel may be produced from a wide variety of feed stocks, this section will focus on those that are either in widespread use, or are frequently discussed as feed stocks with very high potential for producing biofuels. Feed stocks for the BTL process will not be discussed, as any biomass source can be used for this process. The following feed stocks are specific to the lipid conversion technologies discussed in this chapter.

7.7.1 Soybeans

The United States is the world's largest producer of soybean oil (Sheehan 1998), producing approximately 10 million metric tons in 2006 (USDA June 2007). Worldwide production of soybean oil is 35 million metric tons (Rupilius and Ahmad 2007). Soybean oil is typically produced by cracking the soybeans and extracting the oil with a solvent such as hexane. Finished soybean oil is widely used as cooking oil, in various processed foods, and for the production of biodiesel.

Relative to other oil crops, productivity of oil from soybeans is low. Soybean yields in 2006 in the U.S. amounted to 2871 kg/ha (USDA January 2007). At a typical soybean oil yield of 18%, this would have produced an average oil yield of 0.52 tons/ha. The average yield in Brazil, another major producer of soybean oil,

---

[13] Companies actively involved in developing Fischer-Tropsch technology include Shell, operating a GTL facility in Bintulu, Malaysia since 1993; Sasol, with CTL and GTL experience in South Africa; and ConocoPhillips and Syntroleum, both with GTL demonstration plants in Oklahoma.

has been reported at 0.40 tons/ha.[14] These oil yields are far below reported yields of other oil crops such as rapeseed, palm oil, or coconut.

While the oil yields are low, soybean oil does have an advantage over many bio-oil crops. Soybeans are capable of atmospheric nitrogen fixation, so they can be grown with little or no nitrogen fertilizer inputs (Pimentel and Patzek 2005). Because nitrogen-based fertilizers are energy intensive to produce, the energy balance for the agricultural step should be much more favorable than for crops requiring nitrogen fertilizer. This also means that soybeans will contribute less water pollution in the way of fertilizer runoff into waterways.

The expansion of soybean cultivation is not without controversy. In Brazil, critics have charged that soybean cultivation is a major driver of deforestation in Amazonia, resulting in multiple negative impacts on biodiversity (Fearnside 2001). Some researchers also argue that the potential for drought is increasing due to the increased reflectivity of the cleared land (Costa et al. 2007). In the United States, use of genetically-modified soybeans is common. This has resulted in criticism from various countries and environmental groups opposed to the practice.

7.7.2 Rapeseed

Whereas biodiesel in the U.S. is produced primarily from soybean oil, rapeseed oil, also sometimes called canola,[15] is the feedstock of choice for European biodiesel (Thuijl et al. 2003). Like soybean oil, rapeseed oil is edible. Rapeseed oil yields are about 1 ton/ha – double those of soybean oil. Rapeseed is produced mainly in China, Canada, the Indian subcontinent, and Northern Europe (Downey 1990). Rapeseed oil was the first vegetable oil used for transesterification to biodiesel, and remains the most widely-utilized vegetable oil in the production of biodiesel (Puppan 2002). The most common biodiesel produced from rapeseed oil is called Rapeseed-Methyl-Ester, or RME. RME has a slightly higher energy density than most biodiesels, and produces lower NOx and CO emissions than biodiesel produced from soybean oil (EPA 2002).

The primary disadvantage of rapeseed relative to some oil crops is that it has high nitrogen fertilizer requirements. Some life cycle analyses have shown a relatively small environmental benefit from RME relative to petroleum diesel, and a higher energy input than soybean oil, primarily because of the fertilizer requirements (De Nocker et al. 1998, Zemanek and Reinhardt 1999).

---

[14] Unlike the U.S., Brazil does not utilize genetically modified organisms (GMOs) in the production of soybeans (Mattsson et al. 2000).

[15] Rapeseed oil with less than 2% erucic acid content is trademarked as canola by the Canadian Canola Association.

7.7.3 *Palm Oil*

Palm oil is an edible oil extracted from the fruit of the African Oil Palm. In 2006, worldwide palm oil production surpassed soybean oil to become the most widely produced vegetable oil in the world. In 2006, palm oil production was 37 million tons and accounted for just over 25% of all biological oil production (Rupilius and Ahmad 2007). This is a substantial oil yield relative to other lipid crops. For perspective, total distillate usage (diesel and fuel oil) in the United States was approximately 208.5 million tons[16] in 2006 (EIA 2007).

By far the most productive lipid crop, palm oil is the preferred oil crop in tropical regions. The yields of up to five tons of palm oil per hectare can be ten times the per hectare yield of soybean oil (Mattson et al. 2000). Palm oil is a major source of revenue in countries like Malaysia, where earnings from palm oil exports exceed earnings from petroleum products (Kalam and Masjuki 2002).

Palm oil presents an excellent case illustrating both the promise and the peril of biofuels. Driven by demand from the U.S. and the European Union (EU) due to mandated biofuel requirements, palm oil has provided a valuable cash crop for farmers in tropical regions like Malaysia, Indonesia, and Thailand. The high productivity of palm oil has led to a dramatic expansion in most tropical countries around the equator (Rupilius and Ahmad 2007). This has the potential for alleviating poverty in these regions.

But in certain locations, expansion of palm oil cultivation has resulted in serious environmental damage as rain forest has been cleared to make room for new palm oil plantations. Deforestation in some countries has been severe, which negatively impacts sustainability criteria, because these tropical forests absorb carbon dioxide and help mitigate global warming (Schmidt 2007). Destruction of peat land in Indonesia for palm oil plantations has reportedly caused the country to become the world's third highest emitter of greenhouse gases (Silvius et al. 2006).

As a result of the potential environmental dangers posed by the expansion of biofuels, the Dutch government is developing sustainability criteria for biomass that will be incorporated into relevant policy decisions (Cramer 2006). The intention is employ life cycle analyses (LCAs) to measure the overall impact from using various biomass sources. For instance, if the developed world mandates large amounts of biofuels, but this come at the price of massive deforestation of tropical rainforests, the LCA will attempt to incorporate those negatives into the overall assessment. The categories that the Dutch group intends to evaluate are (1). Greenhouse gas balance; (2). Competition with food, local energy supply, medicines and building materials; (3). Biodiversity; (4). Economic prosperity; (5). Social well-being; and (6). Environment.

In addition to the Dutch initiative, some other countries are evaluating the sustainability of biofuels (Rollefson et al. 2004). Yet such efforts may be ultimately futile unless a binding, worldwide agreement can be implemented. While

---

[16] See Calculation 3.

slash-and-burn growers may find that the Dutch will not buy their products, they may easily find other buyers for their product in the global marketplace.

7.7.4 Jatropha

Jatropha curcas is a non-edible shrub native to tropical America, but now found throughout tropical and subtropical regions of Africa and Asia (Augustus et al. 2002). Jatropha is well-suited for growing in arid conditions, has low moisture requirements (Sirisomboon et al. 2007), and may be used to reclaim marginal, desert, or degraded land (Wood 2005). The oil content of the seeds ranges from 30% to 50%, and the unmodified oil has been shown to perform adequately as a 50/50 blend with petroleum diesel (Pramanik 2003). However, as is the case with other bio-oils, the viscosity of the unmodified oil is much higher than for petroleum diesel. The heating value and cetane number for jatropha oil are also lower than for petroleum diesel. This means it is preferable to process the raw oil into biodiesel or green diesel.

Jatropha appears to have several advantages as a renewable diesel feedstock. Because it is both non-edible and can be grown on marginal lands, it is potentially a sustainable biofuel that will not compete with food crops. This is not the case with biofuels derived from soybeans, rapeseed, or palm.

Jatropha seed yields can vary over a very large range – from 0.5 tons per hectare under arid conditions to 12 tons per hectare under optimum conditions (Francis et al. 2005). However, if marginal land is to be used, then yields in the lower range will probably by typical. Makkar et al. determined that the kernel represents 61.3% of the seed weight, and that the lipid concentration represented 53.0% of the kernel weight (Makkar et al. 1997). Therefore, one might conservatively estimate that the average oil yield per hectare of jatropha on marginal, non-irrigated land may be 0.5 tons times 61.3% times 53.0%, or 0.162 tons of oil per hectare. Jatropha oil contains about 90% of the energy density of petroleum diesel, so the energy equivalent yield is reduced by an additional 10% to 0.146 tons per hectare. While this is substantially less than the oil production of soybeans, rapeseed, or palm oil, the potential for production on marginal land may give jatropha a distinct advantage over the higher-producing oil crops.

A commercial venture was announced in June 2007 between BP and D1 Oils to develop jatropha biodiesel (BP 2007). The companies announced that they will invest $160 million with the stated intent of becoming the largest jatropha biodiesel producer in the world. The venture intends to produce volumes of up to 2 million tons of biodiesel per year.

Jatropha has one significant downside. Jatropha seeds and leaves are toxic to humans and livestock. This led the Australian government to ban the plant in 2006. It was declared an invasive species, and 'too risky for Western Australian agriculture and the environment here' (DAFWA 2006).

While jatropha has intriguing potential, a number of research challenges remain. Because of the toxicity issues, the potential for detoxification should be studied (Heller 1996). Furthermore, a systematic study of the factors influencing oil yields should be undertaken, because higher yields are probably needed before jatropha can contribute significantly to world distillate supplies.[17] Finally, it may be worthwhile to study the potential for jatropha varieties that thrive in more temperate climates, as jatropha is presently limited to tropical climates.

7.7.5 Algae

Certain species of algae are capable of producing lipids, which can be pressed out and then converted to renewable diesel. Algae-based renewable diesel is an appealing prospect, as this could potentially open up biofuel production to areas unsuitable for farming. Furthermore, the estimates of the oil production potential from algae have been as high as 160 tons/ha – 30 times that of palm oil.

From 1978 to 1996, the U.S. Department of Energy funded a study by the National Renewable Energy Laboratory (NREL) on the feasibility of producing renewable fuels from algae (Sheehan et al. 1998). The study examined a number of strains of algae for potential lipid production, as well as those that could grow under conditions of extreme temperature, pH, and salinity. Researchers examined the molecular biology and genetics of algae, and identified important metabolic pathways for the production of lipids.

While the production of biofuels from a raw material like algae has obvious appeal, the NREL close-out report concluded that there are many technical challenges to be overcome. A major challenge was encountered in the attempts to increase oil yields. Oil concentrations could be increased by stressing the algae and causing it to shift from a growth mode into a lipid production mode, but this resulted in lower overall oil yields because algal growth slowed. The researchers also discovered that contamination was often a problem upon moving from the laboratory into open pond systems.

The close-out report suggested that algae could potentially supply the equivalent of a large fraction of U.S. demand, but costs must come down, and technical challenges must be solved. On the subject of costs, the report noted 'Even with aggressive assumptions about biological productivity, we project costs for biodiesel which are two times higher than current petroleum diesel fuel costs.' Furthermore, because of lack of data on continuous lipid production from algae, the energy return on the process is unknown.

7.7.6 Animal Fats

Total production of animal fats in the U.S. was approximately 4.5 million tons in 2006 (U.S. Census Bureau 2007). This is just under half the mass of soybean oil

---

[17] See Calculation 4.

produced each year in the U.S. It is also the energy equivalent of around 1.5 days of U.S. petroleum demand.

Animal fats contain fewer double bonds than do most vegetable oils (Peterson 1986). This has an influence on the properties of the renewable diesel product. For example, biodiesel properties have been shown to vary depending on whether the biodiesel was produced from animal or plant lipids. In 2002, the EPA compared plant-based biodiesels derived from soybean, rapeseed, and canola oils, to animal-based biodiesels derived from tallow, grease, and lard (EPA 2002). The study found that animal-based biodiesels had a slightly lower energy density, but higher cetane numbers than plant-based biodiesels. The study also found that animal-based biodiesel produced substantially fewer NOx and particulate matter emissions.

Animal fats also respond differently to the hydrotreating process than do vegetable oils. Animal fats are more amenable to the hydrotreating process because double bonds are saturated in the hydrotreating process. Feed stocks like animal fats, with fewer double bonds than vegetable oils, will require less hydrogen to convert the oil to green diesel.

While animal fats are a byproduct of meat processing, there are significant environmental costs associated with industrial animal agriculture. The production of meat is a highly inefficient process. The production of beef requires relatively large inputs of water, grain, forage, and fossil fuels. Production of 1 kilocalorie of beef protein requires a fossil fuel input of 40 kilocalories (Pimentel and Pimentel 2003). This suggests that animal-based biofuels may be legitimately considered recycled fossil fuels.

7.7.7 Waste Biomass

North America and Western Europe combine to produce an estimated 500 million tons of municipal waste (UNEP 2004a). The main contributors to municipal waste throughout the developed world are organic materials such as food waste, grass clippings, waste cooking oils, and paper (UNEP 2004b). Waste biomass that is presently destined for landfills has great appeal as a feedstock for biofuels production, as it is an available biomass source that does not compete with food. Of this waste biomass, the BTL process can potentially convert any of it to liquid fuels. The lipid conversion technologies are however limited to the waste cooking oil fraction.

Waste cooking oils can either be converted to biodiesel via transesterification, or to green diesel via hydrotreating. For the hobbyist, the waste oil feedstock can often be acquired from restaurants at little or no cost. The conversion to biodiesel may be carried out without expending a great deal of capital, meaning that biodiesel can be produced from waste cooking oil at a very low cost.

Businesses are beginning to realize the opportunity in recycling waste cooking oil into transportation fuel. In July 2007, McDonald's UK restaurants announced their intention to run their delivery fleet on the waste cooking oil generated by 900 of their restaurants (McDonald's 2007). A program under way in New York City is on pace to recycle 450 tons of used cooking oil to biodiesel in 2007 (RWA 2007).

7.8 Conclusions

Biofuels can contribute to our energy portfolio, and many different options are available. But some options pose high environmental risks, some compete with food, and some are far more sustainable than others. Each option should be carefully weighed against the overall impact on the environment and society as a whole. Sustainable energy solutions must be pursued, and rigorous life cycle analyses should be undertaken for all of our energy choices.

We live in a world with limited resources, and a declining endowment of fossil fuel reserves. Much of the world aspires to a higher standard of living. The energy policies that we pursue should attempt to balance the needs of all citizens, worldwide. These policies must carefully consider the ecology of the planet, so future generations are not denied opportunities because of the choices we make today.

7.9 Conversion Factors and Calculations

While SI units are used in this chapter, Imperial/UK units are commonly used in the UK and in the U.S. Therefore, a number of common conversion factors are listed here which should enable to reader to convert between SI and Imperial units. A number of measures in the text have been converted from Imperial units, but the conversion factors listed should enable the reader to reproduce all figures.

Also, because different assumptions of physical properties (density, energy content, etc.) will lead to slightly different results, certain assumptions and calculations used in this chapter are provided in this section.

7.9.1 Conversion Factors

> 1 barrel of oil = 42 gallons = 158.984 liters = 0.137 metric tons
> 1 barrel of oil = 5.8 million BTUs of energy = 6.1 gigajoules (GJ)
> 1.0 hectare = 10,000 m^2 = 2.47 acres
>
> The specific gravity of crude oil is 0.88.
> The specific gravity of diesel oils is 0.84.
> The specific gravity of biodiesel is 0.88.
> The specific gravity of ethanol is 0.79.
>
> Lower Heating Values

The lower heating value (LHV) is the heat released by combusting a substance without recovering the heat lost from vaporized water. The LHV is a more accurate representation of actual heat utilized during combustion, as vaporized water is rarely recovered.

The LHV for crude oil is 138,100 Btu/gallon = 38.5 MJ/liter = 45.3 GJ/t
The LHV for distillates is 130,500 Btu/gallon = 36.4 MJ/liter = 42.8 GJ/t
The LHV for biodiesel is 117,000 Btu/gallon = 32.6 MJ/liter = 37.8 GJ/t
The LHV for ethanol is 75,700 Btu/gallon = 21.1 MJ/liter = 26.7 GJ/t

7.9.2 Calculations

In this section, several of the calculations referenced in the text are reproduced.

Calculation 1: Current oil usage in the United States is approximately 21 million barrels per day. The energy value of 1 barrel of oil is approximately 5.8 million BTUs. Ethanol production of 7 billion barrels per year is equivalent to 457,000 barrels per day. This is 2.2% of daily oil usage on a volumetric basis, but ethanol has approximately 76,000 BTUs/bbl, versus 138,000 BTUs/bbl for oil. Therefore, 7 billion gallons of ethanol per year is worth 1.2% of U.S. daily oil consumption. Backing out the energy inputs required to produce the ethanol (fossil fuels for tractors, trucking, fertilizer, pesticides, etc.) drops the net offset to well less than 1% of U.S. daily oil consumption.

Calculation 2: If the energy input is 0.382, then the net energy is (1-0.382) * 3.3 billion tons of rapeseed oil. The balance of 1.26 billion tons would be equivalent to the energy required to produce, process, and distribute the final product.

Calculation 3: In the United States, distillate demand in 2006 was 4.17 million barrels per day. One barrel of oil is equivalent to 0.137 metric tons; therefore distillate demand in 2006 was 0.57 tons per day, or 208.5 tons per year.

Calculation 4: Consider the potential for displacing 10% of the world's distillate demand of 1.1 billion tons per year – 110 million tons - with jatropha oil. Jatropha, with about 10% less energy than petroleum distillates, will require 122 million tons (110 million/0.9) on a gross replacement basis (i.e., not considering energy inputs). On marginal, un-irrigated land the yields will likely be at the bottom of the range of observed yields. At a yield of 0.146 tons per hectare, this would require 836 million hectares, which is greater than the 700 million hectares currently occupied by permanent crops. An estimated 2 billion acres is considered to be degraded and perhaps suitable for jatropha cultivation (Oldeman et al. 1991). There are also an estimated 1.66 billion hectares in Africa that are deemed suitable for jatropha production (Parsons 2005). This could provide a valuable cash crop for African farmers. But, until an estimate is made of the energy inputs required to process and distribute the jatropha-derived fuel on a widespread basis – especially on marginal land – the real potential for adding to the world's net distillate supply is unknown.

Acknowledgments I would like to acknowledge the patience and support displayed by my family as I completed this chapter. I also want to acknowledge the helpful suggestions submitted by readers of The Oil Drum and my blog, R-Squared, regarding specific renewable diesel topics they wanted to see covered. A special thanks goes to David Henson and Ilya Martinalbo from Choren Industries, who provided very useful input on BTL technology. Finally, I would like to thank Professor Pimentel for the opportunity to make this contribution.

References

Altin, R., Cetinkaya, S., & Yucesu, H.S. (2001). The potential of using vegetable oil fuels as fuel for Diesel engines. *Energy Convers. Manage., 42*, 529–538.

Arena, B., Holmgren, J., Marinangeli, R. Marker, T., McCall, M., Petri, J., Czernik, S., Elliot, D., & Shonnard, D. (2006, September). *Opportunities for Biorenewables in Petroleum Refineries* (Paper presented at the Rio Oil & Gas Expo and Conference, Instituto Braserileiro de Petroleo e Gas).

Augustus, G.S., Jayabalan, M., & Seiler, G.J. (2002). Evaluation and bioinduction of energy components of *Jatropha curcas*. *Biomass and Bioenergy., 23*, 161–164.

Boyd, J. (2007). *Biotech breakthrough could end biodiesel's glycerin glut*. Retrieved July 31, 2007 from the Rice University web site Rice: http://www.media.rice.edu/media/NewsBot. asp?MODE=VIEW&ID=9731

BP. (2007). *BP and D1 Oils Form Joint Venture to Develop Jatropha Biodiesel Feedstock*. Retrieved July 14, 2007 from the BP corporate web site: http://www.bp.com/genericarticle.do? categoryId=2012968&contentId=7034453

CIA, Central Intelligence Agency. (2007). *The World Factbook*. Retrieved July 28, 2007 from https://www.cia.gov/library/publications/the-world-factbook/geos/xx.html

ConocoPhillips. (2007). *ConocoPhillips and Tyson Foods Announce Strategic Alliance To Produce Next Generation Renewable Diesel Fuel*. Retrieved July 21, 2007 from the ConocoPhillips corporate web site: http://www.conocophillips.com/newsroom/news_releases/ 2007+News+Releases/041607.htm

Costa, M.H., Yanagi, S.N. M., Souza, P. J. O. P., Ribeiro, A., & Rocha, E. J. P. (2007). Climate change in Amazonia caused by soybean cropland expansion, as compared to caused by pastureland expansion, *Geophys. Res. Lett., 34*, L07706, doi:10.1029/2007GL029271.

Cramer, J. (Project Chair). (2006). *Criteria for Sustainable Biomass Production*. Retrieved July 15, 2007 from http://www.forum-ue.de/bioenergy/txtpdf/project_group_netherlands_criteria_for_ biomass_production_102006bonn.pdf

DAFWA, Department of Agriculture and Food, Western Australia. (2006). *Jatropha Banned in WA*. Retrieved August 3, 2007 from http://www.agric.wa.gov.au/content/sust/bio-fuel/191006jatrophe.pdf

De Nocker, L., Spirinckx, C., & Torfs, R. (1998). *Comparison of LCA and external cost analysis for biodiesel and diesel*, from Proceedings of the 2nd International Conference on Life Cycle Assessment in Agriculture, Agro-Industry and Forestry, Brussels, December 3–4 1998.

Downey, R.K. (1990). Canola: A quality brassica oilseed. (In J. Janick & J.E. Simon (Eds.), *Advances in new crops*. (pp. 211–217). Portland, OR: Timber Press).

EIA, Energy Information Administration. (2004). *World Output of Refined Petroleum Products*. Retrieved July 12, 2007 from the EIA web site: http://www.eia.doe.gov/pub/international/ iea2004/table32.xls

EIA, Energy Information Administration. (2006). *Annual Energy Outlook 2006*. DOE/EIA-0383, 57–58.

EIA, Energy Information Administration. (2007). *Weekly Supply Estimates – Product Supplied Distillate Fuel Oil*. Retrieved July 6, 2007 from the EIA web site: http://tonto.eia.doe.gov/ dnav/pet/pet_sum_sndw_a_epd0_vpp_mbblpd_w.htm

EPA, U.S. Environmental Protection Agency. (2002). *A Comprehensive Analysis of Biodiesel Impacts on Exhaust Emissions*. EPA420-P-02-001.

Fearnside, P. (2001). Soybean cultivation as a threat to the environment in Brazil. *Environ. Conserv., 28*, 23–38.

Fort, E.F. & Blumberg, P.N. (1982). *Performance and durability of a turbocharged diesel fueled with cottonseed oil blends*. (Paper presented at the International Conference on Plant and Vegetable Oils as Fuel, ASAE).

Francis, G., Edinger, R., & Becker, K. (2005). *A concept for simultaneous wasteland reclamation, fuel production, and socio-economic development in degraded areas in India: Need, potential and perspectives of Jatropha plantations Natural Resources Forum, 29*(1), 12–24.

Goering, C.E., Schwab, A. Dougherty, M. Pryde, M. & Heakin, A. (1982). *Fuel properties of eleven vegetable oils*. (Paper presented at the American Society of Agricultural Engineers meeting, Chicago, IL, USA).

Heller, J. (1996). *Physic nut Jatropha Curcas L. Promoting the conservation and use of underutilized and neglected crops*. Institute of Plant Genetics and Crop Plant Research (Gartersleben) and International Plant Genetic Resources Institute: Rome Vol. 1.

Hemmerlein, M., Korte, V., & Richter, H.S. (1991). *Performance, exhaust emission and durability of modern diesel engines running on rapeseed oil*. SAE Paper 910848.

Hodge, C. (2006). *Chemistry and Emissions of NExBTL*. (Presented at the University of California, Davis). Retrieved July 21, 2007 from http://bioenergy.ucdavis.edu/materials/ NExBTL%20Enviro%20Benefits%20of%20paraffins.pdf

Kalam, M.A. & Masjuki, H.H. (2002). Biodiesel from palmoil – an analysis of its properties and potential, *Biomass and Bioenergy, 23*(6), 471–479.

Kinast, J. NREL, National Renewable Energy Laboratory. (2003). *Production of Biodiesels from Multiple Feed-stocks and Properties of Biodiesels and Biodiesel/Diesel Blends*. NREL/SR-510-31460.

Knothe, G. (2001). Historical perspectives on vegetable oil-based diesel fuels. *INFORM, 12*(11), 1103–1107.

Ledford, H. (2006). Liquid fuel synthesis: Making it up as you go along. *Nature, 444*, 677–678.

Lewis, C. (1997). *Fuel and Energy Production Emission Factors*. MEET Project: Methodologies for Estimating Air Pollutant Emissions from Transport. Retrieved July 28, 2007 from http://www.inrets.fr/nojs/infos/cost319/MEETdeliverable20.pdf

Makkar, H., Becker, K., Sporer, F., & Wink, M. (1997). Studies on the nutritive potential and toxic constituents of different provenances of Jatropha curcas. *J. Agric. Food Chem., 45*, 3152–3157.

Mattson, B., Cederberg, C., & Blix, L. (2000). Agricultural land use in life cycle assessment (LCA): Case studies of three vegetable oil crops. *J. Cleaner Prod., 8*, 283–292.

McDonald's Corporation. (2007). *McDonald's Delivery Fleet to Convert to 100% Biodiesel*. Retrieved July 25, 2007 from http://www.mcdonalds.co.uk/?f=y

Neste Oil Corporation. (2007). *Neste Oil inaugurates new diesel line and biodiesel plant at Porvoo*. Retrieved July 21, 2007 from http://www.nesteoil.com/default.asp? path=1,41,540,1259,1260,7439,8400

NREL, National Renewable Energy Laboratory. (2006). *Biodiesel and Other Renewable Diesel Fuels*, NREL/FS-510-40419

Oldeman, L.R.,. Hakkeling, R.T.A., & Sombroek, W.G. (1991). *World Map of the Status of Human-induced Soil Degradation: An explanatory note*. Wageningen, International Soil Reference and Information Centre, Nairobi, United Nations Environment Programme.

Parsons, K. (2005). Jatropha in Africa: Fighting the Desert & Creating Wealth. *EcoWorld*. Retrieved July 14, 2007, from http://www.ecoworld.com/home/articles2.cfm?tid=367

Peterson, C.L. (1986). Vegetable Oil as a Diesel Fuel: Status and Research Priorities, *ASAE Trans., 29*(5), 1413–1422.

Pimentel, D. & Patzek, T.W. (2005). Ethanol Production Using Corn, Switchgrass and Wood; Biodiesel Production Using Soybean and Sunflower. *Nat. Resour. Res., 14*(1), 65–76.

Pimentel, D., & Pimentel, M. (2003). Sustainability of meat-based and plant-based diets and the environment. *Am. J. Clin. Nutr., 78*(3), 660S–663S.

Pramanik, K. (2003). Properties and use of Jatropha curcas oil and diesel fuel blends in compression ignition engine. *Renewable Energy Journal, 28*(2), 239–248.

Puppan, D. (2002). *Environmental evaluation of biofuels, Period Polytech Ser Soc Man Sci., 10*, 95–116.

Rollefson, J., Fu, G., & Chan, A. (2004). *Assessment of the Environmental Performance and Sustainability of Biodiesel in Canada*. Retrieved July 15, 2007 from http://www.studio255. com/crfa/pdf/res/2004_11_NRCBiodieselProjectReportNov04.pdf

Rupilius, W. & Ahmad, S. (2007). Palm oil and palm kernel oil as raw materials for basic oleochemicals and biodiesel. *Eur. J. Lipid Sci. Technol., 109*, 433–439.

RWA Resource Recovery (2007). *June Year to Date Collection Statistics.* Retrieved July 29, 2007 from http://www.rwaresourcerecovery.org/

Schlick, M.L., Hanna, M.A., & Schinstock, J.L. (1988). Soybean and sunflower oil performance in diesel engine. *ASAE, 31*(5).

Schmidt, C. (2007). Biodiesel: Cultivating Alternative Fuels. *Environ. Health Perspect., 115*(2), A86–A91.

Sheehan, J. NREL, National Renewable Energy Laboratory. (1998). *An Overview of Biodiesel and Petroleum Diesel Life Cycles,* NREL/TP-580-24772.

Sheehan, J., Dunahay, T., Benemann, J., & Roessler, P., DOE, U.S. Department of Energy. (1998). *A Look Back at the U.S. Department of Energy's Aquatic Species Program—Biodiesel from Algae.* NREL/TP-580-24190.

Silvius, M., Kaat, A., van de Bund, H., & Hooijer, A. (2006). *Peatland Degradation Fuels Climate Change.* (Wageningen: Welands International).

Sirisomboon, P., Kitchaiya, P., Pholpho, T., & Mahuttanyavanitch, W. (2007). Physical and mechanical properties of Jatropha curcas L. fruits, nuts and kernels, *Biosyst. Eng., 97*(2), 201–207.

The Glycerol Challenge. (2007). Retrieved July 8, 2007 from http://www.theglycerolchallenge.org/index.html

Thuijl, E. van, Roos, C.J., & Beurskens, L.W.M. (2003). *An Overview of Biofuel Technologies, Markets, and Policies in Europe.* Energy Research Centre of the Netherlands, ECN report ECN-C–03-008.

UNEP, United Nations Environment Programme. (2004a). *Projected trends in regional municipal waste generation. In UNEP/GRID-Arendal Maps and Graphics Library.* Retrieved July 18, 2007 from http://maps.grida.no/go/graphic/projected_trends_in_regional_municipal_waste_generation.

UNEP, United Nations Environment Programme. (2004b). *Municipal solid waste composition: for 7 OECD countries and 7 Asian cities. In UNEP/GRID-Arendal Maps and Graphics Library.* Retrieved July 18, 2007 from http://maps.grida.no/go/graphic/municipal_solid_waste_composition_for_7_oecd_countries_and_7_asian_cities

United States Census Bureau. (2007). *Fats and Oils: Production, Consumption, and Stocks: 2006.* Retrieved July 23, 2007 from http://www.census.gov/industry/1/m311k0613.pdf

USDA, United States Department of Agriculture. (2007, January). *2006 Soybean Crop a Record-Breaker.* Retrieved July 15, 2007 from http://www.nass.usda.gov/Newsroom/2007/01_12_2007.asp

USDA, United States Department of Agriculture. (2007, June). *Soybean oil: U.S. supply and disappearance.* Table 3. Retrieved July 15, 2007 from \http://www.ers.usda.gov/Briefing/ SoybeansOilCrops/Data/table3.xls

Van Gerpen, J. (1996). *Cetane Number Testing of Biodiesel.* (Paper presented at the Third Liquid Fuel Conference: Liquid Fuel and Industrial Products from Renewable Resources, St. Joseph, MI).

West, T. (2004). The Vegetable-Oil Alternative. [Electronic version]. *Car and Driver.* Retrieved June 28, 2007 from http://www.caranddriver.com/article.asp?section_id=4&article_id=7818

Wood, P. (2005). Out of Africa: Could Jatropha vegetable oil be Europe's biodiesel feedstock? *Refocus, 6*(4), 40–44.

Zemanek, G. & Reinhardt, G. (1999). Notes on life-cycle assessments of vegetable oils *Lipid-Fett, 101*(9), 321–327.

Chapter 8
Complex Systems Thinking and Renewable Energy Systems

Mario Giampietro and Kozo Mayumi

Abstract This chapter is divided into three parts. Part 1 deals with theoretical issues reflecting systemic problems in energy analysis: (i) when dealing with complex dissipative systems no quantitative assessment of output/input energy ratio can be substantive; (ii) metabolic systems define "on their own", what should be considered as useful work, converters, energy carriers, and primary energy sources; (iii) the well known trade-off between "power" (the pace of the throughput) and "efficiency" (the value of the output/input ratio). This makes it impossible to use just one number (an output/input ratio) for the analysis of complex metabolic systems. Part 2 introduces basic concepts related to Bioeconomics: (i) the rationale associated with the concept of EROI; (ii) the conceptual definition of a minimum threshold of energy throughput, determined by a combination of biophysical and socio-economic constraints. These two points entail that the energy sector of developed countries must be able to generate a huge net supply of energy carriers per hour of work and per ha of colonized land. Part 3 uses an integrated system of accounting (MuSIASEM approach) to check the viability of agro-biofuels. The "heart transplant" metaphor is proposed to check the feasibility and desirability of alternative energy sources using benchmark values: (i) what is expected according to societal characteristics; and (ii) what is supplied according to the energy system used to supply energy carriers. Finally, a section of conclusions tries to explain the widespread hoax of agro-biofuels in developed countries.

✉ M. Giampietro
ICREA Research Professor, Institute of Environmental Science and Technology (ICTA), Autonomous University of Barcelona, Building Q – ETSE - (ICTA), Campus of Bellaterra 08193 Cerdanyola del Vallès (Barcelona), Spain
e-mail: mario.giampietro@uab.cat

K. Mayumi
Faculty of IAS, The University of Tokushima, Minami-Josanjima 1–1, Tokushima City 770-8502, Japan
e-mail: mayumi@ias.tokushima-u.ac.jp

D. Pimentel (ed.), *Biofuels, Solar and Wind as Renewable Energy Systems*,
© Springer Science+Business Media B.V. 2008

Keywords Biofuels · bioeconomics · complex systems · alternative energy sources · renewable energy systems · multi-scale integrated analysis of societal and ecosystem metabolism (MuSIASEM) · EROI (Energy Return On Investment).

8.1 Theoretical Issues: The Problems Faced by Energy Analysis

8.1.1 The General Epistemological Predicament Associated to Energy Analysis

Attempts to apply energy analysis to human systems have a long history starting with Podolinsky (1883), Jevons (1865), Ostwald (1907), Lotka (1922, 1956), White (1943, 1959), and Cottrell (1955). In the 1970's energy analysis got a major boost by the first oil crisis. In that period the adoption of the basic rationale of Net Energy Analysis (Gilliland, 1978) resulted into a quantitative approach based on the calculation of output/input energy ratios. Energy analysis was widely applied to farming systems, national economies, and more in general to describe the interaction of humans with their environment (e.g., Odum, 1971, 1983; Rappaport, 1971; Georgescu-Roegen, 1971, 1975; Leach, 1976; Slesser, 1978; Pimentel and Pimentel, 1979; Morowitz, 1979; Costanza, 1980; Herendeen, 1981; Smil, 1983; 1988). The term energy analysis, rather than energy accounting, was officially coined at the IFIAS workshop of 1974 (IFIAS, 1974). The second "energy crisis" in the 80s led to a second wave of studies in the field (Costanza and Herendeen, 1984; Watt, 1989; Adams, 1988; Smil, 1991, 2003; Hall et al., 1986; Gever et al., 1991; Debeir et al., 1991; Mayumi, 1991, 2001; Odum, 1996; Pimentel and Pimentel, 1996; Herendeen, 1998; Slesser and King, 2003). However, quite remarkably, the interest in theoretical discussions of how to perform energy analysis quickly faded outside the original circle. This was due to both the return to an adequate world supply of oil in the 90s and the lack of consensus in the community of energy analysts about how to do and how to use energy analysis. *"Indeed, the scientists of this field were forced to admit that using energy as a numeraire to describe and analyze changes in the characteristics of ecological and socioeconomic systems proved to be more complicated than one had anticipated (Ulgiati et al., 1998)"* (Giampietro and Ulgiati, 2005).

In this first section we explore the nature of the epistemological impasse experienced in the field of energy analysis, in order to put better in perspective, in the second and third section, our discussion on how to do an effective analysis of alternative energy sources to oil. The main point we want to make here is that such an impasse is generated by the fact that the term "energy" refers to a very generic concept. This generic concept can only be associated, in semantic terms, with "the ability to induce a change in a given state of affairs". However, as soon as one tries to formalize this semantic conceptualization of energy into a specific quantitative assessment or a mathematical formula, there are many possible ways of doing such a contextualization and quantification. The choice of just one of these ways depends on the interests of the analysts, that is, on why one wants to do such a quantitative analysis in the first place. Before performing any quantitative analysis about energy transformations, one has to go through a series of decisions, which translate into the choice of a particular narrative about the change to be quantified. The decisions are:

(1) *what is the relevant change, which must be associated with a relevant task/event for the analysis, on which we want to focus.* This implies individuating a relevant performance of the energy system, which we want to describe using numbers. In this pre-analytical step the relevant task/event has to be expressed, first, in semantic terms (to check the relevance of the analysis) and not in energy term – e.g. making profit by moving goods to the market;

(2) *what is the useful work required to obtain the relevant change/task/event.* This implies coupling the relevant task defined in semantic term to a definition of the final performance of the energy system, this time expressed in energy term – e.g. the mechanical work associated with the movement of the goods to be transported to the market;

(3) *what is the converter generating the useful work.* This implies individuating a structural-functional complex, which is able to convert a given energy input into the required useful work – e.g. either a given truck or a given mule used for the transportation of goods;

(4) *what is the energy carrier required as energy input by the selected converter.* After choosing a converter associated with the supply of the useful work, the definition of an energy input is obliged – e.g. if we select a truck as converter, then gasoline has to be considered as the relative energy input. Had we selected a mule for the transport, then hay would have to be considered as the relative energy input;

(5) *what is the energy source required to generate an adequate supply of the specified energy carrier.* At this point, the definition of an energy source is related to the availability of a biophysical gradient capable of supplying the required energy input to the converter at a specified pace. Also in this case, choice #3 of a converter, defining the identity of the required energy carrier, entails, in last analysis, what should be considered as the relative energy source for this energy system. In our example of the truck, this would be a stock of oil (with an adequate ability to extract, refine and supply gasoline to the truck). Otherwise, it would be a healthy grassland with enough productivity of hay, if the transport is done by mule.

For this reason, energy analysts dealing with sustainability issues must pay due attention to the "transparency" of their work. That is, the unavoidable process of formalization of a given problem structuring in a set of numerical relations should be an occasion to promote a dialogue with stakeholders and policy makers on the choices made. The alternative is to hide the value calls used in such a formalization "under the carpet" and to sell the final output of the analysis as if it were a substantive "scientific output" indicating the truth. Transparency means that scientists should provide the users of the model a plain critical appraisal of: (i) basic assumptions, (the chosen narrative used for issue definition); (ii) the choices made in the implementation of a particular methodology and accounting scheme; (iii) the quality of the data used in the analysis; (iv) the choices of the criteria selected to define performance; (v) the particular selection of a set of indicators and their feasibility domains; (vi) the choice of a scale making it possible to quantify the selected concepts (boundary conditions, initiating conditions, and duration of the analysis); (vii) the choice of the goals determining the relevance of the analysis, (viii) the influence of the socio-political context in which the analysis is performed (political influence of lobbies, sponsors of the study, etc.).

A general discussion of systemic epistemological problems associated with energy analysis when used to tackle sustainability issues is available in: Giampietro and Mayumi, 2004; Mayumi and Giampietro, 2004, 2006; Giampietro, 2003, 2006; Giampietro et al., 2006a,b. We want to focus here only on three points relevant for the discussion of how to do an analysis of the viability and desirability of alternative energy sources to fossil energy.

8.1.2 Point 1 – when Dealing with Complex Dissipative Systems no Quantitative Assessment of Output/Input Energy Ratio can be Substantive

Even though different types of energy forms are all quantifiable using the same unit (Joules) – or using other units which are reducible to the Joule by using a fixed conversion factor (e.g. Kcalories, BTU, KWh) – different energy forms may refer to logically independent narratives about change and in this case they cannot be reduced to each other in a substantive way. This implies that the validity and usefulness of a given conversion ratio, determining an energy-equivalent of an energy form into another energy form, has always to be checked in semantic terms. Such a validity depends on the initial semantics about what should be considered as a relevant change and the relative set of choices used in the quantification. Put in another way, as soon as one tries to convert a quantitative assessment of a given energy form, expressed in Joules, into another quantitative assessment of a different energy form, still expressed in Joules, one has to choose: (A) a semantic criterion, for determining the equivalence over the two energy forms; and (B) a protocol of formalization, to reduce the two to the same numeraire. This double choice introduces a degree of arbitrariness linked to a series of well known problems in energy analysis:

(i) *the impossibility of summing, in a substantive way, apples & oranges –* referring to the fact that any aggregation procedure has to deal with different energy forms having different qualities. Looking for just one of the possible ways to consider them as "belonging to the same category" entails an unavoidable loss of relevance, since different forms can be perceived as belonging to logically different categories.

when deciding to sum apples and oranges the chosen protocol will define the final number and its usefulness. That is, if we decide to calculate their aggregate weight, we will get a number which is not relevant for nutritionists, but for the truck driver transporting them. On the other hand, if we sum them by using their aggregate nutritional content, we will get a number which is not relevant for either an economist studying the economic viability of their production and the truck driver. The more we aggregate items which can be described using different attributes (i.e., energy inputs which are relevant for different tasks, such as power security, food security, environmental security) using a single category of equivalence, the more we increase the chance that the final number generated by this aggregation will result irrelevant for policy discussions" Giampietro, 2006.

"Without an agreed upon useful accounting framework it is impossible to discuss of quantification of energy in the first place (Cottrel, 1955; Fraser and Kay, 2002; Kay, 2000; Odum, 1971; 1996; Schneider and Kay, 1995). That is, the same barrel of oil can have: (a) a given energy equivalent when burned as fuel in a tractor, but no energy equivalent when given to drink to a mule (when using a narrative in which energy is associated with its chemical characteristics which must result compatible with the characteristics of the converter); (b) a different figure of energy equivalent when used as a weight to hold a tend against the wind (when using a narrative in which energy is associated with the combined effect of its mass and the force of gravity, within a given representation of contrasting forces); (c) a different energy equivalent when thrown against a locked door to break it (when using a narrative in which energy is associated with the combined effect of its mass and the speed at which it is thrown, within a given representation of contrasting forces). I hope that this simple example can convince the reader that quantitative assessments of "the energy equivalent of a barrel of oil" cannot be calculated a priori, in substantive terms, without specifying first "how" that barrel will be used as a form of energy (end use) Giampietro, 2006.

(ii) *the unavoidable arbitrariness entailed by the joint production dilemma –* referring to the fact that when dealing with multiple inputs and outputs – which are required and generated by any metabolic system – arbitrary choices, made by the analyst, will determine the relative importance (value/relevance) of end products and by-products. In fact, when describing a complex metabolic system as a network of energy and material flows linking different elements belonging to different hierarchical levels it is possible to generate multiple non-equivalent representations. These different representations will reflect a different issue definition (narrative about the relevant change to be investigates) and therefore will be logically independent. Incoherent representations of the same system cannot be reduced in substantive way to each other. *"The energy equivalent per year of the same camel can be calculated in different ways using different quality factors when considering the camel as: (i) a supplier of meat or milk; (ii) a supplier of power; (iii) a supplier of wool; (iv) a supplier of blood to drink in emergencies in the desert; and (v) a carrier of valuable genetic information"*. Giampietro, 2006.

(iii) *the unavoidable arbitrariness entailed by* the *truncation problem –* referring to the fact, that several non-equivalent descriptions are unavoidable when describing a system operating simultaneously on multiple scales. This fact, by default, entails the co-existence of different boundaries for the same "entity" when perceived and represented at these different scales. In turn, this implies that what should be considered as embodied in the inputs and/or in the outputs depends on the choice of the scale (determining the choice of just one of the possible definition of boundaries) at which the assessment is performed. The final result is that more than one assessment can be obtained when calculating the energy embodied in a given transformation. A famous example of this fact is represented by the elusive assessment of the energetic equivalent of one hour of human labor.

The literature on the energetics of human labor (reviewed by Fluck, 1981, 1992) shows many different methods to calculate the energy equivalent of one hour of labor. For example, the flow of energy embodied in one hour of labor can refer to: (i) the metabolic energy of the worker during the actual work only, including (e.g. Revelle, 1976) or excluding (e.g. Norman, 1978) the resting metabolic rate; (ii) the metabolic energy of the worker including also non-working hours (e.g. Batty et al., 1975; Dekkers et al., 1978; Hudson, 1975); (iii) the metabolic energy of the worker and his dependents (e.g. Williams et al., 1975); or (iv) all embodied energy, including commercial energy, spent in the food system to provide an adequate food supply to the population (Giampietro and Pimentel, 1990); (v) all the energy consumed in societal activities (Fluck, 1981); (vi) finally, H.T. Odum's EMergy analysis

Table 8.1 Examples of non-equivalent assessments of the energy equivalent of 1 hour of human labor found in scientific analyses

Level	Time horizon of assessment	NARRATIVE	Range of values	Energy Type	Factors affecting the assessment
n+3 Gaia	Millennia	EMergy analysis of biogeochemical cycles and ecosystems	10–100 GJ	Embodied solar energy	* Ecosystem type * Choice in the representation * transformities * choice of ecological services included
n+1 society	1 year	Societal metabolism	200–400 MJ	Oil equivalent	* energy sources mix * energy carriers mix * end uses mix * efficiency in energy uses * level of technology * level of capitalization
n household	1 year	Time allocation Technological conversions	2.0–4.0 MJ 20–40 MJ	Food energy Oil equivalent	* quality of the diet * convenience of food products * food system characteristics
n-2 body/organs	1 hour	physiology	0.2–2.0 MJ	ATP/food energy	* body mass size * activity patterns * population structure (age and gender)

(1996) includes in the accounting of the energy embodied in human labor also a share of the solar energy spent by the biosphere in providing environmental services needed for human survival. Thus, the quantification of an energy input required for a given process (or an energy output) in reality depends on the choice made when defining the boundary of that process.

> *Rigorous scientific assessments of the 'energy equivalent of 1 hour of labor' found in literature vary from 0.2 MJ to more than 20 GJ, a range of the order of 100,000 times! This problem did not pass unnoticed, and since the 1970s, there was more than one conference on the topic in the series "Advances in Energy Analysis." Also there was a task force of experts selected from all over the world dedicated to study these discrepancies. Rosen's theory of models, can help explain this mystery. Insight comes from the concepts surrounding possible bifurcations in the meaning assigned to a given label "energy equivalent of 1 hour of labor". As illustrated by Table 8.1, these different assessments of the energy equivalent of 1 hour of human labor are based on non-equivalent narratives.* Giampietro et al. 2006b.

8.1.3 Metabolic Systems Define on Their Own, what Should be Considered as Useful Work, Converters, Energy Carriers, Primary Energy Sources

A first consequence of the peculiar characteristics of metabolic systems is that they define for themselves the scale that should be used to represent their metabolism. That is, what is an energy input for a virus cannot be represented and quantified using the same descriptive domain useful for representing and quantifying what is an energy input for a household or for an entire society. In more general terms we can say that metabolic systems define the semantic interpretation of the categories which have to be used to represent their energy transformation – a self-explanatory illustration of this point (already discussed in Section 8.1.1) is given in Fig. 8.1. This peculiarity of metabolic systems has to do with an epistemic revolution associated with the development of non-equilibrium thermodynamics:

> *living systems and more in general socio-economic systems are self-organizing (or autopoietic) systems which operate through auto-catalytic loops. This means that the energy input gathered from the environment is used by these systems to generate useful work used to perform several tasks associated with maintenance and reproduction. The gathering of an adequate energy input must be one of these tasks in order to make it possible to establish an autocatalytic loop of energy forms (Odum, 1971; Ulanowicz, 1986). Therefore, in relation to this characteristic, the expression "negative entropy" has been proposed by Schröedinger (1967) to explain the special nature of the energetics of living systems. Each dissipative system defines from its own perspective what is high entropy (= bad) and negative entropy (= good) for itself. This implies that living systems and socio-economic systems can survive and reproduce only if they manage to gather **what they define** as "energy input" (negative entropy or "exergy" within a given well defined system of accounting) and to discard **what they consider** "waste" (high entropy or degraded energy). However, what is waste or "high entropy" for a system (e.g. manure for a cow) may be seen as an energy input or "negative entropy" by another system (e.g. soil insects). This seminal idea has been consolidated by the work of the school of Prigogine (Prigogine, 1978; Prigogine and Stengers, 1981) when developing non-equilibrium thermodynamics, a new*

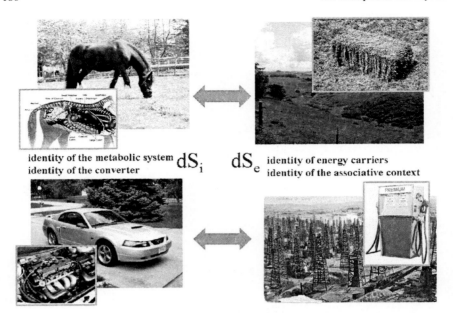

Fig. 8.1 Metabolic systems define for themselves the semantic of energy transformations (energy source and energy carrier)

type of thermodynamic which is compatible with the study of living and socio-economic systems (Schneider and Kay, 1994). However, because of this fact, non-equilibrium thermodynamics of dissipative systems entails a big epistemological challenge. As soon as we deal with the interaction of different metabolic systems defining in different ways for themselves what should be considered as "energy", or "exergy", or "negative entropy", not only it becomes impossible to have a "substantive" accounting of the overall flows of energy, but also it becomes impossible to obtain a "substantive" definition of quality indices for energy forms (Kay, 2000; Mayumi and Giampietro, 2004). Giampietro, 2006.

A second key characteristic of metabolic systems is that their expected identity entails a given range of value for the pace of the consumption of their specific energy input. For example, humans cannot eat for long periods of time either 100 Kcal/day (0.4 MJ/day) or 100,000 Kcal/day (400 MJ/day) of food. If the pace of consumption of their food intake is kept for too long outside the expected/admissible range – e.g. more or less 2,000–3,000 Kcal/day (8–12 MJ/day) depending on the characteristics of the individual – they will die. For all metabolic systems, there is an admissible range for the pace of the various metabolized flows. This expected range of values for the throughput implies that the very same substance of a metabolized flow – e.g. a vitamin – can be good or toxic for the body, depending on the congruence between the pace at which the flow is required and the pace at which the flow is supplied. What is considered as a resource when supplied at a given pace can become a problem (waste) when supplied at an excessive pace. An example of this fact is represented by eutrophication of water bodies (too much of good thing – too

Same flow - human dejections - different perceptions depending on different contexts

Night soil in CHINA -> a resource

It provides nutrients and energy to farmers

Human dejections in USA -> a cost

It requires heavy investments of capital and energy to be dealt with

Fig. 8.2 The relevance of the pace of the throughput

much nutrients for the aquatic ecosystem, which can only handle the metabolism of these nutrients at a given pace). Another example applied to human societies is given in Fig. 8.2. Human dejections can represent a valuable resource in a rural area (determining an energy gain for the system) or a waste problem in a city (determining an energy loss for the construction and operation of the treatment plant).

8.1.4 The well Known Trade-Off Between "Power" (the Pace of the Throughput) and "Efficiency" (the Value of the Output/Input Ratio) Makes it Impossible to Use Just a Number (an Output/Input Ratio) for the Analysis of Complex Metabolic Systems

Very often in conventional energy analysis a single number – e.g. an output/input energy ratio – is used to define the efficiency of an energy system. However, in order to use such a ratio for comparing the performance of different energy systems, we should be, first of all, sure that the two systems to be compared do have the same identity as metabolic systems. That is, do they belong to the same type of energy converter? Do they perform the same set of functions?

A truck moving 100 tons at 60 miles per hour consumes more gasoline that a small motorbike bringing a single person around at 15 miles per hour. But "so what"? Does it means that small motorbikes are "better" in substantive terms than huge tucks? A single output/input assessment does not say anything about the relative efficiency of the two vehi-

cles, let alone their usefulness for society. It is well known that there is a trade-off between energy efficiency and power delivered (Odum and Pinkerton, 1955). Summing energy forms (oranges and gasoline) which are used by different metabolic systems, which are operating at different power levels, using a single overall assessment, implies assuming the same definition of efficiency for different systems that are doing different tasks, while operating at different power levels—bikes and trucks. Again this assumption has only the effect of generating numbers which are simply irrelevant". Giampietro, 2006

It is impossible to compare the mileage of a truck and a motorbike, since they are different types of metabolic systems, having a different definition of tasks, useful work, and also a different definition of constraints on the relative pace of conversion of the energy input into the final useful work. Even willing to do so, the owner of a motorbike cannot move 100 tons at 60 miles per hour. A numerical assessment – e.g. a number characterizing an output/input energy ratio – reflects the chain of choice made by the analyst, when formalizing the semantic concepts associated with the chosen narrative about energy conversions. Metabolic systems having different semantic identities have to be characterized using a different selection of attributes of performance.

8.1.5 The Implications of These Epistemological Predicaments

In conclusion, the epistemological predicament associated with complexity in energy analysis deals with the impossibility of reducing to a single quantitative assessment – an output/input energy ratio: (A) the representation of events taking place simultaneously across different scales; (B) the representation of events which requires the adoption of non-equivalent narratives. This predicament implies that we should abandon the idea that a single index/number can be used to characterize, compare and evaluate the performance of the metabolism of complex energy systems. Discussing the trade-off between energy efficiency and power delivered Odum and Pinkerton (1955) note: *"One of the vivid realities of the natural world is that living and also man-made processes do not operate at the highest efficiencies that might be expected from them"*. Meaning that the idea that the output/input energy ratio should be maximum or a very relevant characteristic to define the performance of an energy system, is not validated by the observation of the natural world. The same basic message associated to an explicit call for the adoption of a more integrated analysis based on multiple criteria and wisdom (addressing and acknowledging the pre-analytical semantic step) was given by Carnot himself more than a century earlier: *"Regarding the need of using a multicriterial approach, it should be noted that in 1824, well before the introduction of the concept of Integrated Assessment, Carnot (1824) stated in the closing paragraph of his Reflections on the motive power of fire, and on machines fitted to develop that power: "We should not expect ever to utilize in practice all the motive power of combustibles. The attempts made to attain this result would be far more harmful than useful if they caused other important considerations to be neglected. The economy of the combustible [efficiency] is only one of the conditions to be fulfilled in heat-engines. In many cases it is only secondary. It should often give precedence to safety, to strength, to the durability of the engine, to the small space which it must occupy, to small cost of installation, etc. To know how to appreciate in each case, at their true value, the considerations of convenience and economy which may present themselves; to know how to discern the more important of those which are only secondary; to balance them properly against each other; in order to attain the best results by the simplest means; such should be the leading characteristics of the man called to direct, to co-ordinate the labours of his fellow men, to make them co-operate towards a useful end, whatsoever it may be" [pag. 59]"*. (Giampietro et al., 2006a).

Following the suggestion of Carnot we present, in the rest of the chapter, an alternative approach to the analysis of the feasibility and desirability of alternative energy sources. This approach is based on the concept of "bioeconomics", which can be used to operationalize the rationale of Net Energy Analysis, and in particular the elusive concept of EROI (Energy Return On the Investment) when dealing with metabolic systems operating over multiple scales.

8.2 Basic Concepts of Bioeconomics

8.2.1 The Rationale Associated with the Concept of EROI

The very survival of metabolic systems entails their ability to gather and process the flow of energy inputs they must consume. This implies that these energy inputs must be used for two different tasks: (i) to keep gathering other energy inputs in the future; and (ii) to sustain additional activities needed for the survival of the metabolic systems such as reproduction, self-repair, and development of adaptability (Rosen, 1958; Ulanowicz, 1986). Therefore, the energy gathered from the environment in the form of a flow of energy carriers cannot go entirely into discretional activities, since a fraction of it must be spent in the process of gathering and processing this energy input. There is a forced overhead on the energy input used by a metabolic system and this unavoidable overhead is behind the concept of Net Energy Analysis. According to this concept we can say that an energy input has a high quality, when it implies a very small overhead for its own gathering and processing. An economic narrative can help getting this concept across. Actually, the use of this economic analogy was proposed by Georgescu-Roegen (1975), exactly to discuss the quality of energy sources: "There certainly are oil-shales from which we could extract one ton of oil only by using more than one ton of oil. The oil in such a shale would still represent available, but not accessible, energy" (ibid, p. 354). His distinction between "available" energy and "accessible" energy can be summarized as follows:

- *available energy* is the energy content of a given amount of an energy carrier. This reflects an assessment which deals only with the characteristics of the energy carrier;

- *accessible energy* is the net energy gain, which can be obtained when relying on a given amount of an energy carrier obtained by exploiting an energy source. This assessment deals with the overall pattern of generation and use of energy carriers in the interaction of the metabolic system with its context.

A well known example of the relevance of this distinction is found in the field of human nutrition. In fact, the energy required to activate and operate the metabolic process within the human body entails an overhead on the original amount of available energy found in the nutrients. This overhead is different for different typologies of nutrient. For example, the energetic overhead for making accessible the available energy contained in proteins is in the range of 10–35%, whereas it is only 2–5% when metabolizing fat (FAO, 2001). Therefore, when calculating the ability to supply energy to humans with a given amount of nutrients it is important to consider that the same amount of available energy in the food – e.g. 1 MJ of energy from protein and 1 MJ of energy from fat – does provide a different amount of accessible energy when going through the metabolic process – e.g. 0.75 MJ out of 1 MJ from proteins versus 0.97 MJ out of 1 MJ from fat.

The example proposed by Georgescu-Roegen to convey the same concept is that of the "pearls dispersed in the sea". These pearls may represent, in theory, a huge economic value when considered in its overall amount. However, the practical value of pearls depends on the cost of extraction. In regard to this example, we cannot avoid to think to the many assessments found in literature of the huge potentiality of "biomass energy" when discussing of the potentiality of biomass as alternative to oil. Like for the pearls dispersed in the ocean, there is a huge amount of biomass dispersed over this planet. The problem is that this analysis seems to ignore the costs for extracting this biomass and converting it into an adequate supply of energy carriers! According to this reasoning, there are also millions of dollars in coins lost in the sofas of US families. Yet no businessman is starting an economic activity based on the extraction of this potential resource. The basic concept of bioeconomics is that it is not the total amount of pearls, biomass or coins that matters, but the ability to generate, using this total amount, a net supply of the required resource at the required pace.

The standard approach used to evaluate an economic investment provides a very effective generalization of this discussion. For example, it is impossible to evaluate an economic investment "which yields 10,000 US$ in a year". This investment may be either very good or very bad. It is very good if it requires 10,000 US$ of fixed investment; or it is very bad if requires 1,000,000 US$ of fixed investment. The economic concept to be used here is the concept of the Return On the Investment, which is extremely clear to anybody when discussing of economic transformations. However, as soon as one deals with the evaluation of energy transformations – e.g. the potentiality of biofuels as alternative to oil – the concept of EROI is very seldom adopted. For example, the well known study of Farrell et al. (2006) on Science, which had the goal to provide a comprehensive review of controversial assessments of biofuels found in literature, has been criticized by many energy analysts for having totally ignored the issue of EROI (Cleveland et al., 2006; Kaufmann, 2006; Patzek, 2006; Hagens et al., 2006).

When applied to energy analysis the EROI index can be defined as:

EROI [Energy Return On the Investment] *the ratio between the quantity of energy delivered to society by an energy system and the quantity of energy used directly and indirectly in the delivery process.*

This index has been introduced and used in quantitative analysis by Cleveland et al., 1984; Hall et al., 1986; Cleveland, 1992; Cleveland et al., 2000; Gever et al., 1991. An overview of the analytical frame behind EROI is given in Fig. 8.3. The figure illustrates two crucial points: (1) the key importance of considering the distinction between primary energy sources, energy carriers, and final energy services, when handling numerical assessments of different energy forms; and (2) a systemic conceptual problem faced when attempting to operationalize the concept of EROI into a single number due to the need of dealing with an internal loop of "energy for energy", which is operating across hierarchical levels. This internal loop entails a major epistemological problem in the quantification of such a ratio (for more see Giampietro and Mayumi, 2004; Giampietro, 2007a).

Still we can say that the total energy consumption of a society depends on its aggregate requirement of useful work or final energy services (on the right of the graph) which is split, according to the overhead associated with the EROI between: (i) Energy for Energy – used for the internal investment within the energy

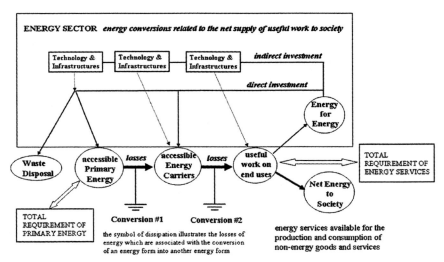

Conceptualization of the complex set of relations over different energy forms in the metabolism of a developed society – the EROI establishes a non-linear link between: (i) Total requirement of primary energy; and (ii) Total requirement of energy services

Fig. 8.3 The complex role of EROI in determining the characteristics of the energetic metabolism of a society

sector needed to deliver the required energy carriers – the energy consumption (or metabolism) of the energy sector; and (ii) Net Energy to Society – used for the production and consumption of "non-energy goods and services" - the energy consumption (or metabolism) of the rest of the society.

In spite of an unavoidable level of arbitrariness in the calculation of EROI, this scheme indicates clearly the tremendous advantage of fossil energy over alternative energy sources (for more see Giampietro, 2007a). In relation to the costs of production of energy carriers, oil has not to be produced, it is already there. Moreover, in the previous century it was pretty easy to get: the EROI of oil used to be 100 MJ per MJ invested, according to the calculations of Cleveland et al. (1984). For this reason, in the community of energy analysts there is an absolute consensus about the fact, that the major discontinuity associated with the industrial revolution in all major trends of human development (population, energy consumption per capita, technological progress) experienced in the XXth century was generated by the extreme high quality of fossil energy as primary energy source (for an overview of this point see Giampietro, 2007a). This means that to avoid another major discontinuity in existing trends of economic growth (this time in the wrong direction), it is crucial that when looking for future alternative primary energy sources, to replace fossil energy, humans should obtain the same performance, in terms of useful work delivered to the economy per unit of primary energy consumed.

As explained earlier a very high EROI means that the conversion of oil into an adequate supply of energy carriers (e.g. gasoline) and their distribution absorbs only a negligible fraction of the total energy consumption of a society. This small overhead makes it possible that a large fraction of the total energy consumptions goes to cover the needs of society, with very little of it absorbed by the internal loop "energy for energy". Moreover, due to the high spatial density of the energy flows in oil fields and coal mines the requirement of land to obtain a large supply of fossil energy carriers is negligible. Finally, waste disposal has never been considered as a major environmental issue, until acid rain deposition and global warming forced world economies to realize that there is also a sink side – beside the supply side - in the biophysical process of energy metabolism of whole societies. As a matter of fact, so far, the major burden of the waste disposal of fossil energy has been paid by the environment, without major slash-back on human economies. Compare this situation with that of a nuclear energy in which uranium has to be mined, enriched in high tech plants, converted into electricity in other high tech plants, radioactive wastes have to be processed and then kept away (for millennia!) both from the hands of terrorists and from ecological processes.

The narrative of the EROI is easy to get across: the quality of a given mix of energy sources can be assessed by summing together the amount of all energy investments required to operate the energy sector of a society and then by comparing this aggregate requirement to the amount of energy carriers delivered to society. By using this narrative it is easy to visualize the difference that a "low quality energy source" can make on the profile of energy consumption of a society. This is illustrated in the two graphs given in Fig. 8.4 (from Giampietro et al., 2007). The upper part of the figure – Fig. 8.4a – provides a standard break-down of the profile of different energy consumptions over the different sectors of a developed economy. Total Energy Throughput (TET) is split into the Household sector (Final Consumption) and the economic sectors producing added value (Paid Work sector – PW). The economic sector PW is split into: Services and Government, Productive Sectors such as Building, Manufacturing, Agriculture (minus the energy sector) and the Energy Sector (ES). The example adopts an average consumption per capita of 300 GJ/year and an EROI > 10/1. This entails that only less than 10% of TET goes into the energy sector. Let's assume now that we want to power the same society with a "low quality primary energy source". For example, let's imagine a system of production of energy carriers with an overall output/input energy ratio of 1.33/1. The lower part of – Fig. 8.4b (right side) – shows that for 1 MJ of net energy carrier supplied to society this energy system has to generate 4 MJ of energy carriers. As mentioned earlier, the huge problem with primary energy sources alternative to oil is that they *have to be produced, and they have to be produced using energy carriers.* That is, a process of production of primary energy sources must use energy carriers which have to be converted into end uses. This fact entails a double energetic cost (to make the carriers that will be used then within the internal loop to produce the primary energy required to make the energy carriers). That is, this internal loop translates into an extreme fragility in the overall performance of the system. Any negative change in this loop does amplify in non-linear way. A small reduction of about 10% in the output/input ratio – e.g. from 1,33/1 to 1,20/1 implies that the net supply of 1 MJ delivered to society would require the production of 6 MJ of energy carriers rather than 4MJ (for more on this point see Giampietro and Ulgiati, 2005).

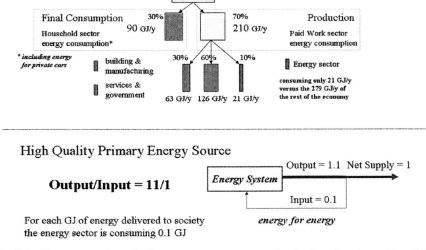

Fig. 8.4a The pattern of metabolism across compartments of a developed society with a "high quality" primary energy source (EROI >10/1)

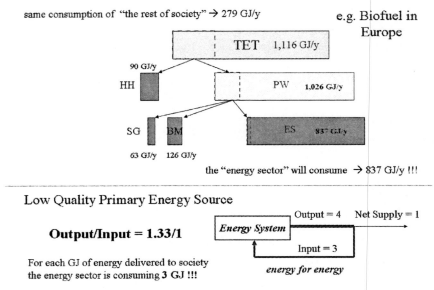

e.g. Biofuel in Europe

Fig. 8.4b The pattern of metabolism across compartments of a developed society with a "low quality" primary energy source (EROI < 2/1)

Let's image now to power the same society illustrated in Fig. 8.4a (a developed society) using a "low quality primary energy source" (EROI = 1.33/1) and keeping the same amount of energy invested in the various sectors (beside the energy sector). The original level of energy consumption per capita for the three sectors described in Fig. 8.4a is 279 GJ/year, which is split into: (i) 90 GJ/year in Final Consumption (residential & private transportation); (ii) 63 GJ/year in Service and Government; and (iii) 126 GJ/year Building and Manufacturing and Agriculture. In this case, the energy sector – when powered by low quality energy sources – would have to consume for its own operations 837 GJ/year per capita. Then, when combining the energy consumed by the rest of society and the energy consumed by the energy sector the total energy consumption of the society would become 1,116 GJ/year per capita – an increase of almost 4 times of the original level! Obviously such a hypothesis is very unlikely. It would generate an immediate clash against environmental constraints, since the industrial and post-industrial metabolism of developed society at the level of 300 GJ/year per capita has already serious problems of ecological compatibility, when operated with fossil energy. However, the environmental impact would not be the only problem. There are also key internal factors that would make such an option impossible. Moving to a primary energy source with a much lower EROI than oil would generate a collapse of the functional and structural organization of the economy. In fact the massive increase in the size of the metabolism of the energy sector would require a massive move of a large fraction of the work force and of the economic investments right now required in the other sectors of the economy. A huge amount of hours of labor and economic investment will have to be moved away from the actual set of economic activities (manufacturing and service sector) toward the building and operation of a huge energy sector, which will mainly consume energy, material and capital for building and maintaining itself.

8.2.2 The Combination of Biophysical and Socio-Economic Constraints Determines a Minimum Pace for the Throughput to be Metabolized

Due to the organization of metabolic systems across different hierarchical levels and scales, there are "emergent properties" of the whole that cannot be detected when considering energy transformation at the level of the individual converter. In socio-economic systems, these "emergent properties" may be discovered only when considering other dimensions of sustainability – e.g. the characteristics of social or economic processes determining viability constraints – which are forcing metabolic systems to operate only within a certain range of power values. To clarify this point let's discuss an example based on an analysis of the possible use of feeds of different quality in a system of animal production. This example is based on the work of Zemmelink (1995).

In the graph shown in Fig. 8.5 numerical values on the horizontal axis (e.g. A1, A2) represent an assessment of the quality of feed (based on nutrient and energy content per unit of mass). They reflect the given mix of possible feed types which are available in a given agro-ecosystem: (i) dedicated crops or very valuable by-products = high quality; (ii) tree leaves = medium quality; and (iii) rice straw = low quality. Therefore, moving on the horizontal axis implies changing the mix of possible feed types. "Very high quality feed" implies that only dedicated crops or very high quality by-products can be used; "very low quality feed" implies that also rice straw can be used in the mix. The points on the curve represent the size of the herd (e.g. S1, S2, on the vertical axis on the right). The diagonal line indicates the relation between levels of productivity (pace of the output) of animal products – i.e. beef – (e.g. P1 and P2 on the vertical axis on the left) and the "quality" of feed used as input for animal production (e.g. the point A1 and A2 on the horizontal axis). When using only animal feeds of a high quality one can get a high level of productivity (boost the output), but by doing so, one can only use a small fraction of the total primary productivity of a given agro-ecosystem. This analysis describes an expected relation between: (i) productivity in time (power level – on the vertical axis on the left); (ii) ecological efficiency (utilization of the available biomass – on the horizontal axis); (iii) stocks in the system (the size of the herd – on the vertical axis on the right) in animal production. This emergent property of the whole determining the viability and desirability of different types of biomass depends on both: (i) the required level of productivity (determined by the socio-economic context) – the economic break-even point on the vertical axis on the left; and (ii) the characteristics of the agro-ecosystem (the set of biological conversions and the ecological context). This study confirms that the need of operating at a high level of productivity implies

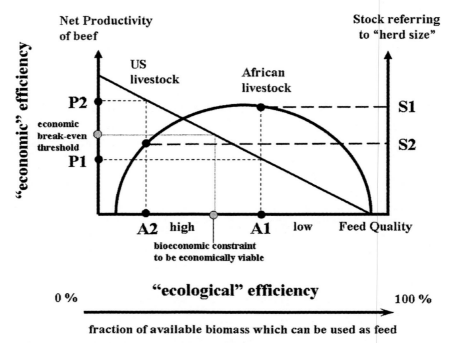

Fig. 8.5 Feed quality and net productivity of animal production

reducing the ecological efficiency in using the available resources. That is, when the socio-economic constraints force to operate at a very high level of productivity, a large fraction of tree leaves and all available rice straw can no longer be considered as feed, but they will result just waste.

This analysis provides a clear example of the need of contextualization for biophysical analysis. That is, when looking only at biophysical variables we can only characterize whether or not a feed input of quality "A1" is an input of "adequate quality" for a system of production of beef operating at a rate of productivity P1. However, the ultimate decision on whether or not the level of productivity P1 is feasible and desirable for the owner of the beef feed-lot cannot be decided using only this biophysical analysis. The viability and desirability of the level of productivity P1 depends on the constraints faced on the interface beef feed-lot/rest of society. This evaluation of desirability has to be done considering a different dimension of analysis. In this case, the acceptability of P1 has to be checked using a socio-economic dimension (the position of the economic break-even point on the vertical axis on the left). This viability check has to do with the evaluation of the pace of generation of added value (linked with the level of productivity P1) required for the viability of the production system.

In conclusion, the very same feed input of quality "A1" can be either: (1) perfectly adequate for that system of animal production in a given social context (e.g. in a developing country); or (2) not acceptable, when moving the same biophysical process from a developing country to a developed country. That is, a change in the socio-economic context can make level P1 no longer acceptable. When forced to operate at a higher level of productivity (e.g. P2) to remain economically viable, the owner of the feed-lot would find the feed input of quality "A1" no longer either viable or desirable. In biophysical terms, the feed input of quality "A1" would remain of an adequate quality for sustaining a given population of cows, but no longer of an "adequate quality" for sustaining, in economic terms, the threshold of productivity, required by the owner of the feed-lot to remain economically viable.

The set of relations described in the graph of Fig. 8.5 is based on well known biological processes for which it is possible to perform an accurate analysis of the biological conversions associated with animal production. Yet, due to the complexity of the metabolic system operating across multiple scales, and due to the different dimensions of analysis which have to be considered, the concept of "quality of the energy input to the whole system" depends on: (1) the hierarchical level at which we decide to describe the system – e.g. the cow level versus the whole beef feed-lot level; and (2) the context within which the system is operating (in this case on the economic side of the animal production system). When considering also socio-economic interactions, there are emergent properties of the whole (the performance based on multiple criteria mentioned by Carnot), which can affect the viability or desirability of an energy input (the minimum admissible feed quality for achieving an economic break-even point). These emergent properties can affect the admissible pace of the metabolism of the whole, and therefore induce a biophysical constraint (the need of reaching a certain threshold of power level) within a particular conversion process (the transformation of feed into beef at the hierarchical level of the whole production system). This can imply that what is an effective energy input, when operating at a lower power level (in this example the mix of feed of quality "A1" in Uganda) is no longer a viable or desirable energy input when operating in the USA. That is, even when the biophysical parameters of the system remain completely unchanged – keeping the same cows, the same set of potential energy inputs for the feed, the same techniques of production – it is the coupling with the external context – beef feed-lot/rest of society – that will affect the biophysical definition of "quality" for what should be considered as a viable energy input.

In conclusion the question: "are crop residues useful feed for a beef feed-lot?" cannot be answered without first checking the biophysical constraints on energy transformations which are determined by the set of expected characteristics of the whole metabolic system. These expected characteristics are determined by its interaction with its context. The question about the viability and desirability of crop residues as alternative feed cannot be answered just by looking at one particular dimension and one scale of analysis. According to the analysis presented in Fig. 8.5 crop residues **may** provide nutritional energy to cows, but their viability and desirability *depends on the severity of the biophysical constraints* determined by the socio-economic characteristics of the whole. Exactly the same answer can be given in relation to the possibility of using biomass for the metabolism of a socio-economic system.

8.2.3 Economic Growth Entails a Major Biophysical Constraint on the Pace of the Net Supply of Energy Carriers (per hour and per ha) in the Energy Sector

Let's image that, in order to reduce the level of unemployment in rural areas of developed countries, a politician would suggest to abandon the mechanization of agriculture and to go back to pre-industrial agricultural techniques requiring the tilling and the harvesting of crops by hand. By implementing this strategy it would be possible to generate millions and millions of job opportunities overnight! Hopefully, such a suggestion would be immediately dismissed by political opponents as a stupid idea. Everybody knows that during the industrial revolution the mechanization of agriculture made it possible to move out from rural areas a large fraction of the work force. This move had the effect to invest human labor into economic sectors able to generate added value at a pace higher than the agricultural sector. This is why, no developed country has more than 5% of its work force in agriculture and the richest countries have less than 2% of their work force in agriculture (Giampietro, 1997a).

As a matter of fact, changes in the structure and the function of socio-economic systems can be studied using the metaphor of societal metabolism. The concept of societal metabolism has been applied in the field of industrial ecology (Ayres and Simonis, 1994; Duchin, 1998; Martinez-Alier, 1987), in particular in the field of matter and energy flow analysis (Adriaanse et al., 1997; Fischer-Kowalski, 1998; Matthews et al., 2000). By adopting the concept of societal metabolism it is possible to show that the various characteristics of the different sectors (or compartments) of a socio-economic systems must be related to each other, as if they were different organs of a human body. In particular it is possible to establish a mechanism of accounting within which the relative size and the relative performance of the various sectors in their metabolism of different energy and material flows must result congruent with the overall size and metabolism of the whole. These two authors have developed a methodological approach – Multi-Scale Integrated Analysis of Societal and Ecosystem Metabolism (MuSIASEM) – originally presented in several publications as MSIASM – e.g. Giampietro, 1997b, 2000, 2001; Giampietro and Mayumi, 2000a,b; Giampietro et al., 1997a, 2001; Giampietro and Ramos-Martin, 2005; Giampietro et al., 2006c, 2007; Ramos-Martin et al., 2007; Giampietro, 2007a – which can be used to perform such a congruence check.

That is, the MuSIASEM approach can be used to check the congruence between: (i) the characteristics of the flows to be metabolized as required by the whole society; and (ii) the characteristics of the supply of the metabolized flows, as generated by individual specialized compartments. An overview of the possible application of this method to the analysis of the quality of energy sources is presented in Giampietro, 2007a; Giampietro et al. 2007. Just to provide an example of the mechanism used to perform this congruence check, we provide in Fig. 8.6 an analysis of the energetic metabolism of a developed society (e.g. Italy) in relation to the profile of use of human activity over 1 year.

Very briefly, when considering the system "Italy" at the hierarchical level of the whole society – considered as a black box (on the right of the figure) – we can

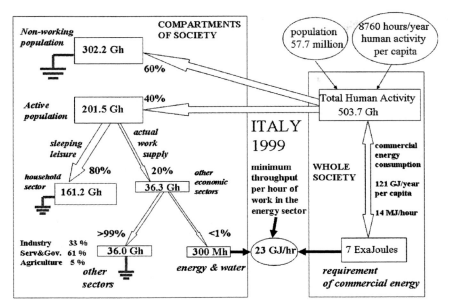

Fig. 8.6 Minimum threshold of energy throughput per hour of labor in the energy sector of a developed country

say that 57.7 millions of Italians represented a total of 503.7 Giga hours (1 Giga = 10^9) of human activity in the year 1999. In the same year they consumed 7 Exa Joules (1 Exa = 10^{18}) of commercial energy. This implies that at the level of the whole society, as average, each Italian has consumed 14 MJ/hour (1 Mega = 10^6) of commercial energy.

Let's imagine now to open the black box and to move to an analysis of the individual sectors making up the Italian economy (moving to the left of the figure). In this way, we discover that the total of human activity available for running a society has to be invested in a profile of different tasks and activities which have to cover both: (i) the step of production of goods and services; and (ii) the step of consumption of goods and services. For example, more than 60% of the Italian population is not economically active – e.g. retired, elderly, children, students. The fraction of human activity associated with this part of the population is therefore not used in the process of production of goods and services (but it is used in the phase of consumption). Furthermore the active population works only for 20% of its available time (in Italy the work load per year is 1,780 hours). This implies that out of the total of 503.7 Giga hours of human activity available to the Italian society in 1999, only 36.3 Giga hours (8% of the total!), were used to work in the economic sectors producing goods and services. In that year, almost 14 hours of human activity have been invested in consuming per each hour invested in producing! Let's now see how this profile of distribution of time use affect the availability of working hours to be allocated in the mandatory task of producing the required amount of energy carriers in the energy sector. This requires looking at what happened within the tiny 8% of the total human activity invested in the productive sector. Out of these 36.3 Giga hours, 60% has been invested in the Service and Government sector. The industrial sector and the agricultural sector have absorbed another 38%, leaving to the energy sector less than one percent (<1%) of the already tiny 8% of the total. This is a well known characteristic of modern developed societies, which are very complex. This complexity translates into a huge variety of goods and services produced and consumed, which, in turn, requires a huge variety of different activities across the different sectors associated with different jobs descriptions and different typologies of expertise (Tainter, 1988).

In conclusion, in Italy in 1999, only 0.0006 of the total (not even 1/1000th!) of the total human activity has been used for supplying the energy carriers associated with the consumption of 7 Exa Joules of primary energy consumed in that country that year. This means that by dividing the total consumption of the "black box Italy" by the hours of work delivered in the energy sector, the performance of the energy sector in relation to the throughput of energy delivered to society per hour of labor in the energy sector has been of **23,000 MJ/hour**.

It should be noted that if rather than considering Italy had we considered USA the consumption per capita would have been much higher (333 GJ/person year or 38 MJ/hour in 2005). After adjusting for a different population structure (50% of the population in the work force) assuming 2,000 hours/year of work load and only 0.007 of the work force – about 1 million workers* – in the sector supplying fossil energy carriers, the resulting throughput of energy delivered to society per hour of labor in the energy sector is **47,000 MJ/hour**. [* this excludes almost 1 million workers in gas stations and trucks needed for transporting liquid fuels, which are not included in the calculation since they are required for the distribution of fuels independently from the energy source used to produce them].

8.3 Using the MuSIASEM Approach to Check the Viability of Alternative Energy Sources: An Application to Biofuels

8.3.1 The "Heart Transplant" Metaphor to Check the Feasibility and Desirability of Alternative Energy Sources

To visualize the type of integrated analysis based on the MuSIASEM approach for linking the characteristics of the energy sector to the characteristics of the whole society, we propose the metaphor of a heart transplant, illustrated in Fig. 8.7 (more details in Giampietro and Ulgiati, 2005; Giampietro et al., 2006c). Let's imagine that the actual energy sector based on fossil energy as primary energy source, is the heart, which, at this very moment, is keeping alive a given person (e.g. a given society). Let's imagine now that we want to replace this heart with an alternative heart (e.g. an energy sector powered by biofuels from agricultural production). Let's imagine that we want to perform this transplant because someone claims that the alternative

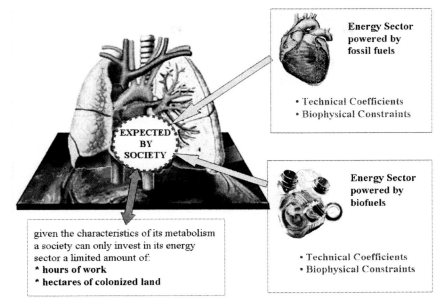

Fig. 8.7 The metaphor of the heart transplant

heart is much better (e.g. it makes it possible to have "zero emission" of GHGs from the energy sector and a total renewability of the supply of energy carriers).

Still, it would be wise, before starting the operation of transplant, to check whether or not such a substitution is: (i) feasible; and (ii) desirable. To do such a check it is necessary to compare the performance of the actual heart with the performance that we can expect from the alternative heart we want to implant. This comparison can be obtained by checking the congruence between: (A) the pace of the required flow of energy carriers determined by the characteristics of the whole society; and (B) the pace of the net supply of energy carriers which can be achieved by the "alternative energy sector" we want to implant. The application of this approach is presented in the next section, which compares the performance of the actual energy sector powered by fossil energy with the performance of an energy sector powered by biofuels. For the sake of simplicity we will focus only on two biophysical constraints on the pace of the flow of energy carriers: (i) "the requirement of hours of labor in the energy sector to generate the required supply" versus "the availability of hours of labor which can be allocated in the energy sector by a given society"; (ii) "the requirement of hectares of land in the energy sector to generate the required supply" versus "the availability of hectares of land which can be allocated to the energy sector by society". With this choice, we ignore additional issues, which are very relevant when checking the viability of biofuels as alternative energy sources. These additional issues should include: water demand, soil erosion, preservation of natural habitat for biodiversity.

8.3.2 Checking the Feasibility and Desirability of Biofuels Using Benchmark Values

8.3.2.1 The Biophysical Constraints Over the Required Flow of Energy Carriers

Let's first define the two benchmarks values to characterize the viability and desirability of the supply of energy carriers from the energy sector operating in a developed society.

In relation to the throughput per hour of labor – that is, according to the analysis described in Fig. 8.6 – within a developed country the throughput of energy per hour of labor in the energy sector has to be in the range of values between 23,000 MJ/hour and 47,000 MJ/hour.

Coming to the benchmarks referring to the spatial density of the energy flow, Fig. 8.8 provides a comparison of the ranges of power density of different primary energy sources (the graph on the left of the figure) against the ranges of power density of different typologies of land use associated with the pattern of metabolism of developed societies (the graph on the right of the figure). In relation to this figure we can immediately detect that the differences in these values are so big to require the use of a logarithmic scale. It is well known that before of the industrial revolution (before the powering of societal metabolism by fossil energy) the number of big cities – i.e. cities above the million people size – was

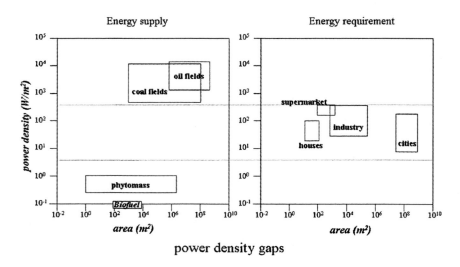

power density gaps

after Vaclav Smil 2003 Energy at the Crossroads, The MIT press
(Fig. 5.2 and Fig. 5.3)

Fig. 8.8 Power density gap between the required and supplied flows of metabolized energy

very small. The percentage of urban population in pre-industrial societies was very low. As a matter of fact, when using biomass as primary energy source one has to rely on a power density of the energy input per square meter which is much lower than the density at which energy is used in typical land uses of urban settling (Giampietro, 2007a). In relation the requirement of a high power density of the net supply of energy carriers, the movement from agricultural biomass to biofuel makes things much worse, because the density of net power supply is heavily reduced by the internal loop of energy carriers consumed within the process generating biofuel.

In conclusion the two benchmark values for a developed country are:

throughput per hour labor in the energy sector: **23,000–47,000 MJ/hour**
power density of fossil energy consumption in urban land uses: **10–100 W/m^2.**

8.3.2.2 The Confusion About the Energetic Assessment of Biofuels

There is a great confusion in literature, when coming to the assessment of the energetic performance of biofuels (e.g. Farrell et al., 2006; Shapouri et al., 2002; Patzek, 2004; Patzek and Pimentel, 2005; Pimentel et al., 2007). This confusion is due to the lack of agreement on how to calculate the net energy supply of biofuel from energy crops. This is a crucial starting point since in a biofuel system energy carriers are produced (e.g. in the form of ethanol or oils), but also consumed (e.g. in the form of electricity and fossil fuels, during the production of the energy crop, transport and in the conversion of biomass into the final biofuel). Obviously, to be considered as an energy source the energy output of this process needs to exceed the energy input. But even more important, in relation to its feasibility and desirability, the requirement of land, labor and capital for generating a net supply of biofuels should not imply a serious interference with the actual functioning of the whole socio-economic system. In relation to this point there are two key issues to be considered: (1) how to handle the implications of net energy analysis – that is, one should acknowledge the crucial distinction between ***gross*** and ***net*** production of biofuel; and (2) how to handle the differences in quality of the different energy forms accounted among the inputs and the outputs of the process.

1 the implication of net-supply of energy carriers – let's imagine to have a biofuel system, fully renewable (not depending on oil for its own functioning) and having zero CO_2 emission, operating with an output/input 1.33/1. The consequences of this fact have been discussed in Fig. 8.4b. This system has to produce 4 barrels of biofuel to supply 1 net barrel to society. It should be noted that by addressing the net supply of energy carriers (a net supply of energy carriers and not a mix of input/output of different energy forms) it is much easier to appreciate the importance of adopting the EROI concept. The distinction between gross production of ethanol and net supply of ethanol to society is crucial, since it implies a strong non-linearity in the requirement of land, labor, capital per unit of net supply (Giampietro et al., 1997b; Giampietro and Ulgiati, 2005).

2 how to handle the different quality of different energy inputs and outputs – As discussed in Part 1, the summing of energy forms of different quality should be performed with extreme care. The problem with the assessment of biofuels is that, not only the vast literature assessing the energetic of "crops/biofuel systems" covers different routes and crop types, but also that different authors use different assumptions and different conversion factors for such a summing. The mentioned chapter of Farrell et al. (2006) reviewed a large number of studies and found that differences in the assessments can be explained by: (i) different technology assumptions; and (ii) differences in the method of accounting for by-products. In relation to the first problem further standardization might help for the accounting of the inputs. But the confusion about the overall output/input energy ratio will still remain since it is the second point – the choice of how to account for by-products (aggregating different energy forms) – which is more relevant in generating differences in the assessments. As a matter of fact, it is important to observe that there is no scientific consensus on whether or not the process producing biofuels in temperate areas (corn-ethanol) has a positive output/input. The estimate of a clear positive return of the production of biofuel from agriculture is due to the system of accounting implemented by the supporters of biofuels. They have chosen a system of accounting in which the wastes generated by the process – e.g. dry distillers grains (DDG) – are calculated as if they were equivalent to a net supply of barrels of biofuel to society (e.g. as done in Shapouri et al., 2002). The explanation for this choice is that the by-products of the production of biofuels can be used as feed. Therefore, according to this rationale, the amount of oil that would be required to generate the same amount of feed obtained using the distillation wastes, should be added in the calculation as if it were an actual supply of energy carriers (the barrel of oil saved in this way). Opponents disagree (e.g. Pimentel et al., 2007) saying that the energy credit given to DDG is too high and that the quality of the feed based on DDG is much lower than the feed they are supposed to replace. But there is another major problem with this accounting method: the rationale backing up the energy credit for by-products feeds does not address the issue of scale (Giampietro et al., 1997b; Giampietro and Ulgiati, 2005). That is, if the production of biofuels were implemented on large scale, the amount of DDG generated by such a production would exceed of several times the demand for feed (an assessment is provided later on in the section dealing with the analysis of the corn-ethanol production in the USA). This implies that they would represent a serious environmental problem, to which analysts should associate an energetic and economic costs and not a positive return (Giampietro et al., 1997b).

8.3.2.3 Benchmark Values for the Net Supply of Energy Carriers (Barrels of Ethanol)

Production of Barrels of Ethanol from Sugarcane in Brazil

We used official data provided by a pro-ethanol institution (UNICA – Sugar Cane Agroindustry Union) in Brazil. Data and technical coefficients taken from the report compiled under the supervision of De Carvalho Macedo (2005) have been checked against several publications assessing technical coefficients of the production of ethanol from sugarcane in Brazil (an overview in Patzek and Pimentel, 2005; Pimentel et al., 2007). Again, also in this case, there are not substantial discrepancies in the assessment of technical coefficients (inputs and outputs); both in phase I (of production of agricultural biomass) and in phase II (fermentation and distillation for producing ethanol). Details on the data set generating the following benchmarks are given in Box 8.1. The resulting benchmarks are:

Box 8.1 Brazilian ethanol production (2004)

GROSS OUTPUT → 83 million liters of ethanol –> 1,766,000,000 MJ of ethanol

GROSS INPUTS → Labor 2,200 full time jobs (of which 73% of them in agriculture)

→ Land in production 13,333 ha –> 133,330,000 m^2

GROSS technical coefficients for biofuel over the whole process.

GROSS OUTPUT → 75,000 kg/ha (12 kg/1 lit) → 6,250 liters (1lt = 21.5 MJ) → 134 GJ/ha

---

Phase 1 – Agricultural Production Sugarcane – GROSS TECHNICAL COEFFICIENTS

INPUT labor → 210 hours/ha/year → 33.6 hours/1,000 liters

land → 6,250 liters/ha → 0.16 ha/1,000 liters

fossil energy → 40 GJ/ha → 6.4 GJ/1,000 liters

Phase 2 – Fermentation/Distillation of Ethanol – GROSS TECHNICAL COEFFICIENTS

INPUT labor → 90 hours/ha/year → 14.4 hours/1,000 liters

land → negligible → negligible

fossil energy → 48 GJ/ha → 7.7 GJ/1,000 liters

Box 8.1 (Continued)

NET technical coefficients for biofuel over the whole process.

TOTAL ETHANOL $\rightarrow$ 133 GJ/ha $\rightarrow$ 21.5 GJ/liter
ENERGY CARRIERS
OUTPUT
TOTAL FOSSIL ENERGY $\rightarrow$ 88 GJ/ha $\rightarrow$ 14.1 GJ/liter
CARRIERS INPUT
OUTPUT/INPUT IN ENERGY $\rightarrow$ 1.5/1 $\rightarrow$ 1.5/1
CARRIERS
NET SUPPLY = 33% of gross supply of ethanol – 3 liters gross ethanol $\rightarrow$ 1 liter net supply

The Net Supply of energy carriers (biofuel) supplied to society by the Brazilian ethanol sector is determined by the relation between: 3 liters of gross supply; 2 liters of gross supply required for internal consumption; 1 liter of net supply: (3–2)/3 = 0.33.

Only 33% of the Gross Output of the ethanol which is produced within the production system represents a net supply of energy carrier for society

Benchmarks related to the net supply delivered by Brazilian ethanol

Net supply $\rightarrow$ 27.7 millions liters (33% of the gross) $\rightarrow$ 588,000,000 MJ (33% of the gross)

Total inputs (aggregate values from UNICA study):
* labor $\rightarrow$ 4,400,000 hours (2,200 full jobs $\times$ 2,000 hours/year)
* land $\rightarrow$ 13,333 hectares

Technical coefficients of the process (per hectare and per liter of ethanol)
Total labor demand gross supply: $\rightarrow$ 48 hours/1,000 liters (300 hours/ha/year)
Total land demand gross supply: $\rightarrow$ 0.16 ha/1,000 liters (6,250 liters/ha)

Throughput per hour of labor in the sugarcane-ethanol production system:
Net supply per hour of labor = *134 MJ/hour* $\rightarrow$ 6.3 liter/hour (using labor data UNICA)
Net supply per hour = *148 MJ/hour* $\rightarrow$ 6.9 liter/hour (using available technical coefficients)

Throughput per unit of land in production in the sugarcane-ethanol production system:
Net supply per unit of land = 45 GJ/ha/year $\rightarrow$ 4 MJ/m^2 $\rightarrow$ *0.1 W/m^2*

Please note that when considering the requirement of fossil energy for the two-step process:

(i) agricultural production of the sugarcane; and (ii) conversion of the sugarcane into ethanol; we assumed as valid the pro-ethanol claim that the burning of the bagasse provides: (1) the entire heat energy consumed in the step of distillation; (2) the entire amount of electricity used in the process; and (3) no pollution costs are generated by this process due to the appropriate recycling of the wastes. Therefore, the assessment of the internal requirement of fossil energy (the requirement of "barrel of ethanol" required in a full self-sufficient process) refers only to the consumption of energy carriers for both the phase of agricultural production (for transportation, production of fertilizers, pesticides, irrigation, the making of steels and the technical infrastructures) and the phase of fermentation-distillation (for transportation and technical infrastructures).

We recall here the benchmark values required by a developed society:

throughput per hour labor in the energy sector: *23,000–47,000 MJ/hour*
power density of fossil energy consumption in urban land uses: *10–100 W/m²*.

The example of ethanol from sugarcane in Brazil, illustrates that even when considering the best possible scenario for biofuel, that is: (i) the use of the sugarcane-ethanol conversion which provides the highest EROI achieved so far in the production of biofuels; and (ii) the situation of Brazil, a country which has enough land to be able to produce sugarcane for energy (a semi-tropical agriculture, which can use a large amount of land not in production of food, because of low demographic pressure); the differences in value from what it would be required to run the metabolism of a developed country and what is provided by a system agricultural production-ethanol is in the order of hundreds of times.

Production of ethanol from corn in the USA

There is a well established data-set for the process corn-ethanol production in the USA, and also in this case, there are not major differences in the physical assessment of inputs and outputs among different studies. This is to say that the differences found in the overall assessment of the output/input energy ratio are basically generated by different choices on how to account for the various inputs and outputs and not by the initial accounting of biophysical inputs and outputs. Details of our calculations are given in Box. 8.2 (where no energy credit is given to the by-products in the form energy carriers). The two resulting benchmarks are:

Box 8.2 Production of ethanol form corn in USA (2004)

GROSS technical coefficients for biofuel over the whole process.

GROSS OUTPUT → 8,000 kg/ha (2.69 kg/1 lit) → 3,076 l/ha (1lt = 21.5 MJ) → 66.13 GJ/ha

STEP 1 – *Agricultural Production of Corn – GROSS TECHNICAL COEFFI-CIENTS*
INPUT   labor → 12 hours/ha/year → 4 hours/1,000 liters
        land → 3,076 liters/ha   → 0.32 ha/1,000 liters
  fossil energy → 29.3 GJ/ha   → 9.5 GJ/1,000 liters

STEP 2 – *Fermentation/Distillation of Ethanol – GROSS TECHNICAL COEF-FICIENTS*
INPUT   labor → 14.76 hours/ha/year → 4.8 hours/1,000 liters
        land → negligible   → negligible
  fossil energy → 31.9 GJ/ha   → 10.4 GJ/1,000 liters

The assessment of labor demand for the phase of agricultural production is from Pimentel (2006), whereas the labor requirement for fermentation/distillation is based on two different assessments:
1 USDA 2005a suggests for an average plant with a capacity of 40 million gallons year (155 million liters/year) the requirement of 41 full jobs in the plant, and 694 indirect jobs related to the operation of the plant. This would be equivalent to an input of 1.5 million hours (9.5 hours/1000 liters);
2 USDA 2005b suggests 17,000 jobs in the ethanol industry per each billion gallons of ethanol produced. This would be equivalent to an input of 34 million hours per 3,870 million liters/year (8.8 hours/1,000 liters).
   Since it is not clear whether or not the hours of agricultural production are already included in these assessments, for safety (in favor of the biofuel option) we took out the 4 hours of agricultural labor from the most favorable of the two assessments.

NET technical coefficients for biofuel over the whole process.

TOTAL ETHANOL ENERGY   → 66.1 GJ/ha → 21.5 GJ/liter
CARRIERS OUTPUT
TOTAL FOSSIL ENERGY   → 61.2 GJ/ha → 19.9 GJ/liter
CARRIERS INPUT
OUTPUT/INPUT IN ENERGY   → 1.1/1   → 1.1/1
CARRIERS
NET SUPPLY = 9% of the supply of ethanol – 11 liters of gross ethanol → 1 liter net supply

The Net Supply of energy carriers (biofuel) supplied to society by a corn-ethanol production system is determined by the relation between: 11 liters of gross supply; 10 liters of internal consumption; 1 liter of net supply: (11–10)/11 = 0.09.
   Only 9% of the Gross Output of the ethanol which is produced within the production system represents a net supply of energy carrier for society

Box 8.2 (Continued)

Benchmarks related to the net supply delivered by the corn-ethanol production systems
Total labor demand gross supply: → 8.8 hours/1,000 liters → 114 liters/hours
Total land demand gross supply: → 0.32 ha/1,000 liters (3,076 liters/ha/year)
Net supply per hour → 10.4 liters/hour [= 11(gross)/1(net) production]
Net supply per hectare → 277 liters/ha (9% of the gross) → 6 GJ/ha (9% of the gross)

Throughput per hour of labor in the corn-ethanol production system:
Net supply per hour of labor = 10.4 liters/hour → **224 MJ/hour**

Throughput per unit of land in production in the corn-ethanol production system:
Net supply per unit of land = 6 GJ/ha/year = 0.6 MJ/m^2/year → **0.02 W/m^2**

Please note that when considering the requirement of fossil energy for the two-step process:

(i) agricultural production of the corn; and (ii) conversion of the corn into ethanol; we assumed as valid the pro-ethanol claim that the by-products of agricultural production provide the entire heat energy consumption of the step of distillation. Therefore, the requirement of fossil energy refers only to the consumption of energy carriers both for the phase of agricultural production (transportation, production of fertilizers, pesticides, irrigation, the making of steels and technical infrastructures) and the phase of fermentation-distillation (transportation and technical infrastructures).

When comparing the two sets of benchmarks, the US system does better in terms of productivity of labor, since it uses much more capital than the Brazilian system. However, this is paid by a larger internal consumption of energy carriers (an internal loop of "energy for energy") to substitute labor with technical devices. The side effect is a skyrocketing requirement of land per unit of net supply delivered to society.

As a matter of fact, it is the skyrocketing increase in the requirement of primary energy production, due to the internal loop of energy for energy, which makes it impossible to power a developed society with biofuels. For example, let's imagine that biofuels would be used to cover a significant fraction of the actual consumption of fossil energy fuels in a developed country. Let's consider Italy in 1999 with a consumption of 7 EJ/year (1 EJ = 10^{18}J), a moderate level of consumption of energy for a developed country (121 GJ/year per person). This is a little bit more than a third of what is consumed per capita in the USA today. To cover just 10% of this consumption – 0.7 EJ/year – the agricultural sector should provide a net supply of 32.5 billion liters of ethanol, which, assuming a system fully renewable and capturing the CO_2 emitted, requires 358 billion liters of gross production (adopting a ratio 11 gross/1 net).

When using the benchmarks calculated before for ethanol from corn in Box 8.2, we find out that Italy would require: (A) 34 Ghours of labor in biofuel production (this is the 94% of the hours of work supply provided by the Italian work force in 1999); and (B) 117 millions hectares of agricultural land (this would be more than 7 times the 15.8 millions of agricultural area in production in Italy in 1999). Please note that: (i) nobody want to be farmers in Italy anymore, and at the moment, it is difficult to find enough farmers to produce even food; (ii) Italy does not have any surplus of food production (since the food consumed in Italy would already require the double of the arable land which is in production – Giampietro et al., 1998); (iii) an expansion of agricultural production on marginal areas would increase dramatically the requirement of technical inputs – e.g. fertilizers – further reducing the overall output/input energy ratio; (iv) the environmental impact of agriculture (soil erosion, alteration of the water cycle, loss of habitats and biodiversity, accumulation of pesticides and other pollutants in the environment and the water table) is already serious. Any expansion in marginal areas would make it much worse.

So biofuel from agriculture does not make any sense in a crowded developed country, even when the goal is to cover only 10% of the total and the level of energy consumption per capita is low. What about a country, like the USA, with higher consumption, but also with much more land available?

When considering the USA, we adopt a less ambitious goal: to cover just 10% of the fuels used in transportation. That is, the 10% of the 30% of the total of US energy consumption in 2006. With this target, the agricultural sector should generate a net supply of 3 EJ of ethanol – a net flow of 140 billion liters.

As promised, earlier, let's now use the EROI calculated by Shapouri et al. (2002) of 1.3/1 [after assuming a positive energy credit for by-products] for the calculation of the ratio gross/net supply. This is a much favorable ratio than that used in Box 8.2 (1.1/1). But yet, in order to be renewable and "zero emission", this biofuel system should produce 4 liters of ethanol to generate 1 liter of net supply. This would translate into a gross production of 12 EJ of ethanol – the gross production of 558 billions of liters. In turn, this translates into the requirement of: (1) a gross production of 1,500 millions tons of corn – which is 6 times the whole production of corn in USA in 2003 – USDA (2006); and (2) the generation of 500 million tons of DDG by-products – *which is 10 times the total US consumption of high protein commercial feeds* – 51 million tons – recorded in 2003 – USDA (2006). Here the negative effect generated by an enlargement of scale becomes crystal clear. Just to cover 10% of fuel in transportation – that is just 3% of total energy consumption of the USA! – the production of by-products from the system corn-ethanol would reach a size so large to make it invalid the rationale of giving an energy credit for the production of by-products. In fact, when reaching a scale of production of ethanol able to cover 3% of total energy consumption of USA these by-products will represent a serious environmental problem (and a serious energetic cost!), let alone a credit of fossil energy.

But after having proved this point, if we take out the energy credit for by-products used in the calculation of the EROI of Shapouri et al. (2002), we are back to the value of 1.1/1 (11 liters of gross ethanol production per liter of net supply) used for calculating the benchmarks in Box 8.2. Then, when repeating the calculation for the USA with this value we find that the net supply of 3 EJ of ethanol – a net flow of 140 billion liters – would translate into a requirement for a gross production of 33 EJ – 1,540 billion liters. This gross production of ethanol would require: (A) 148 Ghours of labor in biofuel production (this would represent almost 48% of the labor supply which could be provided by US work force after absorbing all the unemployed!); and (B) 5,500 million hectares of arable land (this would represent more than 31 times the 175 millions of arable land in production in USA in 2005).

This total lack of feasibility of a large scale biofuel solution based on a self-sufficient corn-ethanol system able to guarantee independence from fossil energy and zero CO_2 emission, clearly indicates that the actual production of ethanol in the USA is possible only because such a production *is powered by fossil energy fuels*! But IF we drop the motivation of independence from fossil energy and the zero emission, THEN it is the common sense that should suggest to a developed country that it is not wise to: (A) pay a price higher than 100 US\$ to buy a barrel of oil; (B) then add a lot of capital, land and some significant labor – additional production factors that have also to be paid; (C) consume natural resources and stress the environment (e.g. soil erosion, nitrogen and phosphorous in the water table, pesticides in the environment, fresh water consumption); to produce 1.1 barrel of oil equivalent in the form of ethanol.

8.4 Conclusion

8.4.1 "If the People have No Bread, Then Let's Them Eat the Cake..."

The interest in alternative energy sources to oil has been primed in this decade by the explosion of two issues: (1) global warming associated with green-house effect; and (2) peak oil. When combining these two problems, and ruling out the option that humans should consider alternative patterns of development not based on the maximization of GDP, it is almost unavoidable to conclude that what humankind needs is a primary energy source which: (i) does not produce emissions dangerous for the global warming; and (ii) is renewable. For those that are not expert in the field of energy analysis and more in general of the analysis of the metabolism of complex adaptive systems it is natural to come out with the simple sum $1 + 1 = 2$ and therefore conclude that producing biomass to be converted in biofuel is the solution that makes it possible to kill two birds with one stone. For those in love with this idea, the gospel is always the same: (1) producing the biomass used to make biofuels absorbs the carbon dioxide which will be produced when using that biofuel – therefore this is a method which has zero-emissions; and (2) since it uses solar energy, the supply of biofuel from biomass is renewable. The key result of this solution is an ideological one: by substituting "barrels of oil" with "barrels of biofuel" there is no longer the need of questioning the myth of perpetual economic growth (the idea which is possible to maximize the increase of GDP and expand human population for ever). Unfortunately things are not that easy and many birds killed with a single stones (together with magic bullets) work only in the fiction stories or in the promises made by politicians. In this chapter we explained in theory and with numerical examples why $1 + 1$ is not equal to 2 when dealing with the production of biofuel from crops.

When looking at the growing literature on biofuels, and at the many initiatives aimed at supporting the research on alternative energy sources, it looks like that because of the urgency and the seriousness of the energy predicament, now, in the field of alternative energy "everything goes" (for a list of bizarre examples see Giampietro et al., 2006c). In relation to this point, it is important to be aware of the stigmatization used by Samuel Brody (1945!) in the last chapter of his master-piece on power analysis of US agriculture. To those proposing, then, to power the mechanization of the US agricultural sector with ethanol from corn he reminded the famous quote attributed to Marie Antoinette: "if the people have no bread, then lets them eat the cake ... "

As a matter of fact, buying a barrel of oil at a price higher than 100 US$, and then adding capital, labor and land to it (all factors of production which requires additional energy and cost in economic terms) to produce a net supply of 1.1 barrel equivalent of ethanol seems to be not a particular smart move. First, it indicates that something went wrong with the study of energy analysis at the academic level. Second, it is also an indication of the incredible amount of freedom that fossil energy has granted to humans living in developed countries. They can afford (but for short periods of time!) to make impractical choices when deciding about how to use their available resources – "if the people are angry and we are out of bred, then lets' give them the cake...". There is a positive side of this fact, however. The impractical choices of developed countries heavily investing in biofuels from agricultural crops will help those developing countries that are using the valuable resource represented by oil to produce goods and services, to be more competitive on the international market. They will sell goods and services produced using a barrel of oil, to those that use a barrel of oil to make 1.1 barrel of oil-equivalent of ethanol (and paying also a higher cost for their food, because of this choice). A massive production of biofuels in developed countries will help developing countries in reducing the existing gradient of economic development.

8.4.2 Explaining the Hoax of Biofuels in Developed Countries

Before closing we want to answer a last question: How it is possible that developed countries are investing so many resources into such an impractical idea? Answering this question requires combining together three completely different explanations

Explanation 1 – Humans want to believe that there is always an easy solution
Due to the facility with which is possible to make the sum $1 + 1 = 2$ (biofuels
are renewable and they are zero emission) it is extremely easy for the uninformed
public to arrive to the conclusion that biofuels represent the perfect alternative to
fossil energy. Since the dominant western civilization is terrorized by the idea that it
will fall like all the previous dominant civilizations, the "public opinion" expressed
by western civilization needs to believe in the existence of a silver bullet that can
remove such a possibility. Therefore, the myth of biofuels represents a fantastic win-
dow of opportunities both for academic departments looking for funds of research,
and for politicians on the various sides of the political arena looking at an easy
consensus (following the opinion polls). In this situation, everyone has to jump into
the biofuel wagon to avoid to be labeled as being against sustainability. Because it is
about looking for a myth, it really does not matter that many of the discussions about
the economic benefits of the biofuel solution – e.g. the creation of a lot of jobs in
rural areas! – are based on a serious misunderstanding about the biophysical foun-
dations of the economic process. Jobs not only do provide income to families, but
also increases the costs when producing the relative goods or services. Suggesting a
strategy of a massive move of the work force into biofuel production in a developed
country is similar to the idea of suggesting a return to the harvesting of crops by
hands to increase the number of jobs in agriculture. It belongs to the stereotype of
Marie Antoinette reasoning. . .

Explanation 2 – Many talking about biofuels do not know energy analysis After the first oil crisis at the beginning of the 70s there was a boom of studies in
energy analysis. In this period several methods were developed to assess the quality and potentiality of primary energy sources. However, the first generation of
energy analysts that "cried wolf" too early has soon been forgotten together with
the work they generated. Energy analysis has been removed from the scientific
agenda and from academic courses (resisting only in departments of anthropol-
ogy or farming system analysis). As a matter of fact, we happen to be among the
organizers of a conference "Biennial International Workshop Advances in Energy
Studies" http://www.chim.unisi.it/portovenere/ held any other year since 1998. We
can confidently claim that within the historic community of energy analysts it is
impossible to find a single scientist, who believes that the production of biofuels from energy crops can be considered as a viable and desirable alternative to
oil. All those that had the opportunity to study basic principles of energy analysis know very well that the quality of a primary energy source has to be assessed
considering the overall EROI. Other scientists claim that it is just a matter of using
common sense – e.g. work of Cottrell (1955); Smil (1983, 1991, 2001,2003) and
Pimentel and Pimentel (1979) – to conclude that food is more valuable of fossil
fuel for any type of society. There are others that propose elaborated approaches
to account for the differences in quality between energy sources, energy carriers
and end uses. By doing so, energy analysis can explain pretty well the link between energy and economic growth (Ayres et al., 2003; Ayres and Warr, 2005;
Cleveland et al., 1984, 2000; Costanza and Herendeen, 1984; Gever et al., 1991;
Hall et al., 1986; Jorgenson, 1988; Kaufmann, 1992). This literature is extremely

clear and effective in making the intended points. There is no chance to power a developed economy on biofuels. So the real issue to be explained is how it comes that all the existing work in energy analysis is at the moment completely ignored by those proposing to invest large amount of money in the production of biofuel from energy crops. This fact calls for another explanation.

Explanation 3 – Biofuels from energy crops represent the last hope for the agonizing paradigm of industrial agriculture In the third millennium, finally, the crisis of the industrial paradigm of agriculture (called also high external input agriculture) is becoming evident also for those that would prefer ignoring it. High input agriculture is now experiencing what is called in jargon "Concorde Syndrome": technological investments and technological progress have the goal of doing more of the same, even though nobody is happy with that "same". High tech agriculture is only capable of producing agricultural surplus that do not have a demand in developed countries and that are too expensive for developing countries (Giampietro, 2007b). Moreover: (A) one of the original goal of the industrialization of agriculture – getting rid of the farmers as quick as possible, in order to be able to move more workers into the industrial and service sectors – does no longer make sense both in developed and in developing countries (Giampietro, 2007b); (B) the hidden costs associated with industrial agriculture, carefully ignored by those willing to preserve the "status quo" are becoming huge: (i) in relation to the health (obesity, diabetes, cardiovascular diseases, accumulation of hormones and pesticides in the food system); (ii) in relation to the environment (soil erosion, loss of biodiversity and natural habitat, pollution and contamination of the water table, alteration of water cycles, loss of natural landscapes); (iii) in relation to the social fabric, especially in rural areas (loss of tradition, loss of the symbolic and cultural dimension of food, loss of traditional landscapes); (iv) in relation to the economy (subsidies and indirect economic support are becoming more and more needed due to the market treadmill – the costs of production grows faster than sales prices). For all these reasons there is "a spectre haunting the establishment of the agricultural sector". The spectre is represented by the hypothesis that the subsidies to the production of agricultural commodities will be sooner or later phased out. As a consequence of this it will be necessary to negotiate a new "social contract" with the farmers about the new role that agriculture has to play in modern and sustainable societies. This contract will not rely on the massive adoption of the industrial agriculture paradigm.

This is the last explanation for the enthusiasm about the idea of using agriculture to produce biofuels. This would represent a third fat bird to be killed with the same rock (moving to the sum $1 + 1 + 1 = 3$). Not only biofuels are supposed to: (i) replace oil in a renewable way; (ii) generate zero emission, but also (iii) stabilize the "status quo" in the agricultural sector, in face of the agonizing paradigm of industrial agriculture. Putting in another way, by switching to biofuels it would be possible to keep the existing flow of subsidies into commodity production within the industrial paradigm of agriculture with virtually no limits. In fact, a self-sufficient biofuel system consumes almost entirely what it produces in its own operation, so that the supply of energy crops for biofuel will never be too much. For those willing to keep receiving subsidies for industrial agriculture the subsidized production of biofuels is very close to the invention of the machine of perpetual motion!

Acknowledgments The first author gratefully acknowledges the financial support for the activities of the European Project DECOIN – FP6 2005-SSP-5-A: 044428.

References

Adams, R.N. 1988. *The Eighth Day: Social Evolution as the Self-Organization of Energy*. University of Texas Press, Austin.

Adriaanse, A., Bringezu, S., Hammond, Y., Moriguchi, Y., Rodenburg, E., Rogich, D., and Schütz, H. 1997. *Resource Flows: The Material Basis of Industrial Economies*. World Resources Institute, Washington DC.

Ayres, R.U., Ayres, L.W., and Warr, B. 2003. Exergy, power and work in the US economy, 1900–1998. *Energy* 28(3), 219–273.

Ayres, R.U. and Simonis, U.E. 1994. *Industrial Metabolism: Restructuring for Sustainable Development*. United Nations University Press, New York.

Ayres, R.U. and Warr, B. 2005. Accounting for growth: The role of physical work. *Structural Change and Economic Dynamics* 16(2), 181–209.

Batty, J.C., Hamad, S.N., and Keller, J., 1975. Energy inputs to irrigation. *Journal of Irrigation and Drainage Div. ASCE* 101(IR4), 293–307.

Brody, S. 1945. *Bioenergetics and Growth*. Reinhold Publ. Co. New York, pp. 1023.

Carnot, S. 1824. *Reflexions sur la puissance motrice du feu sur les machines propres a developper cette puissance*. Bachelier, libraire. Paris.

Cleveland, C.J. 1992. Energy quality and energy surplus in the extraction of fossil fuels in the U.S. *Ecological Economics* 6, 139–162.

Cleveland, C.J., Costanza, R., Hall, C.A.S., and Kaufmann, R. 1984. Energy and the U.S. Economy: A Biophysical Perspective. *Science* 225(4665), 890–897.

Cleveland, C.J., Hall, C.A.S., and Herendeen, R.A. 2006. Letters – energy returns on ethanol production. *Science* 312, 1746.

Cleveland, C.J., Kaufmann, R., and Stern, S.I. 2000. Aggregation and the role of energy in the economy. *Ecological Economics* 32, 301–317.

Costanza, R. 1980. Embodied energy and economic valuation. *Science 210*, 1219–1224.

Costanza, R. and Herendeen, R. 1984. Embodied energy and economic value in the United States Economy: 1963, 1967 and 1972. *Resources and Energy 6*, 129–163.

Cottrell, W.F. 1955. *Energy and Society: The Relation between Energy, Social Change, and Economic Development*. McGraw-Hill, New York.

Debeir, J.-C., Deleage, J.-P., and Hemery, D. 1991. *In the Servitude of Power: Energy and Civilization through the Ages*. Zed Books Ltd., Atlantic Highlands, NJ.

De Carvalho Macedo I. 2005. Sugar Cane's Energy: Twelve studies on Brazilian sugar cane agribusiness and its sustainability. UNICA – Sugar Cane Agroindustry Union – Berlendis Editores Ltda São Paulo, Brazil.

Dekkers, W.A., Lange, J.M., and de Wit, C.T. 1978. Energy production and use in Dutch agriculture. *Netherlands Journal of Agricultural Sciences* 22, 107–118.

Duchin, F. 1998. *Structural Economics: Measuring Change in Technology, Lifestyles, and the Environment*. Island Press, Washington DC.

FAO. 2001. Human Energy Requirements. Report of a joint FAO/WHO/UNU Expert Consultation Rome 17-24 Ocrober 2001. FAO Food and Nutrition Technical Report Series

Farrell, A.E., Plevin, R.J., Turner, B.T., Jones, A.D., O'Hare, M.O., and Kammen, D.M. 2006. Ethanol can contribute to energy and environmental goals. *Science* 311, 506–508.

Fischer-Kowalski, M. 1998. Societal Metabolism: The intellectual history of material flow analysis part I, 1860–1970. *Journal of Industrial Ecology* 2(1), 61–78.

Fluck, R.C. 1981. Net Energy Sequestered in agricultural labor. *Transactions of the American Society of Agricultural Engineers* 24, 1449–1455.

Fluck, R.C. 1992. Energy of Human Labor In: R.C. Fluck (Editor) *Energy in Farm Production* (Vol. 6 of Energy in World Agriculture) pp. 31–37 Elsevier Amsterdam.

Fraser, R. and Kay, J.J., 2002. "Exergy Analysis of Eco-Systems: Establishing a Role for the Thermal Remote Sensing" In: D. Quattrochi and J. Luvall (Eds.), *Thermal Remote sensing in Land Surface Processes*, Taylor & Francis Publishers.

Georgescu-Roegen, N. 1971. *The Entropy Law and the Economic Process*. Harvard University Press, Cambridge, MA.

Georgescu-Roegen, N. 1975. Energy and economic myths. *Southern Economic Journal* 41, 347–381.

Gever, J., Kaufmann, R., Skole, D., and Vörösmarty, C. 1991. *Beyond Oil: The Threat to Food and Fuel in the Coming Decades*. University Press of Colorado, Niwot.

Giampietro, M. 1997a. Socioeconomic pressure, demographic pressure, environmental loading and technological changes in agriculture. *Agriculture, Ecosystems and Environment* 65, 201–229.

Giampietro, M. 1997b Linking technology, natural resources, and the socioeconomic structure of human society: A theoretical model. In: L. Freese (Ed.), *Advances in Human Ecology*, Vol. 6. JAI Press, Greenwich (CT), pp. 75–130.

Giampietro, M. (guest editor) 2000. Societal metabolism, part 1: Introduction of the analytical tool in theory, examples, and validation of basic assumptions. *Population and Environment*, special issue, 22(2), 97–254.

Giampietro, M. (guest editor) 2001. Societal metabolism, part 2: Specific applications to case studies. *Population and Environment*, special issue, 22(3), 257–352.

Giampietro, M. 2003. *Multi-Scale Integrated Analysis of Ecosystems*. CRC Press, Boca-Raton, FL. 437pp.

Giampietro, M. 2006. Theoretical and practical considerations on the meaning and usefulness of traditional energy analysis *Journal of Industrial Ecology* 10(4). 173–185.

Giampietro, M. 2007a. Studying the "addiction to oil" of developed societies using the approach of Multi-Scale Integrated Analysis of Societal Metabolism (MSIASM) – In: F. Barbir and S. Ulgiati (editors) *Sustainable Energy Production and Consumption and Environmental Costing* – NATO Advanced Research Workshop – NATO Science for Peace and Security Series: C-Environmental Security Springer, The Netherlands.

Giampietro, M. 2007b. The future of agriculture: GMOs and the agonizing paradigm of industrial agriculture. In: S. Funtowicz and A. Guimaraes (Editors) *Science for Policy* Oxford University Press.

Giampietro, M., Allen, T.F.H., and Mayumi, K. 2006b. The epistemological predicament associated with purposive quantitative analysis *Ecological Complexity* 3(4), 307–327.

Giampietro, M., Bukkens, S.G.F., and Pimentel, D. 1997a. Linking technology, natural resources, and the socioeconomic structure of human society: Examples and applications. In: L. Freese (Ed.), *Advances in Human Ecology*, Vol. 6. JAI Press, Greenwich (CT), pp. 131–200.

Giampietro, M. and Mayumi, K. 2000a. Multiple-scale integrated assessment of societal metabolism: Introducing the approach. *Population and Environment* 22(2), 109–153.

Giampietro, M. and Mayumi, K. 2000b. Multiple-scales integrated assessments of societal metabolism: Integrating biophysical and economic representations across scales. *Population and Environment* 22(2), 155–210.

Giampietro, M. and Mayumi, K. 2004. Complex systems and energy. In: C. Cleveland, (Ed.), *Encyclopedia of Energy*. Vol. 1, pp. 617–631. Elsevier, San Diego, CA.

Giampietro, M., Mayumi, K., and Bukkens, S.G.F. 2001. Multiple-scale integrated assessment of societal metabolism: an analytical tool to study development and sustainability. *Environment, Development and Sustainability* 3(4), 275–307.

Giampietro, M., Mayumi, K., and Munda, G. 2006a. Integrated assessment and energy analysis: Quality assurance in multi-criteria analysis of sustainability. *Energy* 31(1), 59–86.

Giampietro, M., Mayumi, K., and Ramos-Martin, J. 2006c. Can biofuels replace fossil energy fuels? A multi-scale integrated analysis based on the concept of societal and ecosystem metabolism: Part 1. *International Journal of Transdisciplinary Research* 1(1), 51–87 [accessibile sul sito http://www.ijtr.org/]

Giampietro, M., Mayumi, K., and Ramos-Martin, J. 2007. How serious is the addiction to oil of developed society? A multi-scale integrated analysis based on the concept of societal and ecosystem metabolism *International Journal of Transdisciplinary Research* 2(1), 42–92 – available on line at: www.ijtr.org

Giampietro, M., Pastore, G., and Ulgiati, S. 1998. Italian agriculture and concepts of sustainability. In: E. Ortega and P. Safonov (Eds.), *Introduction to Ecological Planning Using Emergy Analysis with Brazilian Case Studies*. LEIA-FEA Unicamp, Campinas, Brazil.

Giampietro, M. and Pimentel, D. 1990. Assessment of the energetics of human labor. *Agriculture, Ecosystems and Environment* 32, 257–272.

Giampietro, M. and Ramos-Martin, J. 2005. Multi-scale integrated analysis of sustainability: a methodological tool to improve the quality of the narratives. *International Journal of Global Environmental Issues* 5(3/4), 119–141.

Giampietro, M. and Ulgiati, S. 2005. An integrated assessment of large-scale biofuel production. *Critical Review in Plant Sciences* 24, 1–20.

Giampietro, M., Ulgiati, S., and Pimentel, D. 1997b. Feasibility of large-scale biofuel production: Does an enlargement of scale change the picture? *BioScience* 47(9), 587–600.

Gilliland, M.W., Ed., 1978. *Energy Analysis: A New Policy Tool*. Westview Press, Boulder, CO.

Hagens, N., Costanza, R., Mulder, K. 2006. Letters – Energy Returns on Ethanol Production. *Science* 312, 1746.

Hall, C.A.S.; Cleveland, C.J.; and Kaufman, R. 1986. *Energy and Resource Quality*. New York: John Wiley & Sons.

Herendeen, R.A. 1981. Energy intensities in economic and ecological systems. *Journal of Theoretical Biology* 91, 607–620.

Herendeen, R.A. 1998. *Ecological numeracy: quantitative analysis of environmental issues* John Wiley & Sons, New York.

Hudson, J.C., 1975. Sugarcane: its energy relationship with fossil fuel. *Span* 18, 12–14.

IFIAS (International Federation of Institutes for Advanced Study), 1974. *Energy Analysis*. International Federation of Institutes for Advanced Study, Workshop on Methodology and Conventions – Report No. 6. IFIAS, Stockholm, p. 89.

Jevons, W.S., [1865] 1965. *The Coal Question: an inquiry concerning the progress of the nation, and the probable exhaustion of our coal-mines*. A. W. Flux (Ed.), 3rd ed. rev. Augustus M. Kelley, New York.

Jorgenson, D.W. 1988. Productivity and economic growth in Japan and the United States. *The American Economic Review* 78(2), 217–222.

Kaufmann, R.K. 1992. A biophysical analysis of the energy/real GDP ratio: implications for substitution and technical change. *Ecological Economics* 6, 35–56.

Kaufmann, R.K. 2006. Letters – energy returns on ethanol production. *Science* 312, 1747.

Kay, J. 2000. 'Ecosystems as self-organizing holarchic open systems: narratives and the second law of thermodynamics', In: S.E. Jorgensen and F. Muller (Eds.), *Handbook of Ecosystems Theories and Management*, London: Lewis Publishers, pp. 135–160

Leach, G. 1976. *Energy and Food Production*. I.P.C. Science and Technology Press limited, Surrey, U.K.

Lotka, A.J. 1922. Contribution to the energetics of evolution. *Proceedings Natural Academy Science* 8, 147–154.

Lotka, A.J. 1956. *Elements of Mathematical Biology*. Dover Publications, New York.

Martinez-Alier, J. 1987. *Ecological Economics. Energy, Environment and Society*. Blackwell, Oxford, U.K.

Matthews, E., Amann, C., Fischer-Kowalski, M., Bringezu, S., Hüttler, W., Kleijn, R., Moriguchi, Y., Ottke, C., Rodenburg, E., Rogich, D., Schandl, H., Schütz, H., van der Voet, E., and Weisz, H. 2000. *The Weight of Nations: Material Outflows from Industrial Economies*. World Resources Institute, Washington, DC.

Mayumi, K. 1991. Temporary emancipation from land: from the industrial revolution to the present time. *Ecological Economics* 4, 35–56.

Mayumi, K. 2001. *The Origin of Ecological Economics: The Bioeconomics of Georgescu-Roegen*. Routledge, London, UK.

Mayumi, K. and Giampietro, M. 2004. Entropy in ecological economics. In: J. Proops and P. Safonov (Eds.), *Modeling in Ecological Economics*. Edward Elgar, Cheltenham (UK), pp. 80–101.

Mayumi, K. and Giampietro M. 2006. The epistemological challenge of self-modifying systems: Governance and sustainability in the post-normal science era *Ecological Economics* 57, 382–399.

Morowitz, H.J. 1979. *Energy Flow in Biology*. Ox Bow Press, Woodbridge, CT.

Norman, M.J.T. 1978. Energy inputs and outputs of subsistence cropping systems in the tropics. *Agro-Ecosystems* 4, 355–366.

Odum, H.T. 1971. *Environment, Power, and Society*. Wiley-Interscience, New York.

Odum, H.T. 1983. *Systems Ecology*. John Wiley, New York.

Odum, H.T. 1996. *Environmental Accounting: Emergy and Environmental Decision Making*. John Wiley, New York.

Odum, H.T. and Pinkerton, R.C. 1955. Time's speed regulator: the optimum efficiency for maximum power output in physical and biological systems. *American Scientist* 43, 331–343.

Ostwald, W. 1907. The modern theory of energetics. *The Monist* 17, 481–515.

Patzek, T. 2004. Thermodynamics of the corn-ethanol biofuel cycle. *Critical Review Plant Sciences* 23(6), 519–567.

Patzek, T.W. 2006. Letters – Energy Returns on Ethanol Production. *Science* 312, 1747.

Patzek, T.W. and Pimentel, D. 2005. Thermodynamics of Energy Production from Biomass. *Critical Reviews in Plant Sciences* 24(5–6), 327–364.

Pimentel, D., Patzek T., and Cecil G. 2007. Ethanol production: Energy, Economic, and Environmental losses. *Reviews of Environmental Contamination & Toxicology* 189, 25–41.

Pimentel, D. and Pimentel, M. 1979. *Food Energy and Society*. Edward Arnold Ltd., London.

Pimentel, D. and Pimentel, M. 1996. *Food, Energy and Society* (revised edition) University Press of Colorado, Niwot Co.

Podolinsky, S. 1883. Menschliche arbeit und einheit der kraft. *Die Neue Zeit* (Stuttgart, IHW Dietz), p. 413. (In German).

Prigogine, I. 1978. *From Being to Becoming*. W.H. Freeman and Company, San Francisco, CA.

Prigogine, I. and Stengers, I. 1981. *Order out of Chaos*. Bantam Books, New York.

Ramos-Martin, J., Giampietro, M., and Mayumi, K. 2007. On China's exosomatic energy metabolism: an application of multi-scale integrated analysis of societal metabolism (MSI-ASM). *Ecological Economics* 63(1), 174–191.

Revelle, R., 1976. Energy use in rural India. *Science* 192, 969–975.

Rosen, R. 1958. The representation of biological systems from the standpoint of the theory of categories. *Bullettin of Mathematical Biophysics* 20, 317–341.

Schneider, E.D. and Kay, J.J. 1994. "Life as a manifestation of the second law of thermodynamics". *Mathematical and Computer Modelling* 19, 25–48

Schneider, E.D. and Kay, J.J. 1995. "Order from Disorder: The Thermodynamics of Complexity in Biology", In: Michael P. Murphy, Luke A.J. O'Neill (Eds.), *What is Life:The Next Fifty Years. Reflections on the Future of Biology*, Cambridge University Press, Cambridge, pp. 161–172.

Schrödinger, E. 1967. *What is Life & Mind and Matter.* Cambridge University Press, London.

Slesser, M. 1978. *Energy in the Economy*. MacMillan, London.

Slesser, M. and King, J. 2003. *Not by Money Alone: Economics as Nature Intended* Jon Carpenter Publishing, Charlbury, Oxon.

Rappaport, R.A. 1971. The flow of energy in an agricultural society. *Scientific American* 224, 117–133.

Shapouri, H., Duffield, J., and Wang, M. 2002. *The Energy Balance of Corn-Ethanol: An Update.* Report 813. USDA Office of Energy Policy and New Uses, Agricultural Economics, Washington, DC.

Smil, V. 1983. *Biomass Energies* Plenum Press, New York.

Smil, V. 1988. *Energy in China's Modernization.* M.E. Sharpe, Armonk, New York.

Smil, V. 1991. *General Energetics.* Wiley, New York.

Smil, V. 2001. *Enriching the Earth.* The MIT Press, Cambridge, MA.

Smil, V. 2003. *Energy at the crossroads: Global Perspectives and Uncertainties.* The MIT Press, Cambridge MA.

Tainter, J.A. 1988. *The Collapse of Complex Societies.* Cambridge University Press, Cambridge.

Ulanowicz, R.E. 1986. *Growth and Development: Ecosystem Phenomenology.* Springer-Verlag, New York.

Ulgiati, S., Brown, M., Giampietro, M., Herendeen, R., and Mayumi, K. (Eds.) 1998. *Proceedings of the Biennial International Workshop Advances in Energy Studies (1): Energy flows in Ecology and Economy.* Porto Venere, Italy 26–30 May 1998 – MUSIS (Museum of Science and Scientific Information), Rome, Italy.

USDA 2005a. http://www.ars.usda.gov/research/programs/programs.htm?np_code=307&docid=281&page=1

USDA 2005b. http://www1.eere.energy.gov/biomass/economic_growth.html

USDA 2006 http://www.nass.usda.gov/Publications/Ag_Statistics/2006/CHAP01.PDF (Table 1–45).

Watt, K. 1989. Evidence of the role of energy resources in producing long waves in the US economy. *Ecological Economics* 1, 181–195.

White, L.A. 1943. Energy and evolution of culture. *American Anthropologist* 14, 335–356.

White, L.A. 1959. *The Evolution of Culture: The Development of Civilization to the Fall of Rome.* Mac Graw-Hill, New York.

Williams, D.W., McCarty, T.R., Gunkel, W.W., Price, D.R., and Jewell, W.J., 1975. Energy utilization on beef feed lots and dairy farms. In: W.J. Jewell, (ed.), *Energy, Agriculture and Waste Management.* Ann Arbor Science Publishers, Ann Arbor, pp. 29–47.

Zemmelink G. 1995. Allocation and Utilization of Resources at the Farm Level In: *A Reseacrh Approach to Livestock Production from a Systems Perspective* – Proceedings of the Symposium "A farewell to Prof. Dick Zwart" – Dept. of Animal Production Systems – Wageningen Agricultural University, pp 35–48.

Chapter 9
Sugarcane and Ethanol Production and Carbon Dioxide Balances

Marcelo Dias De Oliveira

Abstract Ethanol fuel has been considered lately an efficient option for reducing greenhouse gases emissions. Brazil has now more than 30 years of experience with large-scale ethanol production. With sugarcane as feedstock, Brazilian ethanol has some advantages in terms of energy and CO_2 balances. The use of bagasse for energy generation contributes to lower greenhouse gases emissions. Although, when compared with gasoline, the use of sugarcane ethanol does imply in reduction of GHG emissions, Brazilian contribution to emission reductions could be much more significant, if more efforts were directed for reduction of Amazon deforestation. The trend however is to encourage ethanol production.

Keywords Sugarcane ethanol · CO_2 mitigation · CO_2 balances · bagasse · Co-generation

9.1 Introduction

When the oil crisis hit Brazilian economy, and raised concerns about national sovereignty in the mid-70's, sugarcane industrialists were quick to perceive in the scenario an opportunity to avoid bankruptcy. After some ups and downs of the Brazilian ethanol program the same sector is taking advantage of another scenario, this time related to growing environmental concerns regarding global warming. Brazil now has jumped on the bandwagon of the environmentally friendly fuel alternative, and is experiencing a revival of the ethanol program, the Pró-alcool, first established in the mid 70's.

Government incentives and subsides established by the Pró-alcool program, let the country to experience a considerable increase of ethanol production and ethanol-fueled automobile passenger fleet. By 1984, 94.4% of the passenger cars in Brazil were fuelled by ethanol. Posterior decline in oil prices associated with increase of

✉ M.D. De Oliveira
Avenida 10, 1260, Rio Claro - SP - Brazil, CEP 13500-450
e-mail: dias_oliveira@msn.com

D. Pimentel (ed.), *Biofuels, Solar and Wind as Renewable Energy Systems*,
© Springer Science+Business Media B.V. 2008

Brazilian domestic production and high prices of sugar contributed to an expressive reduction of ethanol production in the country. By 1999, ethanol-fueled cars fell to less of one percent of total sales (Rosa and Ribeiro, 1998).

Current enthusiasm with Brazilian biofuels, particularly sugarcane ethanol, is motivated by increasing worldwide concerns with climate change. Government, society and scientists talk passionately about the benefits of a "green" energy source and possible Brazilian contributions for the reducing of greenhouse gases (GHG) emissions. The ethanol industry is quickly capitalizing the benefits of these circumstances, and Brazilian government is clearly willing to encourage increases for ethanol production.

The present study analyses the CO_2 balance for Brazilian sugarcane ethanol and its possible contributions for GHG mitigation.

9.2 The "Green" Promise

Biofuels are frequently portrayed as "clean fuel" (Moreira and Goldemberg, 1999; Macedo, 1998) and considered to be carbon neutral, since CO_2 emitted through combustion of motor fuel is reabsorbed by growing more sugarcane rendering the balance practically zero (Rosa and Ribeiro, 1998). Numerous articles advocate for an increase in biofuels production and consumption as an environmentally friendly option (Macedo, 1998; Moreira and Goldemberg, 1999 and Farrel et al., 2006).

Sugarcane ethanol is considered and efficient way of reducing CO_2 emissions of energy production. According to Rosa and Ribeiro (1998), the use of ethanol fuel can have a significant contribution to greenhouse gas mitigation. Moreira and Goldemberg (1999), consider the main attractiveness of the Brazilian ethanol program, the reduction of CO_2 emissions compared with fossil fuels, as a solution for industrialized countries to fulfill their commitments with the United Nations Framework Climate Change Convention (UNFCCC). Beeharry (2001), points out that since the net CO_2 released per unit of energy produced is significantly lower compared to fossil fuels, sugarcane bioenergy systems stand out as promising candidates for GHG mitigation. Feedstock for ethanol production, in this particular case, sugarcane, grows by transforming CO_2 from atmosphere and water into biomass, which is, as mentioned before the reason why such fuel is called carbon neutral. Nonetheless, fossil fuel emissions are always associated with any agricultural activity.

9.3 CO_2 Emissions of Sugarcane Ethanol

It has been a popular misconception that bioenergy systems have no net CO_2 emissions (Beeharry, 2001). Considerable amounts of fossil fuel inputs are required for plant growth and transportation, as well as for ethanol distribution, therefore CO_2 emissions are present during the process of ethanol production. Fertilizers, herbi-

Table 9.1 Carbon Dioxide emissions from the agricultural phase of Brazilian sugarcane production

Constituent per ha	Quantity per ha	CO_2 release per unit of constituent[4]	CO_2 release
Nitrogen	70.0 kg[1]	3.14 per Kg	220.0 kg
Phosphorous (P_2O_5)	23.0 kg[1]	0.61 per Kg	14.0 kg
Potassium (K_2O)	132.0 kg[1]	0.44 per kg	58.1 kg
Lime	1500.0 kg[1]	0.13 per kg	195.0 kg
Herbicides	0.5 kg[2]	17.20 per kg	8.6 kg
Insecticides	3.0 kg[2]	18.10 per kg	54.3 kg
Diesel fuel[α]	350.0 L[3]	3.08 per L	1078.0 kg
Total			1628.0 kg

[1] Grupo Cosan – Brasil.
[2] Pimentel and Pimentel – 1996.
[3] Based on Pimentel and Pimentel – 1996.
[4] West and Marland (2002).
[α] values correspondent to oil consumption of all agricultural activities and transport of sugarcane to distilleries.

cides and insecticides have net CO_2 emissions associated with their production, distribution and application. CO_2 emissions from agricultural inputs of sugarcane production are represented on Table 9.1.

Sugarcane production also results in emissions of other GHG, namely methane and nitrous oxide. Based on Lima et al. (1999), CH_4 and N_2O emissions from sugarcane correspond to 26.9 and 1.33 kg per hectare respectively. Such emissions correspond to, based on Schlesinger (1997), 672 kg and 399 kg respectively of CO_2 equivalent.

As for its distribution, based on Shapouri et al. (2002), 0.44 GJ are required per m^3 of ethanol, assuming diesel fuel is the source of this energy, and based on West and Marland (2002) CO_2 emissions associated with ethanol distribution are of 227 kg. Therefore net CO_2 emissions from ethanol production is 2926 kg CO_2/ha of sugarcane (Table 9.2).

Theoretically, there are no GHG emissions associated with distillery operations. All the energy required comes from the burning of bagasse, which is a residue of the milled sugarcane. In fact the burning of bagasse generates more energy than the distillery requires, resulting in some surplus of energy. Conceptually CO_2 emissions associated with bagasse burning are not accounted for, since where sequestered

Table 9.2 Carbon dioxide emissions from Brazilian ethanol production

Process	CO_2 equivalent emissions per ha
Agriculture	1628 kg
CH_4	672 kg
N_2O	399 kg
Ethanol distribution	227 kg
Total	2926 kg

during sugarcane growth and will be re-absorbed in the next season. The same rationale applies to the ethanol burning in mother vehicles. For accounting purposes a complete combustion is assumed in both cases.

Based on an average production of 80 tons per ha which is representative of the State of São Paulo, (Braunbeck et al., 1999), and ethanol conversion efficiency of 80 L per ton of sugarcane processed (Moreira and Goldemberg, 1999); the amount of ethanol resulting from one ha or sugarcane plantations is $6.4 \, m^3$. Consequently for production of one m^3 of ethanol, GHG emissions account to 457 kg of CO_2eq production and distribution, this corresponds to approximately 19 kg of CO_2 per gigajoule (kg/GJ) of fuel. Comparative values of CO_2 emission of other fuel sources are indicated on Table 9.3.

Estimating the potential for GHG reduction from the use of ethanol derived from sugarcane requires a comparison with the fossil fuel displaced. In Brazil the automobile fleet has basically three fuel options, natural gas, ethanol and gasoline, the last option is actually a mixture of gasoline and ethanol. The proportion of each fuel varies slightly according to government decisions, currently is 75% gasoline and 25% ethanol. Natural gas running automobiles are not manufactured in Brazil, but automobiles can be converted to natural gas at a price ranging from US$ 1200 to US$ 2100.[1] Although conversion to natural gas continues to rise in Brazil stimulated by its fuel economy, currently such vehicles represent only about 5% of the automobile fleet. The main attention in this work will be devoted to the impacts of ethanol substitution for gasoline.

In 2003, Brazil began to produce flex fuel cars, which can run with both gasoline and ethanol in any proportion using the same tank. In that year about 40 thousand of such automobiles were produced, corresponding to only 2.6% of the new cars. In 2006, flex fuel cars corresponded to almost 60% of the new cars with 1.25 million units (Anfavea, 2007). This augment is directly related with a strategy for increasing biofuel consumption in Brazil, where the consumer is stimulated to use ethanol as an environmental responsible option. The differences in price between ethanol and gasoline also contribute for the scenario. Presently in Brazil, ethanol is about 49% cheaper than gasoline, mostly due to heavier incidence of taxes over gasoline. The

Table 9.3 Comparative emissions of different fuels

Fuel	CO_2/GJ (kg)
Sugarcane ethanol (Brazil)	19
Corn ethanol (USA)	56[α]
Gasoline	78[β]
Natural Gas	53[β]
Coal	92[β]
Diesel	80[β]

[α] Dias de Oliveira et al. (2005).

[β] West and Marland (2002).

[1] Based on Dondero and Goldemberg (2005) and considering 1 US$= 2 reais

advantage of flex fueled cars is that owners can trade back and forth between ethanol and gasoline according to the prices at the pump.

9.4 Gasoline Versus Ethanol

To estimate the effectiveness that ethanol fuel has on reducing GHG emissions for Brazilian conditions, a comparison is made considering the fuel economy of flex fuel automobiles when using ethanol or gasoline.

As mentioned before the production and distribution of one m^3 of ethanol results in emissions of 457 kg of CO_2eq. Assuming a kilometerage for Brazilian flex fuelled cars of 11.78 km/L for gasoline and 8.92 km/L for ethanol.[2] A flex fuelled car using one m^3 of pure ethanol can run for 8920 km, to travel the same distance using gasoline as fuel 757 L are necessary. Given that gasoline in Brazil is actually sold as a mixture of 75% gasoline and 25% ethanol, such volume of gasohol corresponds to 568 L of gasoline and 189 L of ethanol. According to West and Marland (2002), production, distribution and combustion of one m^3 of gasoline result in emissions of 2722 kg of CO_2, therefore the 568 L of gasoline will result in 1546 kg CO_2. For the 189 L of ethanol, the amount of CO_2 emitted correspond to 86 kg, consequently total CO_2 emissions add up to 1632 kg. Hence ethanol option represents 1175 kg of CO_2 emissions avoided per m^3 produced. In the hypothesis of pure gasoline being used instead of gasohol, to substitute one m^3 of ethanol used, approximately 673 L of gasoline are required, resulting in total emissions of 1832 Kg, that is, 1375 Kg CO_2 more than the ethanol being replaced.

9.5 Bagasse as a Source of Energy

The bagasse, is the residue of sugarcane after the same is milled. It has approximately 50% humidity and results in amounts of 280 kg/t of sugarcane (Beeharry, 2001).

The burning of bagasse provides heat for boilers that generate steam and produce the energy required for distillery operations. Since the energy generated surpass distillery necessities, this surplus of electricity has potential for being exported, which is usually known as cogeneration, and according to Beeharry (1996), offers the opportunity to increase the value added while diversifying revenue sources for distilleries. According to Rosa and Ribeiro (1998), the utilization of sugar-cane bagasse for electricity generation may become the great technological breakthrough for Pró-álcool in the context of sustained economic development while conserving the environment. They point out that the period of harvest of the sugar cane corresponds to the "dry period" in the Brazilian hydroelectric system, thus making the

---

[2] Average values based on three of the most sold cars in Brazil, Volkswagen Gol, Fiat Palio, and Celta-Chevrolet, according to Paulo Campo Grande - Quatro Rodas.

use of bagasse in the area particularly attractive for complementing hydroelectricity generation.

Brazilian distilleries generate an average surplus of 1.54 GJ (428 kWh) per ha or sugarcane processed (Dias de Oliveira, 2005). This corresponds to boilers producing steam operating at pressures of 20 bar generating small amounts of electricity (15–20 kWh/ton of cane) enough for the needs of the unit (Moreira and Goldemberg, 1999).

According to Beeharry (1996), advanced technologies could result in the generation of 0.72 GJ (200 kWh) per ton of sugarcane milled. Such scenario would result in a value of energy surplus per ha or sugarcane of approximately 54 GJ (15000 kWh) or 8.43 GJ (2342 kWh) per m^3 of ethanol. Intermediate values indicated by Beeharry (1996), result in the generation of 0.45 GJ (125 kWh) of electricity per ton of sugarcane milled, representing a surplus of 32.4 GJ (9000 kWh) per ha of sugarcane or 5.06 GJ (1406 kWh), per m^3 of ethanol.

According to personal communication in a visit to the Center for Sugarcane Technology (CTC) – Piracicaba, boilers operating with pressures of 20 bars are so far the standard in Brazilian operating distilleries, with new plants being equipped with boilers that work at pressures of 60 bars, and are capable of generating a surplus of 0.14 GJ (40 kWh) of energy per ton of sugarcane milled. Still according to CTC, advanced technologies are yet economically unfeasible.

To better illustrate the impacts that the conditions mentioned above would have in terms of CO_2 emissions, a comparison will be made with current Brazilian system of electricity generation. According to Brazilian National Agency of Electricity Energy (ANEEL), electricity generation in Brazil comes from the sources indicated on Table 9.4.

With the dominance of hydroelectricity generation, Brazilian electricity matrix is responsible for relatively low CO_2 emissions per kWh of electricity produced (kWh$_{el}$). Compared with other sources, hydroelectricity has low carbon dioxide intensity (Krauter and Ruthers, 2004; Weisser, 2007; van de Vate, 1997). An important point though, made by Rosa and Schaeffer (1995) and Fearnside (2002), is that emissions from hydroelectric dams can be much higher than usually attributed for this source, mostly owning to methane emissions resulting from anaerobic decomposition of organic matter of the inundated areas in hydroelectric reservoirs.

Considering Brazilian electric energy matrix and based on West and Marland (2002), Krauter and Ruthers (2004), and van de Vate (1997), each kWh$_{el}$ generated

Table 9.4 Brazilian electricity energy matrix

Source	Percentage
Hydroelectricity	80.23
Petroleum	4.54
Gas	11.42
Coal	1.47
Nuclear	2.09
Wind	0.25

Biomass not included.

Table 9.5 Estimated avoided emissions resulted from the use of ethanol as fuel instead of gasoline, and the surplus of electricity generated by distilleries*

Scenario Avoided emissions (kg)	kWh/ton (GJ/ton)	Avoided emissions (kg) per ha use of ethanol fuel	Avoided emissions (kg) per ha surplus electricity	Total
Current	20	7520	59	7579
60 bars boilers	~ 53	7520	445	7965
Intermediate	125	7520	1251	8771
Advanced	200	7520	2085	9605

* Values calculated do not account for energy losses associated with electricity transmission

in Brazil corresponds to net CO_2 emissions of approximately 139 grams, compared with the to 660 g per kWh_{el} of US calculated by West and Marland (2002) or the 530 kg/kWh_{el} and 439 Kg/kWh_{el} of Germany and Japan respectively, as calculated by Krauter and Ruthers (2004).

Consequently the surplus of electricity per ha of sugarcane is responsible for 59 kg of avoided CO_2 emissions per ha of sugarcane or 9 kg per m^3 of ethanol produced. With current Brazilian ethanol production of 16 million m^3, total avoided CO_2 emissions due to electricity generation correspond to 144,000 tons of CO_2 kg/year.

In the hypothesis that advanced technologies usually referred to as biomass integrated gasifier/gas turbine (BIG/GT) were the standard in Brazilian distilleries, the amount of CO_2 emissions avoided per ha of sugarcane would be of approximately 2085 kg or 326 kg per m^3 of ethanol. Intermediate technologies would represent avoided emissions of 1251 kg of CO_2 per ha or 195 kg CO_2 per m^3 of ethanol. Nevertheless, as mentioned before, advanced technologies are not yet economically feasible.

Considering differences in emissions from use of ethanol and gasoline, and the potential electricity generation of distilleries, avoided emissions for the possible scenarios of ethanol production in Brazil are summarized on Table 9.5.

The results above indicated that consumption of ethanol, produced with current practices in Brazil, reduces CO_2 atmospheric emissions by 1184 kg/m^3, when compared with gasoline use. Cardenas (1993), cited by Weir (1998), reports reduction in CO_2 emission of 1594 kg/m^3 of ethanol used in Argentina.

According to Beeharry (2001), the use not only of the bagasse, but also sugarcane tops and leaves can contribute to distilleries potential for electricity exportation; such option however, would imply the elimination of pre-harvest burning and the use of cane residues that would otherwise be left on the soil, contributing to reduce soil erosion.

9.6 Pre-Harvest Burning of Sugarcane and Mechanical Harvest

One aspect very criticized of sugarcane production is its pre-harvest burning, which has a series of negative impacts. The practice is adopted in order to facilitate the manual cut of the sugarcane. According to Kicrkoff (1991), pre-harvest burning is responsible for increasing the levels of carbon monoxide and ozone in areas where it is planted. Godoi et al. (2004) and Cancado (2003), report increases during the harvest season, of respiratory problems in cities neighboring sugarcane plantations. In 2002, legislation was passed in the state of São Paulo aiming to a gradual elimination of the pre-harvest burning; it established a period of 30 years to its complete elimination (Sirvinskas, 2003). Dias de Oliveira et al. (2005) mentions that pre-harvest burning usually reaches native vegetation surrounding sugarcane crops. Criticism and restrictions to the practice keep mounting and the government of Sao Paulo is working an agreement with the distilleries to completely eliminate the practice by the year of 2014.

With elimination of pre-harvest burning, sugarcane harvest will be made mechanically instead of manually, resulting in increase of the fossil fuel use on agricultural phase of ethanol production, and additional CO_2 emissions.

According CTC- Piracicaba, the harvester machines performances account for 1.045 L of diesel per ton of sugarcane harvested. As a result, mechanical harvest would imply in additional use of diesel fuel in a volume of approximately 84 L/ha resulting in an increase of 259 kg of CO_2 released per ha.

9.7 Distillery Wastes

One aspect usually not addressed in energy balances and thus, GHG emissions is the treatment of distillery wastes, the stillage, a liquid that in Brazil is usually called vinasse. Ethanol production results in vinasse amounts of 10–14 times the volume of ethanol. The characteristics of vinasse are its high concentration of nutrients and high biological oxygen demand (BOD), which ranges from 30 to 60 g/l, according to Navarro et al. (2000). The common destiny of this liquid is its application as a fertilizer in the sugarcane plantations. According to Moreira and Goldemberg (1999), the recommended rate of application is 100 m³/ha.

Such practices raise concerns about possible infiltration of vinasse resulting in groundwater contamination. Hassuda (1989) reports changes in groundwater quality due to vinasse infiltration in the Bauru aquifer localized in the state of Sao Paulo. Gloeden (1994), in another study area also report problems of groundwater contamination due to vinasse infiltration. According to Macedo (1998), transport and application of vinasse requires 41.5 L of diesel per ha, resulting in emissions of 128 kg of CO_2.

An alternative is its treatment, which would require one kWh (3.6 MJ) per kg of BOD removed, according to Trobish (1992), cited by Giampietro et al., 1997. Assuming the BOD values cited by Navarro et al. (2000), and the production of 12 L of vinasse per liter of ethanol, between 8.3 and 16.6 GJ (2304–4608 kWh) of energy is required for BOD clean up, leading up to emissions ranging from 320 to 640 kg of CO_2 per ha of sugarcane used for ethanol production or 50–100 kg of CO_2/m^3 of ethanol.

Another destiny for the vinasse could be its use for biogas production. Besides reducing an environmental problem, biogas production from vinasse is portrayed as an approach to increase the energy efficiency of ethanol production, contributing to mitigation of CO_2 emissions and environmental pollution load of distilleries.

Based on personal communication with CTC, the process of biogas production would result in an energy surplus equivalent of 3.9 GJ (1082 kWh) per ha of ethanol produced. However, according to Cortez et al. (1998), vinasse is not completely transformed in the process and still has high concentration of organic material after biogas production. Treatment of the remaining organic matter would require all the additional energy generated by the biogas, practically reducing to zero any benefit in terms of energy or CO_2 emissions. An study conducted by Granato (2003) at a distillery in the state of Sao Paulo reports a much lower potential of electricity generation from anaerobic decomposition of vinasse, about 47 MJ (13 kWh) per m^3 of ethanol produced, resulting in 299 MJ (83 kWh) of surplus per ha of sugarcane devoted to ethanol production.

The use of vinasse as fertilizer implies in additional use of fossil fuel and reduction of N, P, K and lime in the traditional way. The fossil fuel used for vinasse application results in additional emissions 128 kg CO_2 per ha of sugarcane. Reduction of fertilizer applied in the traditional way results also in reduction of CO2 emissions in the amount of 204 kg, based on Azania et al. (2003). The net result is a reduction in emissions of 76 kg of CO_2. There is also little variation regarding the net energy in both options, with or without vinasse application, corresponding to a reduction of just 3.7% in the last option.

9.8 Possible Additional Sources of Methane

As already mentioned before, common practice is the application of vinasse as a fertilizer in sugarcane crops, there is currently little information about CH4 emissions to the atmosphere resulted from vinasse decomposition, which might significantly affect GHG balances.

The increase of mechanical harvest, will result in a significant amount of residues (sugarcane tops and leaves), that would be otherwise burned in pre-harvest, to be left on the field, which can also become a source of methane emissions. A more detailed GHG balance would have undoubtedly to consider such aspects; therefore more research on these issues is essential.

9.9 CO_2 Mitigation

For the different alternative scenarios described above, avoided CO_2 emissions represented by the use of ethanol are summarized on Table 9.6.

Currently Brazil produces 4.2 billion gallons of ethanol or approximately 16 million m^3 per year, requiring around 3 million hectares of land (Goldemberg, 2007). Assuming ethanol conversion efficiency of 80 L per ton of sugarcane, the values above suggest an average yield for Brazil of approximately 67 tons of sugarcane per ha.

Table 9.6 Avoided CO_2 emissions for different scenarios of ethanol production, in terms of hectare of sugarcane planted or m^3 of ethanol produced

Bagasse use technology	CO_2 avoided option 1	CO_2 avoided option 2	CO_2 avoided option 3	CO_2 avoided option 4
Current	7579 (1184)	6939 (1084)	6681 (1044)	7320 (1144)
60 bars boilers	7965 (1245)	7325 (1145)	7066 (1104)	7706 (1204)
Intermediate	8771 (1370)	8131 (1270)	7872 (1230)	8512 (1330)
Advanced	9605 (1501)	8965 (1401)	8706 (1360)	9346 (1460)

Values in parenthesis represent avoided emissions per m^3 and values outside the parenthesis represent avoided emissions per ha.
Option 1 – Ethanol production without BOD treatment and with manual harvest.
Option 2 – Ethanol production with BOD treatment and with manual harvest.
Option 3 – Ethanol production with BOD treatment and mechanical harvest.
Option 4 – Ethanol production without BOD treatment and with mechanical harvest.
Values don't consider biogas production, nor fossil fuel consumption for the transport and application of vinasse in the fields.

The basic assumptions for calculations on this study assume a productivity of 80 tons of sugarcane per ha, and conversion efficiency of 80 L/ton, therefore an optimistic value for average yield, and consequently for energy efficiency and CO_2 emissions.

Based on such assumptions, current rate of ethanol production requires 2.5 million ha of sugarcane and represents avoided GHG emissions of 18.9 million tons of CO_2eq, approximately the amount of CO_2 release for the consumption of 6.9 million m^3 of gasoline.

Nevertheless, forest burning corresponds to 75% of GHG emissions in Brazil (WWF-Brazil, 2006). Based on Kirby et al. (2006), between 1994 and 2003, the average rate of deforestation in the Amazon forest was approximately 1.93 million ha. Fearnside et al. (2001), estimate that the burning of Amazon forest result in CO_2 emissions of 187 tons/ha. Consequently, the rate of deforestation mentioned above represents 361 million tons CO_2 emitted, which is 19 times bigger than calculated avoided emission of ethanol.

Even considering all distilleries in Brazil using boilers operating at 60 bars, deforestation emissions would be 18.1 times bigger than ethanol avoided emissions. This leads to the conclusion that efforts to preserve Amazon could have results, regarding CO_2 emissions almost 20 times more efficient than efforts to produce or subsidize ethanol.

9.10 Variations of CO_2 Emissions Calculations

CO_2 balances are calculated according to a series of assumptions. Aspects like sugarcane yield and ethanol conversion efficiency can influence significantly in the final result.

Table 9.7 Total emissions of sugarcane ethanol production and distribution resulted from different assumptions of input variables

Variable	Range of possible values	CO_2 emissions per m^3 of ethanol (kg)
Sugarcane Yield	67–86 tons/ha	425–546 kg
Ethanol conversion	80–85 L/ton	430–457 kg
Diesel fuel use	300–600 L	433–577 kg

Table 9.8 Best and worst case scenarios of ethanol CO_2 emissions

	Best case scenario	Worst case scenario
Sugarcane Yield	86 ton/ha	67 ton/ha
Ethanol conversion	85 L/ton	80 L/ton
Diesel fuel use	300 L	600 L
CO2 emisson/m3	379 kg	690 kg

During the development of this study, research centers, distilleries, farmers and literature were consulted, and the CO_2 emissions were calculated based in values that the author considered closest to Brazilian reality. The exception was sugarcane yield, which is considerably lower than the 80 tons/ha used. The reason for using a higher value is that it is representative of the state of Sao Paulo, whose companies will likely dominate any possible ethanol expansion in Brazil. From all sources consulted the input value that had the greatest variation is the amount of fossil fuel required for agricultural operations. Table 9.7 illustrates the effect that some variables have individually on CO_2 balances, values of variables where defined within a reasonable range, based on the sources consulted during the development of this study. Best and worst case scenarios are presented on Table 9.8.

9.11 A Trend in the Near Future

Brazilian government is infatuated with biofuel possibilities, so much so, that in march 25th, 2007; Brazilian president, Luiz Inácio Lula da Silva, stated that "Brazil could become the Saudi Arabia of Biofuels". Brazilian press seems to embrace the idea, as is common place to observe magazines, newspapers and television reporting the benefits of ethanol as an environmentally friendly option. It is possible to read statements in the press like "We have oil that everybody dreams about, right here in our orchards. An it is and inexhaustible source".

For the government there is the interest that Brazilian ethanol could reach American and European markets, increasing this way the flux of money to the country. The distilleries of course support the idea.

Marris (2006), reports projections from Brazilian minister of agriculture, for ethanol production of 26 million m^3 in 2010. Avoided emissions of such production would represent 28.7 million tons of CO_2, considering the technology for energy generation from bagasse burning as 60 bars boilers, BOD treatment and mechanical harvest; such value is equivalent to CO_2 emissions from deforestation of 153,476 ha, that is, approximately just 8% of the average deforestation rates between 1994 and 2003.

A more ambitious project is to export by 2025, 200 million m³ of ethanol (Ereno, 2007). This project has the objective of developing enzymatic hydrolysis of cellulose to increase substantially ethanol conversion capacity from sugarcane.

Whether enzymatic hydrolysis can be reached soon or not, production of ethanol in Brazil tends to increase significantly in the next decade.

In late July/2007, the Inter-American Development Bank (IDB), announced the financing of US$ 120 million dollars for ethanol production in the state of Sao Paulo (see http://www.iadb.org).

Until 2012, 86 new distilleries, or amplification of current distilleries, will help increase ethanol production in Brazil. This corresponds to an investment of US$ 19 billion, with US$ 5 billion originating from Brazilian National Bank of Economical and Social Development (BNDES), meanwhile the program sustainable Amazon, which encompasses the plan for combat of deforestation has a budget for 2007 of US$ 11.8 millions.[3]

With ethanol production in Brazil increasing, environmental problems follow suit, and raise concerns if such increase could, among other problems, worsen Brazilian deforestation, despite the fact that most of the sugarcane production areas are far from the Amazon.

9.12 Environmental Impacts Versus CO₂ Emissions

Although ethanol use as fuel results in less CO_2 emissions when compared to gasoline, it is important to notice that avoided emissions comes to a cost in other environmental impacts. Soil erosion, water quantity and quality and loss of biodiversity are some of the environmental concerns associated with ethanol production in Brazil.

Evapotranspiration rates of sugarcane are bigger than natural vegetation, Moreira (2007) report evapotranspiration rates from sugarcane varying between 1500 and 2000 mm/year. The original vegetation cover in areas of Sao Paulo state where currently sugarcane is planted, and in areas where it is still preserved has, according to Almeida and Soares (2005), evapotranspiration rates of 1350 mm year. Considering sugarcane evapotranspiration rate as 1500 mm, the additional water demanded corresponds to 1.5 million liters of water/ha. According to Smeets et al. (2006), to what extend evapotranspiration from sugar cane production contributes to regional water shortages is unknown.

Large amounts of water are also used for sugarcane washing and distillery operations. Dias de Oliveira (2005), reports that washing sugarcane consumes 3.9 m³ of per ton. Additional water is used in other distillery processes like fermentation for instance. According to Moreira (2007), 21 m³ of water are used for each ton of sugarcane processed, however most of this water is reused and the actual rate of

---

[3] http://contasabertas.uol.com.br/noticias/detalhe_noticias_impressao.asp?auto=1554

water collection is of $1.89\,m^3$/ton of sugarcane. The overall result is that for each kg of CO_2 avoided at least 217 L of water are required.

Sugarcane harvest period coincides with dry season in Brazil, and the large amounts of water withdrawn by the distilleries consists in a major ecological problem.

Water quality is also a concern as well, according to Ballester et al. (1997), diffuse run-off in the Corumbatai river basin in the state of Sao Paulo, characterized by sugarcane plantations, contributes significantly to deteriorate the river's water quality.

Soil erosion values reported for sugarcane plantations range from 31 to 61,4 tons/ha (Sparovek and Schung, 2001, and Ortiz Lopez (1997). Such values would correspond to 4.1 and 8.1 kg of soil loss per kg of CO_2 avoided, and of course its consequent deterioration in water quality.

It seems that global benefits of CO_2 sequestration come with a price in local environmental impacts. The question rises of how to compare benefits and impacts. Dias de Oliveira et al. (2005), used the ecological footprint (EF) approach for such comparisons. The conclusion was that benefits in terms of CO_2 emission from ethanol use were counterbalanced by environmental impacts associated with ethanol production.

9.13 Conclusions

It is undeniable that the use of ethanol from sugarcane represents reduction in CO_2 emissions when compared with gasoline. Nevertheless, the importance of such option regarding its role in global warming has been disproportionable optimistic and leads to neglection of important environmental and social aspects.

According to Hoffert et al. (2002), biomass plantations can produce carbon-neutral fuels for power plants or transportation, but photosynthesis has too low a power density for biofuels to contribute significantly to climate stabilization.

As pointed out by Cerri et al. (2007) based on UNFCCC, GHG emissions in tropics are mainly related to deforestation and agricultural intensification, while in temperate regions GHG comes from the combustion of fossil fuel in the transportation and industry sector. Agricultural intensification and deforestation are exactly the possible outcomes from significant increases of ethanol production in Brazil.

The idea of reducing fossil fuel consumption from temperate areas by using sugarcane ethanol is unpractical. In order to contribute to reduction of fossil fuel used in developed countries, the amount of ethanol that Brazil would have to produce would require a significant increase of the agricultural area devoted for such crops.

The increasing use of flex fueled automobiles also represents disadvantages in terms of fuel economy, and consequently CO_2 emissions. The adjustment of such cars is optimal neither for gasoline nor for ethanol, which makes such cars consume more fuel than if they were specified for using one type of fuel only.

Deforestation of Amazon still seems to be the major environmental issue in Brazil, and is also the most important aspect regarding global warming impacts; therefore more effort should be direct towards its preservation than for ethanol production.

References

Almeida, A.C. & Soares, J.V. (2005). Comparação entre uso de água em plantações de *Eucayptus grandis* e floresta ombrófila densa (Mata Atlântica) na costa leste do Brasil. *Revista Árvore*, 27, 159–170.

Aneel. Agência Nacional de Energia Elétrica. www.aneel.gov.br

Anfavea. Associação Nacional de Fabricantes de Veículos Automotores – Brazil (2007). Anuário estatístico. Retrieved July 18, 2007, from http://www.anfavea.com.br/anuario.html

Azania, A.A.P.M., Marques, M.O, Pavani, M.C.M.D. & Azania, C.A.M. (2003). Germinação de sementes de *Sida rhombipholia* e *Brachiaria decumbens* influenciada por vinhaça, flegmaça e óleo de fúsel. Planta daninha, 21, 443–449.

Ballester, M.V.R., Camargo, P.B., Carvalho, F.P., Hornink, S., Martinelli, L.A., Moraes, J.M. & Krusche, A.V. (1997). Spatial and temporal water quality variability in the Piracicaba river basin, Brazil. *Journal of the American Water Resources Association*, 33, 1117–1123

Beeharry, R.P. (1996). Extended sugarcane biomass utilisation for exportable electricity production in Mauritius. *Biomass and Bioenergy*, 11, 441–449

Beeharry, R.P. (2001). Carbon balance of sugarcane bioenergy systems. *Biomass and bioenergy*, 20, 361–370.

Braunbeck, O., Bauen, A., Rosillo-Calle, F. & Cortez, L. (1999). Prospects for green cane harvesting and cane residue use in Brazil. *Biomass and Bioenergy*, 17, 495–506.

Cancado, J.E.D. (2003). A poluição atmosférica e sua relação com a saúde humana na região canavieira de Piracicaba – SP.

Cardenas, G.J. (1993). Ethanol from bagasse as fuel, contribution to lowering of CO_2. *Ingenieria-Quimica*, 25, 113–116.

Cerri, C.E.P., Sparovek, G., Bernoux, M., Easterling, W.E., Melillo, M. & Cerri, C.C. (2007). Tropical agriculture and global warming impacts and mitigation options. *Scientia Agricola*, 64, 83–99.

Cortez, L.A.B., Freire, W.J. & Rosillo-Calle, F. (1998). Biodigestion of vinasse in Brazil, *Internacional Sugar Journal*, 100, 403–409.

CTC – Centro de Tecnologia Canavieira. Personal communication with Hélcio Lamônica on July 20, (2007).

Dias de Oliveira, M.E., Vaughan, B.E. & Rykiel, Jr. E.J. (2005). Ethanol as fuel: energy, carbon dioxide balances, and ecological footprint. *Bioscience*, 55, 593–602.

Ereno, D. (2007). Álcool de cellulose. *Revista Pesquisa Fapesp*. retrieved on line on June 14, 2007, from http://www.revistapesquisa.fapesp.br/?art=3169&bd=1&pg=1&lg

Farrell, A.E., Plevin, R.J.,Turner, B.T., Jones, A.D., O'Hare, M. & Kammen, D.M. (2006). Ethanol can contribute to energy and environmental goals. *Science*, 311, 506–508.

Fearnside, P.M., Graça, P.M.L.A. & Rodrigues, F.J.A. (2001). Burning of Amazonian rainforests: Burning efficiency and charcoal formation in forest cleared for cattle pasture near Manaus, Brazil. *Forest Ecology and Management*, 146, 115–128.

Fearnside, P.M. (2002). Greenhouse gas emissions from a hydroelectric reservoir (Brazil's Tucuruí dam) and the energy policy implications. *Water, Air, and Soil Pollution*, 133, 69–96.

Giampietro, M., Ulgiati, S. & Pimentel, D. (1997). Feasibility of large-scale biofuel production. *Bioscience*, 47, 587–600.

Gloeden, E. 1994. *Monitoramento da qualidade das águas das zonas não saturadas em área de fertilização de vinhaça*. Dissertation, Institute of Geociencies, Universidade de São Paulo.

Godoi, A.F.L., Ravindra, K., Godoi, R.H.M., Andrade, S.J., Santiago-Silva, M., Vaeck, L.V. &Grieken, R.N. (2004). Fast chromatographic determination of polycyclic aromatic hydrocarbons in aerosol samples from sugar cane burning. *Journal of Chromatography A*, 1027, 49–53

Dondero, L. & Goldemberg, J. (2005). Environmental implications of converting light gas vehicles: the Brazilian experience. *Energy Policy*, 33, 1703–1708.

Goldemberg, J. (2007). Ethanol for a sustainable energy future. *Science*, 315, 808–810

Grande, P.C. (2007). *Números Flexíveis*. Edição online of Quatro Rodas magazine. Retrieved on July 10, 2007, from http://quatrorodas.abril.com.br/reportagens/conteudo_141385.shtml.

Granato, E.F. (2003). Geração de energia através da biodigestão anaeróbica da vinhaça. Dissertaion, Universidade Estadual Paulista.

Grupo Cosan – Brasil. Personal communication on 06/04/2003.

Hassuda, S. (1989). Impactos da infiltração da vinhaça de cana no aquífero Bauru. Dissertation, Institute of Geosciencies. University of São Paulo.

Hoffert, M.I., Caldeira, K., Benford, G., Criswell, D.R., Green, C., Herzog, H., Jain, A.K., Kheshgi, H.S., Lackner, S., Lewis, J.S., Lightfoot, H.D., Manheimer, W., Mankins, J.C., Mauel, M.E., Perkins, L.J., Schlesinger, M.E., Volk, T. & Wigley T.M.L. (2002). Advanced technology paths to global climate stability: Energy for a greenhouse planet. *Science*, 298, 981–987.

Kirby, K.R., Laurance, W.F., Albernaz, A.K., Schroth, G., Fearnside, O.M., Bergen, S., Venticinque, E.M. & Costa, C. (2006). The future of deforestation in Brazilian Amazon. *Futures*, 38, 432–453.

Kirchoff, W.M.J.H. (1991). Enhancements of CO and O_3 from burning in sugarcane fields. *Journal of Atmospheric Chemistry*, 12, 87–102.

Krauter, S. & Ruthers, R. (2004). Considerations for the calculation of greenhouse gas reduction by photovoltaic solar energy. *Renewable Energy*, 29, 345–355.

Lima, M.A, Ligo, M.A.V., Cabral, O.M.R., Boeira, R.C., Pessoa, M.C.P.Y. & Neves, M.C. (1999). *Emissao de gases de efeito estufa provenientes da queima de residuos agricolas no Brasil*. (SP- Brazil: Embrapa Meio Ambiente).

Macedo, I.C. (1998). Greenhouse gas emissions and energy balances in bio-ethanol production and utilization in Brazil. *Biomass and Bioenergy*, 14, 77–81.

Marris, E. (2006). Drink the best and drive the rest. *Nature*, 444, 670–672.

Moreira, J.R. & Goldemberg, J. (1999). The alcohol program. *Energy Policy*, 27, 229–245

Moreira, J.R. (2007). Water use and impacts due ethanol production in Brazil. Presented at International conference at ICRISAT Campus, Hyderabad, India, 29–30 January 2007.

Navarro, A.R., Sepúlveda, M. del C. & Rubio, M.C. (2000). Bio-concentration of vinasse from the alcoholic fermentation of sugar cane molasses. *Water Management*, 20, 581–585.

Ortega, E., Ometto, A.R., Ramos, P.A.R., Anami, M.H., Lombardi, G. & Coelho, O.F. (2001). *Emergy comparison of ethanol production in Brazil: traditional versus small distillery with food and electricity production*. (Presented at the Second Biennial Emergy Analysis Research Conference: "Energy Quality and Transformities".. Gainesville – FL).

Ortiz López, A.A. (1997). Análise dos custos privados e sociais da erosão do solo: o caso da Bacia do rio Corumbatai. Doctor's dissertation. University of São Paulo – ESALQ, Piracicaba.

Pimentel, D. & Pimentel, M. (1996). *Food energy and society*. (Colorado: University Press of Colorado)

Rosa, L.P, & Ribeiro, S.K. (1998). Avoiding emissions of carbon dioxide through the use of fuels derived from sugarcane. *Ambio*, 6, 465–470.

Rosa, L.P. & Schaeffer, R. (1995). Global warming potentials: the case of emissions from dams. *Energy Policy*, 23, 149–158.

Schlesinger, W.H. (1997). *Biogeochemistry, an analysis of global change*. (California: Academic Press)

Sirvinskas, L.P. (2003). Manual de Direito Ambiental. (SP- Brazil: Editora Saraiva)

Shapouri, H., Duffield, J.A. & Wang, N. (2002). The Energy Balance of Corn Ethanol: An Update. Washington (DC): Office of Energy Policy and New Uses. Agricultural Economic Report # 814.

Smeets, E., Junginger, M., Faaij, A., Walter, A. & Dolzan, P. (2006). Sustainability of Brazilian bioethanol. Utrecht University. Copernicus Institute. Report NWS-E-2006-110.

Sparovek, G. & Schung, E. (2001). Temporal erosion-induced soil degradation and yield loss. *Soil Science Society of America Journal*, 65, 1479–1486.

Trobish, K.H. (1992). Recent development in the treatment of chemical waste water in Europe. *Water Science and Technology*, 26, 319–322.

van de Vate, J.F. (1997). Comparison of energy sources in terms of their full energy chain emission factors of greenhouse gases. *Energy Policy*, 25, 1–6.

Weir, K.L. (1998). Sugarcane fields: sources or sinks for greenhouse gas emissions? *Australian Journal of Agricultural Research*, 49, 1–9.

Weisser, D. (2007). A guide to life-cycle greenhouse gas (GHG) emissions from electric supply technologies. *Energy*, 32, 1543–1559.

West, T.O. & Marland, G. (2002). A synthesis of carbon sequestration, carbon emissions, and net carbon flux in agriculture: comparing tillage practices in the United States. *Agriculture, Ecosystems and Environment*, 91, 217–232.

WWF-Brazil. (2006). Agenda elétrica sustentável 2020. Retrieved on June 29, 2007, from http://assets.wwf.org.br/downloads/wwf_energia_2ed_ebook.pdf

Chapter 10
Biomass Fuel Cycle Boundaries and Parameters: Current Practice and Proposed Methodology

Tom Gangwer

Abstract A methodology is presented for standardizing Biomass Fuel Cycle (BFC) analysis and evaluation. The Biomass Fuel Cycle Methodology (BFCM) enables eliminating disparities, minimizing differences, and clearly quantifying variations. Standardized templates, modular staging, and normalized analysis formulations are used to disposition technologies, facilities, activities, boundaries, and parameters. The methodology enables presentation of quantification and characterization information in a straightforward standard format applicable across a broad range of BFC's. BFC literature data is used to illustrate the flexibility, clarity, and diversity of the methodology. The types of insights to be gained concerning the limitations of BFC treatments (boundary shortcomings, energy uncertainties, analysis constraints) are discussed.

Keywords Agriculture · biodiesel · biofuel · biomass · biorefinery · biorefinery · boundary · corn · crop rotation · energy · ethanol · fuel production · infrastructure · methodology · model · modular · net energy balance · net energy value · scenario · soybean · switchgrass · template · yield

Acronyms & abbreviations

ae: air emission
BFC: biomass fuel cycle
BFCM: biomass fuel cycle methodology
bpf: biofuel production
C: corn
CR: crop rotation
d: biodiesel
E: energy

GGE: greenhouse gas emissions
HHV: high heat value
L: loss
LHV: low heat value
N: net biofuel production
NEB: net energy balance
NEV: net energy value
S: soybean

✉ T. Gangwer
739 Battlefront Trail, Knoxville, TN 37934, USA
e-mail: tgangwer@chartertn.net

D. Pimentel (ed.), *Biofuels, Solar and Wind as Renewable Energy Systems*,
© Springer Science+Business Media B.V. 2008

e: ethanol

EC: environmental concern

EG: energy gain

EL: energy loss

F: corn mill fraction processed

TEG: total energy gain

TEL: total energy loss

U: area, mass, or volume

UE: usable energy

Y: yield

10.1 Introduction

The US national security driven: energy independence goal, reduction of pollution, and the pursuit of renewable energy source efforts have resulted in government bio-fuels subsidies of $6 billion per year (Koplow, 2006), and industry development of Biomass Fuel Cycles (BFC's). A methodology has been developed to provide unbiased characterization and analysis for use in technology viability evaluation. The selection of the boundaries, parameters, and associated numerical values for a given BFC has a direct impact on the evaluation of that technology's viability, import to energy independence, and renewable energy value. Currently there are significant judgment differences about BFC component import, analysis scope, boundary selection, and parameter values. Opinions differ and modeled scopes vary on topics such as coproduct energy credit, facility fabrication, waste management, environmental, and parameter numerical value (Dias De Oliveira et al., 2005; Farrell et al., 2006a,b; Graboski, 2002; Hammerschlag, 2006; Kim & Dale, 2005; Patzek, 2004; Pimentel, 1991; Pimentel & Patzek, 2005; Pimentel et al., 2007; Shapouri et al., 1995; 2002; 2004; Wang et al., 1997, Wang & Santini, 2000; Wang, 2005). As a result, as illustrated in Fig. 10.1, significant uncertainties in the published Net Energy Value (NEV) data exist.

The biomass fuel cycle methodology (BFCM) presented is intended to assist in avoiding, minimizing, or, at least, clearly quantifying and delineating analysis differences. The BFCM uses templates, modular modeling, scenario definition, and statistical based methods to standardize analyses, establish unbiased boundary assignments, normalize numerical value treatments, treat data uncertainty, and characterize limitations of results. Adding clarity to the understanding of BFC intricacies and analyses is intended to facilitate national level discussions and decisions on development of biomass fuel capabilities such as infrastructure requirements for an expanded ethanol industry (Brent and Yacobucci, 2006). In the present study, the focus is on the energy and environmental aspects of BFC's.

10.2 BFC Analysis Methodology: A Modular Model Approach

The BCFM is structured so as to be applicable to a broad range of BFC's. The methodology's three stage template system, fuel cycle parameters, boundary treatment, and statistical tools are presented. The approach facilitates modeling and analysis of scenarios involving diverse configurations (e.g., stand alone biomass cycles, crop rotation combined BFC's), agricultural variations (e.g., fertilization versus crop

Normal Distribution Presentation of NEV Published Data

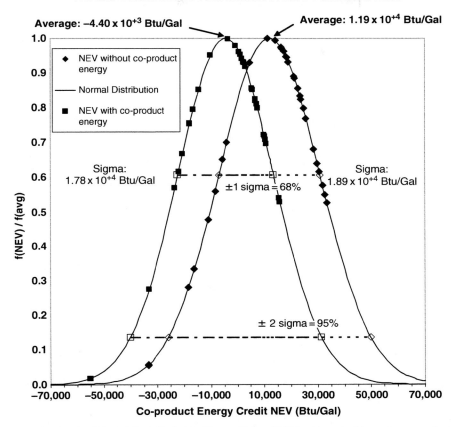

Fig. 10.1 Corn to Ethanol Fuel Cycle Net Energy Value (NEV) with and without the co-product energy (Dias De Oliveira et al., 2005; EBAMM, 2007; Farrell et al., 2006a,b; Graboski, 2002; Hammerschlag, 2006; Patzek, 2004; Pimentel, 1991; Pimentel & Patzek, 2005; Pimentel et al., 2007; Shapouri et al., 1995, 2002, 2005; Wang et al., 1997; Wang & Santini, 2000; Wang, 2005)

rotation, extent of tilling, silage practices/use), biomass to fuel processing variations (e.g., dry versus wet corn milling, cogeneration, cellulous digestion), energy balance consideration, and environmental impact assessment.

10.2.1 BFC General Stages and Templates

The BFCM structures each BFC analysis based on three main analysis stages:

1. Infrastructure (Template 1 given in Table 10.1) – multi-user services/facilities: 70 Sub-activities (59 distinctive + 11 onsite waste management covering 4 waste steam types)

Table 10.1 Template 1 Infrastructure Stage ($j = 1$)

Phase	Sub-phase	Activity: sub-activity	k
Manufacture	Equipment	**Fabricate:** Tractors, Combines, Trucks, Implements, Irrigation systems, Treatment systems (water, waste), Tractor Trailers, Barges, Rail Cars	1
		Onsite: Waste Management[1]	1
Facilities	Biomass Storage (transport: Template 2)	**Physical plant:** Construct, Operations[2], Fuel	2
		Onsite: Waste Management[1]	2
	Barge Terminal	**Physical plant:** Construct, Operations[2], Fuel	3
		Onsite: Waste Management[1]	3
	Rail Terminal	**Physical plant:** Construct, Operations[2], Fuel	4
		Onsite: Waste Management[1]	4
	Seed Plant	**Physical plant:** Construct, Operations[2], Fuel	5
		Onsite: Waste Management[1]	5
	Fertilizer Plant	**Physical plant:** Construct, Operations[2], Fuel	6
		Onsite: Waste Management[1]	6
	Herbicide Plant	**Physical plant:** Construct, Operations[2], Fuel	7
		Onsite: Waste Management[1]	7
	Insecticide Plant	**Physical plant:** Construct, Operations[2], Fuel	8
		Onsite: Waste Management[1]	8
	Lime Plant	**Physical plant:** Construct, Operations[2], Fuel	9
		Onsite: Waste Management[1]	9
	Biorefinery (other operations: Template 3)	**Physical plant:** Construct, Decommission	10
	Fuel Handling Facility (other operations: Template 3)	**Physical plant:** Construct, Decommission	11
	Offsite Water Treatment Plant	**Physical plant:** Construct, Operations/fuel	12
		Source: Biomass Storage, Terminals, Plants, Biorefinery, Fuel handling facility, Farms	
		Onsite: Waste Management[3]	12
	Offsite Waste Facility: Non-aqueous Liquids and Solids	**Physical plant:** Construct, Operations/fuel	13
		Source: Biomass Storage, Terminals, Plants, Biorefinery, Fuel handling facility, Farms	13
		Onsite: Waste Management[3]	13

[1] Wastewater, Non-aqueous liquids, Solids, Air Emissions
[2] includes Maintenance, Repair, Equipment/ Facility Decommissioning
[3] Non-aqueous liquids, Solids, Air Emissions

2. Agriculture (Template 2 given in Table 10.2) – biomass farm activities/facilities: 26 Sub-activities
3. Biofuel Production (Template 3 given in Table 10.3) – biofuel manufacture activities/facilities: 16 Sub-activities

The three general templates detail BFC processes and practices using a Phase, Subphase, Activity, and Sub-activity component structure. These template baselines identify components without consideration of specific BFC potential significance. Component significance will vary both within and across BFC's.

Using the templates, specific BFC modules are established and the cycle boundaries are delineated. Each BFC module Sub-activity is dispositioned (i.e., assigned a parameter/value or justified as not a consideration). Thus each module documents the specifics for use in quantifying and characterizing its' BFC. Introduction into

Table 10.2 Template 2 Agriculture Stage (j = 2)

Phase	Sub-phase	Activity	Sub-activity	k
Land	Growing	Transport to Farm	Seeds	1
			Equipment	1
			Labor	1
			Fertilizer	1
			Lime	1
			Herbicide	1
			Insecticide	1
		Irrigation system & water	Installation	1
			Operations/fuel	1
			Water Pre-application treatment	1
			Maintenance/Repair/Removal	1
		Planting	Pre-planting	1
			Seed Application	1
			Tilling	1
		Field Additives: Operations/fuel	Onsite storage	1
			Fertilizer application	1
			Line application	1
			Herbicide application	1
			Insecticide application	1
	Harvest	Crop and Silage Processing	Operations/fuel	2
			Transport (Storage/Biorefinery)	2
General Items	Full Crop Cycle	Maintain Facilities & Other Equipment Operability	Operations (including Maintenance/Repair)/ fuel	3
		Onsite: Waste Management[1] (includes biomass burning)	Waste dispositioning	3

[1] Wastewater, Non-aqueous liquids, Solids, Air Emissions

Table 10.3 Template 3 Biofuel Production Stage (j = 3)

Phase	Sub-phase	Activity	Sub-activity	k
Biorefinery Plant	Production	Processing to 99.5% Ethanol	Operations/fuel	1
			Maintenance/Repair	1
			Transport of chemicals to Plant	1
			Process water treatment	1
			Co-generation	1
		Onsite: Waste Management[1]	Waste dispositioning	1
Fuel Handling Facility	Fuel Feed Stock	Transport	Operations/fuel	2
		Fuel Blending	Operations/fuel	2
			Maintenance/Repair	2
	Facility Wastes	**Onsite:** Waste Management[1]	Waste dispositioning	2

[1] Wastewater, Non-aqueous liquids, Solids, Air Emissions

a module of new BFC process/practice components or sub-activities to show desired detail is straightforward. This template module approach readily accommodates customization of components while ensuring a standard set of sub-activities is addressed. The module components are analyzed using the standardized analysis and documentation methodologies thereby enabling inter-BFC and intra-BFC comparison.

The application of the three templates to energy and environmental aspects of BFC's is presented in Section 10.4. Although not explicitly addressed, the BFCM could be applied to monetary, production, distribution, regulatory, national security, incentives, and subsidies evaluations through selective expansion of the level of detail in the general templates. Having BFC evaluations linked via these common general templates is advantageous from a continuity, comparison, and clarity perspective.

10.2.2 BFC Parameters and Associated Variability

The BFC variability arises from natural and technological causes. Weather (e.g., wet/dry, temperature, storm damage), location(e.g., farm: soil type/condition, crop disease/pests; biorefinery: infrastructure, economics), transport distance (e.g., from farm to storage/process facility, biofuel distribution distance), seed type, agricultural practice (e.g., crop rotation, fertilization, irrigation), fuel source mix used within cycle (e.g., coal, gas, oil, biomass), biomass type (e.g., corn, soybean, switchgrass), and biofuel process technology (e.g., corn dry/wet mill, cellulose breakdown process) are typical sources of variability. Such viabilities are addressed and quantified by using two different types of parameters. The first is the biomass yield parameters used to quantitatively track the following sources of variability (Section 10.2.2.1):

- Weather, location, seed type, agricultural practice: Crop Yield = Y_{crop}
- Biomass type, biofuel manufacture process: Biofuel Process Yield = Y_{bfp}

The second parameter type is the individual parameters (p_k's and Δ_k's discussed in Section 10.2.2.2) unique to a given module Sub-activity. In the BFCM treatment, Y_{crop} and Y_{bfp} variability relationships are examined separately from the p_k values.

10.2.2.1 Biomass Yield Parameters

For a given BFC:

$$N_{crop\ to\ bfp} = Y_{crop}\ A\ Y_{bfp}$$

Here $N_{crop\ to\ bfp}$ is the BFC net fuel production, Y_{crop} is the agriculture stage biomass crop yield, A is the planted land area, and Y_{bfp} is the biofuel production stage yield. Another BFC general yield and biofuel energy relationship is:

$$E_{biofuel} = N_{corn\ to\ bfp}\ UE_{fuel\ e}$$

Here $E_{biofuel}$ is the BFC created biofuel energy and $UE_{fuel\ e}$ is the biofuel useable energy (see Section 10.3). Combining and rearranging these two equations:

$$E_{biofuel}/A = Y_{crop}\ Y_{bfp}\ UE_{biofuel} \tag{10.1}$$

$E_{biofuel}/A$ is a measure of the BFC crop and biomass fuel production efficiency in creating the biofuel. This equation enables biofuel yield evaluation (see Section 10.4.1) at both the local/regional and national fuel cycle production levels. Clearly gains in crop and process yields mean higher biofuel energy per acre planted.

10.2.2.2 Template Parameters

For each template Activity, there is an assigned k value. This k value is used to index the p_k value assigned to that Activity and it's associated Sub-activities. The p_k value and it's uncertainty Δ_k are specific numerical values used in the analysis. Consider, for example, in Template 1 (Table 10.1) under the Facilities Phase there is the Seed Plant Sub-phase. It's assigned Activity and associated Sub-activities index value is $k = 5$. Therefore it's numerical values used in an analysis are assigned to the p_5 and Δ_5 parameter in the BFCM equations discussed here (see also Section 10.4.2 for specific illustration) The p_k's are used to calculate the $S_{module\ j}$ value of interest:

$$S_{module\ j} = f_j(p_k)$$

and the Δ_k's are used to quantify the uncertainty (Δ_j) associated with that $S_{module\ j}$ (see Section 10.2.4). The $f_j(p_k)$ equations are typically simple summations for the BFC's but can be any mathematical relationship. The detail for a given $S_{module\ j}$ is determined by the BFC scenario and associated module. Both the $S_{module\ j}$ value and its' Δ_j are used to quantifying and characterizing the BFC.

The general relationship applicable to each module is:

$$S_{BFC} = \sum_{j=1}^{m} S_{module\ j}\ U_j\ F_j \qquad (10.2)$$

Here S_{BFC} is the total value (e.g., energy, mass, volume) for the given BFC modeled scenario made up of m modules; U_j is the land area planted, Biorefinery processed biomass, or biofuel volume; and F_j is the scenario specified decimal fraction factor used to evaluate a U_j variation ($F_j = 1$ if U_j held constant). Sections 10.4.2 and 10.4.3 present the application of this equation to energy and environmental treatments respectively.

BFC yields, p_k's, and Δ_k's values, which are annual numbers, are reported in various units in the literature. In order to sum the $S_{module\ j}$'s, the data must be normalize to a common unit. In the current treatment the numerical values are normalized to Btu/Acre. The conversion factors used were: 948.452 Btu/MJ, 0.2520 Kcal/Btu, 3.7854 L/Gal, and 2.471 Acre/Ha. The Biorefinery p_k values were normalized to Btu/Acre using each specific study crop and biofuel yields. The resultant $S_{module\ 3}$ values are thus a function of these specific yields which introduces two sources of variability into the analysis.

10.2.3 BFC Boundaries

A fundamental consideration is the establishment of the given BFC boundaries. As is evident from the results shown in Fig. 10.1, the choice of boundaries can dramatically change results. It is important to clearly and concisely disposition what is included in and excluded from the BFC.

The boundaries for a given BFC are established by using Templates 1, 2, and 3 (see Tables 10.1, 10.2, and 10.3 respectively) as the starting point. The three templates cover a broader range of BFC aspects than typically addressed. Their level of Sub-activity breakout focuses on aspects needing explicate dispositioning. The Sub-activities encompass materials, components, and facilities starting from natural resources through fabrication and usage to disposal. The p_k's quantify aspects such as raw material extraction (e.g., mining of coal and minerals, petroleum drilling), materials fabrication (e.g., steel, fuel, fertilizer, farm equipment), construction (e.g., facilities, roads), operation (e.g., farming, storage, processing, transporting), and waste management (e.g., discharges, emissions, equipment and facility replaced or decommissioned).

The dispositioning (i.e., inclusion or exclusion) of a p_k is a boundary decision. The BFC modules enable capturing the justification, including quantification of the impact, of Sub-activity exclusion. However, as evidenced in Fig. 10.1, Sub-activity exclusion can result in important differences between models. Inclusion has the advantages of simplifying the description, facilitating cross model comparison and evaluation, and minimizing the potential for underestimating (which is inherent to BFC's as a result of their cumulative parameter property).

The energy definitions given in Section 10.3 establish the BFC energy boundaries and accounting of fuel use. Considerations of financial, subsidy, policy, economic, and national security based aspects of a fuel cycle may provide insight into fuel cycle boundaries but should not be used as a basis for disposition because of their introduction of bias.

The end result is the BFC Stage Sub-activities and boundary demarcations are clearly delineated and justified. And the p_k and Δ_k values are presented in a standard format.

10.2.4 Statistical Tools

Use of statistical tools in the BFCM is intended to facilitate error reduction. Sources of imprecision and uncertainty arise from non-random (determinate) and random (indeterminate) errors resulting from method, measurement, estimation, and/or model decisions. Non-random errors can be difficult to detect. Consistent application of the BFCM approach provides one tool of use in avoiding and detecting errors.

The following statistical tools can be used to reduce random error, evaluate p_k and Δ_k significance, identify p_k's and Δ_k's whose refinement will improve $S_{module\ j}$ characterization, assessing boundary dispositions, and minimize introduction of bias.

The present study assumes the following normal distribution relationships apply (Natrella, 1966; NIST, 2006; Skoog and West, 1963):

$$f(p) = \exp\{-[(x - m)^2/2\,\sigma^2]/[\sigma(2\Pi)^{1/2}]\}$$

$$m = \sum_{i=1}^{n}(x_i/n)$$

$$\sigma = \text{standard deviation} = \left\{\left[\sum_{i=1}^{n}(x_i - m)^2\right]/(n - 1)\right\}^{1/2}$$

$$v = \text{variance} = \sigma^2$$

Figure 10.1 is obtained by applying the above equations where p equals the individual NEV values and m is the NEV average value.

Curve fitting data (e.g., linear least squares analysis) is readily accomplished using standard computer spreadsheet program functions.

One can treat the square of the uncertainty (Δ_i^2) associated with each numerical value in a given equation as a variance equivalent and apply absolute and relative deviation addition methods (Skoog and West, 1963) to obtain Δ_k's and Δj's. As an example, for the general relationship:

$$\Delta_j = f_j(\Delta_k)$$

the method first treats sums or differences ($\pm$) using

$$\Delta_{\pm \, \text{equation}} = \left\{ \sum_{k=1}^{n} \Delta_k{}^2 \right\}^{1/2}$$

then multiplications or divisions (x/) using

$$\Delta_{\text{x/equation}} = \left\{ \sum_{k=1}^{n} (\Delta_k/p_k)^2 \right\}^{1/2}$$

as one proceeds from the interior of the function outward. Here n is the number of uncertainty values associated with the numerical values in the $f_j(\Delta_k)$ equation.

10.3 BFC Fuel and Net Energy Balance Definitions

The BFC energy measure of interest is the Net Energy Balance (NEB):
   NEB = Total BFC Energy Gain (EG) – Total BFC Energy Loss (EL)
        = TEG − TEL Concise definition of EG and EL facilitates BFCM boundary dispositioning, energy accounting, and consistency.

10.3.1 Fuel Energy Definitions

When calculating the NEB, the energy gain (i.e., creation of fuel or productive use of BFC biomass or biofuel) and loss (i.e., consumption/expending of non-BFC fuel or energy) accounting needs to be well defined. The energy independence and environmental national goals lead to replacement of fossil fuels (both foreign and domestic) with domestic biomass fuels. BFC energy accounting needs to address all energy consumptions. The BFC energy definitions that follow directly from the above considerations are:

   EL = Energy Loss for given BFC = directly (e.g., burned at given BFC fa-
      cility) or indirectly (e.g., resource extraction/production/refinement, electric-
      ity generation, steam generation, transport) expended fossil (i.e., petroleum,
      coal) fuels, biomass/biofuel, electricity, or energy (e.g., heat) via nuclear/solar/
      water/wind power.
   EG = Energy Gain for given BFC = created biofuels productive combustion
      (e.g., ethanol fuel oxidant in gasoline, ethanol replacement of gasoline,
      biodiesel replacement of petroleum diesel) + biomass or BFC created co-
      products combustion supplying productive heat and/or power (e.g., silage,
      bagasse) + biomass, biofuels, or coproduct conversion to products (e.g.,

biomass digestion resulting in fertilizers, silage composting resulting in lowered field fertilization, conversion of biofuel to pesticides) that displace corresponding products derived from fossil (i.e., petroleum, coal) fuel.

Note both EL and EG include biomass/biofuel used to supply energy to the given BFC. The inclusion in both is needed in order to have the actual total energy value tabulated for the TEL and TEG. In this way both the TEL and TEG values are comprehensive and unencumbered with BFC specific exceptions/treatments. The accounting of the gain resulting from consumed biomass/biofuel displacing fossil fuel is captured in the EG analysis (see Section 10.3.3).

These definitions provide the basis for: excluding through definition the solar energy absorbed in growing the biomass and the caloric energy expended by BFC labor; retention of coproduct energy within the cycle unless some portion of the energy expended to create the coproduct is productively recovered by combustion of the coproduct; treating the use of solid, liquid, or gaseous biomass or biofuel within a given BFC as equivalent to an energy gain (i.e., those biomass fuel consumptions avoid consuming fossil fuels); and treating cogeneration as equivalent to an energy gain (i.e., it avoids consuming fossil fuels). The labor and coproduct aspects are discussed further in Section 10.5.

10.3.2 Fuel Useable Energy

The combustion of a fuel can be simplistically viewed as resulting in energy generation, water (as a gas) containing energy in the form of steam heat, combustion products, and particulates. For fossil, biomass, and biofuel fuels, the relevant energy value is the usable energy realized when a quantity of fuel is burned under normal use conditions:

UE = Useable Energy = fuel High Heat Value (HHV) adjusted for normal use losses (L). HHV is also referred to as the gross heat content of a fuel. Combustion systems differ in their L value due to inefficiencies (e.g., heat leaks, energy transfer, discharge, friction) and operational variations.

For internal combustion engines it is typically assumed the efficiency is the same for all liquid fuels and the main loss is via steam. This L adjusted HHV is commonly referred to as the Low Heat Value (LHV) for the fuel (also called the net heat content) and is commonly used as the UE value. Use of the LHV provides a consistent, common base of comparison. Productive use of L, such as preheater use of boiler system exhaust, increases the UE value with respect to the LHV.

For combustion of solid fuels (e.g., crop biomass such as bagasse), the above assumptions and conditions are not applicable. The L value is much more fuel composition and system efficiency dependent. Capturing BFC energy credit for the use of biomass fuel in place of fossil fuel (e.g., co-generation, pre-heating a process stream) requires consideration of system application specifics.

10.3.3 Fuel Energy Templates and Analysis

When performing the energy EL, EG, and NEB analyses, four templates are used. The Section 10.2.1 Templates 1, 2, and 3 are used to create the BFC specific EL Modules which are then used for the TEL tabulations. The Template 4 given in Table 10.4 is used to create the BFC specific EG Module for the TEG tabulation. In all energy Module tabulations, the applicable UE value should be used.

Table 10.4 Template 4 Energy Gain Stage (j = 4)

Stage	Activity
External-to-Given BFC	Combustion of BFC Created Fuels: Biofuel, Biomass Combustion of Biomass or coproducts for Heat and/or Power Fossil feedstock based products Displacement by Biomass, Biofuel, or coproduct
Infrastructure	Manufacture Operations Fuels: Biofuel, Biomass Facilities Operations Fuels: Biofuel, Biomass
Agriculture	Operations Fuels: Biofuel, Biomass
Biofuel Production	Biorefinery Plant Operations Fuels: Biofuel, Biomass Fuel Handling Facility Operation Fuels: Biofuel, Biomass

Applying the equation 10.2 relationship to the Modules, where we hold U constant, define $S_{\text{module } j} = E_{\text{module } j}$, and calculate the EL's and EG's on a per unit area basis, gives the general BFCM equations:

$$TEL_{BFC} = \sum_{j=1}^{q} E_{\text{module } j}$$

$$TEG_{BFC} = \sum_{j=1}^{1} E_{\text{module } j}$$

Here $E_{\text{module } j}$ is the template derived assessment for module j of the EL or EG value and q and 1 are the number of module values that form the basis for the cited value. Section 10.4.2 presents the NEB analysis for several BFC's.

10.4 BFC Models

The following application of the BFCM to energy and environmental scenario models uses representative as opposed to all inclusive literature data. The purpose is to illustrate the use of the methodology for a few BFC data sets. In the present treatment, the parameters of interest are specified using British thermal unit (Btu), Acre, Gallon (Gal), and Bushel (Bu) units.

10.4.1 Analyzing Yield Aspects

The two main BFC liquid biofuels products are ethanol (e) and biodiesel (d). Consider the created ethanol fuel energy per acre for the corn to ethanol BFC where the portion F of corn processed through the wet versus dry milling is varied. Based on equation 10.1 the energy-yield relationship is:

$$E_e/A \, (\text{Btu/Acre}) = Y_C \, [Y_D \, F + Y_W \, (1 - F)] E_{\text{biofuel e}}$$

Here Y_D is the Y_{bfp} for corn to ethanol Dry mill processing, Y_W is the Y_{bfp} for corn to ethanol Wet mill processing, F is the fraction of ethanol corn Dry mill processed, and $E_{\text{biofuel e}}$ is the ethanol UE fuel value. Figure 10.2 shows the E_e/A linear least square fit results for some corn and ethanol production yields.

From a local/regional and national perspective, the potential gain from BFC improvement is an important consideration. The equation 10.1 E_e/A yield relationship provides insight into such considerations. Large variations in corn yields occur as the result of soil, weather, and crop management practices: 85–245 Bu/Acre (Dobermann and Shapiro, 2004). For biorefinery yields in the 2.6 Gal/Bu range, a region producing at 140 Bu/Acre will attain E_e/A values 25% higher than a region

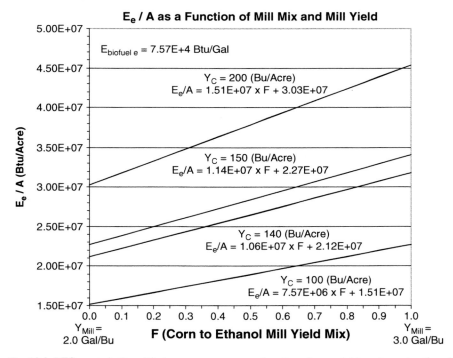

Fig. 10.2 BFC created ethanol fuel energy per acre as a function of crop yields and corn to ethanol mill processing yields

producing 112 Bu/Acre. Alternatively, processing the 112 Bu/Acre region corn at a 2.8 Gal/Bu biorefinery achieves 8% higher E_e/A value over the 2.6 Gal/Bu facility. A subset of this is Wet versus Dry mill utilization considerations illustrated in Figure 10.2. The BFCM facilitates such local/regional Y_C and Y_{bfp} coupled evaluations which may be of value to National energy considerations.

For the soybean to biodiesel BFC the created biodiesel energy per acre is:

$$E_d/A \ (Btu/Acre) = Y_S \ Y_d \ E_{biofuel \ d}$$

Combining the corn and soybean crop rotation and fuel production BFC's:

$$E_{ed}/A \ (Btu/Acre) = Y_C \ CR \ [Y_D \ F + Y_W \ (1-F)] \ E_{fuel \ e} + Y_S \ (1-CR) \ Y_d E_{fuel \ d}$$

Here E_{ed}/A is the combined energy content of ethanol and biodiesel fuel produced and CR is the crop rotation cycle fraction for corn planting (e.g., alternating plantings: $CR = 0.5$; 2 out of every 3 plantings: $CR = 0.67$). Figure 10.3 shows some of the possible correlation plots. For current yield conditions, annual crop rotation gives an E_{ed}/A of $1.73 \times 10^{+7}$ Btu/Acre while corn only (i.e., no rotation) gives $5.50 \times 10^{+7}$ Btu/Acre for the comparable 2 year period. Examination of the left (100% soybean) and right (100% corn) axes shows optimization of the corn to ethanol parameters holds the greater promise for improving biofuel production efficiency, despite $E_{biofuel \ d}$ being 1.55 times $E_{biofuel \ e}$. However, this result does not address the NEB aspects (Section 10.4.2). Nor does it factor in the need for conservation measures to deal with such aspects as soil depletion, crop diseases, and crop pests.

The CR needed to achieve an equal energy gain from each crop in the corn-soybean BFC is given by the relationship:

$$CR = Y_S Y_d E_{fuel \ d} / [Y_C Y_{Mill} E_{fuel \ e} + Y_S Y_d E_{fuel \ d}]$$

Here $[Y_D \ P + Y_W \ (1-P)]$ is defined as the corn to ethanol effective processing yield Y_{Mill}. To achieve parity under the 'current yields' (Fig. 10.3) requires a 5 plantings crop rotation sequence comprised of 1 corn planting for every 4 soybean plantings. The alternate year crop rotation sequence approaches parity for the low corn and high soybean yields. Again the analysis does not include NEB aspects.

10.4.2 BFC Energy Scenario Models and Analysis

The structure of the energy relationships follows directly from the associated modular configuration of the BFC scenario. Templates 1, 2, and 3 (Section 10.2.1) were used to construct the Modules 1 – 9 EL tabulations given in Tables 10.5–10.13. Template 4 (Section 10.3.3) was used to construct the EG Modules 100–102 given in Tables 10.14–10.16. Each Module lists the Sub-activity k assignment (see

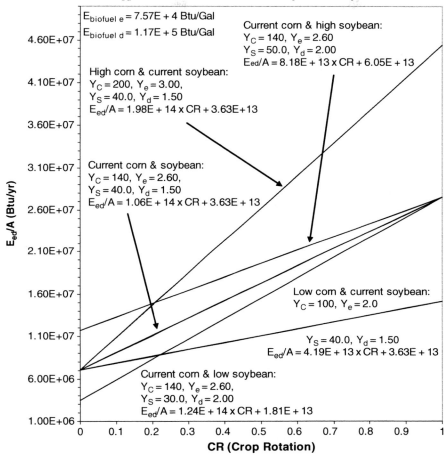

Fig. 10.3 BFC created ethanol-biodiesel fuel energy per acre as a function of yields and crop rotation

Section 10.2.2.2) and the number of literature data points used to obtain p_k, along with the available Δ_k values.

Based on Section 10.3.3, the NEB equation is:

$$NEB_{BFC} = TEG_{BFC} - TEL_{BFC} = \sum_{i=1}^{1} EG_i - \sum_{i=1}^{q} EL_i$$

The 1 and q values are established by the modeled scenario. Table 10.17 lists the BFC module $E_{module\ j}$ relationships which were used to obtain the Table 10.18 BFC scenarios.

Table 10.5 Module 1 Infrastructure for Corn energy loss EL data (EBAMM, 2007) in Btu/Acre $(j = 1)$

Phase	Sub-phase	Activity	Sub-activity	k^*	n^{a^*}	p_k^*	Δ_k^*
Manufacture	Equipment	Fabricate	Tractors, Combines, Trucks, Implements Irrigation, Treatment (water, waste)	1	3	$1.36 \times 10^{+6}$	$1.13 \times 10^{+6}$
Facilities	Seed Plant	Physical Plant	Operations/ fuel	5	2	$4.66 \times 10^{+5}$	$3.89 \times 10^{+5}$
	Fertilizer Plant	Physical Plant	Operations/ fuel	6	23	$3.69 \times 10^{+6}$	$3.43 \times 10^{+5}$
	Herbicide Plant	Physical Plant	Operations/ fuel	7	7	$4.07 \times 10^{+5}$	$2.63 \times 10^{+5}$
	Insecticide Plant	Physical Plant	Operations/ fuel	8	7	$1.09 \times 10^{+5}$	$1.55 \times 10^{+5}$
	Lime Facility	Physical Plant	Operations/ fuel	9	5	$2.13 \times 10^{+5}$	$1.80 \times 10^{+5}$
	Biorefinery	Physical Plant	Construct	10	1	$1.65 \times 10^{+5}$	nv
	Offsite Water Treatment Plant	Treatment of: Water or Wastewater	Operations/fuel	12	1	$3.57 \times 10^{+5}$	nv
			Total EL$_{IC}$ & Δ_{IC}:			$6.77 \times 10^{+6}$	$1.29 \times 10^{+6}$

* With respect to k, n, p_k, and Δ_k, see Section 10.2.2.2 for definitions and Section 10.2.4 for detailed illustration on usage in calculations.
a values obtained by using only non-duplicated data from cited reference
nv: no value

The following illustrates the BFCM module notation and analysis. First consider the Seed Plant Sub-phase in Module 1 $(j = 1)$ shown in Table 10.5. It's $k = 5$ indexed Activity: 'Physical Plant' and associated Sub-activity: 'Operations/fuel' p_5 and Δ_5 values are based on two literature values. This is captured by the $n = 2$ designation in Module 1. In terms of the Section 10.2.2.2 equation:

$$S_{\text{module } j} = f_j(p_k)$$

we have for Module 1:

$$S_{\text{module } j} = E_{\text{module } 1} = f_1(p_k) \equiv EL_{IC}$$

where the $f_1(p_k)$ is a summation of 8 p_k terms $(t = 8)$:

$$EL_{IC} = f_1(p_k) = \sum_{t=1}^{8} p_{k,t}$$

Table 10.6 Module 2 Corn Agriculture energy loss EL data (EBAMM, 2007) in Btu/Acre ($j = 2$)

Phase	Sub-phase	Activity	Sub-activity	k^*	n^{a^*}	p_k^*	Δ_k^*
Land	Growing	Transport to Farm	Seeds	1		In Equipment value	
			Equipment	1	7	$1.66 \times 10^{+5}$	$9.54 \times 10^{+4}$
			Labor	1	1	$1.11 \times 10^{+5}$	nv
			Fertilizer	1		In Equipment value	
			Lime	1			
			Herbicide	1			
			Insecticide	1			
		Irrigation system & water	Operations/fuel	1	3	$2.20 \times 10^{+5}$	$2.60 \times 10^{+5}$
			Pre-planting	1		In Tilling value	
			Seed	1			
		Planting	Tilling	1	33	$3.03 \times 10^{+6}$	$9.42 \times 10^{+5}$
		Field	Fertilizer	1		In Tilling value	
			Line	1			
			Herbicide	1			
			Insecticide	1			
	Harvest	Crop and Silage Processing	Operations/fuel	2		In Tilling value	
			Transport: Storage, Biorefinery	2	6	$1.35 \times 10^{+6}$	$1.15 \times 10^{+6}$
General Items	Full crop Cycle	Facilities & Other Equipment	Operations/fuel	3		In Tilling value	
			Total EL$_C$ & Δ_C:			$4.88 \times 10^{+6}$	$1.51 \times 10^{+6}$

* With respect to k, n, p_k, and Δ_k, see Section 10.2.2.2 for definitions and Section 10.2.4 for detailed illustration on usage in calculations.
[a] values obtained by using only non-duplicated data from cited reference
nv: no value

In the above equation the Seed Plant 'Operations/fuel' Sub-activity we are deals with the second item in Module 1 (i.e., $t = 2$ in the above summation) of Table 10.5 and there are two literature values to sum ($n = 2$):

$$p_{k,2} = \sum_{n=1}^{2} p_{k,i} = 4.66 \times 10^{+5} \, \text{Btu/Acre}$$

Analogous calculations give the other seven Module 1 p_k values. All 8 p_k's are summed to yield the Module 1 energy loss value $6.77 \times 10^{+6}$ Btu/Acre designated EL_{IC} in Table 10.5. The Corn to Ethanol BFC total energy loss is comprised of Modules 1, 2, and 3 (Tables 10.5, 10.6, and 10.7). Thus from the above general BFCM equation, $q = 3$, so:

$$TEL_{Ce} = \sum_{i=1}^{3} EL_j = EL_{IC} + EL_C + EL_{Ce} = 3.025 \times 10^{+7} \, \text{Btu/Acre}$$

Table 10.7 Module 3 Corn to ethanol Production EL data (EBAMM, 2007) in Btu/Acre ($j = 3$)

Phase	Sub-phase	Activity	Sub-activity	K*	n[a]*	p_k*	Δ_k*
Biorefinery Plant	Production	Processing to 99.5% Ethanol	Operations/fuel	1	12	$1.64 \times 10^{+7}$	$2.63 \times 10^{+6}$
			Transport of chemicals to Plant	1	1	$1.82 \times 10^{+6}$	nv
			Process water treatment	1	1	$3.93 \times 10^{+5}$	nv
			Total EL$_{Ce}$ & Δ_{Ce}:			$1.86 \times 10^{+7}$	$2.63 \times 10^{+6}$

* With respect to k, n, p_k, and Δ_k, see Section 10.2.2.2 for definitions and Section 10.2.4 for detailed illustration on usage in calculations.
[a] values obtained by using only non-duplicated data from cited reference
nv: no value

There is only one energy loss term (see Table 10.14), $l = 1$, so $EG_{Ce} = TEG_{Ce}$. The net energy balance equation for this BFC scenario is thus:

$$NEB_{Ce} = TEG_{Ce} - TEL_{Ce} = 2.75 \times 10^{+7} - 3.03 \times 10^{+7} = 2.8 \times 10^{+6}$$

The Table 10.18 presentation:

$$NEB_{Ce} = EG_{Ce} - EL_{IC} - EL_C - EL_{Ce}$$

captures the modular make up of the scenario. The calculation of the Δ values given in the Module Tables and Table 10.18 is performed at each step of the above

Table 10.8 Module 4 Infrastructure for Soybean energy loss EL data (Pimentel & Patzek, 2005) in Btu/Acre ($j = 1$)

Phase	Sub-phase	Activity	Sub-activity	k*	n*	p_k^*	Δ_k^*
Manufacture	Equipment	Fabricate	Tractors, Combines, Trucks, Implements Irrigation, Treatment (water, waste)	1	1	$5.78 \times 10^{+5}$	nv
Facilities	Seed Plant	Physical Plant	Operations/fuel	5	1	$8.90 \times 10^{+5}$	nv
	Fertilizer Plant	Physical Plant	Operations/fuel	6	3	$4.22 \times 10^{+5}$	nv
	Herbicide Plant	Physical Plant	Operations/fuel	7	1	$2.09 \times 10^{+5}$	nv
	Lime Facility	Physical Plant	Operations/fuel	9	1	$2.17 \times 10^{+6}$	nv
	Biorefinery	Physical Plant	Construct	10	3	$3.93 \times 10^{+5}$	nv
			Total EL$_{IS}$ & Δ_{IS}:			$4.66 \times 10^{+6}$	nv

* With respect to k, n, p_k, and Δ_k, see Section 10.2.2.2 for definitions and Section 10.2.4 for detailed illustration on usage in calculations.
nv: no value

Table 10.9 Module 5 Soybean Agriculture EL in data (Pimentel & Patzek, 2005) Btu/Acre (j = 2)

Phase	Sub-phase	Activity	Sub-activity	k*	n*	p_k*	Δ_k*
Land	Growing	Transport to Farm	Seeds	1			In Equipment value
			Equipment	1	1	$6.42 \times 10^{+4}$	nv
			Fertilizer	1			In Equipment value
			Lime	1			
			Herbicide	1			
			Insecticide	1			
		Irrigation system & water	Operations/fuel	1			In Equipment value
			Pre-planting	1			In Tilling value
			Seed	1			
		Planting	Tilling	1	4	$1.23 \times 10^{+6}$	nv
		Field Application	Fertilizer	1			In Tilling value
			Line	1			
			Herbicide	1			
			Insecticide	1			
	Harvest	Crop and Silage Processing	Operations/fuel	2			In Tilling value
			Transport: Storage, Biorefinery	2			In Equipment value
General Items	Full Crop Cycle	Maintain Facilities & Equipment Operability	Operations/fuel	3			In Tilling value
			Total EL$_S$ & Δ_S:			$1.29 \times 10^{+6}$	nv

* With respect to k, n, p_k, and Δ_k, see Section 10.2.2.2 for definitions and Section 10.2.4 for detailed illustration on usage in calculations.
nv: no value

calculation sequence. Since there are only sums and differences for each equation in the calculation sequence, the square of the uncertainty (Δ_k^2) for each term in the equation is analyzed using the $\Delta_{\pm k}$ relationship given in Section 10.2.4.

Table 10.18 documents each scenario, characterizes each module with respect to the number of template Sub-activities dispositioned (e.g., the Table 10.14 corn to ethanol Module 100 Disposition is 1 out of the 8 Overall Template 4 Sub-activities

Table 10.10 Module 6 Soybean to biodiesel Production EL data (Pimentel & Patzek, 2005) in Btu/Acre (j = 3)

Phase	Sub-phase	Activity	Sub-activity	k*	n*	p_k*	Δ_k*
Biorefinery Plant	Production	Processing to 99.5% Ethanol	Operations/fuel	1	5	$2.27 \times 10^{+6}$	nv
			Process water treatment	1	1	$1.23 \times 10^{+5}$	nv
			Total EL$_{Sd}$ & Δ_{Sd}:			$2.39 \times 10^{+6}$	nv

* With respect to k, n, p_k, and Δ_k, see Section 10.2.2.2 for definitions and Section 10.2.4 for detailed illustration on usage in calculations.
nv: no value

Table 10.11 Module 7 Infrastructure for Switch Grass energy loss EL data (EBAMM, 2007) in Btu/Acre ($j = 1$)

Phase	Sub-phase	Activity	Sub-activity	k^*	n^{a*}	$p_k{}^*$	Δ_k^*
Manufacture	Equipment	Fabricate	Tractors, Combines, Trucks, Implements Irrigation, Treatment (water, waste)	1	2	$5.07 \times 10^{+5}$	$5.44 \times 10^{+5}$
Facilities	Seed Plant	Physical Plant	Operations/fuel	5	2	$1.89 \times 10^{+5}$	nv
	Fertilizer Plant	Physical Plant	Operations/fuel	6	5	$1.75 \times 10^{+6}$	$1.08 \times 10^{+6}$
	Herbicide Plant	Physical Plant	Operations/fuel	7	2	$2.67 \times 10^{+5}$	$3.04 \times 10^{+5}$
	Biorefinery	Physical Plant	Construct	10	1	$8.67 \times 10^{+5}$	nv
	Offsite Water Treatment Plant	Treatment of: Water or Wastewater	Operations/fuel	12	1	$5.72 \times 10^{+5}$	nv
			Total EL$_{ISG}$ & Δ_{ISG}:			$4.15 \times 10^{+6}$	$1.25 \times 10^{+6}$

* With respect to k, n, p_k, and Δ_k, see Section 2.2.2 for definitions and Section 10.2.4 for detailed illustration on usage in calculations.
[a] values obtained by using only non-duplicated data from cited reference
nv: no value

Table 10.12 Module 8 SwitchGrass Agriculture EL data (EBAMM, 2007) in Btu/Acre ($j = 3$)

Phase	Sub-phase	Activity	Sub-activity	k^*	n^{a*}	$p_k{}^*$	Δ_k^*
Land	Growing	Transport to Farm	Seeds	1		In Equipment value	
			Equipment	1	1	$1.37 \times 10^{+4}$	nv
			Fertilizer	1		In Equipment value	
			Herbicide	1			
		Planting	Pre-planting	1		In Tilling value	
			Seed	1			
			Tilling	1	5	$1.67 \times 10^{+6}$	nv
		Field Application	Fertilizer	1		In Tilling value	
			Herbicide	1			
	Harvest	Crop and Silage Processing	Operations/fuel	2		In Tilling value	
			Transport: Storage, Biorefinery	2	2	$1.59 \times 10^{+6}$	$4.83 \times 10^{+5}$
General Items	Full Crop Cycle	Maintain Facilities & Other Equipment Operability	Operations/fuel	3		In Tilling value	
			Total EL$_{SG}$ & Δ_{SG}:			$3.27 \times 10^{+6}$	$4.83 \times 10^{+5}$

* With respect to k, n, p_k, and Δ_k, see Section 10.2.2.2 for definitions and Section 10.2.4 for detailed illustration on usage in calculations.
[a] values obtained by using only non-duplicated data from cited reference
nv: no value

Table 10.13 Module 9 Switch Grass to ethanol Production EL data (EBAMM, 2007) in Btu/Acre (j = 3)

Phase	Sub-phase	Activity	Sub-activity	k*	n[a]*	p_k*	Δ_k*
Biorefinery Plant	Production	Processing to 99.5% Ethanol	Operations/fuel	1	4	$7.19 \times 10^{+7}$	$5.90 \times 10^{+7}$
			Process water treatment	1	1	$5.72 \times 10^{+5}$	nv
			Total EL$_{SGe}$& Δ_{SGe}:			$7.25 \times 10^{+7}$	$5.90 \times 10^{+7}$

* With respect to k, n, p_k, and Δ_k, see Section 10.2.2.2 for definitions and Section 10.2.4 for detailed illustration on usage in calculations.
[a] values obtained by using only non-duplicated data from cited reference
nv: no value

Table 10.14 Module 100 Corn to ethanol EG data (Wright et al., 2006) in Btu/Acre

Stage	Activity	n*	EG*	Δ_{EG}*
External-to-Given BFC	Combustion of BFC Created Fuels : Ethanol	1	$2.75 \times 10^{+7}$	nv
	Total: EG$_{Ce}$.& Δ_{Ce}:		$2.75 \times 10^{+7}$	

* With respect to n, EG, and Δ_{EG}, see Section 10.3.3 for definitions and Section 10.2.4 for detailed illustration on usage in calculations.
nv: no value

Table 10.15 Module 101 Soybean to biodiesel EG data (Wright et al., 2006) in Btu/Acre

Stage	Activity	n*	EG*	Δ_{EG}*
External-to-Given BFC	Combustion of BFC Created Fuels: Biodiesel	1	$6.94 \times 10^{+6}$	nv
	Total: EG$_{Sd}$.& Δ_{Sd}:		$6.94 \times 10^{+6}$	

* With respect to n, EG, and Δ_{EG}, see Section 10.3.3 for definitions and Section 10.2.4 for detailed illustration on usage in calculations.
nv: no value

listed in Table 10.4 have been quantified), defines the NEB equations for the indicated scenario, and presents the analysis quantitative results.. The Δ_j values cited are 'lowest estimate' values since Sub-activities are not fully dispositioned and some of the p_k values do not have Δ_k values. The scope and asymmetry in the NEB data is reflected in the Table 10.18 Disposition and Overall values. The limitations of the scenario scope and NEB analysis are thus characterized and documented.

Figure 10.4 shows the corn – soybean crop rotation BFC scenario results. From a NEB perspective, as opposed to the E_{ed}/A production efficiency perspective of

Table 10.16 Module 102 SwitchGrass to ethanol EG data (EBAMM, 2007; Wright et al., 2006) in Btu/Acre

Stage	Activity	n^*	EG^*	Δ_{EG}^*
External-to-Given BFC	Combustion of BFC Created Fuels: Ethanol	1	$2.75 \times 10^{+7}$	nv
Biofuel Production	Biorefinery Plant Operations Fuels: Biomass	1	$5.19 \times 10^{+7}$	nv
	Total: EG$_{SGe}$& Δ_{SGe}:		$7.94 \times 10^{+7}$	

* With respect to n, EG, and Δ_{EG}, see Section 10.3.3 for definitions and Section 10.2.4 for detailed illustration on usage in calculations.
nv: no value

Section 10.4.1, optimization of the soybean to biodiesel parameters would appear (see uncertainty discussion below) to hold the greater promise.

The NEB is a difference based result: NEB = TEG − TEL. As such, it is sensitivity to $S_{module j}$ uncertainty and variation which increases as the TEG and TEL values approach numerical equivalency. The NEB values in Table 10.18 illustrate this limitation. The corn to ethanol BFC data, as illustrated in Module 1, 2, and 3 (see Tables 10.5, 10.6, and 10.7 respectively), have reported p_k and Δ_k values such that some limited statistical insight across reported results can be explored. The uncertainty values are generally of the same order of magnitude as their p_k value. The ethanol UE value reported in the literature also varies. The 7% variance estimate used below is on the low side of the literature range. These uncertainties result in this NEB having a large uncertainty. The data uncertainty impact is also clearly reflected by the NEV results in Fig. 10.1.

The soybean to biodiesel and switchgrass to ethanol BFC's data sets selected were too limited to calculate Δj values. However, both BFC's illustrates the same

Table 10.17 BFC module $E_{module\ j}$ equations

Template	Module j	Module Stage	$E_{module\ j}$
1	1	Infrastructure for Corn EL	EL_{IC}
2	2	Corn Agriculture EL	EL_C
3	3	Corn to ethanol Production EL	EL_{Ce}
1	4	Infrastructure for Soybean EL	EL_{IS}
2	5	Soybean Agriculture EL	EL_S
3	6	Soybean to biodiesel Production EL	EL_{Sd}
1	7	Infrastructure for SwitchGrass EL	EL_{ISG}
2	8	SwitchGrass Agriculture EL	EL_{SG}
3	9	SwitchGrass to ethanol Production EL	EL_{SGe}
4	100	Corn to ethanol EG	EG_{Ce}
4	101	Soybean to biodiesel EG	EG_{Sd}
4	102	SwitchGrass to ethanol EG	EG_{SGe}

Table 10.18 BFC Scenarios, NEB Relationships, and Analysis Results

BFC	Scenario	Components			NEB Equation & Value $\pm\Delta$
		Module & Templates	Disposition[a]	Overall[b]	
Corn to ethanol	Dry vs. Wet milling: $E_{Ce} =$ $E_{DCe} + E_{WCe}$	100	1	8	$NEB_{Ce} =$ $EG_{Ce} - EL_{IC} -$ $EL_C - EL_{Ce} =$ $-2.8 \pm 3.8 \times 10^{+6}$ Btu/Acre
		1	9	70	
		2	18	27	
		3	2	14	
		1	9	70	
		2	18	27	
		3	2	14	
Soybean to Diesel	Soybean only	101	1	8	$NEB_{Sd} = EG_{Sd} -$ $EL_{IS} - EL_S - EL_{Sd} =$ $-1.4 \times 10^{+6}$ Btu/Acre
		4	7	70	
		5	17	27	
		6	2	18	
SwitchGrass to Ethanol	Switchgrass only	102	2	8	$NEB_{SGe} =$ $EG_{SGe} - EL_{ISG} -$ $EL_{SG} - EL_{SGe} =$ $-5.0 \times 10^{+5}$ Btu/Acre
		7	7	70	
		8	12	27	
		9	1	18	
Corn to ethanol + Soybean to Diesel with Crop Rotation	CR = fraction of full crop rotation schedule that corn is grown; (1 − CR) = fraction of full crop rotation schedule that soybean is grown	100	1	8	$NEB_{CeSd:CR} =$ $CRNEB_{Ce} + (1 -$ $CR)NEB_{Sd} =$ See Fig. 10.4
		101	1	8	
		1	9	70	
		2	18	27	
		3	2	14	
		4	7	70	
		5	17	27	
		6	2	18	

[a] number of Sub-activities dispositioned in the Module
[b] total number of Sub-activities in the template

NEB difference problem due to comparable EG and EL values. For the switchgrass to ethanol BFC the 7% ethanol UE uncertainty is 3.9 times the EG − EL difference.

10.4.3 BFC Environmental Scenario Models and Analysis

The environmental aspects are captured in the general Templates 1, 2, and 3 (Section 10.2.2.2) under the Waste Management Sub-activities. The number of potential Environmental Concern (EC) source terms are wastewater − 16, solid waste − 17, non-aqueous liquids − 17, and air emissions − 14. The type, composition, and concentration of environmental pollutant considerations depend on the source activity/process, fuel, and chemicals involved (EPA, 2007b; USDA, 2007b).

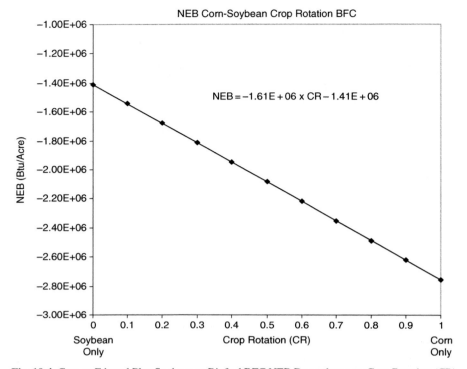

Fig. 10.4 Corn to Ethanol Plus Soybean to Biofuel BFC NEB Dependence on Crop Rotation (CR)

Consider the potential source term air pollutants (EPA, 2007a; USDA, 2007d). Applying the equation 10.2 relationship, where we hold U constant, define $S_{\text{module } j} = EC_{\text{module } j}$, and calculate the EC_{BFC} on a per unit area basis, gives the general BFCM equation:

$$EC_{\text{BFC}} \text{ (mass or volume / Area)} = \sum_{j=1}^{n} EC_{\text{module } j}$$

Here $EC_{\text{module } j}$ is the template derived assessment for module j of the EC value in Btu/Acre and n is the number of literature values that form the basis for the cited value. Using the Air Emissions (AE) aspects of the templates as an example, the AE general relationship is:

$$AE = EC_{\text{module } 1} + EC_{\text{module } 2} + EC_{\text{module } 3} = \sum_{k=1}^{12} ae_{1,k} + ae_{2,3} + \sum_{k=1}^{2} ae_{3,k}$$

Here $ae_{j\,k}$ is the Stage j, Sub-activity k specific pollutant mix. Analysis of the $EC_{\text{module } j}$ Greenhouse Gas Emissions (GGE: $CO_2 + CH_4 + N_2O$) subset using the CO_2 equivalent values reported for the ethanol to corn BFC given in Table 10.19,

Table 10.19 Greenhouse gas emission (GGE) data in g CO_{2e}/Gal (EBAMM, 2007)

Stage	j	k	n^a	$GGE_{jk}{}^b$	$\Delta (GGE_{jk})^b$	Number of quantified GGE_{jk} values
Infrastructure	1	10	2	$7.55 \times 10^{+0}$	nv	1 out of 12
Agriculture	2	3	14	$3.33 \times 10^{+3}$	$6.29 \times 10^{+2}$	1 out of 1
Biofuel Production	3	1	13	$7.84 \times 10^{+2}$	$1.06 \times 10^{+2}$	2 out of 2
	3	2	1	$1.12 \times 10^{+2}$	nv	
Net Greenhouse Gas Emission:			48	$4.23 \times 10^{+3}$	$6.38 \times 10^{+2}$	4 out of 15

[a] values obtained by using only non-duplicated data
[b] factors used to convert data: 2.471 Acre/Hectare and Fig. 10.3 current corn yield values.
nv: no value

shows the estimated Net Greenhouse Gas Emission is $4.29 \pm 0.70 \times 10^{+3}$ (g CO_{2e}/Gal) with 4 out of 15 potential air emission source terms quantified. The impacts, if any, of the other 11 source terms are unspecified in this particular scenario. The asymmetry in the data is further reflected in the cited n values. Thus the BFCM results in Table 10.19 clearly delineate the scope and limitations of the results.

The BFCM template approach can also be used for environmental evaluation of farm conservation measures such as (USDA, 2007a; 2007c) crop rotation, crop residue management, contouring, grade stabilization, soil quality management (erosion and condition), and nutrient/pest/disease management.

10.5 Other Considerations

The differences in interpretation of the BFC boundaries have resulted in disagreement in the literature with respect to the NEV. The energy aspects of coproducts, facility construction, and labor are main issues. While it is desirable to have a positive NEB, the NEB result is not the only consideration. National security, energy independence, financial, and environmental aspects are part of the decision mix which might trump NEB considerations. The BFCM, through definition and methodology, maintains the TEL and TEG parameters as stand alone energy terms which yields an unencumbered NEB.. This enables straightforward cross BFC comparisons without the need to track specific energy exceptions or adjustments.

The consideration and justification of coproduct energy credit or labor caloric aspects is not eliminated by the BFCM, it is just excluded from the NEB analysis. Such adjustments of the NEB would be a post-NEB step.

Reported studies have addressed various Stage activities. The templates incorporate and expand upon these scopes. Consideration of the infrastructure, which includes facility construction, and waste management aspects impacted by BFC growth is an integral part of BFC analysis. The energy to construct storage, seed processing, soil additive, terminals, and waste handling facilities needs to be addressed, particularly in light of the cumulative nature of the NEB. The waste management aspects listed in the Infrastructure Template 1 (see Table 10.1) might appear to be far a field. However, inclusion of such aspects is justified considering the past corn to ethanol BFC (Reynolds, 2002) expansion (annual US ethanol production: $1.75 \times 10^{+6}$ Gal in 1980 to $3.9 \times 10^{+9}$ Gal in 2005) and the hypothesized ($9.8 \times 10^{+9}$ Gal in 2015) growth (Urbanchuk, 2006).

There is a need for standardized p_k and Δ_k estimating methods and establishment of set UE values for the biofuels so the number of significant figures in the UE value is sufficient to yield NEB values with reasonable uncertainties. This, in combination with the BFCM, will enable improvement in the BFC analysis and reduction of the uncertainty of the results. Finally, as demonstrated by the opposed E_{ed}/A and NEB results for the corn − soybean crop rotation, pursuit of multiple BFC aspects would be of value in moving forward on the BFC technologies.

References

Brent D. & Yacobucci, B. D. (2006, March 3). Fuel Ethanol: Background and Public Policy Issues. (CRS Report for Congress, Received through the CRS Web, Order Code RL33290)

Dias De Oliveira, M. E., Vaughan, B. E., & Rykiel, E. J. J. (2005, July). Ethanol as Fuel: Energy, Carbon Dioxide Balances, and Ecological Footprint, *BioScience, 55*, 593

Dobermann, A. & Shapiro, C. A. (2004, January). Setting a Realistic Corn Yield Goal. UNL NebGuide G481 From http://elkhorn.unl.edu/epublic/live/g481/build/#yield

EBAMM (2007). *EBAMM, ERG Biofuel Analysis Meta-Model.* (n.d.). Retrieved February 15, 2007, from http://rael.berkeley.edu/ebamm/

EPA (2007a). *Clean Air Act, U.S. Environmental Agency.* (n.d.). Retrieved April 2, 2007, from http://www.epa.gov/agriculture/lcaa.html#Summary

EPA (2007b). Quick Finder, U.S. Environmental Agency. (n.d.). Retrieved April 2, 2007, from http://www.epa.gov/

Farrell, A. E., Plevin, R. J., Turner, B. T., Jones, A, O'Hare, M, & Kammen, D. M. (2006a, July 13). Ethanol Can Contribute To Energy and Environmental Goals. (Energy and Resources Group (ERG), University of California − Berkeley, ERG Biofuels Analysis Meta-Model (EBAMM), Supporting Online Material, Version 1.1.1, Updated)

Farrell, A. E., Plevin, R. J., Turner, B. T., Jones, A, O'Hare, M, & Kammen, D. M.(2006b, January). Ethanol Can Contribute to Environmental and Energy Security. *Science, 311*, 506–508

Graboski, M. S. (2002, August). Fossil Energy Use in the Manufacture of Corn Ethanol. (Prepared for the National Corn Growers Association)

Hammerschlag, R. (2006). Ethanol's Energy Return on Investment: A Survey of the Literature 1990-Present *Environmental Science Technology, 40*, 1744–1750

Kim, S. & Dale, B. (2005). Environmental aspects of ethanol derived from no-tilled corn grain: nonrenewable energy consumption and greenhouse gas emissions. *Biomass Bioenergy, 28*, 475–489

Koplow, D. 2006. Biofuels at what cost? Government support for ethanol and biodiesel in the United States. The Global Studies Initiative (GSI) of the International Institute for Sustainable development (IISD). http://www.globalsubsidies.org/IMG/pdf/biofuels_subsidies_us.pdf (2/16/07)

Natrella, M. G. (1966, October). *Experimental Statistics. National Bureau of Standards Handbook 91.* (Washington, D.C.: U.S. Government Printing Office)

NIST (2006). *NIST/SEMATECH e-Handbook of Statistical Methods.* (Updated: 7/18/2006). from http://www.itl.nist.gov/div898/handbook/index.htm

Patzek, T. (2004) Thermodynamics of the corn-ethanol biofuel cycle. Critical *Reviews in Plant Sciences, 23*(6), 519-567

Pimentel, D. & Patzek, T. (2005). Ethanol production using corn, switchgrass, and wood and biodiesel production using soybean and sunflower. *Natural Resources and Research, 14*(1), 65–76

Pimentel, D. (1991). Ethanol Fuels: Energy Security, Economics, and the Environment. *Journal of Agricultural and Environmental Ethics, 4*, 1–13

Pimentel, D., Patzek, T. & Cecil, G. (2007). Ethanol Production Energy Economic, Energy, and Food Losses. *Reviews of Environmental Contamination and Toxicology, 189*, 25–41

Reynolds, R. E. (2002, January 15). Infrastructure Requirements for an Expanded Fuel Ethanol Industry. (Prepared by Downstream Alternatives Inc., South Bend, IN for Oak Ridge National Laboratory Ethanol Project)

Shapouri, H., Duffield, J. A., & Graboski, M. S. (1995, July). Estimating the Net Energy Balance of Corn Ethanol. (U.S. Department of Agriculture, Economic Research Service, Office of Energy. Agricultural Economic Report No. 721)

Shapouri, H., Duffield, J. A., & Mcaloon. A. (2004, June 7–9). The 2001 Net Energy Balance of Corn-Ethanol. (Paper presented at the Corn Utilization and Technology Conference, Indianapolis, IN)

Shapouri, H., J. A., Duffield, J., A. & M. Wang, M. (2002, July). The Energy Balance of Corn Ethanol: An Update. (U.S. Department of Agriculture, Office of the Chief Economist, Office of Energy Policy and New Uses. Agricultural Economic Report No. 814)

Skoog, D. & West, D. M. (1963). *Fundamentals of Analytical Chemistry*. Pages 54–57. (New York: Holt, Rinehart, and Wilston)

Urbanchuk, J. M. (2006, Feburary 21). Contribution of the Ethanol Industry to the Economy of the United States. (Prepared for the Renewable Fuels Association)

USDA (2007a). *Farm Conservation Solutions, U.S. Dept. Agriculture Natural Resources Conservation Service.* (n.d.). Retrieved April 2, 2007, from http://www.wi.nrcs.usda.gov/programs/solutions/

USDA (2007b). *Helping People Help the Land, U.S. Dept. Agriculture Natural Resources Conservation Service.* (n.d.). Retrieved April 2, 2007, from http://www.nrcs.usda.gov/

USDA (2007c). *Resource Management System Quality Criteria, U.S. Dept. Agriculture Natural Resources Conservation Service.* (n.d.). Retrieved April 2, 2007, from http://www.id.nrcs.usda.gov/technical/quality_criteria.html

USDA (2007d). *USDA-Agricultural Air Quality Task Force, U.S. Dept. Agriculture Natural Resources Conservation Service.* (n.d.). Retrieved April 2, 2007, from http://www.airquality.nrcs.usda.gov/AAQTF/Documents/index.html

Wang, M. & Santini, D. (2000, February 15). Corn-Based Ethanol Does Indeed Achieve Energy Benefits (Center for Transportation Research, Argonne National Laboratory)

Wang, M. (2005). *15th International Symposium on Alcohol Fuels*, 26–28

Wang, M., Saricks, C., & Wu, M. (1997, December 19). Fuel-Cycle Fossil Energy Use and Greenhouse Gas Emissions of Fuel Ethanol Produced from U.S. Midwest Corn. (Center for Transportation Research, Argonne National Laboratory, Prepared for Illinois Department of Commerce and Community Affairs)

Wright, L., Boundy, B., Perlack, B., Davis, S. & Saulsbury B. (2006, September). *Biomass Energy Data Book Edition 1, ORNL/TM-2006/571.* (Oak Ridge, Tennessee: Oak Ridge National Laboratory)

Chapter 11
Our Food and Fuel Future

Edwin Kessler

Abstract During the past century, inexpensive fuels and an outpouring of new science and resultant technology have facilitated rapid growth and maintenance of human populations, infrastructures, and transportation. Developed countries are critically dependent on the liquid fuels required by present day transportation of goods and services and by agriculture and are dependent on various fuels for generation of electricity. Authorities and the media present physical growth as an economic and social need, but consumption and its growth ultimately cause declining availability and increasing price of fuels and energy. Increased burning of carbon fuels with increase of carbon dioxide in Earth's atmosphere is the principal cause of increasing global warming, which is well-measured and a probable source of future disruption of world ecosystems.

Regrettably for humanity, the power of new technologies has not yet been accompanied by vitally needed political and cultural developments in the U.S. and in many other countries. The political system in the U.S. seems unable to mitigate processes that contribute to global warming nor adequately address declining supplies of liquid fuels, nor does it discourage social pressures for continued physical growth.

Search for alternative sources of liquid fuels for the transportation sector in developed countries and in the United States in particular produce strong connections among energy supply, food supply, and global warming. Various current U.S. programs are examined and none appear effective toward prevention of a future disaster in human terms. The social organism is not ready now to sacrifice for future gain or even for sustainability.

Keywords Energy sources, alternative · energy sources, traditional · batteries · biodiesel · coal · ethanol · geothermal energy · global warming · hydropower · natural gas · nuclear fission · nuclear fusion · petroleum · political and social conditions · solar power · wind · rivers and tides

✉ E. Kessler
1510 Rosemont Drive, Norman, OK 73072
e-mail: kess3@swbell.net

D. Pimentel (ed.), *Biofuels, Solar and Wind as Renewable Energy Systems,*
© Springer Science+Business Media B.V. 2008

11.1 Introduction

Connections among energy supply, food supply, global warming, and political campaigns have become strong in the United States during first years of the 21st century. Liquid fuels derived from petroleum are of enormous importance in developed countries because they are a principal support of the transportation industry (and petroleum- and coal-derived hydrocarbons are also critical ingredients in the chemical industry). Demand for liquid fuels continues to increase, but discoveries are tapering off, and sharply increased price is stimulating search in the U.S. and other nations for sources other than the traditional oil industry, which involves a dependence on foreign suppliers of uncertain reliability. The search for suitable alternatives is influenced and befuddled by powerful established interests whose primary goals are their own economic benefits rather than societal welfare. Several of the programs are examined in detail in following pages, and it should be borne in mind that numerous proposals reflect wishes of special interests more than conclusions from rational analysis. Controversy abounds.

11.2 Price and Availability of Traditional Fuels

Traditional energy sources, i.e., those that produce a substantial amount of the power currently used, include coal, oil, natural gas, hydropower, and nuclear fission. Non-traditional sources, i.e., emerging sources, some on trial or subjects of significant experiments, include wind, tides and river currents, solar, hydrogen, biomass, geothermal, and nuclear fusion. Brief comments on all of these energy sources follow, with much of the presented data obtained from the U.S. Energy Information Administration (see EIA website).

11.2.1 Coal

Coal burning produces about half of all the electrical energy[1] produced in the United States, a ratio that has remained nearly constant for the past twenty-five years, even as electricity usage has increased 70%. Coal is usually said to be so abundant in the United States that its use as an energy source here will endure for centuries. Next to hydropower, it is the cheapest source of energy, and about 85% of the 1.1 billion tons produced and consumed annually in the United States is bituminous coal and is used within the country to generate electricity.

---

[1] Total electric energy produced in the United States in 2005 was 4.05 billion megawatt hours. This would be produced with average generation of 460 thousand megawatts for one year. EIA presents the generating capacity during the 2005 summer, when demand is maximal, as 978 thousand megawatts – in other words, capacity is about twice the average generation. The efficiency of power production in coal-burning plants is in the range 30–40%. In other words, about 30–40% of the heat energy in coal is manifested in the electricity produced.

Coal burning in the U.S. produces annually about 2.1 billion metric tons of carbon dioxide,[2] the major contributor to global warming. The carbon dioxide is emitted to the atmosphere and it is buried permanently (sequestered) only in rare situations where, under high pressure, it enhances tertiary recovery of petroleum. Coal burning has increased 19% since 1990 but was down nearly 1% between 2005 and 2006 because the average U.S. winter in 2006 was milder and the summer cooler than in 2005.

According to EIA data, the price of coal as delivered to power plants in the United States is significantly variable with region, costing much more in New England (~$65/ton in 2005), for example, than in the Midwest (~$20/ton), and owing to increasing world demand, the price is rising as this chapter is developed. In 1975 there was a temporarily doubled price that was largely caused by the Arab oil embargo of 1973, and this peak was followed by a slow decline of coal price.

An important way of looking at the price of coal is through energy content – a typical minehead price in 2005 was about $1.15 per million BTU, or about $20/ton for coal with a 50% carbon content and the delivered price was about $45/ton, but variable depending on the distance from mine to user.

Past sulfurous emissions from coal-burning power plants have been widely associated with "acid rain", which causes corrosion and has altered the pH and ecology of some lakes, especially in northeast U.S. The Shady Point power plant at Panama, Oklahoma, which started in commercial operation in 1991, avoids sulfurous emissions by mixing local high-sulfur coal with limestone, also mined locally. As the limestone is heated, it emits carbon dioxide and combines with the sulfur, producing calcium sulfate, which in another form is known as gypsum. Some of the slag finds a use in neutralizing pollution and some finds use as a road stabilizer, though most goes to land-fill sites.

The Shady Point power plant produces its maximum 320 megawatts throughout 24-hours during June-August while burning daily about 3000 tons of Oklahoma coal mixed with about 1000 tons of limestone. The average sulfur content of the coal is about 3% and its carbon content is variable from about 55% to 70%, depending on mine origin. Its carbon dioxide emissions during summer, based on 60% carbon in the coal, are thus about seven thousand tons daily with about 6% of that from the limestone, and 200 tons/day are extracted from the flue gas as food-grade CO_2. The augmentation of CO_2 by limestone seems unimportant in view of the large ongoing emissions from other coal-burning power plants. (Personally communicated, 2007; also see Shady Point website).

Most actual reductions of sulfur emissions in the U.S. have resulted from use of low-sulfur coal from Wyoming instead of coals with higher sulfur content from

---

[2] Each ton of burned carbon, molecular weight 12, produces 3.66 tons of carbon dioxide, molecular weight 44. Consider a model 1000-megawatt electric power plant operating at 35% efficiency, which burns all contents of a 110-car coal train every day, about 12 thousand tons of coal with a carbon content near 70%. It thereby emits about 30,000 tons of carbon dioxide. See also the table in Section 11.4.

Oklahoma and eastern U.S. Particulate emissions from coal-burning power plants, another cause of "acid rain", have also been greatly reduced in recent years.

Emissions from coal burning include mercury and other heavy metals including arsenic, uranium, and thorium. During 1999–2003, the U.S. Environmental Protection Agency collected and analyzed fish tissue from 500 ponds and lakes across the United States for a wide range of elements and organic toxic chemicals. Levels of mercury or arsenic exceeding EPA screening levels for human health were found in many of them. This contamination is attributed to coal burning, though it seems that this attribution has not been proved. Of twenty one sites sampled in Oklahoma, nine had levels of mercury or arsenic that exceeded EPA screening levels[3] (Environmental Protection Agency, 2007), and many states have issued directives concerning permissible limits on eating fish so contaminated. Questions have been raised about prospects for the enduring use of coal owing to environmental concerns, possible exaggeration of reserves amenable to economical extraction, and probable increased future costs of transportation (Schneider, 2007a).

Further concerning the environment, coal mining in the U.S. state of West Virginia has become very controversial because whole mountain tops have been moved into adjacent valleys in order to expose coal seams. This has caused marked deterioration of water quality and other environmental abominations. Mine safety also continues as a major issue with strident public calls for additional regulation by the U.S. federal government.

China and the United States in 2007 emit nearly equal amounts of carbon dioxide, and further major development of the coal industry in China's Shanxi Province was outlined in a special supplement to *China Daily*, published September 18, 2007. Substantially increased production of raw coal, liquid fuels from coal (usually, Fischer-Tropsch process), and coalbed methane (see following section) were projected during the Taiyuan[4] International Coal and Energy New Industry Expo 2007. This development is seen in China as essential to improved prosperity of the country and its people.

The indicated environmental negatives diminish as advanced technologies are applied. Coal combustion seems destined to remain for decades as a major source of electrical power. However, in spite of promulgation of State policies toward energy conservation and emission controls such as presented by the Shanxi Minister of Commerce, serious concerns persist because coal burning and coal conversion are major producers of carbon dioxide, the principal contributor to global warming (see the Table 11.1 in Section 11.4).

11.2.2 Natural Gas

At the start of the 20th century, natural gas was a little-desired byproduct of the petroleum industry and sold for as little as five cents per thousand cubic feet at

---

[3] And all but two had toxic levels when organics used in industrial agriculture are included.

[4] Taiyuan, in northwest China, is the capitol of Shanxi Province.

the wellhead. During the 1970s, price rose from 17 cents to $1.20 per thousand cubic feet, and during the 1980s and 1990s, natural gas was irregularly priced, but sometimes above $2.50. A substantial price rise to 2007 levels fluctuating between $5 and $7 per thousand cubic feet began about the year 2000. Improved technologies of horizontal drilling and fracturing in tight rock formations have enabled gas production in areas of shale and coal formations in the United States, and the high cost of production is supported by high price of the product. Regrettably, modern methods of extraction often degrade soil and water.

Natural gas is widely used today for home heating and for standby power generation, and gas-to-liquids technologies are being proposed for production of liquid fuels. Gas production and consumption in the United States has been nearly steady at about 24 trillion cubic feet annually since the mid-1990s, and challenges to maintain that level of usage in the presence of an ultimate decline of U.S. supplies have led to proposals for importation of liquefied (strongly cooled) gas (LNG) from the Middle East. However, proposed LNG terminals are often opposed by local groups apprehensive of explosion dangers.

Natural gas is also used for production of the fertilizer bases, ammonium nitrate and urea. As the price of natural gas has risen, its preferred use for home heating and power generating facilities has led to closure of about 40% of U.S. fertilizer production capacity since 1999 and to increasing importation of nitrogen fertilizer from regions where natural gas is much less costly than in the U.S. Imports now account for a little more than half of total U.S. nitrogen supply, which has remained nearly steady at twenty million product tons since 1998.

A recently developed controversy within the United States involves proposed new facilities for electric power generation, with natural gas interests pointing to the lower carbon dioxide emissions associated with natural gas, and coal advocates indicating lower costs with coal.[5] In any case, creation of new power plants, whether gas- or coal-powered, to accommodate continued physical growth leads to increased CO_2 emissions and exacerbation of the global warming phenomenon (see Section 11.4).

It is conceivable that further research will lead to a vast expansion of natural gas supplies and, perhaps, to a medium for the more effective storage of hydrogen (see section 11.3.3) than is available today. Such advances could involve clathrate hydrates, which are abundant below permafrost and along continental margins in and beneath waters whose temperatures are near water's freezing point. Clathrate hydrates are solid combinations of hydrocarbons, especially methane, or carbon dioxide with water. It is estimated that several times the known traditional resources of natural gas are so combined, and there is concern that global warming will lead to release to the atmosphere of vast quantities of clathrate methane. This would be especially important because methane is about 20 times the greenhouse gas that is carbon dioxide. While many clathrate deposits have been identified, an effective technology for methane extraction has not been developed. Mao, et al. (2007)

---

[5] Natural gas is principally methane, CH_4, and coal contains very little hydrogren. When natural gas is burned, its large hydrogen component produces only water.

describe the situation in desirable detail, and their article contains a substantial list of references.

11.2.3 Petroleum

A direct use of oil is for home heating, especially in northeastern United States, and oil refined to gasoline and diesel fuel provides more than 95% of the energy used in the U.S. transportation industry. Oil production in the U.S. peaked at 9.5 million barrels per day in 1970, in close agreement with a prediction of M. King Hubbert.[6] Since 1985, U.S. crude oil production has declined every year, and in 2005 was 5.2 million barrels per day. And, as a result of both declining domestic production and increasing demand, crude oil imported to the United States increased from 5.8 million barrels per day in 1991 to 10.1 million barrels per day in 2005[7]. Total U.S. consumption of crude oil and other imported petroleum products continues to rise about 1% annually, and totaled 20.8 million barrels per day in 2005.

In the early 1970s, the inflation adjusted price hovered near $10/barrel, but it is near $90 and rising irregularly as this article is completed at the end of October 2007. The price of crude oil is reflected in the price of refined products, and gasoline in June 2007 cost as much as $4/gallon in some U.S. markets, and more than $3/gallon on average nationwide.[8] Dependence of the U.S. for oil from foreign sources of uncertain reliability, rising prices, and concern for competition and projected future scarcity (e.g., Simmons, 2005[9]; Ghazvinian, 2007) are stimulating search for alternative motor fuels, discussed further below. But a major concern arises because all carbonaceous fuels produce carbon dioxide emissions that contribute to global warming, and emissions by the U.S. transportation sector are about one third of the total.

A striking example of conflict between efforts to gain access to new oil and the greenhouse problem (discussed in Section 11.4) is provided by the tar sands of northern Alberta. Economically recoverable reserves of heavy oil there are estimated to well exceed one hundred billion barrels, which would supply the whole world for several years at the present rate of consumption (about 30 billion barrels annually). But the extraction process is very energy intensive, involving mining of the sands, their transport in huge trucks to crushing and heating facilities, and costly refinement and transport of a still tarry product via pipelines. In situ heating with large use of water is also implemented for recovery of oils at depth. These energy

---

[6] Hubbert's Peak, so-called.

[7] Only in the year 2002 during this period was there a slight decline of imports from the previous year. The importation of 10 million barrels of oil daily at a price of $80 per barrel is a contribution of $800 million daily to the U.S. deficit in international trade.

[8] The retail price of gasoline in Europe has long tended to be this high and higher, because of much higher taxes.

[9] Simmons presents a comprehensive discussion of oil history and industry in Saudi Arabia, and concludes that the quantity of Saudi Arabian oil reserves is greatly exaggerated in recent announcements.

intensive processes produce much greater release of carbon dioxide than is released during recovery of lighter oils by traditional methods.

The processes for recovery of tarry oil are described at length in a supplement to *E&P Oil and Gas Investor* (Hart Energy Publishing, 2006), which includes a list of companies and their plans to invest $80 billion in Alberta oil sands by the year 2014.[10] Discussion of advanced technologies for extraction and refinement of tarry oil has also been presented (Hart Energy Publishing, 2007).

11.2.4 Hydropower

Most dams are built for flood control and irrigation, but hydropower provides about 7% of all the electricity produced in the United States. The largest hydroelectric facility in the U.S., Grand Coulee Dam, serves multipurposes while providing average power of about 2300 megawatts, the equivalent of two or three ordinary coal-burning plants. In the U.S., it is not expected that additional hydropower can be provided in quantity sufficient to replace other energy shortfalls, but in China, the Three Gorges Dam is scheduled for completion about 2010 and should provide 18 thousand megawatts of electricity.

Dams do have negative effects. Thus, sediment tends to accumulate behind dams, reduced sediment in downstream flows usually fails to compensate for erosion of river deltas, and there are often adverse effects on fisheries.[11] For such reasons and others, especially the destruction of agricultural areas flooded by impounded waters, the construction of hydroelectric facilities produces controversy, and some existing dams have even been proposed for removal.

11.2.5 Nuclear Fission

Studies in astrophysics and atomic physics subsequent to presentation of Einstein's special and general theories of relativity in 1905 and 1916 showed paths for producing enormous energies by conversion from matter. Heavy elements, including uranium, are produced during the collapse of stars much more massive than Sun, and the products of the radioactive decay or fission of the heavy elements are less massive than their sources. The mass difference appears as energy.

Uranium is widely present on Earth, its average concentration is near three parts per million, and it is over ten times more abundant than silver, for example. It consists mainly of the isotope ^{238}U, with about 0.7% ^{235}U, which is principal reactor fuel. For purposes of power generation ^{235}U is concentrated to about 3% by an

---

[10] The 2006 Annual Report of Chevron indicated plans by that company to invest $2 billion in the tar sands. My inquiry as a stockholder about the implications of this investment for carbon dioxide emissions was not answered.

[11] A river dolphin of China has recently been reported extinct, and the principal cause of extinction is believed to be the Three Gorges Dam, under construction at this writing.

energy-intensive gaseous-diffusion process that takes advantage of the slight difference of atomic weights among isotopes. During typical reactor operation, atoms of ^{235}U absorb neutrons and then split into other elements with release of energy and neutrons. The reaction is initiated by stray neutrons and maintained by those released. Materials that absorb neutrons are arranged to maintain a concentration of neutrons that produce heat at the desired rate. The energy statistics are startling: Fission of one kilogram of ^{235}U produces as much energy as combustion of about 40 million kilograms of TNT and without any greenhouse gases.

As in other power plants, the heat generated by controlled fission is used to boil water and create steam that drives turbines to generate electricity. At this writing, nuclear fission provides about 19% of all electricity in the U.S., 16% worldwide, 30% in Japan, and maximally 78% in France. According to the U.S. Energy Information Agency, there were 436 operating reactors in 30 countries worldwide during May 2007, including 103 operating reactors in the United States. There is little question that nuclear reactors could provide abundant electricity but their future is clouded by risk of accidents that degrade wide areas, such as occurred at Chernobyl, by risks from terrorism, and by risks attendant to disposal of highly radioactive nuclear waste for hundreds of thousands of years. Possible effects of seismicity and volcanism at the proposed U.S. disposal site at Yucca Mountain, Nevada, have been examined by Hinze, et al. (2008).

And use of breeder reactors, so-called, which convert uranium of molecular weight 238 to fissionable plutonium of weight 239 and could provide a nearly endless energy supply, is inhibited by fears that the process of separating plutonium from the mix would be adapted to bomb making. Although more than thirty new nuclear plants are under construction in twelve countries as this chapter is prepared, new construction in the United States has been strongly inhibited by negative public opinion. However, the combination of conditions described in preceding sections, coupled with reactor designs that are much improved with respect to simplicity and safety may well lead to a resurgence of fission reactor construction in the U.S. (e.g., *The Economist,* September 8–14, 2007, pp. 13 & 71–73).

In this matter, a paper on net energy (Tyner[12] 2002), should be examined. Owing to energy requirements for construction, operation, waste disposal, and ultimate dismantling of nuclear power plants, Tyner concludes, "any expectation that Nuclear Power will be a viable substitute for fossil fuels is, at best, questionable". There is also the matter of carbon dioxide releases that attend manufacture of the cement and steel needed for reactor construction and the mining and refinement of nuclear fuel. Details are complex and this author proposes that the matter of net consequences be carefully examined. In any event, while electric power however generated is a poor direct substitute for liquid fuel for transportation in 2007, electrical energy can be used for the manufacture of liquid fuels.

---

[12] Gene Tyner, Sr. piloted U.S. aircraft during the Viet Nam war, and, after his retirement from the U. S. Air Force, he gained a doctorate in economics at the University of Oklahoma. Subsequently he consulted on energy issues. He died in 2004.

11.3 Alternative Sources of Energy

As already noted, the high and rising price of oil and its derivative fuels is a principal accelerant to search for alternative fuels. Another motivation for this search lies in concerns about global warming, produced by increasing emissions of carbon dioxide during transportation, power generation and during manufacturing processes attendant to production of steel and cement, for examples. As shown below, it will be difficult to develop an alternative fuel pathway that supports either generation of electricity without excessive carbon dioxide emissions or an automotive industry with markedly reduced usage of petroleum and its products. Further, the programs so far implemented in the United States appear to be means for accumulation of wealth by a relatively small number of beneficiaries who have both the power to control legislation and ability to create a public perception that realistic steps are being taken when the fact is opposite. The incorrect public perception allows business to proceed as usual even though collapse may be just around the corner.

We first discuss several suggested alternate energy sources that may be contributing in a small way, and then we consider possibilities whose successful future application must depend on research results so-far elusive. Then we take up nationally empowered programs involving biologically based fuels.

11.3.1 Wind, Rivers, and Tides

Wind has been used for thousands of years for sailing and for grinding grains, and decades ago in the United States there were, beyond the range of utility lines, many small windmills that powered a few light bulbs and radios. Small windmills are still widely used in western United States to pump water for livestock. Modern wind energy units are especially valuable in remote communities where electricity is otherwise supplied by small diesel-fueled installations, which can be very costly.

According to the Energy Information Administration, wind began to be a significant source of electricity in the United States about 1990.[13] Wind power technology has advanced steadily and large machines now deliver up to five megawatts each during favorable winds. Use of wind power has advanced with particular rapidity in Europe, and Denmark, an acknowledged global leader in wind energy, derives approximately 20% of its electricity from wind turbines and plans for an increase to 50% in 2030. The increase in wind energy production since about 1980 in Denmark has enabled that country to stabilize its carbon dioxide emissions.

Technological advances have greatly reduced the price of power from wind, and land-based wind turbines now cost from $1500 to $3000 per kilowatt, nearly

---

[13] Your author operated one of the first commercial windmills produced by the Bergey Windpower Company of Norman, Oklahoma, a one-kilowatt device, on his farm from 1981 to 1984. A report of its operation (Kessler and Eyster, 1987) is included in the references, and is a fair primer on wind energy technology. The Bergey Windpower Company is a leading producer of small turbines, 1.5–50 kW.

competitive with coal-burning power plants. According to the American Wind Energy Association (2007), the most efficient wind generators in windy places can deliver power at a cost of five to ten cents per kilowatt hour. This is similar to the charge imposed by most utilities in the U.S., but wind power in the U.S. is still subsidized with a federal tax credit of 1.5 cents per kWh.[14]

Electricity is produced by wind with no gaseous emissions at all, though emissions occur during manufacture of the steel, concrete, and other items used in fabrication and erection of the turbines. Where winds are favorable, the overall payback is large, however, and is still increasing with technological advances. The great height, several hundred feet, of modern machines places them above the layer where friction with the ground causes a strong diurnal variation of wind – at the greater height the average wind is nearly constant throughout the average day. Since the rate of electrical power generation is proportional to the cube of the wind speed, site selection is very important. Site selection in Oklahoma has been aided by a network of over one hundred weather-reporting stations within the State (Kessler, 2000; Oklahoma Mesonet, 2007).

The capacity of electricity production from wind is increasing in the U.S., with approximately 5000 megawatts added during the two-year period 2004–05. Subsequent additions brought the total U.S. wind power capacity to 12,634 megawatts as of June 30, 2007, more than one percent of the U.S. total of about one million megawatts (See footnote 2). Production of electricity from wind does seem to be a good, but, as noted elsewhere (e.g., Tyner, 2002), "... even if wind machines were constructed everywhere it is practical to erect wind machines in the United States they would only be able to provide a pitifully small fraction of the net energy compared to that needed to power the industrial economy of the United States..." This seems true in Oklahoma, although five wind farms have been installed and others are planned. Installed wind capacity in Oklahoma totaled 690 megawatts in August, 2007, about three percent of Oklahoma's electric generating capacity (American Wind Energy Association, 2007; Oklahoma Wind Power Initiative, 2007).

Capacity and capacity factors can be confusing. Because wind is highly variable, the average generation by a wind farm is almost always less than half of its capacity with optimum wind, and one third is often taken as a standard. This means that Oklahoma wind farms can presently provide, on average, about 1% of the power that can be provided by traditional facilities. Furthermore, since electricity cannot be economically stored,[15] no amount of wind power installation allows reduction of the number of power plants fueled by coal, natural gas, or nuclear fission, except to the extent that consumers agree to interruptible power supply. Of course, during windy periods, power generators that use fossil fuels can be cut back, thereby reducing emissions and saving non-renewable fuels.

---

[14] Some utilities charge much more for electricity, and the price is sometimes varied substantially with time of day in phase with overall load, to encourage conservation.

[15] Battery technology is advancing but is still a very expensive means for storing large quantities of electricity. Other means such as compressing air for later release to a turbine, pumping water uphill and then letting it down, are also costly. See also Section 11.3.2.

At this writing, wind farms have been proposed offshore Cape Cod, Massachusetts, and offshore south Texas in the United States, but are attended with uncertainties in both costs and esthetics. Research at the Massachusetts Institute of Technology (MIT) envisages anchoring systems for wind farms offshore that would withstand the force of wind and wave in hurricanes at a distance beyond objections from onshore landowners (Anthony, 2007). Average wind at sea is much stronger than on land, and power generation offshore could reverse Tyner's findings. Associated costs and other results of this research remain to be seen.

Utilization of river and tidal flows for energy generation is closely related to wind power technology. Some experiments in Europe were undertaken forty years ago, and there is more activity today, both in Europe and North America. Newspapers have discussed additions of turbines to an experiment ongoing in the East River, New York, and there are proposals for major installations in San Francisco Bay and elsewhere. The sea and rivers harbor enormous energies in waves and flows, but practical utilization is very challenging. Further experiments with river and tidal flows will probably be encouraged and developed with reasonable government assistance.

11.3.2 Solar Power

The diameter, D, of Earth is 12,750 kilometer, and its cross-section is $\pi D^2/4 = 1.28 \times 10^{14}$ square meters. Solar radiation on a flat plate perpendicular to the rays outside Earth's atmosphere is 1.4 kilowatts per square meter.[16] Thus, Earth intercepts 1.8×10^{17} watts of solar energy, i.e., 1.8×10^5 terawatts, which is about fourteen thousand times the rate at which humankind produces energy from a combination of fossil fuels, nuclear, hydropower, and wood and other biomass.

Use of solar energy is prima facie attractive because there is so much of it and because its use has little environmental impact. It may be used in two distinct ways: conversion to electricity and direct heat. The former is presently about ten times more costly than production of electricity by traditional means. An average of ten percent of U.S. electricity would be produced from solar panels of ten per cent efficiency on sunny days from an area of about 180 square kilometers (67 square miles). While this is a very small fraction of Earth's surface, it is a large area in human terms. Power generation would be maximum during the day and zero at night, and unless means were provided for storing produced power and distributing it to meet variable demand, it would be a back-up facility on sunny days to reduce demand for power generated by other means.

The energy and research sides of conversion of solar radiation to electricity are well discussed and explained in *Physics Today* (Crabtree and Lewis, 2007) and, with other energy discussion, in *Science* (Special Section, 2007). Current

---

[16] With atmospheric scattering and absorption, about 1 kW per square meter of normal incidence solar radiation is received at the ground on a clear day.

research and development suggest that efficiencies for conversion of solar radiation to electricity may be doubled within a few years. Even with low conversion efficiencies, communication is much enabled today with panels that produce a few tens of watts for radio links in many field applications without need for connections to a utility's grid, and small solar electric units at reasonable prices maintain electric fences on farms and ranches where access to utility lines is not easily available.

As a direct source of heat, solar radiation does have important practical applications today in water heating, and the design of solar collectors for that purpose has been recently improved with vacuum components manufactured in China (Apricus.com, 2007). Solar water heaters allow avoidance of use of electrical energy for heating, but in cold climates some regrettable complexity is needed in the form of heat exchangers to prevent damage incident to freezing. Solar cookers can be quite effective when Sun is high and skies are clear; your author enjoyed such for several years at his home on an Oklahoma farm and saw several in use in a monastery during a trip to Tibet.

Major solar installations of both the photovoltaic and direct heat types are on line in California and Nevada, USA. For direct heat, known as concentrated solar power (CSP), hundreds of mirrors track Sun and reflect its energy to a tower where the concentrated solar radiation flashes water to pressurized steam at 250C for driving turbines. Another direct heat technology, uses a series of parabolic troughs that focus Sun's energy on a central pipe and thereby heat oil therein to about 400C. The oil flows to a steam generator connected to a turbine for generation of electricity. A new CSP facility is currently under construction near Las Vegas, Nevada and a photovoltaic facility is expected to be on line at the end of 2008 with fourteen megawatts for Nellis Air Force Base, also near Las Vegas.

Use of solar direct heat is being realized in experimental new power plants in Spain and in Algeria (Trade Commission of Spain, 2007). The two methods noted above are subjects of major experiments by a subsidiary of Abengoa, a holding company. A heat storage mechanism involving troughs 18-feet wide with 28 thousand tons of liquid salt is also being developed in Spain. Planned for completion in 2012, the so-called Sanlúcar La Mayor Solar Platform should generate more than 300 megawatts of solar power with both of these technologies and photo-voltaic panels as well.

The government of Algeria plans to invest in solar power some of its revenues gained from exports of oil and natural gas, about $55 billion annually at this writing. The firm, New Energy Algeria, established in 2002 to exploit renewable resources, has partnered with Abengoa for construction of a 150 megawatt power plant that combines the solar resource abundant in the Sahara desert with generation of electricity by natural gas. It is reported that the company hopes to produce six thousand megawatt capability by the year 2020 and export that to Europe via cables under the Mediterranean Sea. The first Algeria facility is projected to use cogeneration with natural gas to fill gaps at night and during occasional cloudy periods.

11.3.3 Hydrogen and Batteries

Numerous research challenges and prospects for a U.S. hydrogen economy have been detailed by Crabtree, et al. (2004), and widely discussed by media. It is not expected that hydrogen would be used directly as an automotive fuel because pure hydrogen is very difficult to store in quantity. But use of hydrogen is attractive because the product of hydrogen oxidation in fuel cells is simply water, and there is no attendant environmental contamination. Perhaps the most important of present applications of hydrogen as a fuel are in the U.S. space program, and there are automotive trials in a fuel cell program that is highly experimental. The fuel cell is properly regarded as an energy storage device, as is a battery.

Basic to development of a hydrogen economy would be economical means for production of hydrogen in much larger amounts than produced in the present chemical sector of the U.S. economy. Hydrogen is almost ubiquitous but is tightly bound in water and other substances. In addition to the research that would be essential to development of acceptably economic means for hydrogen production, infrastructures for storage and transport of hydrogen would have to be created.

The amount of energy used for hydrogen production is several times the energy of the hydrogen produced. Partial justification for expansion of a hydrogen production industry might be found in the burning of abundant low-cost coal as a source of the electrical energy needed for hydrogen production by disassociation of water, but greenhouse gas emissions with coal burning are inhibiting. Of course nuclear power could also be used, but expansion of the nuclear industry is inhibited by concerns for contamination and disposal of nuclear waste. Expanded use of solar power may represent an ultimate good source of energy for hydrogen production.

The challenges for hydrogen lie in development of economies in all of production, storage, and distribution, and numerous research efforts are underway.

If batteries could be developed to the point that they would safely and economically provide the range, power and rapid "plug in" recharge that automobile users want from their automobiles, there could be significant savings of liquid fuels. Batteries used in laptop computers during the year 2007 have very high energy densities but have had safety problems. If safety were assured along with achievement of economic gains through further research and large scale production, electric automobiles powered by numerous laptop batteries could become a reality, as discussed by Schneider (2007b). Further background is available on numerous web sites.

11.3.4 Geothermal

Earth's interior heat has been used for human needs for thousands of years. Hot springs have been used for baths, and today in Iceland, a volcanic area, geothermal sources provide 40% of Reykjavik's hot water! In addition, there are about 20 hectares of geothermally heated greenhouses in Iceland for production of fruit, flowers, and vegetables. However, expansion of greenhouse production in Iceland is inhibited by low levels of natural illumination, which leads to implementation of artificial lighting. More important, Iceland's self-sufficiency is presently impeded by the availability of lower-priced imports, which provide about 75% of Iceland's fruits and vegetables.

Use of geothermal heat for electric power generation dates from 1904 at Lardarello, Italy, where local volcanism provides heat sources near Earth's surface. In the United States, some twenty power plants at the Geysers, north of San Francisco, California, provide 850 megawatts of power from dry steam – this comes from strata less than three thousand meters below the surface, and the total amount of electrical energy produced is similar to that provided by one typical coal-burning facility.

MIT professor Jefferson Tester recently noted that Earth's interior heat, if accessed much more widely for power generation, could provide humankind's demand for power generation for thousands of years (Bullis, 2006). And Roach (1998) has noted that about 99% of Earth's total mass is at temperatures between 1000 and 5000C. However, the necessary heat must be found in a thin surface layer within which the average rise of temperature with depth is about 25C/km. Temperatures near 200C are necessary for viable power generation from geothermal heat, and, owing to spatial variations in the rate of temperature rise with depth, there are many places where wells to depths of about five km find the desired temperatures. Possible applications of geothermal heat are becoming more promising owing to major advances in the drilling technologies applied to recovery of oil and natural gas, particularly in the technologies of horizontal drilling and rock fracturing.

An important geothermal experiment ongoing at this writing near Basel, Switzerland, illustrates both potential and pitfalls (Häring, et al. 2007). In addition to a field of monitoring wells, three principal wells for the facility were planned initially in Basel, one for water injection and two for production of hot water. It was planned to deliver about 3.5 megawatts of electrical power to the grid and the equivalent of about 5.5 megawatts of heat for local heating. However, initial tests were accompanied by earth tremors sufficient to produce significant apprehension in the local population and a flurry of claims for minor damage, and at this writing (September 2007) the project has been stopped pending further assessments.

As this is written, only about 1500 megawatts of electricity is provided globally from geothermal sources – this is comparable to the production of one large coal-burning plant or two ordinary facilities.

11.3.5 Nuclear Fusion

Fusion, in contrast to fission, involves combination of light elements to make more massive elements whose atoms weigh less than the sum of those used for their creation. As with creation of the fission element, uranium, this is a process that takes place in massive stars. Under extreme conditions of temperature and pressure, light elements beginning with hydrogen are fused into heavier elements, ending with collapse of the star and creation of elements heavier than iron, including uranium and some highly radioactive transuranic elements. Elements lighter than iron produce energy when fused; heavier elements produce energy when split.

Hydrogen, consisting of one proton and one electron, constitutes about 74% by mass of the known universe, and most of the balance consists of helium, with only about 2% represented by all other elements. On Earth, hydrogen is about 11% of the mass of the oceans, with deuterium (hydrogen of mass 2) comprising about 1/70% by mass of the total hydrogen. A third isotope of hydrogen, tritium, with two neutrons and one proton, is of importance because of a prospect of its use in a fusion process that may someday be perfected on Earth.

While energy production by fission of uranium is well-established world-wide, energy production by fusion of hydrogen, akin to a controlled hydrogen bomb, is still in its infancy and may never be feasible on Earth. However, effective fusion technology is much sought because it would produce no long-lived radioactive aftermath nor carbon dioxide, and does not, per se, have implications for nuclear war. And centrally important, if the technology for energy production via fusion were perfected, the production of electricity sufficient for any purpose of humankind could be limited only by the number and power of fusion reactors constructed.

Recent history and technical challenges facing the international fusion program have been presented in *Science* (Clery, 2006). The effort toward power by fusion began in several countries during the 1950s. In 1985, programs in separate countries began to be internationalized after a summit conference at Geneva produced agreements between Russian premier Gorbachev and U.S. President Reagan. The program is known as ITER – International Thermonuclear Experimental Reactor. Its latest manifestation is an agreement among seven governments[17] to construct an experimental reactor in Cadarache, in southern France, at a cost presently estimated near $12 billion over ten years. After construction, the facility would be run for twenty years to develop improved knowledge of a proper subsequent design. It will be enormous and very unlike any existing power plant on Earth.

It is presently believed, on the basis of numerous ongoing experiments, that this greatly scaled-up facility will demonstrate net generation of power, but the technical challenges are awesome. Basically, the problem is to replicate on Earth the very high pressure and high temperature conditions in stellar interiors. This would be accomplished with strong electric currents that produce a strong magnetic force and a pinch effect.[18] The zone of extreme temperature must be held away from the walls of the facility because contact would reduce temperature by conduction, the magnetic fields must be controlled to prevent instabilities in the toroidal active zone and the materials used must resist embrittlement by radiation.

ITER fuel consists of a mixture of deuterium and tritium, the former separated from water by distillation and the latter produced in the reactor itself. At sufficient temperature and pressure the velocity of the hydrogen atoms becomes large enough

---

[17] China, the European Union, India, Japan, South Korea, Russia, and the United States.

[18] The pinch effect is manifested during thunderstorms on Earth by narrowness of lightning channels and by crushing of thin-walled cylindrical objects struck by lightning. It is also seen in the filamentary nature of solar prominences.

to overcome the electrostatic repulsion of the nuclear protons, and helium and energetic neutrons are created.

ITER construction is scheduled to begin in 2008, and orders being placed at this writing include such costly items as superconducting magnets. The outcome is uncertain, but potential reward is enormous, and "nothing ventured, nothing gained". Electricity satisfies many needs and can provide the energy needed for manufacture of liquid fuels.

11.3.6 Biofuel Research, Ethanol and Biodiesel

Search for a biological base to alternative fuels is wide-ranging. In 2007, the U.S. Department of Energy provided $375 million over five years to establish bioenergy research centers at the Lawrence Berkeley National Laboratory in California, the University of Wisconsin at Madison, and at Oak Ridge National Laboratory. Efforts at these centers will be focused on devising biological processes to convert cellulose to liquid fuel. The research presumes that success could be followed by viable harvesting of cellulosic materials of forest products, grasses, and crop residues, but as mentioned again in the last paragraph of the next section, impacts on agricultural practice and land use may be unsustainable.

In related research at the J. Craig Venter Institute in Rockville, Maryland, some studies are focused on creating bacteria that contain the genomes for making biofuels from cellulose (Pennisi, 2007).

Whether or not research such as described in the preceding two paragraphs is "successful", both it and its possible future applications will assuredly be controversial. Humankind already consumes a large fraction of the energy represented in annual biological growth,[19] and our search seems directed toward new modes of exploitation rather than toward carefully planned elimination of waste and reduction of demands on non-renewable resources.

The ethanol and biodiesel programs described in the following two sections, except for conceptual production of ethanol from cellulose, use already developed technology for production of liquid fuels from the biosphere.

11.3.6.1 Ethanol from Corn, Sugar, and Cellulose

Much of the following discussion is well presaged in a pamphlet distributed twenty-seven years ago from the Federal Reserve Bank of Kansas City (Duncan and Webb, 1980). The FRB report appears to have been prompted by concerns arising from the embargo placed on export of Arab oil to the United States in the 1970s. Concerns with prospective declines of petroleum-based gasoline also led to a more

---

[19] Indeed, Pimentel and his students found that the American population uses annually more than three times the amount of solar energy that is incorporated into the growth of all green plants in the U.S. (personally communicated)!

formal examination of conversion of biomass to ethanol (Energy Research Advisory Board, 1980 & 1981). Despite the substantial negative energy conversion ratio presented by these reports, interest in ethanol production as a substitute for gasoline has increased and, in the United States, has culminated in Congressional legislation which calls for production of 36 billion gallons of biofuels by 2022. But will this be achieved, and should it be achieved? At this writing in mid-2007, production of ethanol from corn in the United States is at a rate of about six billion gallons annually, having increased from one billion in 1990. The United States among nations thus leads annual production of ethanol, having recently replaced Brazil.

Ethanol from corn is produced by first mixing finely ground corn with water and adding enzymes alpha amylase and glucoamylase to the warmed mixture for conversion of the starch to glucose. Ethanol is then produced from this simple sugar by fermentation with yeast, and the ethanol is concentrated by distillation. Well over one hundred ethanol plants have been built during the past few years at a cost of more than $50 million each in the United States and several tens more are planned. A typical plant consumes about fifty thousand bushels of corn daily, 20 million bushels annually, and produces about one million barrels (@ 42 gallons) of ethanol annually, which is roughly equivalent to 5% of U.S. oil consumption for one day. The total investment in ethanol plants is thus about $6 billion and the yield of six billion gallons of ethanol is equivalent to 4.5 billion gallons of gasoline, equivalent to five or six days supply of oil in the United States.

Every day, the public is swamped by media presentations pro and con, reflecting intense controversy. There are a host of arguments against this program, and, in your author's opinion, this program and several others have gone forward either because lobbyists effectively buy legislation with contributions to legislators or instill fear among candidates that elections will be lost if programs desired by special powerful interests are not supported. The ethanol program will ultimately prove to be destructive. Consider the following.

First, the net energy argument concerning corn-to-ethanol: Prominent contradictory analyses have been presented (Pimentel, et al., 2007, and Shapouri, et al., 2002). The former, in agreement with earlier studies, finds that more energy is required to grow, harvest, and transport corn, ferment it to ethanol, and distill the ethanol to increase its purity to 90% or more, than is obtained from the ethanol. The latter finds the opposite to be true. The ratios of input to output energies presented by the two studies are within the limits 1.5:1 and 0.5:1. It is important to note that Pimentel, et al., includes some energy inputs that are admittedly omitted in the analysis by Shapouri, et al. It is critically important to observe that western societies cannot function in the manner to which they have become accustomed with either ratio, because they are drastically unfavorable with respect to historic oil, said to have been 0.01:1 during early days of discovery and exploitation, and increased to about 0.05:1 today, owing to high costs of recovery in hostile environments.

Further in connection with the energy ratios, recall that any conversion process involves energy loss. For example, the energy in gasoline is less than that in the oil from which it is refined. But we make gasoline from oil because gasoline has higher uses than crude oil. Similarly, it may be argued that we make ethanol from corn because the ethanol has an important use as motor fuel, and corn has been in surplus.[20] Argued in a different way, the inputs of energy toward production of ethanol involve, for example, heat for distillation, which may be produced from coal-burning power plants or even by the burning of coal within the ethanol plant itself, and we need ethanol more than coal. However, as previously indicated, an inhibiting quality of coal burning is its implication for global warming (more on this in Sections 11.4 and 11.5, below).

Second, several studies have shown that use of ethanol as a motor fuel increases emissions of nitrous oxide precursors of ozone and air pollution, which are already serious causes of asthma and allergies in several U. S. cities. We note in passing that this matter is also controversial, but it appears that those who claim that ethanol reduces harmful emissions benefit personally from ethanol manufacture.

Third, the fermentation process produces 2.7 gallons of ethanol per fifty-six pound bushel of corn. This means, for example, that two billion bushels of corn, about 20% of the U.S. corn crop, can produce 5.4 billion gallons of ethanol. Because the energy in ethanol on a volume basis is about 70% of that in gasoline, this is equivalent in gasoline to less than 4 billion gallons or 100 million barrels. As noted above, this is only five days of U.S. oil consumption!

In these rough calculations, we see truth in part of a statement released by U.S. Senator John McCain (2003): ". . . ethanol does nothing to reduce fuel consumption, nothing to increase energy independence, and nothing to improve air quality". Regrettably, Senator McCain as a candidate in 2007 for the Republican presidential nomination in 2008 is now supporting the national ethanol program because the nature of the U.S. political system gives inappropriate power to interests that benefit from the program. Some other candidates for political office in the United States have similarly switched their positions.

A fourth aspect of the ethanol program is its impact on the availability of corn for feed, owing to diversion of a portion of the crop for manufacture of auto fuel. At an extreme, in reference to a perceived looming shortage of animal feeds and human food, the conversion of foods to fuels and especially the ethanol programs have been labeled "The Internationalization of Genocide" by the Cuban publication *Granma* (Castro, 2007). Strong general condemnation in this publication also notes the small fraction of fuel needs to be provided by conversion of large amounts of grain for "voracious automobiles". Certainly, the price of corn and other feeds is being increased by increased demand for corn and by planting to corn of land formerly used to grow other feeds. Between 1980 and 2006, the price of U.S. corn fluctuated considerably but, with few exceptions, remained below $2.50/bushel (56 pounds). At this writing in September 2007, the price of corn is about $3.75 per bushel[21] and

---

[20] With hunger stalking a third of Earth's human population, no food item may be thought to be in surplus.

[21] And the price of wheat surged to $9/bushel during 2007, more than double historic values. Much of the price surge has been attributed to failure of the wheat crop in Australia, owing to drought. Price increase is also a result of the transfer of cultivation from wheat to corn. Soybean price has been similarly affected.

this with other related price increases is receiving most of the blame for a reduction of U.S. food aid by more than half since the year 2000.

This surge in the price of corn has a direct impact on the cost of animal feeds and hence on the price of beef, chicken, and pork, and newspaper articles have carried many indications of related concerns. This has carried over to demonstrations in Mexico, for example, since the price of corn relates directly to the price of tortillas, a dietary staple there that consists almost wholly of corn. However, in the United States, the impact on many items bought in stores may be minimal because the overwhelming part of the price of typical packaged foods reflects value-added processing and costs of packaging and distribution following purchase of the raw commodity. For example, a 14-ounce package of corn tortilla chips, which sold in U.S. supermarkets for about $2.35 in August 2007, contains less than 4.5 cents of farmers' share with corn prices at $3/bushel. A doubling of the price of corn would raise the price of the tortilla chips only 4.5 cents! Somewhat more significant would be the impact on a four-pound package of corn flour, selling for $2.50 in U.S. supermarkets. Farmers' share here is about eighteen cents, so a doubling of the corn price would raise the price to consumers by eighteen cents. Of course, these simple calculations do not account for ripple economic effects.

A possible positive international benefit of a higher corn price lies in improved competitiveness of corn grown by traditional methods in less industrialized countries. Thus, the historical low price of corn grown in the U.S. by industrial methods and exported to Mexico under the North American Free Trade Agreement has reduced the marketability of corn grown on small farms in Mexico, and this may change with a higher price of U.S. corn. Another small plus is distillers grain, the high protein product that remains after fermentation of starch. This product can be fed to cattle during the finishing stages of their fattening for slaughter in our industrial agriculture.

A fifth important negative impact of both the ethanol program and biodiesel program (see below) is reduction of already stressed water supplies, especially in western United States. This concern has been widely publicized in the United States during fall 2007, and is treated in detail by the U.S. National Academy of Sciences (2007).

The U.S. ethanol program is subsidized at the federal level by a nominal tax credit of 51 cents per gallon, and further supported by a tariff on importation of ethanol. The tax credit has been shown in a report by the Congressional Research Service to amount in actuality to 68 cents/gallon owing to the manner in which the credit is administered (Congressional Research Service, 2005). The federal subsidy is augmented in Oklahoma by legislation granting an additional tax credit of twenty cents/gallon.

Often overlooked in the corn-to-ethanol program are heightened general negative impact of increased corn production on ecosystems and high cost of transporting ethanol. Land planted to corn in 2007 totaled about 93 million acres, the highest since 1933, and recent yield of about 155 bushels/acre is nearly double that typical of thirty years ago. The increased acreage is a response to the ethanol program and the increased yield reflects large fertilizer inputs, which involve energy-intensive production of fertilizer with dark implications for hydrocarbon inputs and emissions of greenhouse gases. There are also serious implications for erosion of land in increased production of corn, because soil erosion under corn far exceeds replacement.[22]

Regarding transport of ethanol to markets, existing pipelines cannot be used because ethanol is a strong solvent and would become contaminated with pipeline residues while causing corrosion to the pipelines themselves. Therefore, pending solution to these problems, more expensive truck and rail transport is necessary, and these factors have not been accounted for in the federally supported program.

The corn-to-ethanol program is also causing a large increase in the price of farmland, which, according to articles in The New York Times on August 10, 2007, is increasingly shutting out beginning farmers with limited capital. In spite of loan programs such as those provided by the U.S. Farm Service Agency, the average age of the U.S. population that actually farms has been increasing for years, and efforts to facilitate entry of young people to farming have been increasingly assumed by individual States and by such pro-bono organizations as the Center for Rural Affairs (2007).

Much touted is an ethanol program in Brazil, which provides about 25% of auto fuel there. During 2007, Brazil achieved independence from imported oil owing to a combination of its ethanol program with a significant discovery in an offshore oil field. Brazilian ethanol is made from sugar, which is easier to ferment than corn, and about 4 billion gallons is produced annually from sugar cane grown on about six million hectares of farmland (10% of farmland in Brazil). It is much easier to satisfy Brazil's automotive fuel demand than U.S. demand, because the area of Brazil is 8.5% larger than the U.S.' "lower 48" while its automotive fuel use is only 10% of that in the U.S. It has been reported that the farmland devoted to sugar cane in Brazil was formerly used to grow fruit and vegetables and that no appreciable amount of rainforest has been removed in order to accommodate demand for sugar (Lagercrantz, 2006). However, we wonder whether some of those displaced from horticulture will clear present jungle for new farms.

Finally, research toward conversion of cellulose to ethanol is now subject to much discussion. Will cellulosic conversion be a successor to corn-to-ethanol in the United States? Grasses, especially switchgrass,[23] are commonly portrayed as a viable future rootstock, and development of effective conversion technology is widely publicized as imminent. Extensive research on this subject is underway, with the U.S. Dept. of Energy awarding hundreds of millions of dollars for development of pilot plant s for experimentation with several technologies. Knowledgeable botanists have expressed reservations, noting that serious implications of continuous

---

[22] With improved tillage methods and the Conservation Reserve Program (see footnote 24) soil erosion has been recently declining in the United States, but is not yet at levels consistent with sustainability of fertile topsoil.

[23] Switchgrass is one of the four climax grasses identified with the U.S. tall grass prairie. The others are big and little bluestem and indiangrass. There are hundreds of grass species in the U.S. prairie.

monocropping for net energy consumption, pesticide usage, erosion, water use, and reconversion of the conservation reserve[24] have not been well explored and that little note is being taken of the large amount of cellulose that would be required to replace just a few percent of current U.S. oil consumption. In short, some think that much use of switchgrass in a monoculture and other cellulose for ethanol production could produce an industry resembling that for corn, and some field experiments to clarify these issues are being planned (Wallace, 2007). Also to be considered is the impact of a cellulosic industry on maintenance of domestic herbivores.

11.3.6.2 Biodiesel

In a diesel engine, the fuel air mixture is compressed so much that the accompanying rise of temperature causes self-ignition of the fuel. The higher compression and temperature in a diesel engine than in the usual internal combustion engine produces higher fuel efficiency, i.e., increased mileage with a vehicle. Diesel engines are desirable for this reason, and also because of their simplicity associated with absence of spark plugs and distributor. Diesel fuel is less volatile than gasoline and can be made from both petroleum, with declining availability, and from animal and plant fats and oils. Diesel fuel with a recent biological origin is known as biodiesel.

Glycerin in animal and plant fats and oils must be removed before the lipids can be used in diesel engines. The usual refining process, known to chemists as transesterification for removal of glycerin, involves a reaction of the oils with an alcohol, addition of some water and later heat to remove the water, and various other stages including addition of catalysts, which are recovered for reuse. Several somewhat similar refining processes can be used effectively and are well established. The removed glycerin has a market in soap manufacturing and a few other applications.

In the United States, soybeans are presently the principal source of the oils used to produce biodiesel, and about 73 millions acres of cropland have been devoted to soybean production. With a generous estimate of production at 40 bushels (each 60 pounds)/acre, and 1.4 gallons biodiesel/bushel, total biodiesel production would be 4 billion gallons or 100 million barrels, if all of the soy beans now grown in the U.S. were used for oil production. Corresponding to the preceding analysis of ethanol, this would replace only the amount of petroleum that the U.S. presently consumes in five days! Note that strongly negative conclusions are implied even though no consideration has been given here to other negatives associated with energy inputs required to grow, harvest, and transport soybeans.

In parallel with the ethanol analysis, large-scale production of diesel fuel from oil seeds would have unintended undesirable consequences on markets and the

---

[24] The CRP enrolls landowners to remove highly erodible or environmentally sensitive lands, up to 40 million acres nationally, from agricultural production for contract periods up to 10 to 15 years. In return for incentive payments, the land is planted in grasses, legumes and trees for management as wetlands, wildlife habitat, windbreaks, etc.

conservation preserve. It is reported by George Monbiot[25] that a rush to produce biodiesel from palm oil in Malaysia is causing great losses to primitive rainforest that already represents only a remnant wildlife habitat. The palm oil industry is cognizant of such criticisms and is planning a conference during fall 2007 to review its practices.

Other sources of biodiesel are waste oil at restaurants and homes and animal fat produced by meat processors. During summer, 2007, Tyson Foods, Inc. announced contracts that would produce 175 million gallons (four million barrels) of biodiesel fuel from 25% of Tyson's fat production. This is just 20% of U.S. petroleum supply for one day. While having little effect on U.S. petroleum dependence, this diversion is expected to cause significant price rises in the soap industry.

In the presence of serious unintended consequences and far-flung ripple effects, production of biodiesel fuel in the U.S. is presently subsidized by $1 per gallon at the federal level and also receives subsidies in many States, including 20 cents/gallon in Oklahoma.

Actual U.S. production of biodiesel from soybeans in 2007 is about 300 million gallons or seven million barrels per year, about a third of petroleum products used in one day. In April 2006, a biodiesel facility was opened in Durant, Oklahoma, and was slated to sell the soybean derivative under the brand name, BioWillie, given by famous singer Willie Nelson. On July 13th, 2007, it was reported that a group of note holders had filed an involuntary Chapter 7 bankruptcy petition against the Dallas-based owner of the biodiesel production plant. According to the petition filed in the U.S. Bankruptcy Court for the District of Delaware, Earth Biofuels had not been paying its debts as they became due. However, on January 21, 2008, it was reported that the petition for involuntary bankruptcy had been dismissed by the Court and that Earth Biofuels had consummated an agreement with Alliance Processors to purchase waste grease collected at restaurants in Texas. Up to 400 thousand gallons of grease per month is expected to be supplied.

This is a commendable program. After all, "Waste not, want not", but it should be recognized that if each gallon of grease makes nearly a gallon of diesel fuel, the grease collection is equivalent to about nine thousand barrels per month, or less than one-twentieth of one percent of the petroleum used in the United States each day! The first article quoted the Chair and CEO of Earth Biofuels, "The biofuels industry and other alternative fuels are absolutely essential to our nation's energy security and our ability to maintain economic independence. The goal of energy independence won't be achieved through use of a single technology."

Your present author does not agree with the first part of the quote but believes that the second part is probably true.

Finally, algae are still another source of biodiesel, theoretically very promising. Again, however, the practical challenges are very great and the ultimate outcome of

---

[25] Monbiot is author of numerous media presentations and of the important book, Poisoned Arrows, which is about his somewhat covert travels in Indonesia, where he reports that poorly regulated copper mining is devasting the lives and culture of indigenous tribes.

research in this area is speculative. At this writing, several companies are involved, and Greenfuel Technologies of Cambridge, Massachusetts, U.S.A., is partnering with Arizona Public Service at one of the latter's power plants to develop a system that would feed algae with the plant's carbon dioxide emissions.

11.4 Greenhouse Warming and its Connections

We discuss global warming because it carries grave implications for the future human condition and because it is being caused by human activities, mostly by the burning of carbon-containing fossil fuels for transportation and for generation of electrical power.[26] The global warming issue is thus tightly connected to the fuel-decline issue as illustrated in a short article by your author (Kessler, 1991) and elsewhere. Extraction and burning of carbon fuels since the start of the industrial revolution and particularly the burning of a substantial fraction of extractible resources since World War II has been the source of the present developed economies with high levels of material well-being, and naturally there is a wish to preserve and enhance this condition.[27] In this connection it is important to have in mind both the relative amounts of carbon dioxide produced by the combustion of basic fuels and their heats of combustion. As shown in Table 11.1, with each million BTU produced by combustion of carbon, about 119 kilograms of carbon dioxide are produced. Coal used as fuel is the largest emitter of carbon dioxide in relation to energy produced.

Continued present political reality carries implications for changed weather and climate, for rapid changes in agricultural practice, for substantial rises of sea level, and for changes of oceanic flora and fauna in response to oceanic uptakes of carbon dioxide with resultant increase in oceanic acidity.

There are at least four aspects of global warming with public interest. First, How enduring will global warming be as presently measured? Second, Are human beings a principal cause? Third, Is global warming important for human beings? Fourth, Can it be mitigated by humankind? Although global warming continues to have outspoken deniers in 2007, a proper answer to each of these questions is a resounding "yes", but it is necessary to add that mitigation of global warming and its effects presents to humankind a challenge unprecedented in its magnitude. It is not at all clear at this writing that the challenge will be well met.

The first and second questions above are addressed in this section, and the third and fourth are addressed in Section 11.5.

---

[26] Roughly one third of U.S. carbon dioxide emissions are attributable to each of transportation, buildings, and industry.

[27] "Naturally", perhaps, but "material well being" is not pursued by all cultures, nor is it clearly "good". For example, the Amish tend to reject less manual labor and television brought by advanced technology. Pursuit of "material well being" brings increased leisure to many but not all, and may enhance problems of societal health including obesity, juvenile delinquency, hectic family life, and justice not explored here.

Table 11.1 Approximate heats of combustion and CO_2 emissions for common fuels[28]

Fuel	MJ/kg	Mcal*/kg	BTU/lb	BTU/kg	CO_2/BTU**
Carbon[#]	32.6	7.8	14021	30916	119
Coal[+]	36	8.6	15445	34056	97
Diesel	45	11	19300	42600	73
Ethanol	30	7	12800	28500	66
Gasoline	47	11	20400	44600	69
Hydrogen	142	34	61000	135000	zero
Methane	55	13	23900	52500	49
Natural gas	54	13	23000	51200	49
Propane	50	12	21500	47400	63

M = one million; J = joules; 1kg-cal = 3.96BTU; 1g-cal = 4.19 joules; 1kg = 2.205lb; 1 million joules = 0.278 kilowatt-hours
*gram calories; **grams CO_2/1000 BTU or kg CO_2/MBTU; [#]Graphite [+]Bituminous, 90% Carbon, 5% hydrogen
No significant difference between methane and natural gas is shown here.

11.4.1 The Reality of Global Warming

The Intergovernmental Panel on Climate Change has issued impressive documentation (more than fifteen hundred pages) including both technical details and accounts readily understood by laypersons. These accounts are available on the internet (IPCC, 2007) and are an excellent source of details concerning the following account.

Global warming is world-wide and given the immense variability of weather, no local phenomenon taken by itself proves or disproves global warming. Aggregation of many local effects can be evidence of global warming.

The temperature record at Oklahoma City from January 2004 is a small piece of evidence for global warming. Thirty-four of forty-eight months from January 2004 through December 2007 had above-normal temperatures and thirteen experienced below normal. The largest above normal was +11.0Fahrenheit and the largest below normal was −4.6F. The overall average was +1.93F above normal. Also, during this period, 29 high temperature date records were tied or broken (either the maximum temperature for a particular date or the highest minimum temperature for a particular date) and five low temperature records were tied or broken. Fortunately for the local inhabitants, while Oklahoma winters during 2004–2007 tended to be mild, summers there, usually very hot, were cooler than the long-term average in 2004 and 2005 and not excessively warmer than average in 2006 and 2007.

Sometimes skepticism about global warming is produced by other extreme local conditions. Such was especially the case during the weekend of April 7–8, 2007, in North America, when a severe cold wave covered eastern sections and some low temperature records were broken. However, a figure from the National Oceanic and Atmospheric Administration that depicts temperature anomalies over the whole of

---

[28] This table reflects a variety of sources: *Handbook of Chemistry and Physics*, published by the CRC Press, Wikipedia, personal calculations, EIA and other internet data, and input from a friend.

Earth (NOAA, 2007) shows that our planet as a whole was experiencing above normal temperatures at that time. With a few minor exceptions, eastern North America was the only place on this third planet from Sun that was experiencing temperatures substantially below seasonal averages, and temperatures well above long-term averages prevailed over most of Earth, especially in Arctic regions. And almost all global anomaly charts during 2006 and 2007 are similar in showing a larger area of Earth with above normal temperatures than with below normal temperatures. Of course, pattern details change constantly.

Another indicator of global warming is in a report from the National Oceanic and Atmospheric Administration on March 15, 2007. This states that overall on Planet Earth, the average temperature during three winter months in the northern hemisphere, December 2006 through February 2007, was the warmest recorded since such record keeping began a little more than one hundred years ago. And the eleven warmest years of record on a global basis have occurred during the past twelve years.

Consider conditions in Europe. In an article (*Weather*, 2007) published by the Royal Meteorological Society in the United Kingdom, it is stated that the 12-month period from March 2006 through February 2007 was the warmest ever recorded in the 350-year period of the central England temperature (CET) record. The CET record is the longest instrumental temperature series on Planet Earth. Furthermore, records during the past several years, documented monthly in *Weather* show that practically every month has had above normal temperatures overall in both the U.K. and in continental Europe, and readers will well remember the heat wave of summer 2003 in Europe, when up to 35 thousand deaths were attributed to record-breaking high temperatures. There were heat waves in Europe in 2006 and 2007 also, though of lesser intensity (but 45C in Greece and some Balkan states, with devastating forest fires in 2007), and there has been a substantial increase in the frequency of heat waves in Europe.

A report in EOS (Komar, 2007) documents a convincing increase of wave height since 1985, as measured by buoys near the southeastern coast of the United States. The increased wave height is presented as indicative of increasing storm intensities, a consequence of rising ocean temperatures. There has also been technical documentation indicating increased frequency of drought and flood, and possible increased frequency and severity of hurricanes. Flooding in central England during summer 2007 and record-breaking floods in parts of India during 2006 and 2007 are not proof of global warming, but are suggestive.

During August 2007, observations showed that Arctic sea ice had retreated to a record minimum. Melting was particularly prominent north of the Arctic coasts of Alaska and Siberia. By September 2007, the Arctic ice limit had retreated northward at some longitudes more than 500 miles further from its distance from the Siberian coast on same dates in 2006,[29] much more than expected. In this connection, a

---

[29] On the Greenland side, the ice cover in September 2007 was similar to that in 2006, but the number of melt days on the Greenland ice cap was also a record high during summer 2007.

chapter by Hansen et al., (2007), seems important. For about 20 years, Hansen has been a principal spokesperson for the climate change science community. In its indications that the IPCC documents are conservative estimates of the rate at which climate change is proceeding and of the rate at which remedial action must be taken to avoid passage of a point of no return, this chapter presaged the remarkable 2007 retreat of Arctic ice.

11.4.2 Climatic Fluctuations

There are several causes of climatic fluctuations. Diminution of solar radiation during the Middle Ages is thought to have contributed to global cooling at that time, the so-called Maunder Minimum. Earth's orbit and inclination to the ecliptic are perturbed by the gravitational influence of other planets, particularly Jupiter and Saturn, as analyzed by Milankovich 100 years ago, and some of the major historical ice ages and subsequent warmings are attributed to these variations. Volcanism with strong emissions of carbon dioxide and particulates are believed to influence climate, with particulates tending to reduce temperature and carbon dioxide tending to increase it. Depending on the cause, some climatic fluctuations are opposite on northern and southern hemispheres, and some are synchronous.

Concerning the present climatic fluctuation, it has been shown that geothermal heat associated with volcanic eruptions, black smokers on the ocean floor, etc., are not contributors (Roach, 1998). And although some blame solar variation for climate change, the present oscillation with the sunspot cycle is less than 2 watts/meter^2 in a total radiance of 1370 watts/meter^2 and cannot be a significant factor.

Present concerns are principally related to carbon dioxide, which, next to Sun, of course, and water vapor, is the principal regulator of temperature on Earth.[30] The heat trapping effects of carbon dioxide have been known for at least one hundred years, and were well taught at MIT and elsewhere fifty years ago. Increase of atmospheric carbon dioxide causes a diminution of heat transfer by radiation from the lower atmosphere to the upper atmosphere, an increase of temperature in lower atmospheric layers, and a compensating increase of heat transfer by atmospheric convection (mass motion, as in water boiling on a stove).

Climatic temperature fluctuations during the past 850 thousand years have been deduced from analysis of ice cores obtained in Greenland and Antarctica. Atmospheric gases in the ice essential to these analyses include carbon dioxide and oxygen isotopes ^{18}O and ^{16}O. Particulates are also in the ice, which shows annual

---

The distribution of Arctic ice can be tracked daily at the following website maintained by The Meteorological Service of Canada: http://www.weatheroffice.gc.ca/analysis/index_e.html.

[30] Molecule for molecule, methane is a much more potent greenhouse gas than carbon dioxide. However, the methane content of the atmosphere has stabilized at a low value. Certain chlorofluorocarbons are also potent greenhouse gases and are implicated in the "ozone hole", which is persistent at this writing, especially in the southern hemisphere. The elimination of production of certain chlorofluorocarbons mandated by the Montreal Protocol may be evaded in some countries.

striations. It is important that although the maximum atmospheric carbon dioxide content during this historical period was about 300 parts per million by volume (ppmv), the 2007 content is about 385 ppmv and increasing by about 2 ppmv each year.

Precision measurement of atmospheric CO_2 was begun by Charles Keeling in 1958, and is now monitored at stations around the world. Records show annual increase every year and a within-year variation that is attributed to the cycle of plant growth in the northern hemisphere. During the 1960s, annual increase was only about one half ppmv per year, and the four-fold rate of increase since then corresponds closely with the increasing rate at which humans are burning fossil fuels.

Globally, carbon burning has increased from about two billion tons annually during the late 1950s to more than seven billion tons annually today, with total present-day emissions of carbon dioxide about 25 billion tons annually (Marland, et al, 2005). U.S. facilities that generate electrical power typically burn every day all the coal contained in railroad trains of even more than one hundred cars. Each car may contain about seventy tons of carbon in coal, and each power plant thereby produces about thirty thousand tons of carbon dioxide every day. More coal-burning power plants are being built here and elsewhere, one per week in China, where there are awesome environmental consequences of its rapid industrialization (Kahn and Yardley, 2007).

As noted above, the 2007 carbon dioxide content of Earth's atmosphere is about 385 ppmv, 30% above 300 ppmv, which was the approximate maximum during the pre-industrial 850 thousand years for which atmospheric values can be accurately determined by analyses of gases trapped in polar ice. The present extraordinary content of carbon dioxide is believed to be the significant cause of rising global temperatures.

11.5 Political and Social Conditions, Especially in the United States

Political and Social Conditions in the United States are determinants of all of the legislation passed in the U.S. Congress and in state Legislatures. Of course, we are here concerned with legislation related to U.S. dependence on foreign suppliers for energy and the intertwined problems of global warming and agriculture. Very regrettably, serious deficiencies in rational attention to science, to unintended consequences, and to long-term issues are prominent in the politics of the U.S. government. The shape of legislation is very much determined by moneyed interests that work through lobbyists. Lobbying is an important and needed source of information, but it seems beyond proper control in the United States. Numerous publications from pro bono organizations such as the October 2007 issue of *National Voter* from the U.S. League of Women Voters inform the public of moneyed and corrupt influences that hurt this country, but public power and even public interest are so far inadequate to stem related bad practice sufficiently. Much of the U.S. public seems focused on entertainment. Even the U.S. President, though faced with a war in Iraq, said, "Go to Disneyland". A scholarly and comprehensive discussion of the U.S. political system (and some other systems) has been presented by Vago (1981).

As previously noted, the United States is the world's largest emitter of carbon dioxide, just a bit ahead of China, which is nearly caught up in the year 2007. Although large emitters, both the U.S. and China have been among the least inclined to control their warming emissions. China notes that although it will soon pass the United States in total emissions, its per capita emissions are only about one-third those of the United States, and the average standard of living of its people lags seriously. Compounding the condition of present large emissions, there are continuing strident calls in the United States and elsewhere for further economic growth, which can only increase demand for electrical energy and liquid fuels, both of which associate with increased emissions of carbon dioxide. If physical growth and associated demand growth continue, ultimate demand for fuels would increase along with the emissions therefrom, regardless of any measures directed toward conservation or improved efficiency.

In the United States, search for replacement of petroleum-derived liquid fuels reflects ardent wishes to preserve and even continue to enhance the automotive economy. The search involves investigation of alternative fuels as described in previous sections of this chapter and recovery of energy resources via activities made economically feasible by the high and rising price of traditional sources. Thus, for example, there are immensely expensive oil recovery projects in deep waters of the Gulf of Mexico, where oil rig leases now cost up to a million dollars daily.

Extraction of oil from tar sands in western Canada is of special concern. As discussed in Section 11.2.3, this expanding industry anticipates investments of about $80 billion during the next seven years to provide liquid fuels for the automotive industry. This industry produces substantially more carbon dioxide per unit of oil that is extracted, refined from its tarry beginnings, and delivered to users than the traditional oil industry. As traditional liquid fuels become scarcer, there are also calls for their production from coal and natural gas. This would also enhance emissions of carbon dioxide.

Sequestration (permanent burial) of carbon dioxide for sufficient reduction of global warming is costly. There has been considerable discussion of sequestration in the U.S. press, but the only significant practice, so far, occurs where injection of carbon dioxide enhances recovery of petroleum (tertiary recovery).

Proposals to reduce carbon emissions through a tax on carbon burning have been implemented in a few European countries and others, but not in the United States, owing to opposition from special interests. Similarly, although the cost of limiting mercury emissions from coal has been reported to be less than 0.3 cents per kilowatt-hour, installation of such emission control is being implemented on a time line longer than ten years (Srivastava, et al. 2006).[31]

---

[31] Pollution is much better controlled in the United States than in China, referenced in the penultimate paragraph of preceding Section 11.4.2. Differing political and social conditions in different

Resistance to change is illustrated in the U.S. State of Oklahoma with striking examples of efforts to continue to expand highway travel while ignoring opportunities for provision of improved public transportation and freight service via rail, which is much more energy-efficient and if better utilized would significantly reduce both the threats of global warming and the dependence of the United States on petroleum. For example, powerful highway interests have been intent on replacing the Oklahoma City Crossstown Highway (U.S. Interstate Route 40) at a cost of more than half a billion dollars for less than four miles of new road. This program, started in the mid-1990s, proposes a new large highway that is not a public need on a route that would destroy the Union Station rail yard, owned by Oklahoma City. Union Station, in excellent condition and on the U.S. Historical Register, was a multimodal transportation center fifty years ago, and was purchased in 1989 for announced use for public transportation. Tracks and rights of way to all parts of the State are owned by the State and converge at Union Station, although all tracks are not in good condition at this writing.

The Crosstown replacement proposal illustrates the immense power of the U.S. automobile and truck lobbies and related special interests. If implemented, this proposal would increase truck travel through Oklahoma City and increase ozone and related health problems there while reducing prospects for economical, energy efficient public transportation and freight service throughout the State. This would occur with Oklahoma already behind many other U.S. states and cities in provision of public transportation. While many Oklahomans can hardly afford to buy and maintain the private cars necessary there for travel to work or cannot drive for reasons of health,[32] a variety of other reasons encourage efficient transportation of passengers and freight by rail, and retention of a facility that could be a hub for both freight and passenger service as it once was.

Lack of sufficiently effective programs in the United States is also a consequence of a cultural condition described in the *Harvard Divinity Bulletin* (Weiskel, 1990). Many individuals think in terms of anthropocentrism, e.g., "Earth is made for Man", and policies along this line are too often manifested in government. We should also be concerned with exceptionalism, the notion that humankind, owing to large brains, is exempt from the laws of nature applicable to other living beings. Both concepts have a basis in the Abrahamic religions established in both western and Islamic societies. Culture wars in the U.S. and elsewhere often pit these traditional concepts against new ideas about humankind's proper place. The new ideas spring partly from a torrent of new science about the cosmos ranging from the infinitesimal to the farthest galaxies. Regrettably, the new ideas also bring a kind of new religion with a new exceptionalism. The new religious beliefs hold that problems as they develop will inevitably be solved by new science and technology, and some government support of research stems from this attitude. Indeed, in speeches from

---

countries around the world are highly relevant to associated environmental problems and to their address.

[32] The average annual cost of car ownership and use in the United States is now estimated to exceed $7000.

highest levels of government, it is often proclaimed that our problems will be solved by research, even when the speakers have little knowledge of either science or its natural limitations.[33]

Much legislation provided by the political system in the United States is an exchange for financial contributions to campaigns. Will our system (and others, too) remain inadequate to deal with the global warming and energy decline phenomena? If it does remain inadequate, it will not be because the U.S. system is vastly different than it used to be – although there has been concentration of control of media by narrow interests, this control over news delivery has been somewhat offset by democratizing effects of the Internet. Historically, our political system has frequently supported powerful groups that sacrifice the good of a large sector for personal short-term benefits. This author thinks that the last times that populace and government rose to needed heights was when the critical nature of conditions related to WW II became more than obvious. And subsequent to WW II there was the good Marshall Plan.

In the United States and elsewhere, many research programs are well funded. As noted in a short article (Kessler, 1991), the political establishment is pleased to provide the wherewithal, in part because the hope for favorable outcomes is a basis for postponement of actions that are politically difficult to implement even though they could be immediately effective. And, of course, research must be encouraged; a plethora of research outcomes in every field of study are the principal basis for our industrial and postindustrial worlds, and further highly favorable results seem inevitable.

For example, a recent helpful outcome in Japan has produced light emitting diodes (LEDs) that are about 50% efficient in their production of light from electrical energy, and the cost of LED production is being reduced rapidly. LEDs may be on track to replace both incandescent lights with efficiency about 5% and fluorescents, 25%. The U.S. Dept. of Energy has estimated that about 22% of electricity production is devoted to lighting, so the new products may lead to both reduced CO_2 emissions and better lighting around the world, including in communities remote from utility power (Ouellette, 2007).

Important developed differences between now and decades ago are more in the nature of our times than in qualities of our political system. General demand has risen and continues to rise with increasing world population, and some basic resources that are essential to maintenance of infrastructure and provision of essentials are not as plentiful as formerly and are more expensive to obtain. The immense power of tools created by spectacular advances in science and technology means that malfeasance in the application of those tools leads to increasingly harmful consequences. Thus, private automobiles have provided unprecedented and very welcome mobility to many, but they are still being promoted even though they are principal contributors to carbon dioxide emissions and decline of liquid fuels. While products

---

[33] Of course, some problems are solved by research, but many of the political pronouncements about expectations from scientific research reflect more faith than science.

of advanced science and resultant technologies are essential to most of our daily lives, many more people in the United States than in Europe seem to reject findings and implications of science when those findings conflict with historical matters of faith or call for specific short-term sacrifice for dimly-perceived benefits in the long term.

Science and technology are seen as the major source of means for tapping the wealth of Earth. To what extent may further advances lead to means for marked reduction of our impacts? Such favorable developments will depend much more on scientific guidance to research directions than on political guidance!

Geometrical orientations of Earth to Sun are projected to rule out global cooling and recurrence of glaciations for another 30 thousand years, and this means that global warming will continue inexorably unless emissions of greenhouse gases are greatly diminished or there is an unexpected diminution of Solar radiation or extensive volcanism on Earth. Therefore, it may well be that within a few decades, humans on Earth will have to accommodate powerful forces that will make early adjustments seem easy by comparison. New problems may well include migrations of millions of people forced to leave submerging habitats, shortages of water in areas now dependent on glacial runoff, hotter summers, fluctuations of food supply following intensified droughts and floods, and increased social unrest. There are solutions to global warming problems, but none is easy, and most political systems are inhibiting. Will we humans meet this immense challenge to our established ways and cultures? Delay compounds difficulty and cost of solutions.

11.6 Conclusions

The United States has not yet a single program effective toward reduction of its dependence on foreign sources for liquid fuels or toward mitigation of the looming disaster represented by global warming. If existing programs were effective, we would expect that imports of petroleum products would be declining, but such imports are continuing to increase. And the existing biofuels programs are already damaging the agricultural economy. In large part, the programs in place are a consequence of a political system whose legislation is too-much based on contributions from the already rich and powerful, and insufficiently responsive to conditions and findings from advanced and still burgeoning science and technology. Overall, the situation is a consequence of the human condition, little changed during thousands of years.[34]

Such programs as improved insulation of existing houses, new construction of "green" buildings, and facilitation of transportation alternatives such as bicycling, are steps in right directions and have won grass-roots support, but all are far too

---

[34] Characterized in part in *Sophocles*, "No thing in use by man, for power of ill, can equal money. This lays cities low, this drives men forth from quiet dwelling-place, this warps and changes minds of worthiest stamp, to deeds of baseness, teaching men all shifts of cunning, and to know the guilt of every impious deed... By base profit won, you will see more destroyed than prospering..."

small. The major programs, ethanol from corn and sugar cane and biodiesel from palm oil, soybeans, and canola are deceptive responses. They provide short-term profit to special interests and they do provide fuels, but even the aggregate amount of fuels produced in these programs is a trivial proportion of present consumption and, the production processes yield, at best, no net reduction of carbon dioxide emissions. The alternative fuels programs damage the agricultural economy by causing increases in the price of corn and other human foods and livestock feeds, losses of already diminished habitat including tropical rainforests and wildlife, and losses of topsoil and increased stress on water supplies.

As noted above, unless carbon dioxide emissions are quickly reduced, global warming will be a very serious matter for future generations and will force large adjustments in ecosystems worldwide. Concern rises because in the United States and in rapidly developing countries such as China and India, policies remain strongly oriented toward economic and even physical growth with increasing emissions of carbon dioxide.

What should be done in the United States, for example, beyond such programs as tightening CAFÉ[35] standards, weatherizing homes and utilizing energy-saving construction in new work, installing solar heating, and expanding use of time-of-day pricing of electricity, all of which are or would be good though inadequate? A proper practical course is difficult to identify, and an effective course may be impossible to identify. In other words, it may be too late to avoid serious damages from global warming and to preserve social order in face of fuel declines. But, we must keep trying, and it is clear enough that in order to confront consequences of global warming and decline of liquid fuels, societies in developed (and developing) countries must practically be turned on their heads! And if they do not turn themselves soon, they will be turned later by large forces beyond human control.

As a first step, the notion of continuous economic growth must be abandoned,[36] and global population, which has increased threefold in your author's lifetime, must be much reduced. Whatever else is done, if population growth proceeds, all other saving actions will be nullified and even overwhelmed owing to increased demand. Abplanap's succinct statement (1999) applies, necessary changes being made, to physical growth of many entities in the presence (or absence) of technological advances: "... Any kind of agricultural 'green revolution' which is not accompanied by effective population control merely resets the limiting parameters at higher levels and enables countries with a large proportion of starving citizens to increase the absolute numbers of starving people".

Is population reduction feasible? Population is sustained with an average birth number near 2.1 per female inhabitant. If this average were reduced to 2.0 the impact on individuals would be very minor but the eventual impact on world population would be major. If world population were to decline just one percent per year,

---

[35] Corporate Average Fuel Economy, i.e., average automotive mileage as mandated by federal legislation.

[36] And replaced by increased learning, cultural growth, equity and justice. A tall order!

numbers would be reduced by half in 70 years and again by half in another 70. In 2007, this must be seen as only a utopian dream, since the large proportion of young people in the present world population guarantees substantial growth of the global population in the near term.[37] Further, strong diverse forces, even the U.S. government at this writing, offer little or no support for birth control,[38] and Chambers of Commerce all across America promote growth among the highest of their priorities. Of course, population matters are very different in different economies, demographies, and cultures, and associated problems, including treatment and education of females, are not explored here.[39]

Second, it would be helpful in the United States to have a massive shift in funding from highway building to construction of a national rail system for both passenger travel and improved freight transport. Such a system, emulating that already in place and still under rapid development in Europe and somewhat too in Asia, would be inherently more energy efficient than automobiles and truck travel on highways, and even further emission reductions would be achieved to the extent that trains become more fueled with electricity from overhead wires or from liquefied natural gas in place of diesel fuel.

Such a transportation alternative in the U.S. might be paid for in part by an increased federal tax on gasoline and diesel fuels. If rail were more emphasized, U.S. highways would be less burdened with cars and trucks, highway maintenance costs would decline, and emissions of carbon dioxide and health-threatening gases from the automotive sector in this leader country would decline. And decline of truck traffic would quicken if trucks were taxed in relation to the maintenance costs they impose – road damage is proportional to the fifth power of axle weight.[40] Groups of citizen-activists are working in these directions, but during 2007 in the United States, there is little official interest in such programs – indeed, such programs lack substantial support from the federal level in the United States and are opposed by highway and automotive lobbies. In 2007 there is still strong political support toward expansion of the highway system.

Third, further enhancement of already burgeoning communication technologies may proceed to a level that somewhat reduces energy-consumptive travel.

The three items above could be resource-conserving approaches in a relatively short term. But for true sustainability in terms of geological age, we should, barring success with nuclear fusion as a source of electrical energy, begin to explore development of a very broad solar economy, because only solar energy is projected to endure much as at present for billions of years. This means that solar power plants would be built with help from fossil or nuclear fuels to support an economy with

---

[37] Barring more serious war or pestilence, of course.

[38] China has learned the hard way, and brutality properly opposed is a sometime component of birth control efforts in China, but the United States government declines to acknowledge the seriousness of population numbers even when those numbers strain the food supply.

[39] Nor have we discussed abatement of terrorism and war and spread of justice internationally.

[40] In Oklahoma, the tax on diesel fuel as this document is prepared is three cents/gallon less than on gasoline.

fewer human numbers indefinitely, and the solar power would be used to maintain and enhance the power system itself. This vision of a farther future is mentioned by Patzek on his website and a possible solar path has been detailed by Zweibel, et al. (2008).

So, in summary, What is our food and fuel future? It is highly problematic, and a decent future for humans is much dependent on rationalization of decision-making at all levels to findings and implications of science and technology! The rapid pace of change in this 21st century also calls for a much more rapid response of proper decision making to major findings of science and technology.

Will humanity on Earth be a "flash in the pan"? Consider a 30-volume encyclopedia, each volume with one thousand pages, each page with an average one thousand words. Let these thirty volumes present a linear history of Life on Earth since multi-celled organisms became prevalent perhaps one billion years ago, with the start of accumulation of the fossil fuels that we humans use today. How much space is devoted to the sixty-five years since World War II, during which we humans have extracted about half of Earth's readily extractable liquid fossil fuels and much coal, and caused an astonishing increase in atmospheric content of carbon dioxide? Is the answer disturbing? Only two words on the last page of the last volume! How long will we endure and how much space might describe our future post-industrial society?

Acknowledgments Thanks to Marjorie Bedell Greer and Richard Hilbert for suggestions based on their readings of an early typescript, to Hilbert and to Charles Wright for sociological insights and to Tom Elmore for imparting some of his encyclopedic knowledge of the railroad history of Oklahoma. David Sheegog contributed to the discussion of ethanol, and Steve Shore helped with the table in Section 11.4. Before semi-retirement, Dr. Greer was a professor of anatomy at the Oklahoma University Health Sciences Center in Oklahoma City, Dr. Hilbert was Chair of the Sociology Dept. at the University of Oklahoma in Norman, and he continues to lecture, and Charles Wright is an attorney and sociologist. Tom Elmore is Executive Director of the North American Transportation Institute, Moore, Oklahoma, David Sheegog is a psychologist and rancher, and Steve Shore is a professor of chemistry at Oklahoma City Community College. Thanks also to David Pimentel for several important suggestions.

References

Abplanap, P. L. (1999). A letter to *Technology Review*, Sept–Oct.

American Wind Energy Association (2007). http://www.awea.org/projects/, retrieved August 28, 2007.

Anthony, R. (2007). Safe at Sea, *Spectrum, Massachusetts Institute of Technology, XVIII, X,* 17.

Apricus.com (2007) See this webpage, http://www.Apricus.com, Retrieved Dec. 3, 2007.

Bullis, K. (2006). Abundant Power from Universal Geothermal Energy, http://www/ technologyreview.com/Energy/17236/, retrieved Oct. 11, 2007

Castro, F. R. (2007). The Internationalization of Genocide, *Granma Internacional*, April 3.

Center for Rural Affairs. (2007). Monthly Newsletters from P.O. Box 136, Lyons, Nebraska 68038–0136.

Clery, D. (2006). ITER's $13 Billion Gamble, *Science, 314, 5797,* 238–242.

Congressional Research Service (2005). Alcohol Fuels Tax Incentives, CRS Order Code RL2979.

Crabtree, G. W., Dresselhaus, M., & Buchanan, M. V. (2004). The Hydrogen Economy. *Physics Today, 57, 12*, 39ff.

Crabtree, G. W. & Lewis, N. S. (2007). Solar Energy Conversion. *Physics Today, 60, 5*, 37–42.

Duncan, M. & Webb, K. (1980). *Energy and American Agriculture*. From the Research Division of the Federal Reserve Bank of Kansas City, U.S.A., Thomas E. Davis, Senior Vice President. 41pp.

Environmental Protection Agency. (2007). National Lake Fish Tissue Study, Retrieved 1 Sept. and earlier from www.epa.gov/waterscience/fishstudy/. (Much detail has been available on the web sites, and your author has been told that a formal summary report is in review and may be available during 2008.)

E&P. (2007). *Coalbed Methane, 80, 6*, 41–55. (A series of presentations on new and developing technologies). (E&P = Exploration and Production, from Hart Energy Publishing, 1616 S. Voss Road, Houston, Texas, 77057.)

Ghazvinian, J. (2007). *Untapped – The Scramble for Africa's Oil*. (New York, Harcourt) 320pp.

Hansen, J., Sato, M., Kharecha, P., Russell, G., Lea, D. W., & Siddall, M. (2007). Climate Change and Trace Gases. *Philosophical Transactions of the Royal Society A, 1925–1954*.

Häring, M.O., Ladner, F., Schanz, U., & Spillmann, T. (2007). Deep Heat Mining Basel, Preliminary Results. Retrieved August 5, 2007 from website: http://www.geothermal.ch/ downloads/dhm_egc300507.pdf

Hart Energy Publishing. (2006). Unleashing the Potential of Heavy Oil. A supplement to E & P Oil and Gas Investor (Principally a description of facilities and investments in the tar sands of Alberta, Canada.) 1616 S. Voss, Ste 1000, Houston, Texas 77057.

Hart Energy Publishing. (2007). Unleashing the Potential of Heavy Oil. A supplement to E & P Annual Reference Guide (A discussion of new technologies.) 1616 S. Voss, Ste 1000, Houston, Texas 77057.

Hinze, W. J., Marsh, B. D., Weiner, R. E., & Coleman, N. M. (2008). Evaluating Igneous Activity at Yucca Mountain. *EOS, 89, 4*, 29–30.

Intergovernmental Panel on Climate Change. (2007). Numerous reports available on the Internet, http://www.ipcc.ch/

Kahn, J. & Yardley J. (2007). As China Roars, Pollution Reaches Deadly Extremes, *The New York Times*, August 26.

Kessler, E. (1991). Carbon Burning, the Greenhouse Effect, and Public Policy, *Bulletin of the American Meteorological Society, 72, 4*, 513–514.

Kessler, E. (2000). Wind power over central Oklahoma, *Report* prepared for the Bergey Wind Power Company, Norman, Oklahoma. 2000, x + 25 pp. + 46 figures. January.

Kessler, E. & Eyster, R. (1987). Variability of wind power near Oklahoma City and implications for siting of wind turbines. *Final Report* on DOE Interagency Agreement No. DE-A1-6-81RL 10336. Pacific Northwest Laboratory, Richland, Washington. September, 74 pp. + appendices. [This report was reprinted by the Oklahoma Climatological Survey, Norman, Oklahoma, in 1994 in a condensed format with small editorial adjustments and some additional notes.]

Komar, P.D. (2007). Higher Waves Along U.S. East Coast Linked to Hurricanes. *EOS, 88, 30*, 301.

Lagercrantz, J. (2006). Ethanol Production from Sugar Cane in Brazil. Retrieved August 10, 2007, from http://www.gronabilister.se/file.php?REF=39461a19e9eddfb385ea76b26521ea48&art= 376&FILE_ID=20060511084611.pdf.

Mao, W. L., C. A. Koh, & E. D. Sloan. (2007). Clathrate hydrate under pressure, *Physics Today, 60, 10*, 42–47.

Marland, G., T.A. Boden, & R.J. Andres. (2005). Global, Regional, and National Fossil Fuel CO2 Emissions, *Carbon Dioxide Information Analysis Center*, Oak Ridge National Laboratory, U.S. Department of Energy, Oak Ridge, Tenn., U.S.A.

Mayes, J. (2007). Warmest 12 months in British Isles instrumental records, *Weather, 62, 4*, 86.

McCain, J. (2003). Statement of U.S. Senator John McCain on the Energy Bill. (November 21st).

National Academy of Sciences. (2007). Water Implications of Biofuels Production in the United States. October, 86pp. Summaries and the complete report are available on the Internet: http://www.nationalacademies.org/morenews/20071010.html

NOAA (U.S. National Oceanic and Atmospheric Administration). (2007). http://www.cdc.noaa.gov/map/images/rnl/sfctmpmer_01b.rnl.html

Oklahoma Mesonet (2007). http://www.mesonet.org/public/

Oklahoma Wind Power Initiative (2007). http://www.ocgi.okstate.edu/owpi/

Ouellette, J. (2007) White LEDs poised for global impact. *Physics Today, 60, 12,* 25–26.

Pimentel, D., Patzek, T. W. & Gerald, C. (2006). Ethanol Production: Energy, Economic, and Environmental Losses. *Reviews of Environmental Contamination & Toxicology, 189,* 25–41.

Pennisi, E. (2007). Replace Genome gives Microbe new identity. *Science, 316, 5833,* 1827.

Roach, W. T. (1998). Can Geothermal Heat Perturb Climate? *Weather, 53, 1,* 11–19.

Schneider, D. (2007a). Coal Futures. *American Scientist, 95, 4,* 314–315.

Schneider, D. (2007b). Who's Resuscitating the Electric Car? *American Scientist, 95, 4,* 403–404.

Shady Point. (2007). Retrieved October 17, 2007: http://www.CO2captureandstorage.info/project_specific.php?project_id=22

Shapouri, H., Duffield, J. A. & Wang, M. (2002). The Energy Balance of Corn Ethanol: An Update. United States Department of Agriculture (USDA), *Agricultural Economic Report Number 813.*

Simmons, M. R. (2005). *Twilight in the Desert.* (New York, Wiley) 428pp.

Special Section: Sustainability and Energy (2007). *Science, 315, 5813,* 781–813.

Srivastava, R. K., Hutson, N., Martin, B., Princiotta, F., & Staudt. J. (2006). Control of Mercury Emissions from Coal-fired Electric Boilers. *Environmental Science and Technology, March 1,* 1385–1391.

Trade Commission of Spain (2007). Solar Energy in Spain. *Technology Review, 110, 5,* S1–S10.

Tyner, G., Sr. (2002). *Net Energy from Nuclear Power.* Retrieved April 3, 2007 from Minnesotans for Sustainability website: http://www.mnforsustain.org/nukpwr_tyner_g_net_energy_from_nuclear_power.htm

Tyner, G., Sr. (2002). *Net Energy from Wind Power.* Retrieved April 3, 2007 from Minnesotans for Sustainability website: http://www.mnforsustain.org/windpower_tyner_g_net_energy.htm

Vago, S. (1981). Law and Society (Englewood Cliffs, New Jersey, Prentice Hall) xi + 372pp. (See esp. pp. 132–135)

Wallace, Linda, L. (2007). Switchgrass is no energy panacea. Essay in *The Norman Transcript,* on Page 4, October 11, and personal conversation. Prof. Wallace is with the Dept. of Botany and Microbiology at the University of Oklahoma, Norman.

Weiskel, T. C. (1990). The Need for Miracles in the Age of Science. *Harvard Divinity Bulletin, XX, 2.* 5ff.

Zweibel, K., Mason, J., & Fthenakis, V. (2008). A Solar Grand Plan. *American Scientist, 298, 1,* 64–73.

Chapter 12
A Framework for Energy Alternatives: Net Energy, Liebig's Law and Multi-criteria Analysis

Nathan John Hagens and Kenneth Mulder

Abstract Standard economic analysis does not accurately account for the physical depletion of a resource due to its reliance on fiat currency as a metric. Net energy analysis, particularly Energy Return on Energy Investment, can measure the biophysical properties of a resources progression over time. There has been sporadic and disparate use of net energy statistics over the past several decades. Some analyses are inclusive in treatment of inputs and outputs while others are very narrow, leading to difficulty of accurate comparisons in policy discussions. This chapter attempts to place these analyses in a common framework that includes both energy and non-energy inputs, environmental externalities, and non-energy co-products. We also assess how Liebig's Law of the minimum may require energy analysts to utilize multi-criteria analysis techniques when energy may not be the sole limiting variable.

Keywords Net energy · EROI · EROEI · liebig's law · ethanol · biophysical economics · oil · natural gas

12.1 Introduction

Human energy use, ostensibly the most important driver underpinning modern society, may soon undergo a major transition of both kind and scale. Though numerous energy technologies are touted as alternative supplies to fossil fuels, scientists and policymakers continue to lack a meaningful and systematic framework able to holistically compare disparate energy harvesting technologies. Net energy analysis attempts to base decisions largely on physical principles, thus looking a step ahead

✉ N.J. Hagens
Gund Institute for Ecological Economics, University of Vermont, 617 Main St., Burlington, VT 05405, USA
e-mail: Nathan.Hagens@uvm.edu

K. Mulder
Green Mountain College, Poultney VT, USA

D. Pimentel (ed.), *Biofuels, Solar and Wind as Renewable Energy Systems*,
© Springer Science+Business Media B.V. 2008

of political and/or market based signals distorted by fiat monetary data. The importance of net energy has been overlooked, primarily as a result of confusing and conflicting results in energy literature. In this chapter, we (a) provide an introduction to the history, scale and scope of human energy use (b) reiterate the role of net energy analysis in a world of finite resources, (c) establish a two dimensional net energy framework synthesizing existing literature and (d) illustrate (via the example of corn ethanol) why multi-criteria analysis is important when energy is not the only limiting variable.

12.2 Net Energy Analysis

Energy, along with water and air, completes the trifecta of life's most basic needs. Organisms on the planet have a long history of successfully obtaining and using energy, mostly represented as food. Indeed, some have suggested that the harness of maximum power by both organisms and ecosystems from their environments is so ubiquitous it should be considered the Fourth Law of Thermodynamics (Odum 1995). Cheetahs, to use one example, that repeatedly expend more energy chasing a gazelle than they receive from eating it will not incrementally survive to produce offspring. Each iteration of their hunting is a behavior optimized to gain the most energy (calories in) for the least physical effort (calories out), thus freeing up more energy for growth, maintenance, mating and raising offspring. Over evolutionary time, natural selection has optimized the most efficient methods for energy capture, transformation, and consumption. (Lotka 1922) This concept in optimal foraging analysis extrapolates to the human sphere via *net energy analysis*, which seeks to compare the amount of energy delivered to society by a technology to the total energy required to transform that energy to a socially useful form. Biophysical minded analysts prefer net energy analysis to standard economic analysis when assessing energy options because it incorporates a progression of the physical scarcity of an energy resource, and therefore is more immune to the signals given by market imperfections. Most importantly, because goods and services are produced from the conversion of energy into utility, surplus net energy is a measure of the potential to perform useful work for social/economic systems.

12.3 An Introduction to EROI – Energy Return on Investment

Knowing the importance of energy in our lives, how do we compare different energy options? Unfortunately, the word 'renewable' does not automatically connote 'equality' or 'viability' when considering alternatives to fossil fuels. In assessing possible replacements for fossil fuels, each alternative presents special trade-offs between energy quantity, energy quality, and other inputs and impacts such as land, water, labor, and environmental health (Pimentel et al. 2002, Hill et al. 2006). When faced with these choices, energy policymakers in business and government will require a comprehensive and consistent framework for accurately comparing all aspects of an alternative fuel.

Many criteria have historically been used to assess energy production technologies based on both absolute and relative yields and various costs (Hanegraaf et al. 1998). Many assess economic flows (e.g. Bender 1999, Kaylen 2005) while others focus on energy (e.g. Ulgiati 2001, Kallivroussis et al. 2002, Cleveland 2005, Farrell et al. 2006) or emissions (e.g. EPA 2002). With the recent acceptance of global climate change as a problem, energy analyses favoring low greenhouse gas emissions are becoming more frequent (Kim and Dale 2005, Chui et al. 2006). Though not yet widely accepted by market metrics, some other analyses have attempted to include environmental and social inputs as well as energy costs. (e.g. Giampietro et al. 1997, Hanegraaf et al. 1998, Pimental and Patzek 2005, Reijnders 2006).

The objective of an energy technology is to procure energy. A common measure combining the strength/quality of the resource with its procurement costs is the ratio of energy produced to energy consumed for a specific technology/source. This concept has many labels in energy literature including the energy profit ratio (Hall et al. 1986), net energy (Odum 1973), energy gain (Tainter 2003), and energy payback (Keoleian 1998). In this chapter, we focus on Energy Return on Investment (EROI) (Hall et al. 1986, Cleveland 1992, Gingerich and Hendrickson 1993) EROI is a ratio and is equal to 'net energy +1'. Total energy surplus is EROI times the size of the energy investment, minus the investment. We will use the terms energy gain, net energy and EROI interchangeably, throughout this chapter.

12.4 Humans and Energy Gain

Ancestral humans first major energy transformation came from the harnessing of fire, which provided significant changes to daily tribal life by providing light, warmth and eventually the ability to work metals, bake ceramics, and produce tools. (Cleveland 2007). More recently, the energy gain of agriculture further transformed human culture. Though the per unit energy gain of widespread agriculture was actually lower than many hunting and gathering practices, a large amount of previously unused land was brought under cultivation, thus freeing up substantially larger energy surplus for society as a whole. (Smil 1991) This is a first example of how an energy return combines with scale to determine an overall energy gain for society. Much more recently, the development of the steam engine catapulted mankind into the fossil fuel era by leveraging the embodied energy in coal deposits. The high energy gain of coal rippled its way through the economy akin to a deposit in a fractional banking system, and the industrial revolution had its first power source. In the 19th century, modern humans learned to unlock the hydrocarbon bonds in the higher quality fossil fuels of crude oil and natural gas, freeing up orders of magnitude more energy than our evolutionary forbears even dreamed about. The changing size of this subsidy, how to measure it and meaningfully compare it to potential

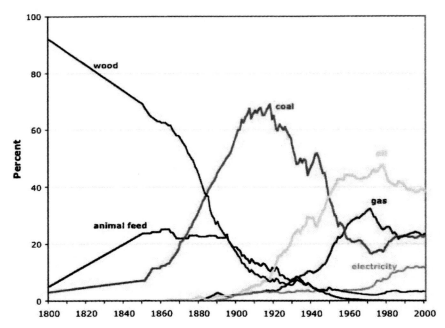

Fig. 12.1 *Composition of US energy by* (Cleveland 2007)

energy substitutes that will be required to power future society is the subject of this chapter (Fig. 12.1).

12.5 Current Energy Gain

The current scale of our energy gain is unprecedented. When coal, oil and natural gas are included, the average American uses 57 barrel of oil equivalents per year (BP 2005). Each barrel of oil contains 6.1178632×10^{9} Joules of energy. An average man would need to work about 2.5 years to generate this amount of heat work[1]. Multiply it by 57, and the average American uses a fossil fuel subsidy equal to over 150 annual energy slaves. But the quality of oil is also fantastic – liquid at room temperature and highly dense – oil possesses energy quality that human labor cannot.

An important nuance underlying the concept of net energy analysis, is that fossil fuel production is itself cannibalistic, as oil production uses a great deal of natural gas (and some oil) to procure. Coal production, wind turbine creation, solar photo-voltaic panels, etc. all require liquid transportation fuels to generate their products

---

[1] An 'average' worker utilizes 300 calories per hour. At 8 hours per day, 5 days per week and 50 weeks per year this is 600,000 calories per year. $(6.1178632 \times 10^{9}$ Joule) per barrel / (600,000 Calories $\times$ 4,184 joules required work energy per year) = **2.44 years/barrel.**

in a modern economy. In fact, over 90% of world transportation is accomplished using liquid fuels. (Skrebowski 2006).

The scale of remaining recoverable crude oil is a topic under much debate, with many analysts saying we are already past peak production (Deffeyes), and others (IEA, Cambridge Energy Research Associates) saying we will reach a broad plateau by 2030–2040. A large number of analysts believe a peak in oil production will occur sometime in the next decade. However, few if any of these analysts look at how much of future oil and gas production nets down to the societal use phase after the energy costs have been accounted for. Nor is there a distinction made in 'crude oil' statistics between actual crude oil, ethanol, coal-to-liquids, etc. all of which not only have disparate energy costs, but different BTU contents as well.

The Hubbert curve of resource extraction is roughly Gaussian in shape, and the energy surplus (or lack thereof) drops down dramatically after its peak (see Hall et al., 1986 for an example on Louisiana). If oil is peaking soon, asking how much is still in the ground is not the most important question. How much can be brought to market at one time? How much energy is left after energy companies use what they require internally to procure the harder to find, deeper, more sulfurous, more environmentally and socially sensitive drilling locations, etc.? These questions ultimately address how much of our remaining fossil resources will be available for non-energy, non-government society.

12.6 An Energy Theory of Value

There is a rich history over many decades of the concept of an energy theory of value, dating back to Howard Scott and the Technocrats who stated that 'A dollar may be worth – in buying power – so much today and more or less tomorrow, but a unit of heat is the same in 1900, 1929,1933 or 2000' (Berndt 1983). In the 1970s, Senator Mark Hatfield argued that 'Energy is the currency around which we should be basing our economic forecasts, not money supply.' His efforts resulted in the passing of (now defunct) Public Law 93.577 which stipulated that all prospective energy supply technologies considered for commercial application must be assessed and evaluated in terms of their 'potential for production of net energy'. (Spreng 1988) And in a still broader sense, ecological analysts have long stated that money does not properly account for externalities – ecologist Howard Odum stated 'Money is inadequate as a measure of value, since much of the valuable work upon which the biosphere depends is done by ecological systems, atmospheric systems, and geologic systems.'

12.7 Why is Net Energy Important?

This 'work' Professor Odum alluded to requires an energy surplus. (Odum 1994) In a world where energy is likely to become scarcer, net energy analysis is more forward looking than conventional economic analysis, and as such can be an important tool for policymakers. Net energy is important because we need energy to accomplish work. The surplus energy of a system, or society, is what allows it to continue growth, maintenance, repair and leisure. Energy technologies can be stock or flow based. Stocks are depletable and non-renewable on human time scales. Flow-based resources are renewable, provided the infrastructure that supports them is renewable. There is only so much low entropy energy present in fossil fuel stocks and solar/tidal flows that can be accessed at a meaningfully positive energy return. If society has collectively become dependent on a certain aggregate energy gain system and attempts to replace it with a lower energy gain portfolio, while keeping all other inputs equal, then a larger % of societies resources (labor, capital, land, water, etc) would have to be devoted to energy procurement, leaving less available for hospitals, infrastructure, science, etc.

So in one sense, the Energy Return on Investment is a story of demand, and how a civilization uses their BTU endowments. A doubling in efficiency of use, or a doubling of conservation efforts, are equivalent to a doubling of an energy surplus. *But if efficiency and conservation do not occur, we are left trying to maintain a high gain system from new energy supply as original stocks of resources deplete.* Historian Joseph Tainter has shown, with both examples from the animal kingdom and historical human societies (Rome), that high energy gain systems undergo social upheaval and ultimately collapse if they cannot maintain the energy gain that their infrastructure is built upon (Tainter 2003). The more energy required to harvest, refine and distribute energy to society, (assuming we're at maximum scale), the less will be left over for non-energy sectors. This is especially important in a society that has built its infrastructure around high-energy-return inputs (Smil 1991). Our modern situation, the energy density required for our shopping centers, hospitals, high rises, etc. is orders of magnitude higher than that of biomass and other renewables. (Smil 2006).

12.8 Net Energy and Energy Quality

In a human system, the desirability of a resource derives both from its absolute energy gain as well as from its utility to a unique sociocultural system. (Tainter 2003) Thermal energy quantity is important from a thermodynamic standpoint. However, a human society does not use or value energy based on its heat component alone. Prehistoric man would have viewed a horse as a source of meat, not as an animate converter of cropland or as a riding steed. Similarly, an ancient Yibal tribesman in Saudi Arabia would have little use for the high energy density oil bitumen just under the sands surface, but enormous use for the energy conversion capacity of a healthy horse. Today's shopping centers and hospitals could not be powered by meat calories or horsepower, but require the dense energy concentrated in fossil fuels. Thus, energy quality is a definition dependent on the context of a society.

When Watt was developing his steam engine, the heat value and liquid form of petroleum were of little use, because the new technologies of that day required wood or coal. And, unlike other mammals, humans have evolved to utilize exosomatic energy, and build and expand society around specific inanimate converters, earlier the steam engine and more recently the internal combustion engine. In this fashion, energy 'quality', as defined by an energy sources ability to perform economic or other work valued by society, can and does depart from a straight thermal assessment of the energy. Coal does not make a refrigerator work, and natural gas does not have the density to run a computer printer; these fuels must first be transformed into higher quality energy, at a thermal loss.

When assessing the quality of an alternative energy, the following factors need to be considered: energy power and density, timing, energy quality, environmental and social impacts of energy procurement and use, geographic and spatial scales, volatility, and the potential scale of the resource (energy surplus). We will now briefly discuss this first set of objective energy quality criteria. The majority of the chapter will deal with the penultimate societal energy metric; the scale of the energy surplus, and its EROI.

Energy density refers to the quantity of energy contained per unit mass or volume. The lower energy density of biomass (12–15 MJ/kg) compared to crude oil (42 MJ/kg) means that replacing the latter with the former will require a larger infrastructure (labor, capital, materials, energy) to produce an equivalent quantity of energy. (Cleveland 2007) The energy carrying molecule hydrogen, has very low energy per unit volume, creating many technical hurdles to a 'hydrogen economy', even were cheap abundant hydrogen fuel stocks available.

Due to the enormous amount of geologic energy invested in their formation, fossil fuel deposits are an extraordinarily concentrated source of high-quality energy, commonly extracted with power densities of 100 to 1000 Watts/m^2 for coal or hydrocarbon fields. (Cleveland 2007). This implies that very small land areas are currently used to supply enormous energy flows. In contrast, biomass energy production has densities well below 1 Watt/m^2, while densities of electricity produced by water and wind are commonly below 10 Watt/m^2. In effect, as power dense fossil resources deplete, less power dense energy must be secured from more of the earth's surface to match the gross amount available from the concentrated high-gain sources (Smil 2006).

Bioenergy made from annual crops will also undergo unexpected volatility from periodic droughts or floods, whereas oil production can provide gasoline and its energy services continuously (or at least until a well runs dry). On a shorter time scale, the intermittency (or fraction of time that an energy source is usable to society), is low for wind and solar technologies as neither the sun nor the wind give us energy twenty four hours a day. This is potentially important with modern electricity generation systems that need to combine power generated from multiple sources and locations to supply electricity '24/7.' A derivative concept of intermittency is the dispersion over time of a source. In economics and finance, investors care greatly about the 'shape' of portfolio returns. A portfolio returning 10% consistently is much preferred to an investment that averages 15% but has periodic negative years. In effect, investors preferences are measured by a 'risk adjusted return' which is the mean return divided by the standard deviation. Energy too, has a risk adjusted return, and constantly flowing and storable fossil fuels have built a society that depends on smooth flows of energy services. Going back to ecosystem services to procure energy may have higher standard deviations of energy availability.

All natural resources show distinct geographical gradients. In the case of oil and natural gas more than 60% of known resources are in the Middle East. Just as with stored ancient sunlight, renewable energy from current sunlight (solar, wind, etc.) is geographically diffuse. This implies that significant investments (of dollars and energy) into new infrastructure will be required to concentrate, store and distribute energy over distance in order to procure useful amounts of energy services to human population centers.

Historical human energy transitions occurred when the human population was small, and had technology that was much less powerful than today. Environmental impacts associated with energy occurred locally but did not exhibit the current global impact. But the future of energy and the environment are linked, as there are numerous ecological constraints. Our future energy systems must be designed and deployed with environmental constraints that were absent from the minds of the inventors of the steam engine and internal combustion engines (Cleveland 2007).

12.9 Energy Return on Investment – Towards a Consistent Framework

Though all of the above are important factors in assessing renewable energy technologies, perhaps the most critical metric is the actual size of energy surplus freed up for society. Once an energy output becomes truly scarce – large sums of dollars won't improve its scarcity, and all the dollars in the world wont change (quickly) the demand system and energy infrastructure dependent on its energy gain. High energy gain can arise from using a resource that is of high intrinsic quality but untapped, or from technological development that allows an increase in the net energy of a previously used resource. The energy gain of mining deep coal, for example, increased greatly after Watt's engine was widely used (Wilkinson 1973). Conversely, energy gain can decline from exploiting a resource that can yield only small returns on effort under any technology, or from having depleted the most accessible reserves of a once abundant resource (Tainter 2003).

Energy Return on Investment (EROI) is an oft-confused controversial but important cousin to energy gain. EROI is basically a combined measure of how high of quality/density the original energy source is with the energy cost that the composite of harvesting technologies uses to deliver the energy to the consumptive stage. EROI is strictly a measure of energy and its 'harvesting' costs in energy terms, not the efficiency of its use or it's transformation to another energy vehicle. For example, once coal is procured out of the ground at a particular energy return, the decision, and subsequent efficiency loss to turn it into electricity or Fischer-Tropsch diesel, are both part of the consumption choices of society *after* the primary fuel is obtained.

The efficacy of EROI analysis is limited by one of its basic assumptions—that all forms of energy are fungible with a statistic determined by their thermal content (Cleveland 1992). This ignores the fact that the quality of an energy source can be the key determinant of its usefulness to society. A BTU of electricity is of higher value to society than a BTU of coal, a fact reflected by the price differential between these two energy sources as well as our willingness to convert coal into electricity at a significant energy loss. Some would argue that a technology with a low EROI should be given stronger consideration if the energy outputs have a higher quality than the energy inputs—an argument raised by Farrell et al. (2006) in support of corn ethanol which has the potential to convert coal and corn (low quality) into a liquid fuel (high quality). Cleveland (1992) has proposed a variant of EROI methodology that incorporates energy quality. Quality-adjusted economic analysis can even support sub-unity EROI energy production depending on context.

The EROI concept has been specifically used in only a small percentage of national energy analyses, but is implicit in any study that uses a form of net energy as a criterion. Recently it was used as a synthesizing concept for multiple comparisons of biofuels (Farrell et al. 2006, Hammerschlag 2006). It has been used to examine nuclear energy (Tyner et al. 1988, Kidd 2004), ethanol (Chambers et al. 1979, Pimentel 2003, Hu et al. 2004, Farrell et al. 2006, Hammerschlag 2006), other biofuels (Baines and Peet 1983, Giampietro et al. 1997, Kallivroussis et al. 2002), wood energy (Baltic and Betters 1983, Potter and Betters 1988, Gingerich and Hendrickson 1993), and other alternative energies (Crawford and Treloar 2004, Berglund and Borjesson 2006, Chui et al. 2006). Ongoing analysis continues on the EROI of various fossil fuels (Cleveland 1992, 2005, Hall, 2008).

At first blush, the calculation of EROI as the ratio of energy outputs to inputs seems straightforward. However, the concept has never expanded into common usage (Spreng 1988). Even with a recent resurgence of interest in this topic due to escalating oil prices, there is still not a widely accepted methodology for calculating either the numerator (the energy produced) or the denominator (the energy consumed) in the EROI equation. While attempting to use this important criteria to compare energy technologies, different researchers are using different methods to arrive at widely disparate notional EROI numbers, thereby diluting the policy value of this energy statistic. The ongoing heated debate over the viability of grain ethanol is a relevant example. A recent publication (Farrell et al. 2006) suggests that previous analyses of the EROI of grain ethanol are errant because of outdated data and faulty methodology. The analysis attempted to standardize previous studies and introduce modifications of the EROI methodology including measuring energy produced per unit of *petroleum* energy invested. However, because a standardized well-defined EROI formula does not exist, nor is there wide acceptance on the reasons why net energy analysis is important, the Farrell et al chapter has not ameliorated the polarization of the debate but rather heightened it (Hagens et al., 2006). At the very least, this lack of precision and consensus has negative implications for the utility of EROI analysis, in particular as a tool for decision makers. At the worst, it leaves the methodology open to manipulation by partisans in the debate over a given technology.

Furthermore, emphasis is being placed on whether or how much the energy return of a proposed technology exceeds unity, without addressing the shortfall in energy return of the segment of energy services it is trying to replace. Corn ethanol advocates and proponents spend a huge amount of resources and time honing and refining the corn-ethanol energy balance – whether it's slightly negative or slightly positive seems to be of great policy significance. At 1.5:1, which is at the high end of the latest range, corn ethanol's energy return remains an order of magnitude below the fossil energy it purports to replace (Cleveland 2001). Unless society makes large scale changes on the consumption/efficiency side, it will need to address the variance between its current energy surplus and what can be expected with the combination of lower quality fossil stocks and less energy dense renewable infrastructure in the future. Due to differences in demand, and the geographic dispersion of high energy gain renewables, there may be a variety of answers to this question at the local/regional level and at the national/global level. Since fossil fuels power a global society, global energy gain, a function of EROI times scale for all energy sources, will be of central importance in the coming decades. In the following pages, we review the various usages of EROI in the literature and place them into a consistent schematic framework. This allows comparison of the different methodologies in use by clarifying both their assumptions and their quantitative components. We then synthesize the different methodologies into a two-dimensional classification scheme with terminology for each version of EROI that will hopefully yield consistent and comparable results between studies going forward.

Figure 12.2 is a theoretical aggregate of EROI and scale. D = direct energy costs, C = indirect energy costs, and B = externality costs (converted to energy). The area under the outer curve represents the total gross energy production X = A + B + C + D. A is the leftover 'net energy'. Since the most efficient areas of productions are usually developed first (e.g. best cropland, best wind sites, etc. (Ricardo 1819) the annual energy gain tends to decline while energy costs tend to rise with scale of development. Externalities also tend to increase.

At time T1 in Fig. 12.2, there is no surplus energy (A or B) leftover after direct and indirect energy costs (C and D) have been accounted for, meaning this 'source' X, is now an energy sink. If we also translate environmental externalities into energy terms (B), we then are faced with an energy sink shortly after time T2. In effect, if we include all costs, direct, indirect, and non-energy parsed into energy, the green shaded area A is the amount of net resource available under the entire graph. The graphic also illustrates that the peak energy gain in terms of net benefits to society is reached more quickly than the peak in gross energy.

It is important to note that unless the energy output and input are identical types, energy extraction can still continue at an energy loss – but these joules needs to come from elsewhere in productive society. One can envision a summation of all energy technologies used globally. If we aggregate all the 'A's' (Or A+B's if we ignore environmental externalities) of all planetary energy sources, we have a sum total of energy gain for society which is able to do useful work and create human utility (beyond the sun warming us and the wind drying our laundry, and other fixed natural flows not considered in the global 500 quadrillion BTUs of annual energy

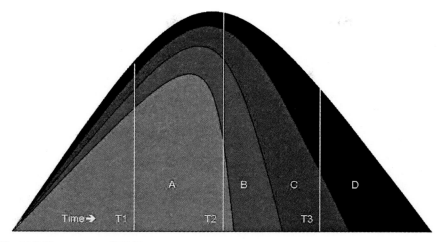

Fig. 12.2 Net energy and EROI as a resource matures over time

use). The surplus energy of a system, or society, is what allows it to continue growth, maintenance, repair and leisure. If our energy sources required equal amounts of energy input in order to obtain an energy output, we would have no surplus energy left for other work (Gilliland 1975). If we had a very small energy surplus, we would only be able to consume at a low level.

EROI has an eventual trade-off with scale – at low scale, EROI can be very high, as the best first principles apply. At higher and higher scale, EROI eventually declines as more resources (energy and other) are needed to harvest the more difficult parts of the original resource. Indeed, analysis of the EROI of US oil and gas exploration shows that we had over 100:1 in the 1930s, when the large oil fields were discovered and put into production. By 1970 the Energy Return on Investment had declined to 30:1 and down to a range of 10–17:1 by 2000. (Cleveland 2001, Hall 2003). Anecdotally, from 2005 to 2006, the finding and production costs of the marginal barrel of oil in the US went from \$15 to \$35 per barrel. (Herold 2007), and offshore in the Gulf of Mexico increased from \$50 to over \$69 per barrel (EIA 2007). Though these are financial increases as opposed to energy, it suggests the high return oil has been found, and increasing amount of dollars (and energy) will be needed to extract the remainder.

12.10 A Framework for Analyzing EROI

Imagine the physical flows of an energy producing technology (T) e.g. a corn ethanol plant. Energy (ED_{in}) and other various inputs ($\{I_k\}$) are taken into the plant and combined or consumed to produce energy output (ED_{out}) as well as possibly other co-products ($\{O_j\}$) i.e. $T(ED_{in}, \{I_k\}) = \{ ED_{out}, O_j \}$. In its narrowest (and least informative) form, EROI (minus 1) is similar to the economic concept of financial

Return on Investment but uses energy as the currency while treating non-energy inputs as negligible. This simple definition yields EROI = ED_{out}/ED_{in}. EROI is rarely used in this simple form (examples being Southwide Energy Committee 1980, Gingerich and Hendrickson 1993), but EROI statistics are frequently published regarding different technologies that ignore the energy costs associated with infrastructure and non-energy inputs (American Wind Energy Association 2006).

12.11 Non-Energy Inputs

EROI rarely conforms to the above simplistic formulation. Depending on the definition of T, the energy inputs, ED_{in} generally do not account for additional and significant energy requirements important to the production process. This energy is embodied in the non-energy direct inputs (Odum 1983), for example the agricultural energy required to grow oilseeds for biodiesel (Hill et al. 2006). Precise calculation of the energy embodied in non-energy inputs is nearly impossible – (e.g. do we include the calories consumed by the farmer for breakfast before he goes to harvest corn? How much energy is the oil field managers expertise worth? etc.). This may be resolved either through an input-output matrix framework or by semi-arbitrarily drawing a boundary beyond which additional, (and presumably negligible), energy inputs are ignored (Spreng 1988). The latter is the accepted approach for Life Cycle Analyses (LCAs – International Standard Organization 1997). A typical EROI formulation applies an appropriate methodology to evaluate the embodied energy costs for the non-energy inputs, which are termed the indirect energy inputs. For a given production process, this should yield a specific set of coefficients, $\{\gamma_k\}$, that give the per-unit indirect energy costs of $\{I_k\}$ (e.g. MJ per tonne soybean). This gives the following version of EROI:

$$EROI = ED_{out}/(ED_{in} + \Sigma \gamma_k I_k). \tag{12.1}$$

Some analyses arbitrarily include the indirect energy costs for certain inputs while excluding the energy cost of others, something that clearly creates difficulty of comparison between studies (Pimentel and Patzek 2005, Farrell et al. 2006). The embodied energy costs of labor in particular are difficult to define but can be a significant component of the energy cost. (Costanza 1980, Hill et al. 2006).

Though energy return analysis obviously treats energy as a critical limiting variable, there are potentially numerous other limiting inputs to a production process. In addition to the direct and indirect energy requirements of an energy technology, important inputs such as land, time, and water, are difficult (some would argue impossible) to accurately reduce into energy equivalent measures. In this chapter we refer to these as *non-energy requirements* so as to distinguish them from *non-energy inputs* (which can be parsed into energy terms). Non-energy requirements can have embodied components as well (Wichelns 2001). For example, the biodiesel conversion process requires labor and water. Similarly, the oilseeds used to produce biodiesel require inputs such as land, labor, and water in addition to direct and indirect energy requirements (Pimentel et al. 1994, Pimentel 2003). The standard assumption underlying past EROI analyses is that all non-energy requirements are held constant and negligible. In a globally connected world of potentially numerous limiting inputs, energy systems analysis will benefit from a relaxing of this assumption.

The direct and indirect non-energy requirements can be handled two different ways. The first method is to identify key, potentially limiting resources and treat them completely separate from energy inputs. This would create a new indicator of efficiency for each resource tracked e.g. EROLI(Land) measured in MJ/ha, or EROWI(Water) measured in MJ/gallon. In particular, for non-energy requirement X, EROXI is given by:

$$EROXI = ED_{out}/(\Sigma \pi_{X,k} I_k) \qquad (12.2)$$

where $\pi_{X,k}$ gives the direct and indirect per-unit requirements of X into I_k.

While this method increases the complexity, it also has advantages. First, it provides a metric of energy harvesting efficiency that could be included in a broader energy systems analysis. In combination with other technologies that require different array of resource inputs, this type of metric can be informative on the scaling capacity of a renewable energy portfolio. Second, this type of multicriteria approach allows for contextual assessment of a technology. Different geographic and political will be limited in their growth by different resources (Rees 1996), a Liebig's law of the minimum for economic growth (Hardin 1999). Some resources like water may be equally if not more limiting than energy (Barlow 2002). An ideal energy technology would optimize on scarce resource X (high EROXI) thus deemphasizing the return necessary on abundant resource Y (lower EROYI).

Another way to deal with non-energy primary inputs is to convert them into energy equivalents via some set of coefficients ($\{\psi_X\}$) for all non-energy requirements X. A justification for this is that in order for any energy procurement process to be truly sustainable, it must be able to regenerate all resources consumed (Patzek 2004). An approach adopted by Patzek (2004) and Patzek and Pimentel (2005) is to assign energy costs based on a resource's exergy (Ayres and Martinas 1995, Ayres et al. 1998), approximately defined as the ability of a system to perform work and equated with its distance from thermal equilibrium. This can also be viewed as the amount of energy necessary to reconstitute a given level of thermodynamic order.

The above set of coefficients yields the following measure for EROI:

$$EROI = \frac{ED_{out}}{\left(ED_{in} + \sum_k \gamma_k I_k + \sum_X \sum_k \psi_X \pi_{X,k} I_k \right)}. \qquad (12.3)$$

Assuming consensus around the validity of the energy equivalents, this measure of EROI provides for complete commensurability by reducing all inputs to a single currency.

12.12 Non-Energy Outputs

Just as consideration of non-energy inputs yields a fuller, and more complex EROI statistic, so too can non-energy outputs be incorporated to provide a more complete indicator of the desirability of a process. Firstly, many technologies yield co-products in addition to a primary energy product. Most studies assume that a credit should be given for these co-products which increases the EROI by reducing the numerator for the process. Mathematically, each co-product O_j is assigned a per-unit energy equivalency coefficient (v_j) indicative of its value relative to the energy product.

The most straightforward method is to assign co-products an explicit energy value based on their thermal energy content (Pimental and Patzek 2005) or their exergy (Patzek and Pimentel 2005). However, co-products are seldom used for their energy content (bagasse in sugar cane ethanol being an exception). If energy is the limiting variable to be optimized, a full energy credit for dry distiller grains or milk, may be aggressive, and the EROI of a technology giving full allocation to co-products will decline as the co-products scale beyond their practical use (e.g millions of tons of DDGs). Energy values can also be assigned according to the energy that would be required to produce the most energy-efficient replacement (Hill et al. 2006). Economic value and mass are two non-energy metrics that are used to establish relative value, both of which are frequently used in life cycle analyses (International Standard Organization 1997, deBoer 2003).

Once the energy equivalency coefficients have been established, the EROI formulation is modified to the following:

$$EROI = \frac{ED_{out} + \sum v_j O_j}{ED_{in} + \sum \gamma_k I_k}. \tag{12.4}$$

For example, when procuring biodiesel from soybeans, the soybean meal is a valuable co-product often used as a source of protein for livestock. An energy credit can be assigned to this co-product based on its actual thermal content (Pimentel and Patzek 2005), its market value (e.g. Mortimer et al. 2003), or its mass (e.g. Sheehan et al. 1998). The fact that calculated EROI can vary by a factor of 2 or more depending on allocation method gives insight that EROI, though much more so than dollars, is not a purely physical concept.

12.13 Non-Market Impacts

We have considered inputs and outputs that are currently recognized by the market system. However, many energy production processes create outputs that have social, ecological, and economic consequences external to the market. As we are all part of a planetary ecosystem, to properly include energy externalities should provide us with more accurate information of the desirability of an energy procuring technology (Hill et al. 2006). Negative externalities can include loss of topsoil erosion, water pollution, loss of animal habitat, and loss of food production capacity (Hanegraaf et al. 1998, Pimentel et al. 2002). Externalities can also be positive such as the creation of jobs and the maintenance of rural communities (Bender 1999).

As with non-energy requirements, these externalities can be incorporated into our framework in one of two ways—as separate indicators in a multicriteria framework or through conversion into energy equivalents. Thus, if topsoil is lost or nitrous oxide is emitted as part of the life cycle of the technology, we can measure EROI (*Topsoil*) or EROI(*Nox*). Studies that include such externalities have been published by the US Department of Energy (1989a, 1989b), Giampietro et al. (1997). Such measures are useful for assessing the scalability of a process within a given context by indicating what resources (e.g. waste sinks) might become limiting under increased production.

Negative externalities also can be assigned energy equivalency coefficients equal to the energy required to prevent or remediate their impacts (Cleveland and Costanza 1984, Pimental and Patzek 2005, Farrell et al. 2006). If we assume a set of externalities $\{E_i\}$ with energy equivalency coefficients $\{v_i\}$, then we must add into the denominator of the EROI calculation the term $\sum v_i E_i$. Not many studies have attempted this approach, however and pursuing this strategy has the drawback of parsing important non-reducible criteria into one metric.

12.14 A Summary of Methodologies

Table 12.1 lists all of the different formulations of EROI (or net energy analysis) presented above based on the formulation of the denominator. For each, we've cited one or more studies that have employed that specific variation. While all the works surveyed fall within the same methodological framework, as outlined above,

Table 12.1 Exisiting EROI Formulations in the Literature

Cost category	Direct	+ Indirect	+ Allocation
	$Cost = ED_{in}$	$Cost = (ED_{in} + \sum \gamma_k I_k)$	Numerator = $ED_{out} + \sum v_j O_j$
Energy	Wood Biomass[a] Wood to Electric[b] $Cost = X$	Soy/Sunflower Biodiesel[c] Solar Cells[d] $Cost = \sum \pi_{X,k} I_k$	Corn Ethanol[e] Soy Biodiesel[f] Numerator = $ED_{out} + \sum v_j O_j$
Primary Input(X)	Hydroelectric, $X = Land^b$ Various Technologies, $X = Water^g$	Corn Ethanol, $X = $ Various Inputs[c,h] Rapeseed Biodiesel, $X = $ Various Inputs[g]	Soy Biodiesel, $X = $ Various Inputs[f] Rapeseed Biodiesel, $X = Water^i$

Table 12.1 (continued)

Cost category	Direct	+ Indirect	+ Allocation
	Cost = E	Cost = $\sum \pi_{E,k} I_k$	Numerator = $ED_{out} + \sum v_j O_j$
Externality (E)	Wind, E = Emissions[j] Various Technologies, E = Soil Loss[g]	Various Technologies, E = Emissions[k] Wind, E = Emissions[l]	Biodiesel, E = Emissions[f] Ethanol, E = GHG[m]
Energy Equivalents	(1) Conversion of externalities into energy: Cost = $ED_{in} + \sum \gamma_k I_k + \sum v_i E_i^{c,h}$ (2) Conversion of primary inputs into energy: Cost = $ED_{in} + \sum \gamma_k I_k + \sum \psi_X \pi_{X,k} I_k^{c,h}$		

Citations:
[a] (Gingerich and Hendrickson 1993)
[b] (Pimentel et al. 1994)
[c] (Pimentel and Patzek 2005)
[d] (Pearce and Lau 2002)
[e] (Farrell et al. 2006)
[f] (Sheehan et al. 1998)
[g] (Hanegraaf et al. 1998)
[h] (Patzek 2004)
[i] (DeNocker et al. 1998)
[j] (American Wind Energy Association 2006)
[k] (European Commission 1997)
[l] (Schleisner 2000)
[m] (Mortimer et al. 2003)
(Table and accompanying text adapted from Mulder et al. 2008)

assumptions and terminology vary significantly among studies resulting in conflicting results that make them difficult to compare.

12.15 A Unifying EROI Framework

If net energy analysis is to produce results that are clear, and comparable across studies, and be of practical use to researchers and policy-makers, it will be necessary for the methodology to become uniform and well-specified. Such standards exist in the area of life cycle analyses (International Standard Organization 1997). However, unlike LCA, it is probably not possible or even desirable that EROI be restricted to a single meaning or methodology. The different levels of energy and environmental analysis outlined above are relevant to different problems, contexts, and research objectives. *The problem heretofore has arisen when the same term is used for methodologies with different assumptions and different goals.*

We propose a two-dimensional framework for EROI analyses (with accompanying terminology) that clarifies the major assumptions in an analysis. In the first dimension, we identify three distinct levels of analysis that can be distilled from the above examples. These levels differ in terms of *what* they include in their analysis.

The first level deals with only the direct inputs (energy and non-energy) and direct energy outputs. We term this *Narrow Boundary EROI* as, while it can offer more precise EROI calculations, it is also the most superficial, restricting the analysis to simple inputs and thus missing many critical energy costs (as well as ignoring co-products). The next level, *Intermediate Boundary EROI*, involves incorporating indirect energy and non-energy inputs as well as crediting for co-products. This is the methodology used by Life Cycle Analysis to estimate the EROI of an energy technology. Intermediate Boundary EROI requires two assumptions that must be made clear: (1) What allocation method is used for the co-products (thermal content, price, mass, exergy etc.); and (2) What boundaries are used for determining indirect inputs. Finally, *Wide Boundary EROI* incorporates additional costs (and possibly benefits) for the externalities of the energy technology. Admittedly, this is the most imprecise but also the most relevant of the EROI measures in that it presents the fullest measure of the net energy available to society.

	Basic EROI	**Total EROI**	**Multicriteria EROI**
Narrow Boundary	$\dfrac{ED_{out}}{ED_{in}}$	$\dfrac{ED_{out}}{ED_{in} + \sum_k \psi_k I_k}$	$\dfrac{ED_{out}}{I_k}$
Intermediate Boundary	$\dfrac{ED_{out}}{\alpha\left(ED_{in} + \sum_k \gamma_k I_k\right)}$	$\dfrac{ED_{out}}{\alpha\left(\begin{array}{l}ED_{in} + \sum_k \gamma_k I_k \\ + \sum_k \psi_k \pi_{X,k} I_k\end{array}\right)}$	$\dfrac{ED_{out}}{\alpha \sum_k \pi_{X,k} I_k}$
Wide Boundary		$\dfrac{ED_{out}}{\alpha\left(\begin{array}{l}ED_{in} + \sum_k \gamma_k I_k \\ + \sum_k \psi_k \pi_{X,k} I_k \\ + \sum_i \nu_i E_i\end{array}\right)}$	$\dfrac{ED_{out}}{\alpha \sum_k \pi_{E,k} I_k}$

Fig. 12.3 Methodological framework for net energy analysis. The side axis determines *what* to include (direct inputs, indirect inputs, and/or externalities). The top axis dictates *how* to include non-energy requirements (ignore, convert to energy equivalents, or treat as separate inputs.) Note that since basic EROI ignores non-energy inputs, it does not have a wide boundary form that accounts for externalities. (Table and accompanying text adapted from Mulder et al. 2008)

Once it has been determined what can (and should) be included in the analysis, the second dimension in our framework dictates *how* to include these inputs. We delineate three choices for handling of the non-energy requirements and externalities. They can be ignored, yielding *Basic EROI*, or converted to energy equivalents, yielding '*Total EROI*', or handled as separate components yielding '*Multi-criteria EROI*'.

Our framework is presented in Fig. 12.3. Note that while the grid is 3×3, it yields only 8 meaningful formulations. The different levels of analyses are nested hierarchically. The computation of a wider boundary EROI for an energy production process should easily yield all other forms of EROI found below it. That is to say, the necessary data will have been compiled and it is merely a decision of which components to include in the calculation. Similarly, a Total EROI calculation will use the same data set as a Multi-criteria EROI with the addition of energy equivalency coefficients. This means that more comprehensive studies should yield results at least partially comparable with less comprehensive studies as seen in a meta-study of ethanol by Farrell et al. (2006).

12.16 Liebig's Law, Multi-Criteria Analysis, and Energy from Biofuels

Though it is becoming apparent that energy will be a limiting variable for society going forward, it is easy to envision other equally limiting variables as the planetary population increases its demand on ecosystems. Water, land, and carbon sinks are only three examples of inputs and impacts of renewable energy production that could limit the potential of a technology (Giampietro et al. 1997, Hagens et al. 2006, Hill et al. 2006). These should be included explicitly in a net energy analysis or else their cost in terms of energy should be estimated.

Liebig's Law of the minimum states that the production of a good or resource is limited by its least available input. In layman's terms something is only as good as its weakest link. This form of ecological stoichiometry will loom large in the procurement of energy alternatives to fossil fuels. Water, land, soil, greenhouse gas emissions, and specific fossil inputs themselves will potentially limit scaling of alternative energy.

Though EROI is generally measured as the ratio of the gross energy return to the amount of energy invested, it has been argued this can give a false indicator of the desirability of a process due to the increasing cost of non-energy requirements as EROI approaches 1. Following Giampietro et al. (1997), let $\omega = EROI/(EROI - 1)$ be the ratio of gross to net energy produced. ω equals the amount of energy production required to yield 1 MJ of net energy. From an energy perspective, all costs have been covered. However, for non-energy requirements the perspective and the implications, change.

Let $EROXI$ be the energy return for 1 unit of non-energy requirement X. Then $1/EROXI$ is the number of units of X required for 1 MJ gross energy production. From the above, it is easily seen that $\omega/EROXI$ units of X are required, or more generally, the net energy yielded per unit of X is equal to $EROXI/\omega$. Since ω increases non-linearly (approaching infinity) as EROI approaches 1, a relatively small change in EROI can produce a large decrease in the 'net EROI' for non-energy require-ments. For energy production processes with significant non-energy requirements such as biofuels, this suggests a low EROI can imply strong limitations on their ability to be scaled up (Giampietro et al. 1997, Hill et al. 2006).

If we assume the Intermediate Boundary EROI for non-cellulosic ethanol from corn is in the neighborhood of 1.34 (Farrell et al. 2006), this implies net energy of .34 for every 1 unit of energy input. The corn-based ethanol Energy Return on Land Invested (EROLI) = 11,633 MJ/ha gross energy production (equivalent to 3475 l per hectare). However, the net energy per unit of land is only 2,908 MJ/ha. At 2004 levels of gasoline consumption for the United States, this is equivalent to consuming the net energy production of 42 ha of cropland per second. If the EROI of ethanol is reduced to 1.2, a decrease of only 10%, the net return on land decreases by 33% while the amount of land required to achieve this same net yield increases by 50%. Conversely, an oil well requires equipment access, roads, etc. but pulls its bounty out of a comparatively small land area. This contrast has significant implications for the potential scale of biofuel production (Giampietro et al. 1997). In effect, due to significant power density differentials, replacing energy-dense liquid fuels from crude oil with less power dense biomass fuels will utilize 1,000- to 10,000-fold increases in land area relative to our existing energy infrastructure (Cleveland 2007).

Though land is one limiting factor, water may be another. In a forthcoming paper, we use Multicriteria EROI analysis to define and quantify the EROWI (Energy Re-turn on Water Invested) for various energy production technologies. Since water and energy may both be limiting, we care about the 'Net EROWI', which is a combined measure of EROI and EROWI for each technology. With the exception of wind and solar which use water only in indirect inputs, the 'Net EROWI' of biofuels are one to two orders of magnitude lower than conventional fossil fuels. We also determined that approximately 2/3 of the world population (by country) will have limitations on bioenergy production by 2025, due to other demands for water (Mulder et al. *In press*).

Nitrogen, a byproduct of natural gas via ammonia, is essential to a plant's ability to develop proteins and enzymes in order to mature. The importance of nitrogen fertilizers to U.S. agriculture, particularly corn and wheat, is evidenced by its ac-celerated use over the last 50 years. From 1960 to 2005, annual use of chemical nitrogen fertilizers in U.S. agriculture increased from 2.7 million nutrient tons to 12.3 million nutrient tons (Huang 2007). This increase is considered to be one of the main factors behind increased U.S. crop yields and the high quality of U.S. agricultural products (Huang 2007). Furthermore, biofuels, especially the ethanols, require large amounts of natural gas for pesticides, seedstock and primary electricity to concentrate the ethanol. In areas that have natural gas fired electricity plants (as opposed to coal), fully 84% of the energy inputs into corn ethanol are from natural gas (the nitrogen, a portion of the pesticides, and the electricity). (Shapouri 2002). Ethanol proponents, other than optimizing 'dollars' (making money), are presuming that 'domestically produced vehicle fuel' is the sole item in short supply. Were the math on corn ethanol somehow scalable to 30% of our national gasoline consump-tion, in addition to land and water, we would use *more than the entire yearly amount of natural gas currently used for home heating* as an input.

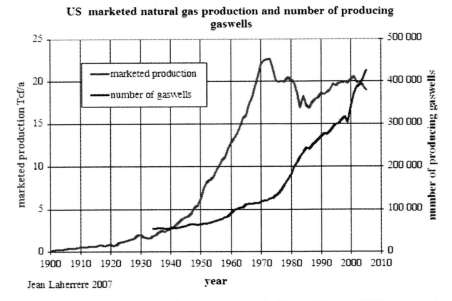

Fig. 12.4 Natural gas production vs. # of natural gas wells (Source Laherrere 2007)

Though many biofuel studies imply that fertilizer, and therefore natural gas, are more abundant and cheaper than petroleum, we are actually on a 'natural gas treadmill' in North America and low prices are being kept down only by 2 consecutive mild winters and summers with no hurricanes. In 1995 the average new gas well in North America took 10 years to deplete. A new gas well in 2007 takes under 10 months. More and more drilling of new gas wells is necessary just to stay at constant levels of production. As can be seen in Fig. 12.4, US production peaked in 1973 followed by another peak in 2001. The second peak required 370% more wells to produce the same amount of gas. Furthermore, the energy/$ effort on Canadian natural gas production implies a decline in EROI from 40:1 to 15:1 from 2000 to 2006, with an extrapolated energy break even year circa 2014. (CAPP 2007, methodology Hall and Lavine 1979). The falling EROI makes it impossible for natural gas production to maintain both low costs and current levels of production. When US oil peaked in 1970, we made up our oil demand shortfall by imports. Natural gas can also be imported (as LNG), but it must first be liquefied at a high dollar and energy cost. It requires over 30% of its BTU content to be transported overseas – another energy loss. In this sense, studies that show energy use on petroleum invested are perhaps overlooking natural gas as a limiting input.

So corn ethanol, and other biofuels requiring both natural gas for fertilizers and pesticides, as well as for electricity to steam the ethanol solution, are essentially turning 3 scarce resources: water, land, and natural gas, into liquid fuels, at an energy gain an order of magnitude lower than what societal infrastructure is currently adapted to. What will the strategy and metrics to measure it become when natural gas too, is recognized as limiting input?

12.17 Conclusion

At some point in the near future, those reading this chapter will witness a forced change from the fossil fuel mix that has powered society smoothly for decades. In a perfect world, all information about externalities and an accurate balance sheet of the size and quality of our resources would be available to decision-makers. In reality however, accurate information about the reliability of upcoming resource flows is opaque beyond a few months. Only 6% of the worlds (stated) oil reserves are owned by public companies subject to SEC requirements, leaving the NOCs and private companies each individually knowing only their own share of the oil pie. It is unlikely the market will respond in time once critical limiting variables to society become apparent. Unfortunately, this cannot be empirically proven until after the fact. To have a framework in hand that anticipates such problems is a first but important step.

New energy technologies require enormous capital investments and significant lead time as well as well-defined research and planning. Aggregating decisions surrounding alternative energy technologies and infrastructure will be both difficult and time sensitive. As a growing population attempts to replace this era of easy energy with alternatives, net energy analysis will reassert its importance in academic and policy discussions. Alongside ecological economics, it is one of the few methods we can use to attempt to measure our 'real' wealth and its costs. As such, it will be advantageous to adhere to a framework that is consistent among users and attempts to evaluate correctly the complex inputs and outputs in energy analysis in ways that are meaningful. Accounting for the subtle and intricate details in net energy analysis is difficult. However, in a growing world constrained by both energy and increasingly by environmental concerns, adherence to a common framework will be essential for policy-makers to accurately assess alternatives and speak a common language.

Perhaps the biggest misconception of net energy analysis, particularly in its most popular usage referring to corn ethanol, is the comparison on whether or not something is energy positive – this myopic focus on the absolute, ignores the much larger question of relative comparisons – what happens to society when we switch to a lower energy gain system? While net energy analysis outcomes will not guide our path towards sustainable energy with the precision of a surgical tool, they are quite effective as a blunt instrument, helping us to discard energy dead-ends that would be wasteful uses of our remaining high quality fossil sources and perhaps equally as important, our time. Ultimately when faced with resource depletion and a transition of stock-based to flow-based resources, EROI will function best as an allocation device, marrying our demand structure with our supply structure, thus guiding our high quality energy capital into the best long term energy investments. Finally, analysts and policymakers may use net energy analysis not only to compare the merits of proposed new energy technologies, but also as a roadmap for possible limitations on demand, if global energy systems analysis points to declines in net energy not adequately offset by conservation, technology or efficiency. A framework like the one presented above, may also be useful for analyses involving limiting inputs in addition to energy.

References

American Wind Energy Association. 2006. Comparative air emissions of wind and other fuels. Retrieved on Jan 27 2007 from http://www.awea.org/pubs/factsheets.html.

Ayres, R., L. Ayres, and K. Martinas. 1998. Exergy, waste accounting, and life-cycle analysis. Energy 23:355–363.

Ayres, R., and K. Martinas. 1995. Waste potential entropy: The ultimate ecotoxic? Economie Appliquees 48:95–120.

Baines, J., and M. Peet. 1983. Assessing alternative liquid fuels using net energy criteria. Energy 8:963–972.

Baltic, T., and D. Betters. 1983. Net energy analysis of a fuelwood energy system. Resources and Energy 5:45–64.

Bender, M. 1999. Economic feasibility review for community-scale farmer cooperatives for biodiesel. Bioresource Technology 70:81–87.

Berglund, M., and P. Borjesson. 2006. Assessment of energy performance in the life-cycle of biogas production. Biomass & Bioenergy 30:254–266.

Barlow, M., Clarke T. 2002. "Blue Gold: The Battle Against Corporate Theft of the World's Water" Stoddart, Toronto.

Berndt, E. 1983. From technocracy to net energy analysis: Engineers, economists, and recurring energy theories of value. Pages 337–366 in A. Scott, editor. Progress in Natural Resource Economics. Clarendon, Oxford.

British Petroleum. 2005. Quantifying Energy: BP Statistical Review of World Energy 2005.

Canadian Assoc of Petroleum Producers (CAPP). "Wells and Meters Drilled in Canada 1981–2006", 2007.

Chambers, R., R. Herendeen, J. Joyce, and P. Penner. 1979. Gasohol: Does it or doesn't it produce positive net energy. Science 206:789–795.

Chui, F., A. Elkamel, and M. Fowler. 2006. An integrated decision support framework for the assessment and analysis of hydrogen production pathways. Energy & Fuels 20:346–352.

Cleveland, C. 1992. Energy quality and energy surplus in the extraction of fossil fuels in the U.S. Ecological Economics 6:139–162.

Cleveland, C. 2001. "Net Energy from Oil and Gas Extraction in the United States, 1954-1997". Energy, 30: 769-782.

Cleveland, C. 2005. Net energy from the extraction of oil and gas in the United States. Energy 30:769–782.

Cleveland, C. 2007 "Net Energy Analysis", Retrieved 3/16/2007 from http://www.eoearth.org/article/Net_energy_analysis

Cleveland, C. 2007 "Energy Transitions Past and Future", The Encyclopedia of Earth, Retrieved 7/29/2007 from http://www.eoearth.org/article/Energy_transitionspastandfuture

Cleveland, C., and R. Costanza. 1984. Net Energy Analysis of Geopressured Gas-Resources in the United States Gulf Coast Region. Energy 9:35–51.

Cleveland, C., R. Costanza, C. Hall, and R. Kaufmann. 1984. Energy and the United States Economy – A Biophysical Perpsective. Science 225:890–897.

Costanza, R. 1980. Embodied energy and economic valuation. Science 210:1219–1224.

Crawford, R., and G. Treloar. 2004. Net energy analysis of solar and conventional domestic hot water systems in Melbourne, Australia. Solar Energy 76:159–163.

deBoer, I. 2003. Environmental impact assessment of conventional and organic milk production. Livestock Production Science 80:69–77.

DeNocker, L., C. Spirinckx, and R. Torfs. 1998. Comparison of LCA and external-cost analysis for biodiesel and diesel. in Proceedings of the 2nd International Conference LCA in Agriculture, Agro-industry and Forestry. VITO, Brussels.

Energy Information Agency. 2007. "Performance Profiles of Major Energy Producers 2006", December 2007

EPA. 2002. A Comprehensive Analysis of Biodiesel Impacts on Exhaust Emissions. EPA420-P-02-001, US Environmental Protection Agency.

European Commission. 1997. ExternE, Externalities of Energy, Methodology Report, Vol. 1. European Communities.

Farrell, A., R. Plevin, B. Turner, A. Jones, M. O'Hare, and D. Kammen. 2006. Ethanol can contribute to energy and environmental goals. Science 311:506–508.

Giampietro, M., S. Ulgiati, and D. Pimental. 1997. Feasibility of Large-Scale Biofuel Production. Bioscience 47:587–600.

Gilliland, M. 1975. Energy analysis and public policy. Science 189:1051–1056.

Gingerich, J., and O. Hendrickson. 1993. The theory of energy return on investment – A case-study of whole tree chipping for biomass in Prince Edward Island. Forestry Chronicle 69:300–306.

Hagens, N., R. Costanza, and K. Mulder. 2006. Energy Returns on Ethanol Production. Science 312:1746.

Hall, C. A. S., and M. Lavine. 1979. "Efficiency of Energy Delivery Systems:1. An Economic and Energy Analysis", Environmental Management, vol 3, no 6, pp 493–504.

Hall, C. A. S., C. J. Cleveland, and R. Kaufmann. 1986. Energy and resource quality: The ecology of the economic process. Wiley, New York.

Hall, C., Tharakan, P. Hallock, J. Cleveland, C. and Jefferson, M. 2003. "Hydrocarbons and the Evolution of Human Culture", Nature 426:318–322, 20 November 2003.

Hammerschlag, R. 2006. Ethanol's energy return on investment: A survey of the literature 1990 – Present. Environmental Science & Technology 40:1744–1750.

Hanegraaf, M. C., E. E. Biewinga, and G. Van der Bijl. 1998. Assessing the ecological and economic sustainability of energy crops. Biomass & Bioenergy 15:345–355.

Hardin, G. J. 1999. The ostrich factor: Our population Myopia. Oxford University Press, New York.

Herold and Co. Upstream Performance Summary 2007.

Hill, J., E. Nelson, D. Tilman, S. Polasky, and D. Tiffany. 2006. Environmental, economic, and eneretic costs and benefits of biodiesel and ethanol biofuels. Proceedings of the National Academy of Sciences 103:11206–11210.

Hu, Z., F. Fang, D. Ben, G. Pu, and C. Wang. 2004. Net energy, CO2 emission, and life-cycle cost assessment of cassava-based ethanol as an alternative automotive fuel in China. Applied Energy 78:247–256.

Huang, W. 2007. Impact of Rising Natural Gas Prices on U.S. Ammonia Supply, US Department of Agriculture, August 2007

International Standard Organization. 1997. Environmental Management – Life Cycle Assessment – Principles and Framework. ISO, Geneva.

Kallivroussis, L., A. Natsis, and G. Papadakis. 2002. The energy balance of sunflower production for biodiesel in Greece. Biosystems Engineering 81:347–354.

Kaylen, M. 2005. An economic analysis of using alternative fuels in a mass burn boiler. Bioresource Technology 96:1943–1949.

Keoleian, G. 1998. Application of Life Cycle Energy Analysis to Photovoltaic Design. Progress in Voltaics 5.

Kidd, S. 2004. nuclear: Is there any net energy addition? Nuclear Engineering International 49:12–13.

Kim, S., and B. Dale. 2005. Environmental aspects of ethanol derived from no-tilled corn grain: Nonrenewable energy consumption and greenhouse gas emissions. Biomass & Bioenergy 28:475–489.

Laherrere, J. 2007, "North American natural gas discovery & production", August 2007, ASPO France, pg 15.http://aspofrance.viabloga.com/files/JL_NAmNG07.pdf

Lotka, A., 1922. "Contributions to the Energetics of Evolution", Biology. 8:147-151.

Mortimer, N. D., M. A. Elsayed, and R. Matthews. 2003. Carbon and Energy Balances for a Range of Biofuel Options. Resources Research Unit, Sheffield Hallam University, Sheffield, England.

Mulder, K., and Hagens N. 2008."Energy Return on Investment – Towards a Consistent Framework", AMBIO Vol. 37, no 2, pp. 74–79 March 2008.

Odum, H. T. 1973. Energy, Ecology, and Economics. Ambio 2:220–227.

Odum, H. T. 1983. Systems ecology: An introduction. Wiley, New York.

Odum, H.T. 1994. "Ecological and General Systems: An Introduction to Systems Ecology", University Press of Colorado, Revised Edition of Systems Ecology, 1983, Wiley.

Odum, H. T. 1995. 'Self-Organization and Maximum Empower', in C.A.S.Hall (ed.) Maximum Power: The Ideas and Applications of H.T.Odum, Colorado University Press, Colorado.

Patzek, T. 2004. Thermodynamics of the corn-ethanol biofuel cycle. Critical Reviews in Plant Sciences 23:519–567.

Patzek, T., and D. Pimentel. 2005. Thermodynamics of energy production from biomass. Critical Reviews in Plant Sciences 24:327–364.

Pearce, J., and A. Lau. 2002. Net energy analysis for sustainable energy production from silicon based solar cells. in Proceedings of Solar 2002, Reno, Nevada.

Pimentel, D., and T. W. Patzek. 2005. Ethanol production using corn, switchgrass, and wood; biodiesel production using soybean and sunflower. Natural Resources Research 14:65–76.

Pimentel, D. 2003. Ethanol Fuels: Energy Balance, Economics, and Environmental Impacts are Negative. Natural Resources Research 12:127–134.

Pimentel, D., M. Herz, M. Glickstein, M. Zimmerman, R. Allen, K. Becker, J. Evans, B. Hussain, R. Sarsfeld, A. Grosfeld, and T. Seidel. 2002. Renewable energy: Current and potential issues. Bioscience 52:1111–1120.

Pimentel, D., and T. W. Patzek. 2005. Ethanol production using corn, switchgrass, and wood; biodiesel production using soybean and sunflower. Natural Resources Research 14:65–76.

Pimentel, D., G. Rodrigues, T. Wang, R. Abrams, K. Goldberg, H. Staeker, E. Ma, L. Brueckner, L. Trovato, C. Chow, U. Govindarajulu, and S. Boerke. 1994. Renewable energy – economic and environmental issues. Bioscience 44:536–547.

Potter, L., and D. Betters. 1988. A net energy simulation model – Applications for domestic wood energy systems. Forest Products Journal 38:23–25.

Rees, W. 1996. Revisiting carrying capacity: Area-based indicators of sustainability. Population and Environment 17:195–215.

Reijnders, L. 2006. Conditions for the sustainability of biomass based fuel use. Energy Policy 34:863–876.

Ricardo, D. 1819. On the principles of political economy, and taxation, 1st American edition. J. Milligan, Georgetown, D.C.

Schleisner, L. 2000. Life cycle assessment of a wind farm and related externalities. Renewable Energy 20:279–288.

Shapouri H., J. Duffield and M. Wang, "The Energy Balance of Corn Ethanol – An Update", USDA 2002.

Sheehan, J., V. Camobreco, J. Duffield, M. Graboski, and H. Shapouri. 1998. An Overview of Biodiesel and Petroleum Diesel Life Cycles. NREL/TP-580-24772, National Renewable Energy Laboratory, Golden, CO.

Skrebowski, Chris, Oil Depletion Analysis Center (ODAC Newsletter May 2006).

Smil, V. 1991. General Energetics: Energy in the Biosphere and Civilization. John Wiley, New York.

Smil, V. 2006. "21st Century Energy - Some Sobering Thoughts." OECD Observer 258/59: 22–23, December 2006

Southwide Energy Committee. 1980. Petroleum product consumption and efficiency in systems used for energy wood harvesting. American Pulpwood Association, No. 80-A-10.

Spreng, D. T. 1988. Net-energy analysis and the energy requirements of energy systems. Praeger, New York.

Tainter, J. A. 2003. Resource transitions and energy gain: Contexts of organization. Conservation and Ecology 7.

Tyner, G., R. Costanza, and R. Fowler. 1988. The net energy yield of nuclear power. Energy 13:73–81.

Ulgiati, S. 2001. A Comprehensive Energy and Economic Assessment of Biofuels: When Green is not Enough. Critical Reviews in Plant Sciences 20:71–106.

US Department of Energy. 1989a. Energy Systems Emissions and Material Requirements. Prepared by the Meridian Corporation, Washington, DC.

US Department of Energy. 1989b. Environmental Emissions from Energy Technology Systems: The Total Fuel Cycle. R L San Martin, Deputy Assistant Secretary for Renewable Energy, Washington D.C.

Wichelns, D. 2001. The role of 'virtual water' in efforts to achieve food security and other national goals, with an example from Egypt. Agricultural Water Management 49:131–151.

Wilkinson, R. G. 1973. "Poverty and progress: an ecological model of economic development". Methuen, London, UK.

Chapter 13
Bio-Ethanol Production in Brazil

Robert M. Boddey, Luis Henrique de B. Soares, Bruno J.R. Alves and Segundo Urquiaga

Abstract In this chapter the history and origin of the Brazilian program for bioethanol production (ProÁlcool) from sugarcane (*Saccharum* sp.) are described. Sugarcane today covers approximately 7 Mha, with 357 operating cane mills/ distilleries. The mean cane yield is 76.6 Mg ha^{-1} and almost half of the national production is dedicated to ethanol production, the remainder to sugar and other comestibles. The mean ethanol yield is 6280 L ha^{-1}. An evaluation of the environmental impact of this program is reported, with especial emphasis on a detailed and transparent assessment of the energy balance and greenhouse gas (CO_2, N_2O, CH_4) emissions. It was estimated that the energy balance (the ratio of total energy in the biofuel to fossil energy invested in its manufacture) was approximately 9.0, and the use of ethanol to fuel the average Brazilian car powered by a FlexFuel motor would incur an economy of 73% in greenhouse gas emissions per km travelled compared to the Brazilian gasohol. Other aspects of the environmental impact are not so positive. Air pollution due to pre-harvest burning of cane can have serious effects on children and elderly people when conditions are especially dry. However, cane burning is gradually being phased out with the introduction of mechanised green-cane harvesting. Water pollution was a serious problem early in the program but the return of distillery waste (vinasse) and other effluents to the field have now virtually eliminated this problem. Soil erosion can be severe on sloping land on susceptible

✉ R.M. Boddey
Embrapa-Agrobiologia, BR-465, Km 07, Caixa Postal 75.505, Seropédica, 23890-000, Rio de Janeiro, Brazil,
e-mail: bob@cnpab.embrapa.br

L.H. de B. Soares
Embrapa-Agrobiologia, BR-465, Km 07, Caixa Postal 75.505, Seropédica, 23890-000, Rio de Janeiro, Brazil

B.J.R. Alves
Embrapa-Agrobiologia, BR-465, Km 07, Caixa Postal 75.505, Seropédica, 23890-000, Rio de Janeiro, Brazil

S. Urquiaga
Embrapa-Agrobiologia, BR-465, Km 07, Caixa Postal 75.505, Seropédica, 23890-000, Rio de Janeiro, Brazil

D. Pimentel (ed.), *Biofuels, Solar and Wind as Renewable Energy Systems*,
© Springer Science+Business Media B.V. 2008

soils but with the introduction of no-till techniques and green-cane harvesting the situation is slowly improving. The distribution of the sugar cane industry shows that reserves of biodiversity such as Amazônia are not threatened by the expansion of the program and while there may be no great advantages of the program for rural poor, the idea that it will create food shortages is belied by the huge area of Brazil compared to the area of cane planted. Working conditions for the cane cutters are severe, almost inhuman, but there is no shortage of men (and women) to perform this task as wages and employment benefits are considerably more favourable than for the majority of rural workers. The future will bring expansion of the industry with increased efficiency, more mechanisation of the harvest, lower environmental impact along with a reduction in the number of unskilled workers employed and an increase in wages for the more skilled. This biofuel program will not only be of considerable economic and environmental benefit to Brazil, but also will play a small but significant global role in the mitigation of greenhouse gas emissions from motor vehicles to the atmosphere of this planet.

Keywords Bio-ethanol · Brazil · energy balance · environmental impact · flex-fuel vehicles · greenhouse gas emissions · labour conditions · sugarcane

13.1 Historical Introduction

The present large Brazilian program for bioethanol production is historically derived from the introduction of the sugarcane plant (*Saccharum officinarum*) from the island of Madeira by the Portuguese colonising expedition of 1532 (Machado et al., 1987). At that time Brazil was a Portuguese colony in South America, and its first economic cycle was based only upon natural resources such as brazilwood (*Caesalpinia echinata*), gold and precious stones.

Soon after the exploration of the interior of the country, sugar-cane became the first large-scale plantation crop, and depended on the labour of slaves in the newly-opened wilderness. Until the end of 19th Century, cultures such as rubber (*Hevea brasiliensis*) and coffee (*Coffea arabica*) occasionally eclipsed its economic importance.

In the colonial period, there was a productive rural structure of traditionally mid-to-large-size estates that contributed to populate the interior of the country. The edaphoclimatic conditions in São Paulo and Rio de Janeiro States in the southeast, and Pernambuco State in the northeast, favoured the spread of this crop in these regions. After the abolition of slavery in 1883, the supply of cheap labour to cut cane was initially maintained by the arrival of European immigrants. Consequently, the processing units for sugar production, and later the attached bioethanol distilleries, were closely related to a traditional oligarchy with a resolute and lasting political influence on the country's affairs.

The first trials on the use of ethanol blends in petrol engines took place in the early years of Getulio Vargas dictatorship, soon after the foundation in 1933 of The Sugar and Alcohol Institute (Instituto do Açúcar e do Álcool, IAA). Extensive use of anhydrous bioethanol was attempted during the course of Word War II in order to save oil imports. Later on in 1953, during the democratically-elected second Vargas presidency, the major national oil company, Petrobras, was founded to promote fuel production and industrial development.

When the Oil Crisis of 1973 hit the international fuel supplies, Brazil was importing 72% of its crude oil, and was almost completely dependent on petroleum derivatives for the transport sector. Oil import expenses rose from US$ 600 million that year up to US$ 2.6 billion in 1974. In this period the annual balance of payments changed from a small surplus to a deficit of US$ 4.7 billion. It was against this background that in 1975 the military dictatorship created the National Alcohol Programme (PROÁLCOOL), with the aim of moving towards the introduction of engines fuelled solely by hydrated ethanol. The first automobiles running on ethanol and other bio-fuels were developed at Centre for Aerospace Technology (Centro Técnico Aeroespacial, CTA), a Research Centre of the Brazilian Air Force, located at São José dos Campos, São Paulo State. The motor vehicle industry principally led by the multinational companies Volkswagen, Ford, Fiat and General Motors started large-scale production and new parts and materials were soon developed to resist corrosion and solve the problem of starting the engines from cold. Ethanol production was 500,000 litres per year in 1975 at the beginning of PROÁLCOOL (and reached 3.4 billion litres only five years later – TCU, 1990).

A complete and distinct program of tax and investments was brought out to support PROÁLCOOL, for the industrial sector of new distilleries and enlargements, for sugar-cane farming and for final ethanol consumption. Up to 1990 the investment amounted to more than US$ 7 billion, with almost US$ 4 billion of public resources.

After 1990 no more direct subsidies were supplied by the government but as gasoline was taxed at a much higher rate, cars and other light vehicles were cheaper to run on ethanol and sales from 1983 until 1989 of light vehicles running this fuel outstripped gasoline vehicles. The main problem with the program was that in the late 1980s and through the 1990s crude oil prices declined to below US$ 20 a barrel. Petrobras became very antagonistic to the ethanol program as gasoline was being substituted by ethanol. As a consequence, in order to provide the home market with sufficient diesel and naphtha the company was left with excess gasoline that had to be sold at low prices on the international market. Added to this there were several crises, caused by high international sugar prices and low rainfall that lowered ethanol production, and in some years (1989 and 1990) there were huge queues for ethanol at the gas stations and car buyers lost faith in relying on this biofuel.

It can been seen from the production figures (Fig. 13.1) that in 1988 (when 95% of cars being manufactured were equipped with alcohol engines), hydrated ethanol reached 9.5 billion litres but then varied between 8.7 and 10.7 billion litres until 1999 (9.25 billion litres). By this time very few ethanol-powered cars were being produced and much of the ageing fleet had left the roads, such that in 2000 production fell to less than 7 billion litres, reached a low of just under 5 billion litres in 2001 and only exceeded 7 billion litres again after 2005.

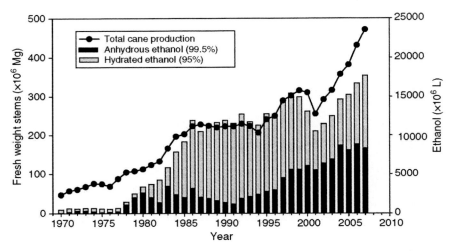

Fig. 13.1 Total sugar cane and anhydrous and hydrated ethanol production in Brazil, 1970–2007.
Data from MAPA (2007) and IBGE (2007)

However, the government could not let the program die, as apart from the pressure from the powerful cane planters lobby, more than 700,000 desperately-needed jobs had been created in the rural sector (TCU, 1990). For this reason in 2001, a law was passed making obligatory to add between 20 and 24% of anhydrous ethanol to all gasoline (Federal Law No. 10,203 of 22nd February). Historically, all over the world tetraethyl lead was added to gasoline to avoid spontaneous combustion before (spark) ignition. This enhancement of octane rating could also be achieved with the addition of ethanol. In fact this was well known, and published in Scientific American a few years before Thomas Midgely in the USA synthesised tetra-ethyl lead in 1922 (see Kovarik, 2005). The change from leaded gasoline to gasohol therefore was perceived to have a beneficial effect on air quality, especially in urban areas, and of course was extremely popular with the sugarcane industry.

However, the great leap forward for Brazilian bioethanol has just begun with the invention and production of ethanol/gasoline FlexFuel Otto cycle engines. Flex-fuel engines were first released in March 2003, a joint project of Volkswagen and Bosch. The compression ration of the engines is between 10:1 and 12.5:1 intermediate between that for gasoline (9–10:1) and ethanol (13–14:1). A carburettor control unit receives two basic signals. A conductivity detector informs the composition of the fuel in the tank and an oxygen probe analyses the concentration of this element in the exhaust vapour. The control unit electronically regulates the air-fuel mixture in order to reach the right stoichiometric rate for optimal burning of any ethanol/gasoline combination. This innovation has coincided with the increase of international crude oil prices, which since 2000 have risen above US$ 30 to between US$ 50 and US$ 100 today.

Until end of July 2006, 2 million FlexFuel powered vehicles were sold and from August 2006 to May 2007 another 1.3 million, totalling 3.3 million (ANFAVEA, 2007). From January to May 2007, 67% of all Otto cycle vehicles sold were Flexfuel, the remainder running on gasohol (20–24% anhydrous ethanol). In June 2007 this proportion reached 89.7%.

13.2 The Sugarcane Crop in Brazil

13.2.1 The Situation Today

With the great international interest in bio-ethanol, the area planted to sugar cane is rapidly expanding. For the 2007 season it is estimated that 7.8 Mha of sugarcane will be planted, an increase of 9.9% over 2006. More than half of the area (55%) planted to cane in Brazil is in the state of São Paulo, and this area increased by 10% over the last year (Table 13.1). While 1.2 Mha was planted in north eastern states, this area has not increased appreciably, and the largest proportional increases have been in the Cerrado (central western savanna) region with an increase of 35% in Mato Grosso do Sul, 20% in Minas Gerais and in the southern state of Paraná (26.5%). São Paulo, and these three states where the area is expanding most rapidly, account

Table 13.1 Area planted to sugarcane in all states and regions of Brazil, the proportional increase in planted area from 2006 to 2007 and mean cane yields[a]

Region	State	Area planted, 2007 (ha × 10³)	% increase in area from 2006	Yield[b] (Mg ha⁻¹)	% area of all sugarcane
North[c]		**19.7**	**−7.4**	**63.0**	**0.25**
	Amazonas	6.0	0.0	58.6	0.08
	Pará	9.0	−20.0	69.5	0.12
	Tocantins	3.7	+5.8	54.4	0.05
North East		**1207.0**	**+1.1**	**56.2**	**15.49**
	Alagoas	400.0	−2.9	60.0	5.13
	Bahia	103.4	−0.5	60.5	1.33
	Ceará	41.3	+2.7	56.8	0.53
	Maranhão	42.2	+3.8	59.7	0.54
	Paraíba	135.3	+16.5	52.5	1.74
	Pernambuco	369.7	−2.1	51.0	4.75
	Piauí	10.1	−1.3	63.1	0.13
	Rio Grande do Norte	61.4	+10.3	55.8	0.79
	Sergipe	43.6	+12.2	61.8	0.56
South East		**5203.2**	**+10.6**	**81.8**	**66.78**
	Espirito Santo	74.4	+6.3	66.5	0.95
	Minas Gerais	637.5	+19.8	77.9	8.18
	Rio de Janeiro	162.9	−0.8	45.3	2.09
	São Paulo	4328.5	+9.9	84.3	55.56

Table 13.1 (continued)

Region	State	Area planted, 2007 (ha × 10³)	% increase in area from 2006	Yield[b] (Mg ha⁻¹)	% area of all sugarcane
Central West		**759.8**	**+11.7**	**76.5**	**9.75**
	Goiás	299.4	−2.8	79.6	3.84
	Mato Grosso	254.0	+15.8	67.5	3.26
	Mato Grosso do Sul	206.4	+35.1	83.0	2.65
South		**601.4**	**+23.7**	**80.6**	**7.72**
	Paraná	547.5	+26.5	84.7	7.03
	Rio Grande do Sul	36.8	+4.2	36.9	0.47
	Santa Catarina	17.1	−5.6	38.7	0.22
	All Brazil	**7790.4**	**+9.9**	**76.6**	**100.0**

[a] http://www.sidra.ibge.gov.br/bda/default.asp?t=5&z=t&o=1&u1=1&u2=1&u3=1&u4=1&u5=1&u6=1&u7=1&u8=1&u9=3&u10=1&u11=26674&u12=1&u13=1&u14=1 accessed 5th June 2007.
[b] Fresh weight of cane stems (predicted).
[c] The Amazonian states of Acre, Amapá, Rodônia and Roriama have no significant area of sugarcane.

for 73.4% of the planted area, and as yields are well above the national average, these states contribute 77.5% of national cane production.

13.2.2 *Sugar and Ethanol Production*

From sugarcane, Brazil produces sugar, hydrous ethanol (5% water) for use in motors adapted for this fuel, anhydrous ethanol (<0.5% water) for mixing with gasoline, and other products such as the alcoholic beverage cachaça and various other products such as molasses and "rapadura" (a traditional sweet cake). FlexFuel motors can function on any mixture of hydrous or anhydrous ethanol with gasoline.

The data from the Ministry of Agriculture (MAPA, 2007) available for the 2005/2006 harvest, only includes sugar and ethanol. Of the "total recovered sugar" exactly 50% was used to produce refined sugar and 50% for ethanol. For the two types of ethanol 49% was anhydrous and 51% hydrated. Predictions were made recently (31st May 2007) by the Ministry of Agriculture that total cane production form the 2007/2008 season would be 582 million Mg, of which 44.8% will be used to produce ethanol fuel, 43.9% for refined sugar and the remaining 11.3% for other products (UOL Economia, 2007). In the 2006/2007 season the production of ethanol was approximately 17.5 billion litres, and for the next year it is estimated at 20.0 billion litres.

The present yield of hydrated ethanol per Mg of cane (fresh weight) is estimated to be 82.0 L (MAPA, 2007) which is close to the value of 85.4 L given by Macedo (1998) for the State of São Paulo. Thus using the mean national value and estimated yield for 2006/2007 of 76.6 Mg cane ha⁻¹ (IBGE, 2007), one ha of sugarcane produced 6281 L of ethanol ha⁻¹.

13.2.3 The Crop Cycle

In São Paulo, and in more productive areas of other states, general practice is to plant cane every 6 years. The first (plant) crop is harvested approximately 18 months after planting, and then there are four subsequent ratoon crops which are harvested at 12-monthly intervals (Macedo, 1998). The land generally lies fallow for the 6 months until the next planting, although occasionally a 'break crop' of groundnut or soybean is grown during this period. This practice is not common as few plantation owners have access to the necessary machinery for planting and harvesting crops other than sugarcane.

13.2.4 Land Preparation (Tillage)

Tillage prior to planting is usually intense, with sub-soiling followed by two or three passes with a heavy disc plough before harrowing and the subsequent formation of the furrows. As cane is planted from setts (lengths of cane stems), the furrows are 20–30 cm wide and approximately the same depth.

With the widespread introduction of zero tillage (ZT) in the mechanised production of grains in Brazil, this practice has recently been adapted for the sugarcane crop. All existing weeds and cane regrowth are treated with herbicide and the only mechanical operation is the furrow making. As the furrows are comparatively wide, a certain proportion of the soil is disturbed, but as spacing is also wide (usually 1.4–1.5 m between rows), this means that 80% or less of the soil surface is not tilled. This should lead to better maintenance of soil structure and aggregate integrity and probably favour soil organic matter accumulation ("C sequestration" – Six et al., 2000), but as this technique has only recently been introduced there do not yet appear to be any studies on the impact of the introduction of ZT on soil carbon stocks.

Virtually all planting is from setts usually produced on-farm and approximately 12 Mg of setts are required per ha.

13.2.5 Fertilisation

At planting the setts are covered with filtercake (from the large filters used to remove suspended material from the cane juice) at between 10 and $20 \, Mg \, ha^{-1}$. Typical nutrient content of this material is given in Table 13.2 and an addition of 10 Mg per ha would amount to an input of 63 kg N, 77 kg P, 15 kg K, 100 kg Ca and 49 kg Mg. In addition best practice (Macedo, 1998) is to add 500 kg of 4-24-24 fertiliser hence adding 20 kg N, 120 kg P_2O_5 and 120 kg $K_2O \, ha^{-1}$. Many agronomists and others have reported that there is very rarely a response of the plant crop to N fertiliser (Azeredo et al., 1986). Ratoon crops do usually respond to N fertiliser but rarely more than $100 \, kg \, N \, ha^{-1}$ are applied. Assuming an application of $20 \, kg \, N \, ha^{-1}$ at planting and $80 \, kg \, N \, ha^{-1}$ for each of the 4 ratoon crops spread over

Table 13.2 Chemical analysis (fresh weight basis) of typical filtercake from Usina Cruangi, Timbaúba, Pernambuco

OM[a]	N	P	K	Na	Ca	Mg	Zn	Cu	Fe	Mn
			$g\,kg^{-1}$					$mg\,kg^{-1}$		
231.1	6.25	7.67	1.50	0.17	10.8	4.90	36	42	3250	300

[a] Organic matter.

a 6 year cycle, the annual mean application becomes $56.7\,kg\,N\,ha^{-1}$. Nearly all other cane-producing countries of the world utilise at least $150\,kg\,N\,ha^{-1}\,yr^{-1}$, and most approximately $200\,kg\,N\,ha^{-1}$.

The reason for much lower use of N fertiliser on cane in Brazil seems to be partially that Brazilian cane varieties, which were first bred in soils of low N fertility, are able to obtain significant inputs of N from association with N_2-fixing bacteria (Lima et al., 1987; Urquiaga et al., 1992; Boddey et al., 2001). There is a considerable amount of literature on this controversial subject and readers are referred to reviews of James (2000), Baldani et al. (2002) and Boddey et al. (2003). However, there is no question that sugarcane has been grown for many decades, often centuries, in many regions of Brazil with no apparent long-term decline in yields or soil fertility, even though it is estimated that more N is removed by export of cane to the mill and trash burning than is added as N fertiliser (Boddey, 1995). The N_2-fixing bacteria that infect the interior of the plant tissues (endophytic diazotrophs) are generally thought to be responsible for most of the input from BNF (Baldani et al., 1997; James, 2000). Sugar cane plants have been found to be infected with very significant numbers of such diazotrophs in other countries such as Australia (Li and Macrae, 1992), Mexico (Muñoz-Rojas and Caballero-Mellado, 2003) and India (Muthukumarasamy et al., 1999, 2002). However, attempts to prove that BNF inputs to sugarcane are of agronomic significance in countries other than Brazil have not been successful (Biggs et al., 2002; Hoefsloot et al., 2005).

13.2.6 Cane Harvesting

Before the 1940s, pre-harvest burning of sugarcane in Brazil was virtually unknown. However, subsequently with the increasing price of labour, pre-harvest burning became almost universal until a few years ago. Until recently, virtually all cane was manually harvested, and one man in one day can manually harvest almost three times as much burned cane as unburned cane. It was the introduction of mechanical harvesting which facilitated the return to trash conservation. A strong lobby of environmentalists, who were especially active in the state of São Paulo, claimed that there were serious human health dangers (respiratory problems) with the annual cane burning (see Section 13.3.3.1). This has led to legislation in this State that mandates that all pre-harvest burning of cane must be phased out by the year 2022. Only on land that has greater than a 12% slope, where machine harvesting is non-viable, will burning be allowed until 2032. Today approximately 20% of sugarcane is not subject to pre-harvest burning (green cane harvesting), and most of this area is in São Paulo (Coelho, 2005).

As one cane harvesting machine can replace 80–100 men, in a country with large pockets of acute rural under- and unemployment, there are considerable negative social consequences of this change in practice. However, from an agronomic point of view the conservation of trash has considerable benefits. Our team at Embrapa Agrobiologia recently completed a 16-year study on the effects of pre-harvest burning versus trash conservation on cane productivity and soil organic matter content (Resende et al., 2006). The study was conducted in Pernambuco, some 100 km from the coast, where rainfall is often sub-optimal for cane production. The results showed clearly that the conservation of trash had most benefit in dry years (Fig. 13.2), and that over the whole 16 years, trash conservation increased cane yields by 25% from a mean of 46 to 58 Mg ha^{-1}. In the same study it was found that soil carbon stocks to 20 or 60 cm depth were not significantly affected by trash conservation. There was a tendency for the unburned plots to have accumulated annually a mean of 90 kg (0–60 cm) to 150 kg C ha^{-1} (0–20 cm).

Relatively short-term studies at two sites in São Paulo, both close to the city of Ribeirão Preto, have been reported, one on an Oxisol (Hapludox) and the other on an Entisol (Quartzipsamment) (Campos, 2004). The accumulation of trash in the unburned cane fields reached respectively 4.5 and 3.6 Mg dry matter ha^{-1} after 4 years without burning. The author concluded that an annual rate could be calculated for C accumulation from these data, but the decomposable fraction was probably achieving steady state by this time. He also observed an increase of approximately 1 Mg C ha^{-1} yr^{-1} in the soil during this period, but there was no replanting

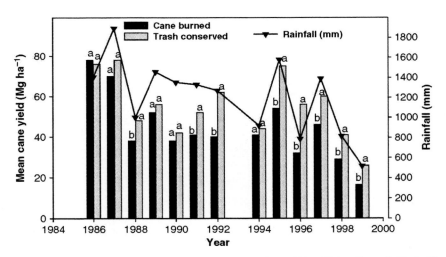

Fig. 13.2 Cane yield and annual rainfall for the period 1984–1999 at Usina Cruangi, Timbaúba, Pernambuco comparing cane managed with pre-harvest burning or trash conservation (green-cane harvesting). Bars surmounted by the same letter in the same year indicate that the effects of burning on cane yield was not significant at P<0.05 (Tukey's HSD, test). From Resende et al. (2006)

of the cane during this period. Based on these data Cerri et al. (2004) and Mello et al. (2006) suggest that the change from pre-harvest burning to trash conservation would promote a mean soil C accumulation of $1.62\,\mathrm{Mg\,C\,ha^{-1}\,yr^{-1}}$. As explained above, when cane is replanted, heavy tillage and deep plowing are used which lead to large mineralization losses of soil organic matter. For this reason, the difference between the SOC stocks under burned and green cane are not likely to reach even $1\,\mathrm{Mg\,C\,ha^{-1}\,yr^{-1}}$ over the long-term (Boddey et al., 2006). On the other hand, the values from the EMBRAPA Agrobiologia experiment in Pernambuco may be lower than a mean for São Paulo, in that this area of Pernambuco often has years with low yields due to lack of rainfall, and mean yields for the region are only about 60% of those for the State of São Paulo.

13.3 Environmental Impact

To consider the environmental impact of the production of bioethanol from sugarcane, and the expansion of this activity, the following items are considered:

A. Global impact: The energy balance of the bio-ethanol production and the impact on greenhouse gas emissions;
B. Local and Regional impact: Atmospheric and water pollution, and soil erosion.

13.3.1 Energy Balance

13.3.1.1 Introduction

Every biofuel requires as least some input of fossil fuel in its manufacture and distribution. Starting with the agricultural operations there are inputs of diesel fuel for ploughing, transporting seeds etc, then for harvesting, factory processing and fuel distribution. To calculate the balance for bioethanol produced from sugarcane in Brazil, we used the most recent available data for all inputs and divided the energy inputs into the following categories:

A. Agricultural operations
B. Transport of cane to mill/distillery and of raw materials from suppliers
C. Factory/distillery operations

Different authors have used different units (often not metric but "Imperial") or for expressing areas, crop yields, fertilisers and units of fuel production. In this chapter we use only SI (metric units) and express energy as joules (MJ or GJ). We have calculated all energy inputs and outputs on a per ha basis. The justification for this is that for agricultural operations the energy used in tillage operations, seeding and harvesting, are usually very similar on a per ha basis regardless of crop yield.

Many authors do not include energy required to build factories and distilleries, and to fabricate farm vehicles and transportation equipment (Sheehan et al., 1998; Shapouri et al., 2002). However, the relevant ISO standard (ISSO 14040–ISO, 2005) for Life Cycle Assessment studies clearly states that "manufacture, maintenance and decommissioning of capital equipment" should be taken into consideration.

13.3.1.2 Fuel for Agricultural Operations

Table 13.3 gives the typical values for fuel consumption (essentially only diesel oil) of the different agricultural machines used in the planting of the sugarcane, and for field operations during growth and regrowth (ratooning). As mentioned above (Section 13.2.3) normal good practise is to plant cane every six years, and subsequently harvest the plant crop and four subsequent ratoon crops. In areas where the

Table 13.3 Consumption of energy as diesel oil in agricultural field operations for sugarcane production in Brazil. Data from Macedo et al. (2003) and http://ftp.mct.gov.br/Clima/comunic_old/coperal5.htm#introdu%E7%E3o

	Field operation	Machine	L/h	ha/h	L/ha	MJ[a]/ha
Plant crop						
	Lime application	MF 290	6.00	1.78	3.37	161.0
	Elimination of old ratoons	Valmet 1280	12.80	1.85	6.92	330.4
	Heavy plough I	CAT D6	27.60	1.98	13.94	665.7
	Subsoiler	CAT D6	26.00	1.16	22.41	1070.4
	Heavy plough II	CAT D6	27.60	2.04	13.53	646.1
	Heavy plough III	CAT D6	27.60	2.04	13.53	646.1
	Harrow	CAT D6	13.00	2.52	5.16	246.4
	Furrow maker	MF 660	11.50	1.26	9.13	435.9
	Distribution of setts	MF 275	3.30	0.79	4.18	199.5
	Closing of furrows and application of insecticide	MF 275	4.80	2.52	1.90	91.0
	Application of herbicides	Ford 4610	4.00	3.30	1.21	57.9
	Interow weeding	Valmet 880	5.50	1.44	3.82	182.4
				Total	99.10	4732.7
Ratoon crop						
	Rowing of trash	MF 275	4.00	1.37	2.92	139.4
	Interow weeding	Valmet 1580	9.20	2.05	4.49	214.3
	Application of herbicides	Ford 4610	4.00	3.30	1.21	57.9
				Total	8.62	411.6
			Annual mean all field operations[b]=		22.3	1064.4

[a] Calorific value of 1.0 L of diesel fuel = 47.73 MJ.
[b] Based on one plant crop and 4 ratoon crops over a 6 year period. (Mean annual fuel consumption = {FcP + (4 × FcR)}/6, where FcP and FcR = fuel consumption for plant crop and ratoon crops, respectively).

plantations are replanted at longer intervals, diesel fuel consumption per year will be lower on an annual basis. For planting cane it is estimated that approximately 99 L of diesel are used per ha (Table 13.3), and as no tillage operations are involved, maintenance of the ratoon crops requires far less fuel (<9 L ha^{-1}). Thus the weighted mean annual diesel consumption is 22.3 L ha^{-1}, which at 47.7 MJ L^{-1} (11.414 Mcal L^{-1} – Pimentel, 1980) gives total input of 1063 MJ ha^{-1} yr^{-1} (Table 13.3).

13.3.1.3 Agricultural Inputs

Introduction

Apart from the diesel fuel energy input computed above, the other fossil energy agricultural inputs are derived from, human labour, industrial fertilisers, seed material, pesticides and the energy utilised to manufacture and maintain the agricultural implements.

Manual labour: The largest labour input is in the harvesting of the cane that is still burned and then cut by hand in approximately 80% of the area in Brazil (Coelho, 2005). A basic "Tarefa" (literal translation = task) for one man to cut cane is 6 Mg per day, and while nearly all workers cut more than 1 tarefa per day, the ratio of manual energy to cut 1 ton of cane is the same. Even if very conservatively we assume that it takes one man 8 hours to cut 6 Mg of cane then for each ha in Brazil which yields a mean of 76.6 Mg fresh cane ha^{-1}, it requires 76.6/6 $\times$ 8 = 102.1 man hours ha^{-1} harvest^{-1}. As there are 5 harvests in 6 years this becomes 85.1 man hours ha^{-1} yr^{-1}. Considering that apart from cane cutting there is manual planting, weeding and many other minor tasks, the estimate of Pimentel and Patzek (2007) of 128 h ha^{-1} yr^{-1} does not seem unreasonable.

Most authors who calculate energy balance for biofuel crops do not count any energy input for human labour. However, as each individual consumes fossil energy to survive and work it seems logical to include their fossil energy consumption as an input to the cane/ethanol production system. Giampietro and Pimentel (1990) estimated that in poor/rural societies such energy inputs range from 25.1 to 62.7 MJ day^{-1} (6000–15000 kcal day^{-1}). Utilising the higher value and assuming that all this energy is utilised in field labour, 1 man hour is equivalent to 62.7/8 = 7.84 MJ. The total energy invested in manual labour thus becomes 1003.5 MJ ha^{-1} yr^{-1} (Table 13.4).

Fertilisers

As calculated above (Section 13.2.5) it is estimated that mean N fertiliser inputs to cane are 56.7 kg N ha^{-1} yr^{-1}. Smil (2001) shows that N fertiliser production has greatly improved in energetic efficiency over the past 50 years from >80 GJ Mg^{-1} NH$_3$ before 1955 to 27 GJ Mg^{-1} NH$_3$ in the most efficient plants operating in the late 1990s. The mean value given by Lægreid et al. (1999, p. 204) is 54 MJ kgN^{-1} for urea production in plants operating in 1999 and this value was adopted giving an overall fossil energy cost of 3062 MJ ha^{-1} yr^{-1}. These same authors give values of 3.19 and 5.89 MJ kg for P and K, respectively. From our mean annual estimates of 16 and 83 kg ha^{-1} of P and K applied, respectively, the energy costs become 51 and 489 MJ ha^{-1} yr^{-1} for these two fertilisers, respectively (Table 13.4).

The other major soil amendment is lime, in that virtually all soils in Brazil used for sugar cane are acidic. Macedo (1998) estimates that as every replanting of the

Table 13.4 Fossil energy input, total energy yield and energy balance of bioethanol produced from sugarcane under present day Brazilian conditions. Energy values expressed on a per ha per year basis. Full explanation given in the text (Section 13.3.1)

Input	Quantity	unit	MJ/unit	MJ/ha/yr
Field operations				
Labour	128.0	h	7.84	1003.5
Machinery	155.4	kg	8.52	1785.6
Diesel	22.3	L	47.73	1064.4
Nitrogen	56.7	kg	54.00	3061.8
Phosphorus	16.0	kg	3.19	51.0
Potassium	83.0	kg	5.89	488.9
Lime	367.0	kg	1.31	478.9
Seeds[a]	2000.0	kg		252.2
Herbicides	3.20	kg	451.66	1445.3
Insecticides	0.24	kg	363.83	87.3
Vinasse disposal	180	m^3	3.64	656.0
Transport of consumables[b]	820.0	kg		276.8
Cane transport[c]	24.7	L	47.73	2058.0
Total transport				*2334.8*
Total field operations				**12709.7**
Factory inputs				
Chemicals used in factory[d]				487.6
Water		L		0.0
Cement	11.5	kg		75.9
Structural mild steel	28.1	kg		841.8
Mild steel in light equipment	23.1			693.5
Stainless steel	4.0	kg		287.1
95% ethanol to 99.5%				225.3
Sewage effluent	0			0.0
Total Factory inputs				**2611.1**
Total all fossil energy inputs				**15320.8**
Output				
Sugarcane yield	76.7	Mg/ha		
Total ethanol yield	6281.0	L/ha	21.45	**134750.4**
Final Energy Balance[e]				**8.8**

[a] This calculated form 2.6% of all field operation inputs.
[b] Transport of Machinery and fuels etc. to plantation/factory.
[c] Transport of cane from field to mill.
[d] Taken from Macedo et al. (2003) Table 13.3.
[e] Total energy yield/fossil energy invested.

cane (every 6 year) a mean of $2 \, Mg \, ha^{-1}$ of lime area added. This value is appropriate for opening up new land (mainly degraded pastures), which have not been limed for several years, but is higher than would be used for a plantation that has been operating for few decades or more. However, using this value of $2 \, Mg \, ha^{-1}$ ($367 \, kg$ lime $ha^{-1} \, yr^{-1}$), and an energy cost of $1.31 \, MJ \, kg$ for lime manufacture (Pimentel, 1980) we estimate a total annual fossil energy cost of $479 \, MJ \, ha^{-1}$ (Table 13.4).

Pesticides

Brazil has probably the largest program (in terms of land area) of any country in the world for biological control of insect pests and this is precisely on the sugarcane crop. The most widespread insect pest of cane is the sugar cane borer (*Diatraea saccharalis*), which can cause serious damage to cane in all regions of Brazil, damage being mainly secondary due to invasion of the tunnels bored in the stems by this pest by fungi. Control was at first by species of flies native to Brazil (*Metagonistlum minense* and *Paratheresia claripalpis*) but now almost universally the introduced wasp (*Cotesia flavipes*) is used, all of which lay their eggs in the larvae of the stem borer (Botelho, 1992). The *C. flavipes* gives the most effective control and is widely used in many cane growing areas by releasing hundreds of thousands of flies/wasps into the fields. At present over 1 M ha of cane are treated with *C. flavipes* to control Diatraea and more than 20 companies are engaged in producing this control agent and the number is growing (Sene Pinto, 2007).

The other main pests are restricted to the northeast region of Brazil and down the coast as far south as Rio de Janeiro, and are the root spittle bug (*Mahanarva fimbriolata)*, which sucks root sap, and the froghopper (*Mahanarva posticata*) which sucks leaf sap. The main damage to the plant is due to toxins injected into the plant phloem at the time of penetration of the insect stylet. Control is by spraying the fields with the fungus *Metarizium anisopliae*, which parasitises the exoskeleton of the sap-sucking pests. This fungus is generally produced on-farm by inoculating sterile boiled rice. A suspension of the fungus/rice (which breaks up into a slurry when vigorously agitated with water) is spayed onto the leaves. This control program is applied on approximately 600.000 ha of cane mainly in the north-eastern region (Sene Pinto, 2007).

There are many other insect pests, but none that have the potential to cause such widespread damage as the stem borer, spittlebug or froghopper. These minor pests are generally controlled by insecticides, and in consequence Brazil's use of insecticides is far lower than that used on other crops such as citrus, coffee or soybean. According to the National Association of Pesticide Manufacturers, in 2006 a total of $1700 \, Mg$ of insecticide (active ingredient – a.i.) of all insecticides were used on 7.1 Mha of cane, a mean of $0.24 \, kg \, ha^{-1}$ (SINDAG, 2007). The same source shows data that only 1 Mg (!!) of fungicide was sold to cane producers. However, as especially in recent years, weed control has become almost universally chemical (mainly glyphosate), herbicide sales for 2006 were $22,851 \, Mg$, a mean of $3.2 \, kg$ a.i. ha^{-1}. According to Pimentel (1980) the fossil energy cost of the insecticides most utilised on sugarcane (Carbofuran, Diuron and Endosulfan) is approximately

364 MJ kg a.i.$^{-1}$ (87 Mcal kg^{-1}). The herbicide glyphosate is cited as having an energy cost of 452 MJ kg a.i.$^{-1}$ (108 Mcal kg^{-1}). We calculate therefore the energy fossil energy inputs for insecticide and herbicide are, respectively, 87 and 1445 MJ ha^{-1} (Table 13.4).

Planting Material

Virtually all sugarcane is planted from setts (stem pieces), and 12 Mg of setts are required per ha every 6 years (Macedo, 1998). The means 2 Mg per year out of 76.6 Mg ha^{-1}, or 2.6%. As the agricultural operations for sett production are the same as for the rest of the cane plantation, the energy input for seeds is regarded as 2.6% of the total agricultural fossil energy input (TAFEI). Thus the fossil energy for seed production = (0.026 × (TAFEI* × 100/(100–2.6)), where TAFEI* is the TAFEI excepting the energy input in the setts. The fossil energy for sett production was estimated to be 252.2 GJ ha^{-1} yr^{-1} (Table 13.4).

Irrigation

Cane is only planted in regions where there is usually sufficient annual rainfall for the crop. Only a very small proportion of the sugarcane area in Brazil is irrigated, but in almost all cases the distillery waste (vinhasse) is applied to the fields. Between 10 and 12 L of vinasse are produced per L of ethanol and the return of this to the fields is valuable source of nutrients, especially potassium. Usually approximately 80 m^3 of vinasse are applied per ha and Resende et al. (2006) calculated that this adds 23 kg N, 8 kg P, 93 kg K and 35 kg S ha^{-1}. The vinasse is mixed with waste water used to wash the cane before grinding, typically giving a total volume of diluted vinasse of 160–200 m^3 ha^{-1} yr^{-1}. According to Dr Rogério P. Xavier (Usina Itamaraty, Mato Grosso) a diesel pump of 125 HP is requires 2 h to irrigate 1 ha with this volume of diluted vinasse, which incurs a fossil energy input of 656 MJ ha^{-1}.

13.3.1.4 Agricultural Machinery

Macedo et al. (2003) gives the density of utilisation of equipment for tractors and harvesters of 41.8 kg ha^{-1}. This would mean for a 20,000 ha plantation there would be 836 Mg of machinery, which certainly does not appear to be an underestimate. For implements towed by tractors he gives 12.4 kg ha^{-1}, and for transport vehicles to haul cane etc., he gives 82.4 kg ha^{-1}.

He cites Pimentel (1980) for the methodology used to calculate energy input from these data as follows:

a) The energy incorporated in the material (steel, rubber for tyres etc.) and for the fabrication, repairs and maintenance are considered. The energy incorporated in this case is essentially in the steel and tyres. The energy for fabrication of the different equipment is given by weight (excluding tyres).

b) The energy for repairs and maintenance is considered to be approximately 1/3 of
the total repair energy cost for the entire life of the equipment. The values utilised
come from the ASEA tables given in Pimentel (1980) (see Appendix 13.1).

c) The useful working life corresponds to 82% of the total life of the equipment and
the energy costs are converted to annual values based on these values.

The energy cost to manufacture steel from iron ore was reviewed by Worrell
et al. (1997). The data for 1991 show that the energy cost in all countries except
China, range from 20 to 30 GJ Mg^{-1}. A later publication by Farla and Blok (2001)
cite the World Energy Council for 1995 with a mean value of 22 GJ Mg^{-1}. For the
purposes of calculating the fossil energy input for steel for agricultural machinery
in this section, as well as for buildings and equipment in cane mills/distilleries we
have adopted the value of 30 MJ kg^{-1}.

A full explanation of how the energy required to manufacture and maintain agricultural machinery are given in Appendix 13.1 and the total value is estimates as
1,785.6 MJ ha^{-1} yr^{-1} (Table 13.4).

13.3.1.5 Transport Costs

The fossil energy cost of transport of the cane from the field to the mill/distillery
depends on the mean distance travelled to and from the mill by the transporting
vehicles, the capacity (Mg cane) of the transport vehicles and the consumption of
diesel fuel per km. The 1990 report by the National Audit Tribunal on the bioethanol
program states that transport of cane a distance of more than 30 km from the mill
is uneconomic (TCU, 1990). Managers answering a quick telephone survey of four
mills in São Paulo and two in Rio de Janeiro States, said their mean radius of transport was between 14 and 20 km. If cane were planted uniformly around a mill to a radius of 30 km, the mean distance to fetch cane would be 22 km, so we have used this
value. The cane transporters are predominantly a truck with a trailer (known in the
business as a "Romeo and Julieta") and they transport between 26 and 30 Mg cane
(Macedo et al., 2003). Using the mean of 28 Mg/transporter, 1 ha of cane (76.6 Mg)
will require 2.74 loads. When loaded their diesel consumption is approximately
1.6 km L^{-1} (Macedo et al., 2003), and we have assumed that when empty this is
3 km L^{-1}. So for a 44 km round trip at a mean of 2.3 km L^{-1}, the diesel consumption to fetch one ha of cane will be 44 × 2.74/2.33 = 51.7 L. There are 5 harvests
every six years so the annual diesel consumption per ha will be 43.1 L ha^{-1} yr^{-1}. At
a energy value of 47.73 MJ L^{-1}, mean fossil energy consumption for transporting
cane to the mill becomes 2058 MJ ha^{-1} yr^{-1} (Table 13.4).

Transport of raw materials to plantation and mill from suppliers: The most important quantities of materials to be transported across the country to the plantations
are lime and fertilisers. While the fertiliser supplier's association does not to provide
data on individual quantities of N. P and K applied to each crop, they do give total
tonnage of fertilisers applied to each crop. For 2006 their data were 3.13 million Mg
of fertilisers added to 7.37 M ha of sugarcane, a mean of 425 kg fertiliser applied
per ha^{-1}. For lime Macedo (1998) assumed that at each replanting (every 6 years)

6 Mg of lime were applied, a mean of 367 kg lime ha^{-1} yr^{-1}. The total (792 kg ha^{-1}) is close to that of the 800 kg ha^{-1} value of Pimentel and Patzek (2007) and to account for the small quantities of inputs to the factories (lubricants, reagents) and pesticides, we use the value of 820 kg ha^{-1} which must be transported. The great majority of mills/distilleries are in the south-eastern region (Table 13.1), predominantly in São Paulo. Fertilisers come from the factories that are situated close to the coast or are imported. In the southeast lime comes predominately from the State of Minas Gerais, but the northeast and other states have their reserves also. The main port of São Paulo State, Santos, is 410 km by road from one of the largest cane growing areas of the state, near Ribeirão Preto. In almost all case, except for a few mills in the Central West region, transport will be predominately by road and less than 500 km distance. Assuming that all transport of raw material (820 kg ha^{-1}) is in trucks hauling 35 Mg for a distance of 500 km and fuel consumption of these vehicles is 2 km L^{-1}, an input of diesel per ha is of 5.8 L or 276.8 MJ ha^{-1} yr^{-1} (Table 13.4).

13.3.1.6 Factory Inputs

Vast amounts of energy are utilised in factory processing, for pumping water, crushing the cane processing the juice, fermentation and distillation of the ethanol. Pimentel and Patzek (2007) estimate that it requires 2.546 Gcal per 1000 L of anhydrous ethanol for steam production for direct heating and to drive the electricity generators. This is equivalent to 16.9 GJ ha^{-1} yr^{-1}. However, all Brazilian mills/distilleries are powered by steam generated from burning the bagasse (crushed cane stems). In fact Macedo (1998) estimated that in most mills there is a surplus of bagasse-derived energy of between 8 and 14%. In some regions (especially São Paulo State) this surplus may be used to generate electricity which is exported to the local grid (in 2005, 350 MW were exported to the grid by cane mills – Coelho, 2005), or the excess bagasse is transported to nearby industries for power generation, or sometimes used to produce fibreboard. Some energy for orange juice extraction plants near Ribeirão Preto (SP), use excess bagasse for power generation. The use of bagasse for all factory inputs means that there is no extra fossil energy required to power the mills/distilleries.

The most important input of fossil energy for the factories is in their construction. The company Dedini S.A. based in Piracicaba, São Paulo State is now responsible for the construction of approximately 80% of the new sugarcane mills/ethanol distilleries in the whole of Brazil. At present (2006/2007) there are 357 mills/distilleries in operation for a total harvested area of 6.72 Mha. This means the average mill has a harvested area of 18,800 ha with an annual production of 1.4 million Mg of cane. As the harvest period is almost universally 180 days, this means average mill throughput is approximately 8,000 Mg of cane per day.

Engineers from Dedini S.A. provided us with construction details of a modern mill/distillery with throughput of 2 million Mg cane year^{-1}. As these mills rarely run at full capacity, this size of mill approximately represents an average size mill in Brazil. Obviously a large new mill will be considerably more energy efficient that smaller and/or older mills, and we take this into account by calculating all energy inputs in the construction of this 2 million ton/day standard mill as if it were functioning only at one third capacity. In other words the total energy involved in its construction will be three times higher per ha, than if we considered that it was functioning at 100% of capacity.

Pimentel and Patzek (2007) calculate this energy input from the energy content of cement, and stainless and mild steel. We follow the same procedure and details of all the buildings, tanks and equipment in the standard mill are given in Tables 13.5 and 13.6. The total cement used in the construction of the mill was estimated as $1,000\,m^3$ ($1,600\,Mg$), total weight of mild steel $4,310\,Mg$ and of stainless steel $410\,Mg$.

Table 13.5 Area of buildings of a modern sugarcane mill/distillery (Dedini S.A., Piracicaba, São Paulo) with a design capacity of 2 million Mg cane year^{-1}. Data provided by Engineers Roberto dos Anjos and Antonio Sesso

	Area (m^2)
Buildings (Total area of factory = 600 × 650 m)	390000
a. Weigh-in/cane reception 18 m^2	18
b. Unloading bay, conveyer house and cane crusher	2160
c. Stores and workshops	110
d. Refectory	137
e. Clinic	108
f. Offices	300
g. Generator shed	900
Other paved/walled areas	
a. Storage tank area	2700
b. Bagasse storage area	7300

Table 13.6 Equipment and storage tanks of a modern sugarcane mill/distillery (Dedini S.A., Piracicaba, São Paulo) with a design capacity of 2 million Mg cane year^{-1}. Data provided by Engineers Roberto dos Anjos and Antonio Sesso

Item	Weight (Mg)
Primary loading conveyer 12 × 13 m	240
Primary cane conveyer (steel)	77
Defibred cane conveyer (rubber)	10.5
Conveyers to feed crusher (4)	60
Conveyer to feed bagasse to furnace	112
Bagasse conveyers (4)	77
Cleaning conveyer and forced air dryers	200
Electricity generator (20 MVA)	120
Turbine to power generator	130
Electricity transformer	15
Furnace	2510
Distillations columns	380
Storage tanks	748
Pipes and tubing	20
Ethanol platform	2

Buildings are regarded as having a useful life of 50 years and a maintenance energy cost of 4% per yr (Macedo, 1997). As the greatest energy input is in cement the energy value of this material is used. As fuel costs are such a large part of manufacturing costs, most cement companies have aggressive energy conservation programs and according to the International Energy Program (IEA, 1999) new manufacturing plants have reduced energy use by between 25 and 40% compared to 10–15 years ago. The report by the IEA (1999) gives a value of 6.61 GJ for the energy required to produced 1 Mg of cement, and other recent reports (Young et al., 2002 and Worrell and Galitsky, 2004) give somewhat lower values of 4.35 and 6.1 GJ Mg^{-1}. We use the former higher value of 6.61 GJ Mg^{-1}. So allowing for a 4% annual maintenance cost the total embodied energy for all buildings over a 50 year period is $(1,600 + (0.04 \times 50 \times 1,600))$ Ú 6.61 GJ which becomes 31,730 MJ or 634.6 MJ yr^{-1}. As we assume that the mill serves to grind one third of 2 million Mg of cane per year, this becomes 0.952 MJ Mg cane milled, or 75.9 MJ ha^{-1} yr^{-1} (Table 13.7).

For the mild steel in the mill/distillery we have assumed that one third is in light equipment and thus subject to more wear and will have a lower useful life

Table 13.7 Energy in the buildings and construction of a standard mill/distillery. Design capacity 2 million Mg year, running at 33% capacity. Methodology for the calculation of the fossil, energy inputs follows that of Pimentel (2007)

	Mass	[a] Useful life[b]	Including maintenance[c]	Including on-site energy utilisation[d]	per year	kg/ha/ year	Total energy
	Mg	yr	Mg	Mg	kg	kg	MJ/ha/yr
Cement in buildings	1600	50	4800.0	5000.0	100000	11.49	75.9
Mild steel (structural)	2873	25	5746.0	6105.1	244205	28.06	841.8
Mild steel in light equipment	1437	10	2011.8	2191.4	201180	23.12	693.5
Stainless steel	410	25	820.0	871.3	34850	4.00	287.1
							1898.3

Basic data on standard cane Factory	
Mg cane harvested by factory	666667 yr^{-1}
Area harvested by factory	8703.2 ha
Energy in cement (MJ/kg)[e]	6.61
Energy in Steel (MJ/kg)[f]	30.0
Energy in stainless steel (MJ/kg)[g]	71.7

[a] Data from Dedini S.A.. Piracicaba. São Paulo.
[b] According to Macedo et al. (2003).
[c] Maintenance energy cost of 4% per year.
[d] 12.5% of mass of each component (Hannon et al. 1978).
[e] From IEA (1999).
[f] From Worrel et al. (1997).
[g] Embodied energy in stainless steel = 2.39 × energy in mild steel (Pimentel and Patzek, 2007).

(10 yrs – Macedo, 1997). The remaining two thirds is considered to be in the structure of the mill, equipment and distillery, and thus will have a longer useful life (25 years). The same calculations have been made in the same way as for the cement in buildings but the embodied energy in mild steel was considered to be $30\,MJ\,kg^{-1}$ as justified in Section 13.3.1.4 above.

The data for the standard mill provided by Dedini S.A. show that 410 Mg of stainless steel was used, mainly in the distillery columns. Pimentel and Patzek (2007) give the embodied energy in stainless steel to be 2.39 times that in mild steel, so the value of $71.7\,MJ\,kg^{-1}$ was used for this material. The useful life of this material was assumed to be 25 years. The energy input for stainless steel in the factory was again calculated using the same procedure as for cement (Table 13.7).

Finally to account for on-site energy utilised in the construction, all values were increased by 12.5% as suggested by Hannon et al. (1978). The total energy requirement for factory buildings and equipment totalled $1898\,MJ\,ha^{-1}\,yr^{-1}$ (Table 13.7).

13.3.1.7 Energy Balance

The details of all fossil energy inputs calculated as described in Sections 13.3.1.1–13.3.1.6 above, are displayed in Table 13.4. The total energy yield of the annual mean per ha ethanol yield of 6,281 L, becomes $134,815\,MJ\,ha^{-1}$ (1 L of ethanol yields $21.46\,MJ\,L^{-1}$ – Pimentel, 1980).

Within the fossil energy inputs in the agricultural operations, fertilisers, especially N fertiliser, are responsible for the largest contributions. The fact that in Brazil N fertiliser use is far lower than in just about any other cane growing area in the world, makes an important economy. If for example $150\,kg\,N\,ha^{-1}\,yr^{-1}$ (typical of most other countries) were used instead of the estimated 56.7 kg, the energy input would rise from 3060 to $8100\,MJ\,ha^{-1}\,yr^{-1}$ increasing the total energy input in agricultural operations (including transport of cane and consumables) by 43%.

Because of their complicated synthesis herbicides are extremely energy intensive and even though only a mean of $3.2\,kg\,a.i.\,ha^{-1}\,yr^{-1}$ are applied, this is the second most important consumables input after fertilisers.

Brazil is fortunate in that most of the country has over 1,000 mm of rainfall a year, and the most productive cane-growing areas have over 1,300 mm of rain. For this reason only a very small area is irrigated so that there is effectively no energy input for irrigation.

The comparatively large input of fossil energy in the manufacture of agricultural machinery, and to a lesser extent, of human labour, show the importance of including these inputs, which is not universal practice in computing such balances (e.g. Sheehan et al., 1998; Shapouri et al., 2002)

As all factory energy is supplied by bagasse, the main fossil energy input (estimated to be $\sim 1,900\,MJ\,ha^{-1}\,yr^{-1}$) is in the infrastructure of the construction and maintenance of structure and equipment of the factory. All factories are built near abundant water supplies (usually rivers) and pumping comes from electricity generated from bagasse, and thus involves minimal fossil energy inputs.

The total energy balance is the Total Energy Yield (TEY) of the biofuel divided by the Fossil Energy Invested (FEI). For today's production levels and practice we calculate this to be approximately 8.8, which is close to the value of 9.2 calculated for ethanol production São Paulo by Macedo (1998), and of 8.3 by Macedo et al. (2003). The main differences between these studies are that (a) we included the manual labour energy input, which was not included in the studies by Macedo and his colleagues, and (b) we used a much more recent estimate for the energy embodied in steel ($30 \, MJ \, kg^{-1}$ – Worrell et al., 1997), rather than that cited by Macedo et al. (2003) of 38–$63 \, MJ \, kg^{-1}$ which are estimates that date from the 1970s.

When the energy balance (TEY/FEI) is high, differences of 1 or 2 units in the this ratio make only small differences in the proportion of energy saved. This is illustrated in Fig. 13.3, which displays the relationship between the economy in fossil energy (% Fossil energy saved) and the energy balance. Thus if a biofuel has an energy balance of 5, this represents an economy in fossil energy of 80%. It might take a lot of ingenuity and expenditure to halve fossil energy inputs to raise the balance to 10, but this would only represent economy in fossil energy inputs of a further 10%.

Pimentel and Patzek (2007) estimated the input of fossil energy to produce Brazilian bioethanol was $13,286 \, MJ \, m^{-3}$ ($3,177 \, Mcal \, m^{-3}$) and a total energy yield of $21,454 \, MJ \, m^{-3}$ ($5,130 \, Mcal \, m^{-3}$). The resulting energy balance of 1.66 is in wide disparity of those calculated by Macedo (1998) and Macedo et al. (2003) and by us in this present study. A comparison of our estimates with those of Pimentel and Patzek (2007) is given in Table 13.8.

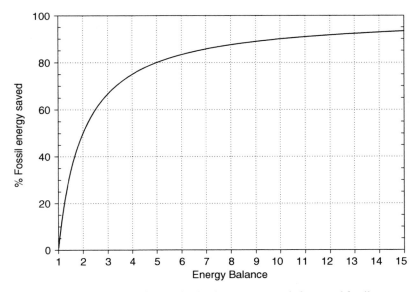

Fig. 13.3 The relationship for biofuel production between energy balance and fossil energy saved

Table 13.8 Comparison of estimates of fossil energy and energy balance computed in this study and that of Pimentel and Patzek (2007)

	This study				Pimentel and Patzek (2007)			
Input	Quantity	unit	MJ/unit	MJ/ha/yr	Quantity	unit	MJ/unit	MJ/ha/yr
Field operations								
Labour	128.0	h	7.84	1003.5	128	h	3.1	656.6
Machinery	136.6	kg	13.07	1785.6	156	kg	112.9	2668.1
Diesel	22.3	L	47.73	1064.4	22.3	L	4.1	1062.2
Nitrogen	56.7	kg	54.00	3061.8	58.3	kg	66.9	3901.8
Phosphorus	16.0	kg	3.19	51.0	16	kg	17.4	250.9
Potassium	83.0	kg	5.89	488.9	83	kg	13.6	1129.1
Lime	367.0	kg	1.31	478.9	367	kg	1.3	476.7
Seeds[a]	2000.0	kg		252.2	21000	kg	5.3	865.7
Herbicides	3.20	kg	451.66	1445.3	5	kg	418.2	2091.0
Insecticides	0.24	kg	363.83	87.3	2	kg	418.2	836.4
Transport of consumables[b]	820.0	kg		276.8	650	kg	3.5	2258.3
Cane transport[c]	76.7	L	15.37	2058.0	77	Mg	499.8	32977.4
Total field operations				**12709.7**				**49174.2**

Table 13.8 (continued)

Input	This study				Pimentel and Patzek (2007)			
	Quantity	unit	MJ/unit	MJ/ha/yr	Quantity	unit	MJ/unit	MJ/ha/yr
Factory inputs								
Chemicals used in factory[d]				487.6				2070.1
Water	11.5	L		0.0	115500	L	0.0	8832.4
Cement	51.2	kg	6.6	75.9	44	kg	200.7	2116.1
Mild steel		kg	30.0	1535.3	22	kg	96.2	3795.2
Stainless steel	4.0	kg	71.8	287.1	16.5	kg	230.0	3795.2
95% ethanol to 99.5%				225.3				188.2
Sewage effluent	0			0.0	110	kg BOD	14.4	1587.1
Total Factory inputs				**2611.3**				**18589.1**
							Distribution	3411.9
Total all fossil energy inputs				15320.8				71175.2
Output								
Sugarcane yield	76.7	Mg/ha			77.0	Mg/ha		
Total ethanol yield	6281.0	L/ha	21.45	**134750.4**	**5499**		**21.45**	**117973.7**
			Final Energy Balance	**8.8**			**Final Energy Balance**	**1,66**

For footnotes see Table 13.4.

There is a huge disparity in the estimates of the energy attributed to transport of consumables (fertilisers and chemicals for the factory) and of hauling cane from the field to the mill, the estimates of Pimentel and Patzek (2007) being respectively 10 and 33 times higher than ours. The consumption of diesel oil estimated by these authors seems totally unrealistic in that if a truck and trailer can carry 34 Mg of cane for a 16 km round trip (their value) then the consumption of diesel at $47.7\,MJ\,L^{-1}$ would be $21.6\,L\,km^{-1}$.

The other large difference is in the specific constants used for the cement and steel used in the construction of the factory, which are, respectively, 30.4 and 3 times greater than those used in our study and justified in Sections 13.1.3.6 and 13.1.3.4.

The energy balance computed by de Oliveira et al. (2005) of 3.7 is also considerably lower than that computed in this present study or Macedo (1998) and Macedo et al. (2003), and again the large difference comes in the utilisation of diesel fuel in the field operations and cane transport. These authors cite a report from the University of Campinas (Unicamp, Campinas, São Paulo State) for a value of 600 L of diesel fuel consumed per ha per year compared to a total of $71.2\,L\,ha^{-1}\,yr^{-1}$ (43.1 L for cane transport, 22.3 L for field operations and 5.8 L for transport of consumables to the plantation/mill) in our study. Substituting our value for diesel consumption in the energy balance of de Oliveira et al. (2005) becomes 7.0.

13.3.2 Greenhouse Gas Emissions

For the ethanol production, fossil fuel is used directly and indirectly for construction of the infrastructure of machinery and consumables together with other chemical and biological processes which are used in sugarcane production. The use of these fossil fuels results in the generation of greenhouse gases (GHGs). The energy data and amounts of material for factories, consumables, machinery, fuels and labour involved in the ethanol life cycle (Table 13.4) were used to estimate GHGs emissions based on emission factors for each component. A summary of the results is displayed in Table 13.9.

Inputs for agricultural operations are calculated from energy in labour, herbicides, insecticides and seeds which come from many different sources and they were assumed to be best represented by crude oil. From IPCC (1996) 1 GJ of crude oil emits $73.3\,kg\,CO_2$, $0.003\,kg\,CH_4$ and $0.0006\,kg\,N_2O$. Estimates from machinery were based on the energy contained in the steel, which was assumed to come from steel factories fuelled by coking coal (1 GJ is equivalent to $94.6\,kg\,CO_2$, $0.001\,kg\,CH_4$ and $0.0015\,kg\,N_2O$). Diesel oil was the energy source for transport of consumables and cane to the factory, fuel for machines and irrigation, which meant each GJ employed emitted $74.1\,kg\,CO_2$, $0.003\,kg\,CH_4$ and $0.0006\,kg\,N_2O$ (IPCC, 2006). Fertilisers and lime complete the components of sugarcane production. In the absence of information regarding the type of lime used in sugarcane areas (proportions of calcitic and dolomitic) emissions of CO_2 from lime addition were estimated by the amount of lime multiplied by the emission factor of 0.75, proposed

Table 13.9 Emissions and avoided emissions of greenhouse gases (CO_2, N_2O and CH_4) during ethanol production phases

Ethanol production phase	Greenhouse gases emitted (per ha)			
	CH_4	N_2O	CO_2	CO_2 eq[a]
	g of CH_4 or N_2O ha^{-1} yr^{-1}		kg ha^{-1} yr^{-1}	
Sugarcane planting[b]	+ 8.9	+ 1.8	+ 718.0	+ 718.7
Crop management[c]	+ 2.7	+ 1,570.5	+ 86.9	+ 573.8
Harvesting[d]	+ 28007.1	+ 381.9	+ 253.8	+ 960.2
Ethanol production[e]	–	–	+ 107.6	+ 107.6
		Total fossil GG emission		+ 2,360.3
Ethanol consumption[f]	–	–	−9580.6	−9580.6
Net greenhouse gas emissions				**−7220.3**

[a] Each mol of N_2O and CH_4 is considered equivalent to 310 and 21 mol CO_2, respectively (IPCC, 2006). Positive values refer to emissions, and avoided emissions when negative.
[b] Machinery and diesel (50% of total), transportation, labour (20% total), herbicide, soil liming, fertiliser addition and planting operation.
[c] Machinery and diesel (10% of total), labour (20% total), insecticides, irrigation and soil emissions.
[d] Machinery and diesel (40% of total) , labour (60% total), emissions from residues after burning to harvest 80% of the area, and transportation.
[e] Ethanol installations and processing.
[f] Assuming ethanol (52% C) is fully burned.

by the IPCC (2006), tier 1. For fertiliser emissions, urea, triple superphosphate and potassium chloride were considered to best represent the NPK formulation used in sugarcane areas. The contribution of each source was estimated by the emission factors proposed by Kongshaug (1998). Assuming the average for this technology in Europe, the production of 1 kg of urea, 1 kg of triple superphosphate and 1 kg of potassium chloride represent 0.61, 0.17 and 0.34 kg CO_2 emitted to the atmosphere, respectively.

After N fertiliser placement (56.7 kg N ha^{-1}) and vinasse application (23 kg N ha^{-1}), it was assumed no NH_4^+ volatilisation occurs, so the total N added was substrate for nitrification and denitrification processes for N_2O emissions. No significant CH_4 production was considered to occur in the sugarcane areas during cropping phase (Macedo, 1998). The harvested area after burning was assumed to be 80% of the whole cropped area. In this case, fractions of 0.005 of total C (5.25 Mg ha^{-1}) and 0.007 of total N (30 kg N ha^{-1}) in burned trash were considered to evolve as CH_4 and N_2O, respectively (IPCC, 2006). For the remaining 20% in unburned areas, the 30 kg N ha^{-1} were considered to be in harvest residues left to decompose in the field which meant a fraction of 0.0125 of this N was emitted as N_2O.

For factory construction and function the emissions coming from cement, steel and chemicals were accounted for, all based on emission factors from the IPCC guidelines (IPCC, 1996). For cement a factor of 0.95 was applied to calculate clinker content from the total cement used. According to Tier 1 this carries an emission

Appendix 13.1 Calculation of fossil energy inputs to agricultural machinery. Data on a per ha per year basis

| Equipment | Mass of equipment[a] | Proportion of steel | Total Steel | Total Tyres | Energy ha^{-1} | | | | | | |
					Steel[b,c]	Tyres[d]	All	Fabrication[e]	Maintenance[f]	Total[g]	per year[h]
	kg ha^{-1}		kg ha^{-1}		MJ ha^{-1}						
Tractors	41.8	0.82	34.3	7.48	1029.5	641.2	1670.8	501.4	496.2	2489	497.8
Implements	12.4	1.00	12.4	0	472.4	0.0	472.4	106.9	146.0	680	85.0
Trucks	82.4	0.94	77.5	4.94	2951.1	1412.5	4363.6	1131.6	881.4	6014	1202.9
	136.6										1785.6

[a] Mass of equipment per ha adapted from Macedo et al. (2003).

[b] Energy intensity of steel in tractors rated at 30 MJ kg^{-1} (See Section 13.3.1.4).

[c] Steel in Implements (ploughs etc.) and trucks rated at 1.27 × 30 MJ (Factor from Pimentel, 1980) = 38.1 MJ kg^{-1}.

[d] Tyres rated at 85.7 MJ kg^{-1} (Macedo et al., 2003).

[e] Fabrication energy = 14.61, 8.62 and 14.61 MJ kg steel for Tractors, Implements and Trucks, respectively.

[f] Maintenance energy cost over working life = 0.297, 0.309 and 0.202 kg total mass for tractors, implements and trucks respectively.

[g] Total adjusted to "reliable life" (TARL) according to Pimentel (1980) is 0.82 × (total embodied + fabrication energy). Hence this Total energy = TARL energy + Maintenance energy.

[h] Working life of Tractors, implement and trucks were rated as 5, 8 and 5 years, respectively.

factor of 0.507, with a 2% correction for cement kiln dust, and this was used to calculate the CO_2 emission. In the case of structural and mild steel, emissions of CO_2 were calculated on the basis of the global average emission factor for iron and steel production (1.06 kg CO_2 kg^{-1} steel produced). For stainless steel, the emission factor for ferrochromium of 1.6 kg CO_2 kg^{-1} steel produced was used (IPCC, 2006). Energy in the production agro–chemicals was considered to be from crude oil for which the emission factors for CO_2, N_2O and CH_4 were mentioned above.

To explain the impact of ethanol from sugarcane produced under Brazilian conditions the agricultural activities were broken down into three different phases: planting, crop management and harvesting, the latter including transportation of cane to mill. The factory phase was also included to close the cycle (Table 13.9).

Emissions of CO_2 predominated at planting and were explained by the fossil fuel energy used in consumables, machinery and transportation of consumables. During plant development N_2O production was derived from fertiliser and vinasse N and nitrification/denitrification gained importance and represented a large share (85%) of the emissions expressed as equivalents of CO_2. Again, the trace gases CH_4 and N_2O represented most of the emissions at harvest, the former was emitted mostly from burning trash at harvest and the latter, partially from burning, but also from decomposition of N in residues in unburned areas (20% of all Brazilian cane). The most important greenhouse gas (GG) emissions are incurred during pre-harvest burning, and amount to 82 kg and 588 kg ha^{-1} yr^{-1} of CO_2 equivalents, as N_2O and CH_4, respectively, 34% of all GG emissions.

The conversion from manual harvesting of burned cane to machine harvesting of green cane would eliminate these emissions as well as approximately 70% of the yearly manual labour input (0.7 × 1004 MJ ha^{-1} or 52 kg CO_2 equivalents ha^{-1}). However, the decomposing trash emits 183 kg CO_2 equivalents ha^{-1} yr^{-1} as N_2O from the 30 kg N left in the cane trash. Furthermore, the harvester (70 Mg of cane harvested per h, machine weight 19 Mg) consumes 40 L of diesel per h (data from Sr. Aureo Tasch, John Deere S.A., Catalão, Goiás) giving a fossil energy input of 2089 MJ ha^{-1} yr^{-1} (155 kg CO_2 equivalents ha^{-1} yr^{-1}). Embodied energy in the machine (effectively 100% steel, 5.5 kg ha^{-1} yr^{-1}) is equivalent to 54.2 MJ (5.1 kg CO_2 equivalents ha^{-1} yr^{-1}). In summary, under manual harvesting annual GG emissions amount to 722 kg CO_2 equivalents ha^{-1} and this falls to 343 kg CO_2 equivalents ha^{-1} if the cane is harvested green with machine harvesting. It is also reported that full ground cover with trash during the year reduces the requirement of herbicide by at least 50% (Antônio Gondim, Usina Cruangi, Timbaúba, Pernambuco; pers. comm.) equivalent to 60 kg CO_2 equivalents ha^{-1} yr^{-1}.

Emissions derived from the factory infrastructure and chemicals for ethanol production from milled cane accounted for less than 5% of the emissions calculated for the whole cycle.

Summing up: all emissions in terms of CO_2 equivalents amount to approximately 2.36 Mg CO_2 ha^{-1} yr^{-1}, close to one fourth of the total emissions avoided whether burning ethanol as a fuel (9.58 Mg CO_2 ha^{-1} yr^{-1}), assuming 100% is converted to CO_2.

13.3.3 Local and Regional Impacts

13.3.3.1 Atmospheric Pollution

Harvesting of cane occurs mainly in the dry season. The pre-harvest burning of the cane facilitates the manual harvest and diminishes the risks of injury to workers from snakes and poisonous spiders. The burning releases inhalable particles with a great number of components, most of all are carbon-rich alumino-silicate based, causing a typical overcast fog-like atmosphere, widespread in the cane districts in this season. However, the occurrence, composition and persistence of the smoke is highly dependent on specific weather conditions and this coincides with admission to hospitals of children and elderly people with respiratory problems (Godoi et al., 2004; Arbex et al., 2007). Some studies state that this effect is regarded as similar to what could be observed in urban areas exposed to industrial and automotive pollution, but also acknowledge that the ethanol addition to gasoline has contributed to decreasing air pollution, at least in the last twenty years, in the urban centres (Cançado et al., 2006).

As mentioned before (Section 13.2.6) biggest cane producer, São Paulo State has passed a law to regulate cane burning since 2003. The law defines what areas are able to use mechanical harvesting due to field slope, and sets a timetable. All the plantations under 12% of inclination should be totally mechanised by 2022. Cane burning in the other areas should be eliminated by 2032, when all areas ought to be harvested without burning.

13.3.3.2 Water Pollution

In the early years most distillery waste was disposed of without treatment into local rivers. As the waste usually contains approximately 1% soluble C and high levels of K, S and N and some P, the results were disastrous. Many rivers became eutrophic, there was massive death of fish and all other aquatic organisms, and the stench of this disposed waste could be scented many kilometres before arriving at a distillery. At first the factory owners were loath to return these wastes to the field as there was often an initial wilting of the cane leaves and signs of damage to the plants (Boddey, 1993). However, it was shown in many experiments that the plants soon recovered and benefited from the extra nutrients, and pumping the waste out onto the fields diluted with other wash water from the mills (which also had significant BOD), was a cheaper source of nutrients than synthetic fertilisers.

Today almost all vinhaça is disposed of by pumping onto the fields, and where State and/or Municipal governments have effective environmental protection agencies, significant water pollution is a thing of the past.

13.3.3.3 Soil Erosion

Several authors have stated that soil erosion in sugarcane fields is major problem. Pimentel and Patzek (2007) write "Sugarcane production causes more intense soil erosion than any crop produced in Brazil because the total sugarcane biomass is harvested and processed in ethanol production". They cite the paper of Sparovek and Schnug (2001) who give a value for soil loss of 31 Mg ha^{-1} yr^{-1}. This estimate was derived from the use of the Universal Soil Loss Equation on just one site near Piracicaba (São Paulo) with a slope of 5–15% and not based on actual soil loss measurements. Fuller details are given of this study in the paper of Bacchi et al. (2000). Actual measurements at this site using the ^{137}Cs radioactive isotope technique (Ritchie and McHenry, 1990, 1995) yielded a mean value for soil loss of 23 Mg ha^{-1} yr^{-1}. This technique yields mean annual values from approximately 1962 to the time of sampling. At the start of the 1960s there was a large increase in ^{137}Cs deposition due to many very large nuclear explosions from H-bomb tests by the USSR and USA.

Only one other study using this technique seems to have been published (Correchel, 2003), and this author reported a mean annual soil loss of 10.8 Mg ha^{-1} yr^{-1} on a 4% slope on an Oxisol.

Lombardi-Neto et al. (1982) conducted a study where actual soil loss was measured from plots in a cane field on a 12.8% slope on a "latosolo roxo" (Typic Haplorthox – US Soil Taxonomy classification) for the plant crop and 2 ratoons. In the year when the soil was deep ploughed for planting losses were high (49 Mg ha^{-1}) but for the subsequent two ratoon crop years losses were minimal (0.20 and 0.01 Mg ha^{-1}) giving a mean loss of 16.4 Mg ha^{-1} yr^{-1}.

Other estimates using the Universal Soil Loss Equation give values between 3.3 and 7.3 Mg yr^{-1} of soil loss on slopes from between 3 and 8%.(de Souza et al., 2005).

In summary, as the crop is only renovated every 6 years, mean annual losses are generally lower than for other crops (e.g. soybean, maize) grown with conventional tillage, but a global figure for Brazil is not available, and where cane is grown on flat land losses will be much lower than in the above studies. Already 30% of the area of sugar cane in São Paulo State is being harvested without burning (green-cane harvesting) and the preservation of the cane trash on the soil surface in these areas will undoubtedly radically reduce erosion losses. While erosion losses are at present are moderate to severe, the increasing use of direct planting (zero tillage) of cane (Section 13.2.4) and green-cane harvesting in the next decade or so, these losses should fall to acceptable levels.

13.3.3.4 Replacement of Food Crops and Invasion of Reserves of Biodiversity

Two major criticisms have been levied against the Brazilian ethanol program that need to be answered:

One can be summarised briefly as "the expansion of the ethanol program will increase the destruction of the Amazon forest". As can be seen from the data presented in Table 13.1, less than 20,000 ha of cane have been planted in this region (0.25% of the cane area) so that the impact on the forest is minute. The government has declared recently that cane factories will not be licensed in reserves of biological diversity such as Amazônia and the Pantanal, and the data indicate that at present such areas are not threatened.

The other criticism is that sugar cane will replace food crops, especially those grown by subsistence farmers, leading to food shortages for the poorer sections of Brazilian society. The expansion of sugar cane is occurring principally onto areas purchased from large landowners/ranchers who have extensive areas of degraded pastures. It is estimated that in the Atlantic coastal region (which includes the States of São Paulo and Paraná where expansion of cane is most rapid) there are perhaps 20 Mha of degraded pastures (Boddey et al., 2003) and in the Cerrado (central savanna) as much as 40 Mha of similar under-utilised, but not infertile land (Sano et al., 2001; Boddey et al., 2004). Brazil has no shortage of land for crop production. The resource-poor and landless require land reform, which is now occurring at an increased pace, but more importantly they need resources and skills to invest in the land and markets for their products and these facilities are only slowly becoming available. A vigorous rural economy fuelled by the intensive production of soybean, maize and sugarcane is more likely to provide employment for the poor, than the vast abandoned tracts of badly managed ranches which dominate the Cerrado region.

13.4 Labour Conditions

In this publication we have restricted comments on the social impact of the Bioethanol program to labour conditions. Approximately 80% of the area of Brazilian sugar cane is still harvested by hand. It is one of the most arduous occupations that exists in any industry. In cutting the burned cane the workers are exposed to the charred residues and they immediately become covered in ash and soot. To protect themselves from the rough cane stalks they must be fully attired with heavy protective clothing and boots and leggings to avoid injury with the sharp heavy cutlasses, while often working in temperatures which can reach 40°C. Even to cut the minimum requirement of one "tarefa" (~6 Mg of cane) requires an immense amount of energy and most workers manage to cut considerably more than this each day. The only reason that workers will accept such employment is because compared to most other work in the rural areas it is relatively well paid. Most workers manage to earn between R$ 600 and R$ 900 (US$ 300–US$ 450 – Globo Rural, 26 August 2007) per month which compares well to the national minimum wage of R$ 380, which is not often attained by the majority of workers in other rural occupations.

According to the last national census conducted by the Brazilian Institute of Geography and Statistics (IBGE, 2004), sugar cane involves the direct employment of more than 251,000 permanent and 242,000 temporary employees, a total of about 493,000 agrarian workforce. Rural labour is almost always the only possible livelihood for the unqualified worker. The number of Brazilian agrarian workers officially registered in the Ministry of Work and Employment is 32.3%, which means full access to public health, working rights and a secured retirement. However permanent and temporary rural employees engaged in the sugar cane industry are 64.9% and 39.7%, respectively. That means even the temporary workers possess guarantees and a formal employment on a wider scale than the remainder of the national agricultural work force (Balsadi, 2007). However, there have been some cases of terrible working conditions and low wages at some mills, with isolated cases where the workers are paid only sporadically or only at the end of the harvest, such that they owe so much money to the "company store" that they are effectively slaves. There is a major campaign of the Federal authorities to eliminate such practices and they levy large fines on these employers, but in such a vast country these abuses still occur.

The cost of harvesting burned cane using manual labour is estimated to be approximately R\$ 1000.oo ha^{-1}, considerably more than machine harvesting which is estimated at R\$ 700.oo ha^{-1}, including the costs of the equipment and maintenance etc. For this reason those companies that have the financial resources are investing in harvesters (cost approximately R\$ 800,000.oo) such that it is estimated that 450 machines will be sold in 2007 (Globo Rural, 26 August 2007). As one machine can harvest approximately 60 Mg h^{-1} (18.7 ha per day – Anon 2005) These machines operate 24 h in three 8 h shifts and the harvest last 180 days. Assuming a operational efficiency of 80%, this year approximately 1.2 Mha will be converted from manual harvesting of burned cane to machine harvesting of green cane. This will not only reduce operational costs, but also promote soil organic matter accumulation (Boddey et al., 2006), reduction in atmospheric pollution and methane emissions from burning, and improve soil fertility. The downside is that for each machine introduced approximately 80 jobs will be lost, increasing rural unemployment by 72,000. While manual cane harvesting may be a terrible task, it is far preferably to being unemployed.

13.5 Conclusions

It is unfortunate that the private companies, sugarcane co-operatives and even research institutions engaged in producing bioethanol, or studying the bioethanol program, rarely publish information in English and much of that in Portuguese is in unpublished reports and other grey literature. For this reason there is a lack of easily-available information on the history, growth and environmental and social impact of this large biofuel program. We have attempted to search out as much information as possible on this subject with special consideration for an international readership.

After relating the history of the program and the present situation today with regard to scale and agronomic and industrial practices, we have attempted to evaluate principally its environmental impact, both local and global, with regard to the ratio of fossil energy inputs to total energy yield of the fuel produced (energy balance), to its role in the mitigation of greenhouse gas emissions and other impacts such at water and atmospheric pollution. For the energy balance and greenhouse gas emissions we have tried to find the latest data not only on field and factory practice but also on the fossil fuel inputs and greenhouse gas emissions.

In agreement with other earlier reports from São Paulo State (Macedo, 1998; Macedo et al., 2003) we conclude that the energy balance is approximately 9:1; i.e. one unit of fossil energy invested produces 9 units of total energy as bioethanol.

As this subject is controversial, we have tried to be as transparent as possible by clearly citing the sources used for our information. The greenhouse gas balance indicates that using bioethanol produced from sugarcane with present practice will result in a 79% abatement of greenhouse gas emissions compared to "pure" gasoline. By "pure" we refer to gasoline as used in the USA and other countries with no ethanol addition, but with MTBE. If a comparison with Brazilian "gasohol" (actually gasoline with a 22–24% ethanol addition) this greenhouse gas abatement becomes about 73%.

With regard to accusations that the expansion of the area of sugarcane in Brazil will induce destruction of the Amazon rain forest or other reserves of biodiversity, the evidence contradicts this. Only 7–8 Mha of land are used as present for sugarcane, compared to over 350 Mha of Amazon forest, and the new areas for cane cultivation are being established in the south-east and central-west regions of Brazil principally in areas of *Brachiaria* and similar pastures used for extensive cattle ranching.

Working conditions for the manually harvesting of cane are extremely severe, almost inhuman, but for the rural poor of many regions of Brazil such work is comparatively well paid and rates of formal employment (with health security and pension provision) are much higher than is general in the remainder of the rural sector.

In the future is seems inevitable that the industry and area planted to cane will grow. Also increased mechanisation is inevitable and far fewer manual workers will be required, which will have negative effects on employment but positive effects on salaries and working conditions within the industry. Increased mechanisation and the consequent abandonment of pre-harvest burning and the introduction of no-till planting will further increase the energy balance, and reduce greenhouse gas emissions as well as other forms of atmospheric pollution and soil erosion. If State and Federal environmental protection agencies become increasing effective and employment laws are enforced, the Brazilian bioethanol program will be of great economic, and environmental benefit to the country, and could play a small but significant global role in the mitigation of greenhouse gas emissions from motor vehicles and reducing the consumption of petroleum.

According to a recent report in the Newspaper Folha de São Paulo on average a Flex fuel car will need 10 L of hydrous ethanol (95% alcohol) to cover the same distance as it would cover with just 7.2 L of Brazilian gasohol. Based on data presented in Table 13.9, the 10 L of ethanol would emit (95% of 3.76 kg of fossil CO_2) 3.57 kg of fossil CO_2. The same distance covered with gasohol would mean an emission of 13.45 kg CO_2 (1.87 kg CO_2 L^{-1} in the mixture gasoline:alcohol (23%) (3.57 kg CO2 L alcohol (our data) and 2.32 kg CO2 L gasoline – IPCC, 2006), even without considering the fossil energy expended in the refining of gasoline.

Acknowledgments The authors thank the Engineers Roberto dos Anjos and Antonio Sesso of Dedini S.A., Piracicaba, São Paulo for information on the dimensions and materials involved in the construction of modern cane factories/distilleries, Dr Rogério P. Xavier of Usina Itamaraty, Mato Grosso, for information on disposal of vinasse, Dr Alexander Resende of Embrapa Agrobiologia for information on the chemical composition of vinasse, Engineer Aureo Tasch, of John Deere S.A., Catalão, Goiás for information on modern cane harvesting machines and Dr Alexandre de Sene Pinto of the Centro Universitário Moura Lacerda, – Ribeirão Preto, São Paulo for information on biological control of insect pests in sugar cane. We are grateful to Embrapa and the International Atomic Energy Agency (IAEA – Vienna) for financing most of our recent work on sugar cane and the valuable co-operation of the students and staff of the Universidade Federal Rural do Rio de Janeiro. The authors RMB, SU and BJRA wish to acknowledge fellowships and grants from the National Research Council (CNPq) and the Rio de Janeiro State Research foundation (FAPERJ).

References

ANFAVEA (2007). Associação Nacional dos Fabricantes de Veículos Automotores – Brasil. Retrieved 24, August, 2007 http://www.anfavea.com.br/tabelas.html

Anon (2005). O maior aliado do setor sucroalcooleiro. *Revista Farm Forum, 4*(15), 4–6. (CNH Latin-America Ltda. Av. Juscelino K. de Oliveira, 11.825, Curitiba, 81450-903, Paraná.).

Azeredo, D. F., Bolsanelli, J., Weber, M., & Vieira, J. R. (1986). Nitrogênio em cana-planta, doses e fracionamento. *Revista da Sociedade dos Técnicos Açucareiros e Alcooleiros do Brasil (STAB), 4*, 26–32.

Arbex, M. A, Martins, L. C., de Oliveira, R. C., Pereira, L. A. A., Arbex, F. F., Cançado, J. E. D., Saldiva, P. H. N., & Braga, A. L. F. (2007). Air pollution from biomass burning and asthma hospital admissions in a sugar cane plantation area in Brazil. *Journal of Epidemiology and Community Health*, 61, 395–400.

Bacchi, O. O. S., Reichard, K., Sparovek, G., & Ranieri, S. B. L. (2000). Soil erosion evaluation in a small watershed in Brazil through ^{137}Cs fallout redistribution analysis and conventional models. *Acta Geologica Hispanica, 35*(3–4), 251–259.

Baldani, J. I., Caruso, L., Baldani, V. L. D., Goi, S. R., & Döbereiner, J. (1997). Recent advances in BNF with non-legume plants. *Soil Biology and Biochemistry, 29*, 911–922.

Baldani, J. I., Reis, V. M., Baldani, V. L. D., & Döbereiner, J. (2002). A brief story of nitrogen fixation in sugarcane reasons for success in Brazil. *Functional Plant Biology, 29*, 417–423.

Balsadi, O. V. (2007). Mercado de trabalho assalariado na cultura da cana-de-açúcar no Brasil no período 1992–2004. *Informações Econômicas, 37*, 38–54.

Biggs, I. M., Stewart, G. R., Wilson, J. R., & Critchley, C. (2002). [15]N natural abundance studies in Australian commercial sugarcane. *Plant and Soil, 238*(1), 21–30.

Boddey, R. M. (1993). Green energy from sugar cane. *Chemistry & Industry*, 19 May 1993 pp. 355–358.

Boddey, R. M., Macedo, R., Tarré, R. M., Ferreira, E., Oliveira, O. C., Rezende, C. P., Cantarutti, R. B., Pereira, J. M., Alves, B. J. R., & Urquiaga, S. (2004). Nitrogen cycling in *Brachiaria* pastures: the key to understanding the process of pasture decline. *Agriculture, Ecosystems and Environment, 103*, 389–403.

Boddey, R. M., Jantalia, C. P., Macedo, M. O., de Oliveira, O. C., Resende, A. S., Alves, B. J. R., & Urquiaga, S. (2006). Potential of carbon sequestration in soils of the Atlantic Forest region of Brazil. In R. Lal, C. C. Cerri, M. Bernoux, J. Etchevers & C. E. P. Cerri (Eds.), *Carbon Sequestration in Soils of Latin America*. (pp. 305–347). New York: Howarth Press.

Boddey, R. M. (1995). Biological nitrogen fixation in sugar cane: A key to energetically viable bio-fuel production. *CRC Critical Reviews in Plant Sciences, 14*, 263–279.

Boddey, R. M., Polidoro, J. C., Resende, A. S., Alves, B. J. R., & Urquiaga, S. (2001). Use of the 15N natural abundance technique for the quantification of the contribution of N2 fixation to sugar cane and other grasses. *Australian Journal of Plant Physiology, 28*(9), 889–895.

Boddey, R. M., Urquiaga, S., Alves, B. J. R., & Reis, V.M. (2003). Endophytic nitrogen fixation in sugar cane: Present knowledge and future applications. *Plant and Soil, 252*, 139–149.

Botelho, P. S. M., (1992). Quinze anos de controle biológico de *Diatraea saccharalis* utilizando parasitóides. *Pesquisa Agropecuária Brasileira, 27,* 255–262.

Campos, D. C. (2004). *Potencialidade do sistema de colheita sem queima da cana-de-açúcar para o seqüestro de carbono.* PhD Thesis, Escola Superior de Agricultura, Luiz Queiroz (ESALQ), Universidade de São Paulo, Piracicaba, SP, Brazil. (in Portuguese).

Cançado, J. E. D., Saldiva, P. H. N., Pereira, L. A. A., Lara, L. B. L. S., Artaxo, P., Martinelli, L. A., Arbex, M. A., Zanobetti, A., & Braga, A. L. F. (2006). The impact of sugar cane-burning emissions on the respiratory system of children and the eldery. *Environmental Health Perspectives, 114,* 725–729.

Cerri, C. C., Bernoux, M., Cerri, C. E. P., & Feller, C. (2004). Carbon cycling and sequestration opportunities in South America: the case of Brazil. *Soil Use and Management, 20,* 248–254.

Coelho, S. T. (2005). *A cana e a questão ambiental: Aspectos socio-ambientais da nova modalidade de energia.* (Seminar presented at the National Development Bank (BNDES), Rio de Janeiro, Nov. 25, 2005).

Correchel, V. (2003). *Avaliação de índices de erodibilidade do solo através da técnica da análise da redistribuição do "FALLOUT" do 137Cs.* DSc. Thesis, Universidade de São Paulo, Piracicaba, SP.

de Oliveira, M. E. D., Vaughan, B. E., & Rykiel, Jr. E. J. (2005). Ethanol as fuel: Energy, carbon dioxide balances and ecological footprint. *BioScience, 55,* 593–602.

de Souza, Z. M., Martins Filho, V. M., Marques Júnior, J., & Pereira, G. T. (2005). Variabilidade espacial de fatores de erosão em LATOSSOLO VERMELHO Eutroférrico sob cultivo de cana-de-açúcar. *Engenharia Agrícola (Jaboticabal, SP) 25,* 105–114.

Farla, J. C. M. & Blok, K. (2001). The quality of energy intensity indicators for international comparison in the iron and steel industry. *Energy Policy, 29,* 523–543.

Giampietro, M. & Pimentel, D. (1990). Assessment of the energetics of human labor. *Agriculture, Ecosystems and Environment, 32,* 257–272.

Globo Rural (2007). Television Program Rede Globo, 26, August, 2007. Full text retreived 27, August, 2007 http://globoruraltv.globo.com/GRural/0,27062,LTO0-4370-298116-1,00.html

Godoi, R. H. M., Godoi, A. F. L., Worobiec, A., Andrade, S. J., de Hoog, J., Santiago-Silva, M. R., & Van Grieken, R. (2004). Characterisation of Sugar Cane Combustion Particles in the Araraquara Region, Southeast Brazil, *Microchimica Acta, 145,* 53–56.

Hannon, B., Stein, R. G., Segal, B. Z., & Serber, D. (1978). Energy and labor in the contruction sector. *Science, 202,* 837–847.

Hoefsloot, G., Termorshuizen, A. J., Watt, D. A., & Cramer, M. D. (2005). Biological nitrogen fixation is not a major contributor to the nitrogen demand of a commercially grown South African sugarcane cultivar. *Plant and Soil, 277,* 85–96.

IBGE (2004). Instituto Brasileiro de Geografia e Estatística, *Pesquisa Nacional por Amostra de Domicílios.* Rio de Janeiro, 24, 2004. 27pp.

IBGE (2007). Instituto Brasileiro de Geografia e Estatística. Retrieved June 5, 2007, from http://www.sidra.ibge.gov.br/bda/default.asp?t=5&z=t&o=1&u1=1&u2=1&u3=1&u4=1&u5= 1&u6=1&u7=1&u8=1&u9=3&u10=1&u11=26674&u12=1&u13=1&u14=1

IEA (1999). International Energy Agency. *The reduction of Greenhouse Gas Emissions from the Cement Industry.* Report No PH3/7, May, 1999, Paris.

IPCC (2006). 2006 IPCC Guidelines for National Greenhouse Gas Inventories. Retrieved July, 2006, http://www.ipcc-nggip.iges.or.jp/public/2006gl/index.htm

ISO 14040 (2005). ISO/DIS 14040 – Environmental management – Life cycle assessment – Principles and framework.

James, E. K. (2000). Nitrogen fixation in endophytic and associative symbiosis. *Field Crops Research, 65,* 197–209.

Kongshaug, G. (1998). Energy Consumption and Greenhouse Gas Emissions in Fertilizer Production. IFA Technical Conference, Marrakech, Morocco, 28 September-1 October, 1998, 18pp.

Kovarik, W. (2005). Ethyl-leaded Gasoline: How a classic occupational disease became an international public health disaster. *International Journal of Occupational and Environmental Health, 11,* 384–397.

Lægreid, M., Bøckman, O. C., & Kaarstad, O. (1999). *Agriculture, fertilizers and the environment.* (Wallingford: CABI).

Li, R. P. & Macrae, I. C. (1992). Specific identification and enumeration of *Acetobacter diazotrophicus* in sugarcane. *Soil Biology and Biochemistry, 24*(5), 413–419.

Lima, E., Boddey, R. M., & Dobereiner, J. (1987). Quantification of biological nitrogen fixation associated with sugar cane using a 15N aided nitrogen balance. *Soil Biology and Biochemistry, 19*(2), 165–170.

Lombardi-Neto, F., Dechen, S. C. F., Castro, O. M. (1982). A cultura da cana-de-açúcar e as perdas de solo e água por erosão. (Paper presented at Congresso Brasileiro Conservação Solo, 4, Campinas, SP. Programa e Resumos. Campinas: Soc. Bras. Ciência Solo, Secretaria Agricultura do Estado de São Paulo e Ministério da Agricultura, 24 p.).

Macedo, I. C., Leal, M. R. L.V., & da Silva, J. E. A. R. (2003). *Greenhouse gas (GHG) emissions in the production and use of ethanol in Brazil: Present situation (2002).* Government of the State of Sao Paulo, Secretariat of the Environment, 47pp.

Macedo, I. C. (1997). *Balanço de Energia na produção de cana-de-açúcar e álcool nas usinas cooperadas: 1996.* Boletim CTC Coopersucar, 23pp.

Macedo, I. C. (1998). Greenhouse gas emissions and energy balances in bio-ethanol production and utilization in Brazil. *Biomass and Bioenergy, 14*(1), 77–81.

Machado, G. R., da Silva, W. M. & Irvine, J. E. (1987). Sugar cane breeding in Brazil: the Copersucar Program. In Copersucar International Sugarcane Breeding Workshop, Cooperativa de Produtores de Cana, Açúcar e Álcool do Estado de São Paulo Ltda, São Paulo, SP, Brazil. pp. 215–232.

MAPA (2007). *Balanço Nacional da Cana de Açúcar e Agroenergia.* Ministério da Agricultura, Pecuária e Abastecimento, Brasília, DF.

Mello, F. F. C., Cerri, C. E. P., Bernoux, M., Volkoff, B., & Cerri, C. C. (2006). Potential of soil carbon sequestration for the Brazilian Atlantic region. In R. Lal, C. C. Cerri, M. Bernoux, J. Etchevers, & C. E. P. Cerri (Eds.), *Carbon Sequestration in Soils of Latin America.* (pp. 349–368). New York: Haworth Press.

Muñoz-Rojas, J. & Caballero-Mellado, J. (2003). Population dynamics of *Gluconacetobacter diazotrophicus* in sugarcane cultivars and its effect on plant growth. *Microbial Ecology, 46,* 454–464.

Muthukumarasamy, R., Revathi, G., & Lakshminarasimhan, C. (1999). Influence of N fertilisation on the isolation of *Acetobacter diazotrophicus* and *Herbaspirillum* spp. from Indian sugarcane varieties. *Biology and Fertility of Soils, 29,* 157–164.

Muthukumarasamy, R., Revathi, G., & Loganathan, P. (2002). Effect of inorganic N on the population, in vitro colonization and morphology of *Acetobacter diazotrophicus* (syn *Gluconacetobacter diazotrophicus*). *Plant and Soil, 243,* 91–102.

Pimentel, D. & Patzek, T. (2007). Ethanol production: Energy and economic issues related to U.S. and Brazilian sugarcane (in press).

Pimentel, D. (Ed.) (1980). *CRC Handbook of energy utilization in agriculture.* (Boca Raton: CRC Press).

Resende, A. S., Xavier, R. P., Oliveira, O. C., Urquiaga, S., Alves, B. J. R., & Boddey, R. M. (2006). Long-term effects of pre-harvest burning and nitrogen and vinasse applications on yield of sugar cane and soil carbon and nitrogen stocks on a plantation in Pernambuco, N.E. Brazil. *Plant and Soil, 281,* 339–351.

Ritchie, J. C. & McHenry, J. R. (1990). Application of radioactive fallout cesium-137 for measuring soil erosion and sediment accumulation rates and patterns: a review. *Journal of Environmental Quality, 19,* 215–233.

Ritchie, J. C. & McHenry, J. R. (1995). *^{137}Cs use in erosion and sediment deposition studies: promises and problems.* Vienna: International Atomic Energy Agency. TECDOC 828.

Sano, E. E., Barcellos, A. O., & Bezerra, H. S. (2000). Assessing the spatial distribution of cultivated pastures in the Brazilian Savanna. *Pasturas Tropicales, 22,* 2–15.

Sene Pinto, A. de, (2007). Centro Universitário Moura Lacerda, - Ribeirão Preto, São Paulo 14076-510. Personal communication.

Shapouri, H., Duffield, J. A., & Wang, M. *The Energy Balance of Corn Ethanol: An Update.* (2002). U.S. Department of Agriculture, Office of the Chief Economist, Office of Energy Policy and New Uses. Agricultural Economic Report No. 814.

Sheehan, J., Camobreco, V., Duffield, D., Graboski, M., & Shapouri, H. *Life Cycle Inventory of Biodiesel and Petroleum Diesel for Use in an Urban Bus.* (1998). U.S. Department of Energy's Office of Fuels Development and U.S. Department of Agriculture's Office of Energy.

SINDAG (2007). Sindicato Nacional da Indústria de Produtos para Defesa Agrícola. Retrieved June 27, from http://www.sindag.com.br/osindag.php

Six, J., Elliott, E. T., & Paustian, K. (2000). Soil macroaggregate turnover and microaggregate formation: a mechanism for C sequestration under no-tillage agriculture. *Soil Biology and Biochemistry, 32,* 2099–2103.

Smil, V. (2001). Enriching the Earth. (Cambridge: MIT Press).

Sparovek, G. & Schnug, E. (2001). Soil tillage and precision agriculture. A theoretical case study for soil erosion control in Brazilian sugar cane production. *Soil & Tillage Research, 61,* 47–54.

TCU (1990). Proálcool, Programa Nacional do Álcool. Relatório de Auditoria Operacional. Tribunal de Contas da União. Brasilia, DF. 116pp.

UOL Economia (2007). Retrieved June 27, from http://noticias.uol.com.br/economia/

Urquiaga, S., Cruz, K. H. S., & Boddey, R. M. (1992). Contribution of nitrogen fixation to sugar cane: Nitrogen-15 and nitrogen-balance estimates. *Soil Science Society of America Journal, 56*(1), 105–114.

Worrell, E., Price, L., Martin, N., Farla, J. C. M., & Schaeffer, R. (1997). Energy intensity in the iron and steel industry: a comparison of physical and economic indicators. *Energy Policy, 25*(7–9), 727–744.

Worrell, E. & Galitsky, C. (2004). Energy Efficiency Improvement Opportunities for Cement Making. An ENERGY STAR® Guide for Energy and Plant Managers. LBNL-54036. U.S. Environmental Protection Agency. Retrieved July, 2007, from http://ies.lbl.gov/ iespubs/ieuapubs.html

Young, S. B., Turnbull, S., & Russell, A. (2002). Towards a Sustainable Cement Industry. Substudy 6: What LCA can tell us about the cement industry. Battelle/World Business Council for Sustainable Development. Retrieved July, 2007, from www.wbcsdcement.org

Chapter 14
Ethanol Production: Energy and Economic Issues Related to U.S. and Brazilian Sugarcane

David Pimentel and Tad W. Patzek

Abstract This analysis employs the most recent scientific data for the U.S. and for Brazil sugarcane production and the fermentation/distillation. These two countries were selected because they are the two largest countries in the world producing ethanol. All current fossil energy inputs used in the entire process of producing ethanol from sugarcane were included to determine the entire energy cost for ethanol production. Additional costs to consumers, including federal and state subsidies, plus costs of environmental pollution and/or degradation associated with the entire production system are discussed. The economic and the broad human food supply issues are evaluated. In addition, other studies are compared.

Keywords Converting biomass · energy costs · environmental costs · subsidization

14.1 Introduction

The supply of "conventional" oil is projected to peak before 2010 and its decline thereafter cannot be compensated fully by other liquid fuels (Youngquist and Duncan, 2003). The United States, Brazil, and other nations critically need to develop liquid fuel replacements for oil in the near future. The present search for alternative liquid fuels has focused on the conversion of biomass into liquid fuels.

Biomass is green plant material, like corn, soybeans, sugarcane, and trees. All biomass converts solar energy into plant material. However, the major difficulty in relying on the use of solar energy collected by plant biomass is that green plants on average, collect only about 0.1% of the solar energy that reaches the land each year

✉ D. Pimentel
College of Agriculture and Life Sciences, Cornell University, 5126 Comstock Hall,
Ithaca, NY 15850,
e-mail: Dp18@cornell.edu

T.W. Patzek
Department of Civil and Environmental Engineering, University of California,
Berkeley, CA 94720, 425 David Hall, MC1716,
e-mail: patzek@patzek.ce.berkeley.edu

D. Pimentel (ed.), *Biofuels, Solar and Wind as Renewable Energy Systems*,
© Springer Science+Business Media B.V. 2008

(Pimentel et al., 2007). In addition, biomass production requires large land areas, suitable soil, nutrients, and freshwater for its production. Then, in the conversion of the biomass into liquid fuel, water, microorganisms, and energy are required.

Professor Mario Giampietro (personal communication, visiting Scholar, Arizona State University, 2007) reported that a person in a developed economy, like the U.S. and Brazil, requires a range of 10–100 W/m^2 of fossil energy in urban land uses. Without including any charges for the fossil energy inputs for sugarcane production in Brazil, the collection of solar energy in Brazil is equal to only 1.58 W/m^2, but with the outcome of producing ethanol, then only 0.4 W/m^2 are captured. Therefore, sugarcane and corn and other biomass resources will not supply a developed economy sufficient energy (Patzek and Pimentel, 2005).

14.2 Energy Inputs in Sugarcane Production

The conversion of sugarcane and other food/feed crops into ethanol by fermentation is a well-known and established technology. In both the U.S. and Brazil, the energy costs for the production inputs and processing inputs were apportioned to each activity. Also, in both countries, the sugarcane bagasse that remains after crushing is burned to provide steam and electricity in the processing activity. The ethanol yield from a large production plant averages about 1 L of ethanol from 12 to 14 kg of fresh sugarcane (Ferguson, 2004; Patzek and Pimentel, 2005; R.M Boddey, Senior Scientist, Empresa Brasileira de Pesquisa Agropecuaria (Embrapa), Brasil, personal communication, 2007).

14.2.1 United States

The production of sugarcane in Louisiana, United States requires a significant energy and dollar investment for the 12 inputs, including labor, farm machinery, fertilizers, pesticides, and electricity (Table 14.1). To produce an average sugarcane yield of 88,000 kg/ha requires the expenditure of about 13 million kcal of energy inputs (mostly oil and natural gas) (Table 14.1). This energy input is the equivalent of about 1,300 kg of oil equivalents expended per hectare of sugarcane. The full production costs total about $2,524/ha for the 88,000 kg/ha or approximately 3 ¢/kg of sugarcane produced.

14.2.2 Brazil

The production of sugarcane in Brazil also requires a significant energy and dollar investments for the 12 inputs (Table 14.2). Energy investment amounts to about 8.8 million kcal, to produce the average sugarcane yield of 77,000 kg/ha or slightly lower than in Louisiana. This is the equivalent of about 393 kg of oil equivalents expended per hectare of sugarcane produced in Brazil (Table 14.2).

Table 14.1 Energy inputs and costs of sugarcane production per hectare in Louisiana

Inputs	Quantity	kcal × 1000	Costs $
Labor	40 h[a]	1,621[b]	520.00[c]
Machinery	70 kg[d]	917[e]	264.44[f]
Diesel	430 L[a]	4,902[g]	860.00[h]
Nitrogen	196 kg[a]	3,136[i]	107.80[j]
Phosphorus	118 kg[a]	444[k]	73.16[l]
Potassium	185 kg[a]	603[m]	57.35[n]
Lime	237 kg[a]	96[o]	4.74[o]
Sulfur	27 kg[a]	53[p]	27.00[h]
Sets	12,000 kg[q]	207[r]	230.00[h]
Herbicides	5.8 kg[a]	580[s]	116.00[h]
Insecticides	2.5 kg[a]	250[s]	50.00[h]
Transportation	715 kg[t]	593[t]	213.80[h]
TOTAL		13,402	$2,524.29
Sugarcane yield	88,000 kg/ha[a]	107,000,000	kcal input:output 1:7.98
Sugar	6,600 kg/ha[a]		

[a] Breaux and Salassi, 2003.
[b] It is assumed that a person works 2,000 h per year and utilizes an average of 8,000 L of oil equivalents per year.
[c] It is assumed that labor is paid $13 an hour.
[d] Energy costs for farm machinery that was obtained from Breaux and Salassi, (2003). Tractors, harvesters, plows and other equipment was assumed to last about 10 years and are used on 160 hectares per year. These data were prorated per year per hectare.
[e] Prorated per hectare and 10-year life of the machinery (Gamble, 2003). Tractors weigh from about 10 tons (DeJong-Hughes, 2005) and harvesters about 10 tons (Taganrog, 2004–2006), plus plows, sprayers, and other equipment.
[f] Hoffman et al., 1994.
[g] Input 11, 400 kcal per liter.
[h] Estimated.
[i] Patzek, 2004.
[j] Cost $0.55 per kg.
[k] Input 3,762 kcal per kg.
[l] Cost $0.62 per kg.
[m] Input 3,260 kcal per kg.
[n] Cost $0.31 per kg.
[o] Pimentel and Patzek, 2005.
[p] Pimentel, 1980.
[q] Patzek and Pimentel, 2005.
[r] R.M Boddey, Senior Scientist, Empresa Brasileira de Pesquisa Agropecuaria (Embrapa), Brasil, personal communication, 2007.
[s] Input 100,000 kcal per kg of herbicide and insecticide.
[t] Goods transported include machinery, fuels, and seed that were shipped an estimated 1,000 km. Input 0.83 kcal per kg per km transported.

Table 14.2 Energy inputs and costs of sugarcane production per hectare in Brazil

Inputs	Quantity	kcal × 1000	
Labor	128 h[a]	157[a]	
Machinery	156 kg[a]	638[a]	
Diesel	22.3 L[a]	254[a]	
Nitrogen	58.3 kg[b]	933[c]	
Phosphorus	16 kg[b]	60[d]	
Potassium	83 kg[b]	270[e]	
Lime	367 kg[a]	114[a]	
Sulfur	2 kg[f]	53[g]	
Sets	21,000 kg[h]	207[h]	
Herbicides	5 kg[i]	500[c]	
Insecticides	2 kg[f]	200[c]	
Transportation	650 kg[j]	540[k]	
TOTAL		3,926	
Sugarcane yield	77,000 kg/ha[l]	94,000,000	kcal input:output 1:23.94
Sugar yield	5,789 kg/ha[l]		

[a] Macedo et al., 2004.
[b] Boddey, 1995.
[c] Pimentel and Patzek, 2005.
[d] Input 3,762 kcal/kg.
[e] Input 3,260 kcal/kg.
[f] Breaux and Salassi, 2003.
[g] Pimentel, 1980.
[h] R.M. Boddey, Senior Scientist, Empresa Brasileira de Pesquisa Agropecuaria (Embrapa), Brasil, personal communication, 2007.
[i] Sartori and Basta, 1999.
[j] Goods transported include machinery, fuels, and seed that were shipped an estimated 1,000 km.
[k] 0.83 kcal/kg/km.
[l] Patzek and Pimentel, 2005.

14.3 Energy Inputs in Fermentation/Distillation

The average costs in terms of energy and dollars for a large (250–300 million liters/year), modern ethanol plant are listed for U.S. and Brazilian sugarcane conversion in Tables 14.3 and 14.4. In the fermentation/distillation process, the sugarcane is crushed and squeezed and approximately 1.5 L of water are added to each kilogram of sugarcane juice. After microbial fermentation, to obtain a liter of 95% pure ethanol from the 10% ethanol to 90% water mixture, 1 L of ethanol must be extracted from approximately 11 L of the ethanol/ water mixture (O. Primavesi, personal communication, Senior Scientist, Embrapa Pecuaria Sujdeste, Brazil, 2007). Although ethanol boils at about 78 degrees C, and water boils at 100 degrees C, the ethanol is extracted from the water in multi-step distillations, which provides 95% pure ethanol (Maiorella, 1985; Wereko-Brobby and Hagan, 1996; S. Lamberson, personal communication, Cornell University, 2000).

In order to be mixed with gasoline in the U.S. and Brazil, the 95% ethanol must be further processed to 99.5% pure ethanol. More water must be removed, requiring additional fossil energy inputs to achieve the 99.5% pure ethanol (U.S. in

Table 14.3 Inputs per 1000 L of 99.5% ethanol produced from U.S. sugarcane[a]

Inputs	Quantity	kcal × 1000	Dollars $
Sugarcane	12, 000 kg[b]	1,828[b]	363.95[b]
Sugarcane transport	12, 000 kg[c]	490[c]	80.00[d]
Water	21,000 L[e]	90[f]	19.04[f]
Stainless steel	3 kg[g]	165[h]	10.60[d]
Steel	4 kg[g]	92[h]	10.60[d]
Cement	8 kg[g]	384[h]	10.60[d]
Steam	2,546,000 kcal[i]	0[j]	21.00[k]
Electricity	392 kWh[i]	0[j]	27.44[k]
95% ethanol to 99.5%	9 kcal/L[l]	9[l]	0.60
Sewage effluent	20 kg BOD[m]	69[n]	6.00
Distribution	331 kcal/L[o]	331	20.00[o]
TOTAL		3,458	$569.83

[a] Output: 1 L of ethanol = 5,130 kcal
[b] Data from Table 14.1.
[c] Calculated for 16 km roundtrip.
[d] Pimentel, 2003.
[e] 1.5 L of water mixed with each kg of sugarcane juice.
[f] Pimentel et al., 2004.
[g] Estimated.
[h] Newton, 2001.
[i] Illinois Corn, 2004.
[j] Bagasse was used as a substitute fuel to generate steam and also as the fuel to generate electricity. The bagasse with 45% to 55% moisture has an energy value of about 1,900 kcal/kg (Liu and Helyar, 2003).
[k] Although there was charge for the fuel, the manipulations of using the fuel to generate steam and produce electricity both cost a small amount for the manipulations.
[l] 95% ethanol converted to 99.5% ethanol for the addition to gasoline (T. Patzek, personal communication, University of California, Berkeley, 2004).
[m] 20 kg of BOD per 1000 L of ethanol produced (Kuby et al., 1984).
[n] 4 kWh of energy required to process 1 kg of BOD (Blais et al., 1995).
[o] DOE, 2002.

Table 14.3). Thus, a total of about 10 L of wastewater must be removed per liter of ethanol produced. This relatively large amount of sewage effluent has to be disposed of at an energy, economic, and environmental cost.

To produce a liter of 99.5% ethanol from sugarcane in the U.S. uses less fossil energy than the energy delivered back as ethanol (input/output ratio 1 kcal: 1.48 kcal) and costs 57¢/per liter ($2.15 per gallon) (Table 14.3). The sugarcane feedstock requires more than 40% of the total energy input in processing. In this analysis the total economic cost, including the energy inputs for the fermentation/distillation process and the apportioned energy costs of the stainless steel tanks and other industrial materials, is $570 per 1000 L of ethanol produced (Table 14.3).

The production of 1,000 L of ethanol under Brazilian conditions uses less energy than in the U.S. (Tables 14.3 and 14.4). With Brazilian ethanol production there is a net return of 2.28 kcal per 1 kcal of fossil energy invested. Yet, economic cost of production is 26¢/liter compared to 57¢/liter in the U.S (Tables 14.3 and 14.4).

Table 14.4 Inputs per 1000 L of 99.5% ethanol produced from Brazilian sugarcane[a]

Inputs	Quantity	kcal × 1000	
Sugarcane	12,000 kg[b]	612[b]	
Sugarcane transport	12,000 kg[c]	490[d]	
Water	21,000 L[e]	90[f]	
Stainless steel	3 kg[g]	165[h]	
Steel	4 kg[g]	92[h]	
Cement	8 kg[g]	384[h]	
Steam	2,546,000 kcal[i]	0[j]	
Electricity	392 kWh[i]	0[j]	
95% ethanol to 99.5%	9 kcal/L[k]	9[k]	
Sewage effluent	20 kg BOD[l]	69[m]	
Distribution	331 kcal/L[n]	331	
TOTAL		2,242	26 ¢/liter[o]

[a] Output: 1 L of ethanol = 5,130 kcal.
[b] Data from Table 14.2.
[c] Calculated for 16 km roundtrip.
[d] R.M Boddey, Senior Scientist, Empresa Brasileira de Pesquisa Agropecuaria (Embrapa), Brasil, personal communication, 2007.
[e] 1.5 L of water mixed with each kg of sugarcane juice.
[f] Pimentel et al., 2004.
[g] Estimated.
[h] Newton, 2001.
[i] Illinois Corn, 2004.
[j] Bagasse was used as a substitute fuel to generate steam and also as the fuel to generate electricity. The bagasse with 45–55% moisture has an energy value of about 1,900 kcal/kg (Liu and Helyar, 2003).
[k] 95% ethanol converted to 99.5% ethanol for the addition to gasoline (T. Patzek, personal communication, University of California, Berkeley, 2004).
[l] 20 kg of BOD per 1000 L of ethanol produced (Kuby et al., 1984).
[m] 4 kWh of energy required to process 1 kg of BOD (Blais et al., 1995).
[n] DOE, 2002.
[o] Calibre, 2006.

14.4 Energy Yield

The largest energy inputs in sugarcane-ethanol production are for producing the sugarcane feedstock, plus the transport energy (Tables 14.3 and 14.4). The total energy input to produce a liter of ethanol in the U.S. is 3,458 kcal and in Brazil 2,242 kcal. A liter of ethanol has an energy value of 5,130 kcal. Thus, in the U.S. there is a positive energy return (1:1.48), while in Brazil the positive energy return is 1:2.28 or slightly larger than in the U.S. (Tables 14.3 and 14.4).

14.5 Economic Costs

Not only does the U.S. ethanol production technology with sugarcane use nearly as much fossil fuel to produce as there is energy in the ethanol produced (Table 14.3), but it also costs substantially more dollars to produce than its energy value is worth on the U.S. market. Clearly, without the more than $3 billion U.S. federal and state government yearly subsidies, U.S. corn or sugarcane ethanol production would be reduced or cease to exist, confirming the basic fact that ethanol production is uneconomical (National Center for Policy Analysis, 2002).

Federal and state subsidies for corn ethanol production total more than $7 per 25 kg (1 bushel), for example. The subsidies are mainly paid to large corporations, like Archer Daniels Midland, and others (McCain, 2003), while corn farmers, for instance, are receiving a maximum of only about 70¢ per 25 kg (1 bushel) for their corn (or about $100 per acre) in the subsidized corn ethanol production system. Senator McCain reports that direct subsidies for ethanol, plus the subsidies for corn grain, amount to 79¢ per liter of ethanol produced (McCain, 2003). Thus, the subsidy per liter of U.S. ethanol is 60 times greater than the current subsidy per liter of gasoline, based on the $5 billion per year subsidy for oil (Gara, 2006; NCGA, 2006; Koplow, 2006).

If the production cost of a liter of ethanol were added to the tax subsidy cost, then the total cost for a liter of sugarcane ethanol would be $1.11. Because of the relatively low energy content of ethanol, 1.6 L of ethanol have the energy equivalent of 1 L of gasoline. Thus, the cost of producing an amount of ethanol equivalent to a liter of gasoline is $1.78 ($6.71 per gallon of gasoline). This is more than the 53¢ per liter, the current cost of producing a liter of gasoline (USCB, 2004–2005).

The subsidies for ethanol produced from sugarcane in the U.S. would have similar effects as subsidies in corn ethanol. Unfortunately the costs to the American consumer are greater than the $8.4 billion/year expended to subsidize and produce the U.S. ethanol because diverting the required corn feedstock from livestock increases corn grain prices for livestock producers. The National Center for Policy Analysis (2002) estimate is that ethanol production is adding more than $1 billion per year to the cost of beef production for consumers (2002). Given that about 78% of the current corn grain harvest is fed to U.S. livestock (USDA, 2004), the doubling or tripling of ethanol production can be expected to increase corn grain prices further for beef production and for other livestock products, including milk and eggs, and ultimately increase costs to the consumer. Therefore, in addition to paying the $8.4 billion in taxes for ethanol and subsidies, consumers are expected to face significantly higher meat, milk, and egg prices, plus higher sugar prices if much sugarcane eventually goes into ethanol production in the market place.

14.6 Land Use in the U.S.

Currently 395,000 ha are devoted to U.S. sugarcane production and most of this is for sugar production (USDA, 2004). Relatively little or no ethanol is being produced from U.S. sugarcane.

Currently, about 18.9 billion liters of ethanol (5 billion gallons) are being produced in the United States each year primarily from corn (DOE, 2005). In contrast, the total petroleum used in the U.S. was about 1,200 billion liters in 2004–2005 (USCB, 2004–2005). To produce the 18.9 billion liters of ethanol, about 5.0 million ha or 20% of U.S. corn land is used. Furthermore, the 18.9 billion liters of ethanol (energy equivalent to 12.5 billion liters of vehicle liquid fuel) provides only 1% of the petroleum utilized by U.S. each year. If corn-ethanol production were expanded to using 100% of U.S. corn production, this would provide only 7% of the petroleum needs!

14.7 Ethanol Production and Use in Brazil

In contrast, Brazil can fuel most of its automobiles and other vehicles with ethanol because Brazilians consume only 9% of the U.S. consumption in petroleum (BP, 2005). Since 1984 the portion of Brazilian sugarcane used for ethanol decreased from a peak of 70% to about 55% in 2000 (Schmitz et al., 2003). During that time the percentage of ethanol cars declined from 94% in 1984 to less than 1% in 1996 (Rosillo-Calle and Cortez, 1998). The difference included gasoline for the cars. Flex cars replaced the ethanol cars and as a result Brazil's oil consumption has increased 42% during the last decade (BP, 2001, 2005).

Proponents of ethanol point to the production of ethanol in Brazil but ignore the fact that the U.S. now produces more ethanol (18.9 billion liters ethanol per year) compared with Brazil that produces about 15.1 billion liters per year (Calibre, 2006). Brazil is fortunate to have the land and climate suitable for sugarcane. Sugarcane is a more efficient feedstock for ethanol production than corn grain (Patzek and Pimentel, 2005). However, because Brazilian energy balance is only slightly positive (1 kcal:2.28 kcal), the Brazilians need to heavily subsidize their ethanol industry as does the U.S.. In the 1980s and 1990s the Brazilian government sold ethanol to the public for 22 ¢ per liter, but it cost the government 33 ¢ per liter to produce (Pimentel, 2003). Because other priorities emerged in Brazil, the government has since abandoned directly subsidizing ethanol (Spirits Low, 1999; Coelho et al., 2002). Now the consumer is paying the subsidy directly at the pump (Pimentel, 2003).

The total Brazilian subsidy is estimated to be about 50% for ethanol production (CIA, 2005). Earlier it was mentioned that it costs 26¢ to produce a liter of ethanol in Brazil that sells for 86¢ per liter (Calibre, 2006). Brazilian gasoline sells for nearly $1.23 per liter or about 43% higher than a liter of ethanol (R.M Boddey, Senior Scientist, Empresa Brasileira de Pesquisa Agropecuaria (Embrapa), Brasil, personal communication, 2007). Thus, higher gasoline prices help subsidize the cost of ethanol production in Brazil (CIA, 2005).

14.8 Environmental Impacts

Some of the economic and energy contributions of ethanol production both in the U.S. and Brazil are negated by the widespread environmental pollution problems associated with ethanol production using sugarcane. Many of the environmental impacts in Brazil associated with sugarcane production also occur in the U.S.

sugarcane production. Sugarcane production causes more intense soil erosion than any crop produced in Brazil because the total sugarcane biomass is harvested and processed in ethanol production. This removal of most of the biomass leaves the soil unprotected and exposed to erosion from rainfall and wind energy. For example, soil erosion with sugarcane cultivation is reported to have the highest soil erosion rate in all Brazilian agriculture, averaging 31 t/ha/yr (Sparovek and Schung, 2001). The 31 t/ha soil loss is 30–60 times greater than sustainability of the soil in agriculture (Troeh et al., 2004; Pimentel, 2006).

In addition, sugarcane production uses larger quantities of herbicides and insecticides and nitrogen fertilizer (Tables 14.1 and 14.2) than most other crops produced in Brazil and these chemicals spread to ground and surface water thereby causing significant water pollution (NAS, 2003).

Relatively large quantities of water are required to produce sugarcane. Because it takes 12 kg of sugarcane to produce 1 L of ethanol, about 7,000 L of water are needed to produce the required 12 kg of sugarcane per liter of ethanol.

Although the Brazilian government has passed legislation to curtail the burning of sugarcane before harvest to reduce air pollution problems, most of the sugarcane in Brazil is still burned and this is resulting in respiratory problems in children and the elderly (Braunbeck et al., 1999; Cancado et al., 2006). The rules need to be enforced to help protect the people from this serious air pollution problem. Additional smoke is released during the removal of forests for sugarcane and other crop production. Between May 2000 and August 2005, Brazil lost more than 132,000 square km of forest, an area larger than Greece (Mongabay, 2006).

The harvesting of sugarcane by laborers is hard and dangerous work, cutting the sugarcane with large knives. As Broietti (2003) reported these are dangerous and miserable conditions under which to work.

All these factors confirm that the environmental and agricultural system in which Brazilian and U.S. sugarcane is being produced is experiencing major environmental problems. Further, it substantiates the conclusion that the sugarcane production system, and indeed the ethanol production system, are not environmentally sustainable now or for the future. Because sugarcane is the raw material for ethanol production in Brazil, it cannot be considered a renewable energy source, considering the production and processing aspects.

Another pollution problem concerns the large amounts of waste-water produced by each ethanol plant. As noted, for each liter of ethanol produced using sugarcane, about 10 L of wastewater are produced. This polluting wastewater has a biological oxygen demand (BOD) of 18,000–37,000 mg/liter depending of the type of plant (Kuby et al., 1984). The cost of processing this sewage in terms of energy (4 kWh/kg of BOD) must be included in the cost of producing ethanol (Tables 14.3 and 14.4).

14.9 Air Pollution

Reports confirm that ethanol use contributes to air pollution problems when burned in automobiles (Youngquist, 1997; Hodge, 2002, 2003, 2005; Niven, 2005). The use of fossil fuels, as well as the use of ethanol in cars, releases significant quantities of pollutants to the atmosphere. Furthermore, carbon dioxide emissions released from burning these fossil fuels contribute to global warming and are a serious concern (Schneider et al., 2002). Additional carbon dioxide is released during the fermentation process. Also, when the soil is tilled serious soil erosion takes place and soil organic matter is oxidized. When all the air pollutants associated with the entire ethanol production system are considered, the evidence confirms that ethanol production contributes to the already serious U.S. and Brazilian air pollution problem (Youngquist, 1997; Hodge, 2002, 2003, 2005; Pimentel and Patzek, 2005; Patzek and Pimentel, 2005).

14.10 Food Security

At present, world agricultural land supplies more than 99% of all world food (calories), while aquatic ecosystems supply less than 1% (FAO, 2002). Worldwide, during the last decade, per capita available cropland decreased 20% and irrigation land 12% (Brown, 1997). Furthermore, per capita grain production has been decreasing, in part due to increases in the world population (Worldwatch Institute, 2001). Worldwide, diverse cereal grains, including corn, make up 80% of the food of the human food supply (Pimentel and Pimentel, 1996).

The current food shortages throughout the world call attention to the importance of continuing U.S. and Brazilian exports of grains and other food crops for human nutrition. The expanding world population that now numbers 6.5 billion, further complicates and stresses the food security problem now and for the future (PRB, 2006). Almost a quarter million people are added each day to the world population, and each of these human beings requires adequate food. Today, the malnourished people in the world number more than 3.7 billion (WHO, 2006). This is the largest number of malnourished people and proportion ever reported in history. Malnourished people are highly susceptible to various serious diseases and this is reflected in the rapid rise in the number of seriously infected people in the world, with diseases like tuberculosis, malaria, and AIDS, as reported by the World Health Organization (Kim, 2002; Pimentel et al., 2006).

14.11 Food versus the Fuel Issue

Using sugarcane, a human food resource, for ethanol production, raises ethical and moral issues (Wald, 2006). Expanding ethanol production entails diverting valuable cropland from the production of food crops needed to nourish people. The energetic and environmental aspects, as well as the moral and ethical issues also deserve serious consideration. In spite of oil and natural gas shortages now facing the U.S., ethanol production is forcing the U.S. to import more oil and natural gas to produce ethanol and other biofuels (Pimentel and Patzek, 2005).

The expansion of ethanol production in the U.S. and Brazil is having negative impacts on food production and food exports (Chang, 2006), and is likely to have further negative impacts on food production and the environment.

Furthermore, increasing oil and natural gas imports in the U.S. and other countries drives up the price of oil and gas. This is especially critical for the poor in developing countries of the world. Even now this is documented by the fact that worldwide per capita fertilizer use has been declining for the last decade because of the increased costs for the poor farmers of the world (Worldwatch Institute, 2001).

14.12 Summary

For a thorough and up-to-date evaluation of all the fossil energy costs of ethanol production from sugarcane in both the U.S. and Brazil, every energy input in the biomass production and ultimate conversion process must be included. In this study, more than 12 energy inputs in average U.S. and Brazilian sugarcane production are evaluated. Then in the fermentation/distillation operation, 9 more fossil fuel inputs are identified and included. Some energy and economic credits are given for the bagasse to reduce the energy inputs required for steam and electricity.

Based on all the fossil energy inputs in U.S. sugarcane conversion process, a total of 1.48 kcal of ethanol is produced per 1 kcal of fossil energy expended. In Brazil, a total of 2.28 kcal of ethanol is produced per 1 kcal of fossil energy expended.

Some pro-ethanol investigators have overlooked various energy inputs in U.S. and Brazilian sugarcane production, including farm labor, farm machinery, processing machinery, and others. In other studies, unrealistic low energy costs were attributed to such energy inputs, as nitrogen fertilizer, insecticides, and herbicides (Corn-Ethanol, 2007).

Both the U.S. and Brazil heavily subsidize ethanol production. The data suggest that billions of dollars are invested in subsidies and this significantly increases the costs to the consumers.

The environmental costs associated with producing ethanol in the U.S. and Brazil are significant but have been generally overlooked. The negative environmental impacts on the availability of cropland and freshwater, as well as on air pollution and public health, have yet to be carefully assessed. These environmental costs in terms of energy and economics should be calculated and included in future ethanol analyses so that sound assessments can be made.

In addition, the production of ethanol in the U.S. and Brazil further confirms that the mission of converting biomass into ethanol will not replace oil. This mission is impossible.

General concern has been expressed about taking food crops to produce ethanol for burning in automobiles instead of using these crops as food for the many malnourished people in the world. The World Health Organization reports that more than 3.7 billion humans are currently malnourished in the world – the largest number of malnourished ever in history.

Acknowledgments We would like to thank the following people for their valuable comments and suggestions on earlier drafts of this manuscript: Andrew B. Ferguson, Optimum Population Trust, Oxon, UK; Mario Giampietro, Istituto Nazionale di Ricerca per gli Alimenti e Nutrizione (INRAN), Rome, IT; Matthew Farwell, Alternative Energy, Energy, Nanotechnology, Palo Alto, CA; Marcelo Dias de Oliveira, University of Florida, Gainesville, FL; Odo Primavesi, Empresa Brasileira de Pesquisa Agropecuária (Embrapa), Brazil; Thomas Standing, San Francisco Public Utilities Commission, San Francisco, CA; Sergio Ulgiati, Department of Chemistry, University of Siena, Italy; Walter Youngquist, Petroleum Consultant, Eugene, OR.

This research was supported in part from a grant from the Podell Emertii award at Cornell University.

References

Blais, J.F., Mamouny, K., Nlombi, K., Sasseville, J.L., & Letourneau, M. (1995). Les mesures deficacite energetique dans le secteur de leau. (In J.L Sassville & J.F. Balis (Eds.), Les Mesures deficacite Energetique pour Lepuration des eaux Usees Municipales: Scientific Report 405, Vol. 3, INRS-Eau, Quebec.).

Boddey, R.M. (1995). Biological nitrogen fixation in sugarcane: a key energetically viable bio-fuel production. *CRC Critical Review in Plant Science, 14*, 263–279.

BP. (2001). *British Petroleum statistical review of the world energy.* June 2001. (New York: Morgan Guaranty Trust Company of New York).

BP. (2005). *British Petroleum statistical review of the world energy.* June 2001. (Providence, RI: J. P. Morgan Chase Bank).

Braunbeck, O., Bauen, A., Rosillo-Calle, F., & Cortez, L. (1999). Prospects for green cane harvesting and cane residue use in Brazil. *Biomass and Bioenergy, 17*, 495–506.

Breaux, J. & Salassi, M.F. (2003). *Projected costs and returns – Sugarcane in Louisiana, 2003.* Louisiana State University Agricultural Center, Louisiana Agricultural Experiment Station. Department of Agricultural Economics and Agribusiness. Bull., No. 211, 38pp.

Broietti, M.H. (2003). *Os assalariados rurais temporaries da Cana.* (San Paulo Plano Editoracao).

Brown, L.R. (1997). *The agricultural link: How environmental deterioration could disrupt economic progress.* (Washington, DC: Worldwatch Institute).

Calibre. (2006). DJ Brazilian sugar millers fix prices as futures jump. Retrieved April 25, 2006 from http:/caliber.mworld.com/m/m.w?lp=GetStory&id=190618291

Cancado, J.E.D., Saldiva, P.H.N., Pereira, L.A.A., Lara, L.B.L.S., Artaxo, P., Martinelli, L.A., Arbex, L.A., Zonobetti, A., & Braga, A.L.F. (2006). Impact of sugar cane-burning emissions on the respiratory system of children and the elderly. *Environmental Health Perspectives, 114* (5), 725–729.

Chang, J. (2006). Difficult road for ethanol in Brazil. *Knight Ridder*, May 2, 2006. 2pp.

CIA. (2005). Brazil natural gas. The Library of Congress Country Studies; CIA World Factbook. Retrieved September 5, 2005 from http://www.photius.com/countries/brazil/economy/brazil_economy_natural_gas.html

Coelho, S.T., Bolognini, M.F., Silva, O.C., & Paletta, C.E.M. (2002). *Biofuels in Brazil: The current situation.* CENBIO –The National Reference Center on Biomass. Technical Texts. Retrieved November 12, 2002 from http://www.cenbio.org.br/in/index.html

Corn-Ethanol. (2007). Corn-based ethanol: Is this a solution? The Oil Drum. Retrieved June 15, 2007 from http://www.theoildrum.com/node/2615

DeJong-Hughes, J. (2005). *Soil compaction: What you can do?* Minnesota Crop e-News, University of Minnesota, Extension.

DOE. (2002). Review of transport issues and comparison of infrastructure costs for a renewable fuels standard: U.S. Department of Energy, Washington, DC. Retrieved October 8, 2002 from http://tonto.eia.doe.gov/FTPROOT/service/question3.pdf

DOE. (2005). Energy efficiency and renewable energy: U.S. Department of Energy. Washington DC. Retrieved January 6, 2006 from http://www1.eere.energy.gov/biomass/ethanol.html

FAO. (2002). *Food balance sheets.* (Rome: Food and Agriculture Organization of the United Nations)

Ferguson, A.R.B. (2004). Sugarcane and energy: *Optimum Population Trust.* July 20, 1999, 9pp.

Gamble, R. (2003). *Lease agreements for farm buildings.* Factsheet. Ministry of Agriculture, Food and Rural Affairs. Ontario, Canada. Retrieved October 24, 2007 from http://www.omafra.gov.on.ca/english/busdev/facts/03-095.htm

Gara, L. (2006). New oil "tax" charges Alaskans for BP pipeline failures; Gives away $5+ billion in state revenue. Alaska State Legislature. Retrieved August 27, 2006 from http://gara/akde,pmcrats/org

Hodge, C. (2002). Ethanol use in US gasoline should be banned, not expanded. *Oil and Gas Journal,* September 9, 20–30.

Hodge, C. (2003). More evidence mounts for banning, not expanding, use of ethanol in gasoline. *Oil and Gas Journal,* October 6, 20–25.

Hodge, C. (2005). Government and fuels: Increased air pollution with the consumption of ethanol in gasoline. Retrieved October 10, 2005 from http://www.arb.ca.gov/fuels/gasoline/meeting/2005/0502052ndopi

Hoffman, T.R., Warnock, W.D., & Hinman, H.R. (1994). Crop enterprise budgets; timothy-legume and alfalfa hay, Sudan grass, sweet corn and spring wheat under rill irrigation; Kittitas County, Washington. Farm Business Reports EB 1173, Pullman, Washington State University.

Illinois Corn. (2004). Ethanol's energy balance. Retrieved August 10, 2004 from http://www.ilcorn.org/Ethanol/Ethan_Studies/Ethan_Energy_Bal/ethan_energy_bal.html

Kim, Y. (2002). World exotic diseases. (In D. Pimentel (Ed.), *Biological invasions: Economic and environmental costs of alien plant, animal, and microbe species* (pp. 331–354). Boca Raton, FL: CRC Press)

Koplow, D. (2006). Biofuels—at what cost? Government support for ethanol and biodiesel in the United States. The Global Initiative (GSI) of the International Institute for Sustainable Development (IISD). Retrieved October 10, 2007 from http://www.globalsubsidies.org/IMG/pdf/biofuels_subsidies_us.pdf

Kuby, W.R., Markoja, R., & Nackford, S. (1984). *Testing and Evaluation of On-Farm Alcohol Production Facilities.* Acures Corporation. Industrial Environmental Research Laboratory, Office of Research and Development, U.S. Environmental Protection Agency, Cincinatti, OH. 100pp.

Liu, D.L. & Helyar, K.R. (2003). Simulation of season stalk water content and fresh weight yield of sugarcane. *Field Crops Research, 82*(1), 59–73.

Macedo, I.C., Leal, M.R.L.V., & da Silva, J.E.A.R. (2004). *Assessment of greenhouse gas emissions in the production and use of fuel ethanol in Brazil.* Government of the State of Sao Paulo, Brazil, Secretariat of the Environment, 36 pp.

Maiorella, B. (1985). Ethanol. (In H.W. Blanch, S. Drew & D.I.C. Wang (Eds.), *Comprehensive biotechnology,* Vol. 3, New York: Pergamon Press).

McCain, J. (2003). Statement of Senator McCain on the Energy Bill. Press Release. Wednesday, November 19, 2003.

Mongabay. (2006). Deforestation in the Amazon. Retrieved August 22, 2006 from http://www.mongabay.com/brazil.html

NAS. (2003). *Frontiers in agricultural research: Food, health, environment, and communities.* (Washington, DC: National Academy of Sciences).

National Center for Policy Analysis. (2002). Ethanol subsidies. Idea House. National Center for Policy Analysis. Retrieved September 9, 2002 from http://www.ncpa.org/pd/ag/ag6.html

NCGA. 2006. Ethanol and coproducts. Oil industry subsidies. Retrieved August 27, 2006 from http://www.ncga.com/Ethanol/publicPolicy/subsidies.asp

Newton, P.W. (2001). Human settlements theme report. Australian State of the Environment Report 2001. Retrieved October 6, 2005 from http://www.environment.gov.au/soe/2001/settlements/settlements02-5c.html

Niven, R.K. (2005). Ethanol in gasoline: Environmental impacts and sustainability. *Renewable and Sustainable Energy Reviews, 9*(6), 535–555.

Patzek, T.W. (2004). Thermodynamics of the corn-ethanol biofuel cycle. *Critical Reviews in Plant Sciences, 23*(6), 519–567.

Patzek, T.W. & Pimentel, D. (2005). Thermodynamics of energy production from Biomass. Critical Reviews in Plant Sciences, 24, (5–6), 327–364.

Pimentel, D. (1980). *Handbook of energy utilization in agriculture.* (Boca Raton, FL: CRC Press)

Pimentel, D. (2003). Ethanol fuels: energy balance, economics, and environmental impacts are negative. *Natural Resources Research, 12*(2), 127–134.

Pimentel, D. (2006). Soil erosion: a food and environmental threat. *Environment, Development and Sustainability, 8*(1), 119–137.

Pimentel, D. & Pimentel, M. (1996). *Food, energy and society.* (Boulder, CO: Colorado University Press).

Pimentel, D. & Patzek, T.W. (2005). Ethanol Production using corn, switchgrass, and wood; biodiesel production using soybean and sunflower. *Natural Resources and Research, 14*(1), 65–76.

Pimentel, D., Berger, B., Filberto, D., Newton, M., Wolfe, B., Karabinakis, E., Clark, S., Poon, E., Abbett, E., & Nandagopal, S. (2004). Water resources: current and future issues. *BioScience, 54*(10), 909–918

Pimentel, D., Cooperstein, S., Randell, H., Filiberto, D., Sorrentino, S., Kaye, B., Nicklin, C., Yagi, J., Brian, J., O'Hern, J., Habas, A., & Weinstein, C. (2006). Ecology of increasing diseases: population growth and environmental degradation. *Human Ecology 35*(6), 653–668.

Pimentel, D., Patzek, T., & Cecil, G. (2007). Ethanol production: energy, economic, and environmental losses. *Reviews of Environmental Contamination and Toxicology, 189,* 25–41

PRB. (2006). *World population data sheet.* (Washington, DC: Population Reference Bureau)

Rosillo-Calle, F. & Cortez, L.A. (1998). Towards proAlcool II – a review of the Brazilian bioethanol program. *Biomass and Bioenergy, 14*(2), 115–124.

Sartori, M.M.P. & Basta, C. (1999). Methodos matematicos para o calculo enegetico da producao de cana-de-acucar. *Energia na Agricultra, 14*(1), 52–68.

Schmitz, T., Schmitz, G.A., & Seale, J.L. (2003). Brazil's ethanol program: the case of hidden sugar subsidies. *International Sugar Journal, 105*(1254), 254–256, 258–265.

Schneider, S.H., Rosencranz, A., & Niles, J.O. (2002). *Climate change policy change.* (Washington, DC: Island Press)

Sparovek, G. & Schung, E. (2001). Temporal erosion-induced soil degradation and yield loss. *Soil Science Society of America Journal, 65,* 1479–1486.

Spirits Low. (1999). Spirits low as Brazil alcohol car in trouble anew. Reuters Limited. Retrieved November 22, 1999 from http://www.climateark.org/articles/1999/alcocaro.htm

Taganrog. (2004–2006). Taganrog combine-harvester factory. Retrieved August 24, 2006 from http://www.tagaonrocity.com/tkz.html

Troeh, F.R., Hobbs, J.A., & Donahue, R.L. (2004). *Soil and water conservation.* (Englewood Cliffs, NJ: Prentice Hall).

USCB. (2004–2005). *Statistical abstract of the United States 2004–2005.* U.S. Census Bureau (Washington, DC: U.S. Government Printing Office)

USDA. (2004). *Agricultural statistics.* (Washington, DC: U.S. Government Printing Office)

Wald, M.I. (2006). Corn farmers smile as ethanol prices rise, but experts on food supplies worry. *New York Times (National),* A, p.13. January 16, 2006. Retrieved October 24, 2007 from http://archives.foodsafetynetwork.ca/agnet/2006/1-2006/agnet_jan_16.htm#story4

Wereko-Brobby, C. & Hagan, E.B. (1996). *Biomass conversion and technology.* (Chichester: John Wiley & Sons)

WHO. (2006). Malnutrition Worldwide. Source: World Health Organization. Retrieved August 27, 2006 from http://www.mikeschoice.com/reports/malnutrition_worldwide.htm

Worldwatch Institute. (2001). *Vital signs.* (New York: W.W. Norton & Company)

Youngquist, W. (1997). *GeoDestinies: The inevitable control of earth resources over nations and individuals.* (Portland, OR: National Book Company)

Youngquist, W. & Duncan, R.C. (2003). North American natural gas: data show supply problems. *Natural Resources Research, 12*(4), 229–240

Chapter 15
Ethanol Production Using Corn, Switchgrass and Wood; Biodiesel Production Using Soybean

David Pimentel and Tad Patzek

Abstract In this analysis, the most recent scientific data for corn, switchgrass, and wood, for fermentation/distillation were used. All current fossil energy inputs used in corn production and for the fermentation/distillation were included to determine the entire energy cost of ethanol production. Additional costs to consumers include federal and state subsidies, plus costs associated with environmental pollution and/or degradation that occur during the entire production process. In addition, an investigation was made concerning the conversion of soybeans into biodiesel fuel.

Keywords Energy · biomass · fuel · natural resources · ethanol · biodiesel

15.1 Introduction

Green plants, such as corn, soybeans, switchgrass and trees, and all other kinds of biomass, convert solar energy into plant material but require suitable soil, nutrients, and freshwater. In the conversion of the biomass into liquid fuel, water, microorganisms, and more energy are required. Andrew Ferguson (2006, personal communication, Optimum Population Trust, Manchester, UK) makes an astute observation that the proportion of sun's energy that is converted into useful ethanol, even using very positive energy data, only amounts to 5 parts per 10,000, or 0.05% of the solar energy.

Some recent papers are claiming returns on ethanol production from corn of anywhere from 1.25 kcal to 1.67 kcal per kcal invested (Shapouri et al., 2004; Farrell

✉ D. Pimentel
College of Agriculture and Life Sciences, Cornell University, 5126 Comstock Hall,
Ithaca, NY 15850,
e-mail: Dp18@cornell.edu

T. Patzek
Department of Civil and Environmental Engineering, University of California, Berkeley,
CA 94720, 425 David Hall, MC1716,
e-mail: patzek@patzek.CE.berkeley.edu

D. Pimentel (ed.), *Biofuels, Solar and Wind as Renewable Energy Systems*,
© Springer Science+Business Media B.V. 2008

et al., 2006; Hill et al., 2006). These excessively high returns are achieved by either omitting several energy inputs, reducing other energy inputs, or giving credits that are too optimistic for the by-products.

15.2 Energy Inputs in Corn Production

The conversion of corn into ethanol by fermentation in a large plant is about 1 liter of ethanol from 2.69 kg of corn grain (approximately 9.5 liters pure ethanol per bushel of corn; see Footnote (a) in Table 15.2) (Pimentel and Patzek, 2005). The production of corn in the United States requires a significant energy and dollar investment for the 14 inputs, including labor, farm machinery, fertilizers, irrigation, pesticides, and electricity (Table 15.1). To produce an average corn yield of 9,400 kg/ha (149 bu/ac) of corn using up-to-date production technologies requires the expenditure of about 8.2 million kcal of energy inputs (mostly natural gas, coal, and oil) listed in Table 15.1. This is the equivalent of about ~930 liters of oil equivalents (~25% of grain energy) expended per hectare of corn. The production costs total about $927/ha for the 9,400 kg/ha or approximately 10¢/kg ($2.54/bushel) of corn produced.

Full irrigation (when there is insufficient or no rainfall) requires about 100 cm/ha of water per growing season. Because only about 15% of U.S. corn production currently is irrigated (USDA, 1997a), only 8.1 cm per ha of irrigation was included for the growing season. On average irrigation water is pumped from a depth of 100 m (USDA, 1997a). On this basis, the average energy input associated with irrigation is 320,000 kcal per hectare (Table 15.1).

15.2.1 Energy Inputs in Fermentation/Distillation

The average costs in terms of energy and dollars for a large (245 to 285 million liters/year), modern drygrind ethanol plant are listed in Table 15.2. In the fermentation/distillation process, the corn is finely ground and approximately 15 liters of water are added per 2.69 kg of ground corn. Some of this water is recycled. After fermentation, to obtain a liter of 95% pure ethanol from the 8–12% ethanol beer and 92–88% water mixture, the 1 liter of ethanol must be extracted from approximately 11 liters of the ethanol/ water mixture. Although ethanol boils at about 78 degrees C, and water boils at 100 degrees C, the ethanol is not extracted from the water in the first distillation, which obtains 95% pure ethanol (Maiorella, 1985; Wereko-Brobby and Hagan, 1996; S. Lamberson, personal communication, Cornell University, 2000). To be mixed with gasoline, the 95% ethanol must be further processed and more water removed, requiring additional fossil energy inputs to achieve 99.5% pure ethanol (Table 15.2). Thus, a total of about 10 liters of wastewater must be removed per liter of ethanol produced, and this relatively large amount of sewage effluent has to be disposed of at an energy, economic, and environmental cost.

To produce a liter of 99.5% ethanol uses 46% more fossil energy than the energy produced as ethanol and costs 45 ¢ per liter ($1.71 per gallon) (Table 15.2). The corn feedstock requires about 32% of the total energy input. In this analysis, the total cost, including the energy inputs for the fermentation/distillation process and the

Table 15.1 Energy inputs and costs of corn production per hectare in the United States

Inputs	Quantity	kcal × 1000	Costs $
Labor	11.4 hrs[a]	462[b]	148.20[c]
Machinery	55 kg[d]	1,018[e]	103.21[f]
Diesel	88 L[g]	1,003[h]	34.76
Gasoline	40 L[i]	405[j]	20.80
Nitrogen	155 kg[k]	2, 480[l]	85.25[m]
Phosphorus	79 kg[n]	328[o]	48.98[p]
Potassium	84 kg[p]	274[r]	26.04[s]
Lime	1,120 kg[t]	315[u]	19.80
Seeds	21 kg[v]	520[w]	74.81[x]
Irrigation	8.1 cm[y]	320[z]	123.00[aa]
Herbicides	6.2 kg[bb]	620[ee]	124.00
Insecticides	2.8 kg[cc]	280[ee]	56.00
Electricity	13.2 kWh[dd]	34[ff]	0.92
Transport	204 kg[gg]	169[hh]	61.20
TOTAL		8,228	$926.97
Corn yield 9,400 kg/ha[ii]		33,840	kcal input:output 1:4.11

[a] NASS, 2003.
[b] It is assumed that a person works 2,000 hrs per year and utilizes an average of 8,000 liters of oil equivalents per year.
[c] It is assumed that labor is paid $13 an hour.
[d] Pimentel and Pimentel, 1996.
[e] Prorated per hectare and 10 year life of the machinery. Tractors weigh from 6 to 7 tons and harvesters 8–10 tons, plus plows, sprayers, and other equipment.
[f] Hoffman et al., 1994.
[g] Wilcke and Chaplin, 2000.
[h] Input 11, 400 kcal per liter.
[i] Estimated.
[j] Input 10,125 kcal per liter.
[k] NASS, 2003.
[l] Patzek, 2004.
[m] Cost $.55 per kg.
[n] NASS, 2003.
[o] Input 4,154 kcal per kg.
[p] Cost $.62 per kg.
[q] NASS, 2003.
[r] Input 3,260 kcal per kg.
[s] Cost $.31 per kg.
[t] Brees, 2004.
[u] Input 281 kcal per kg.
[v] Pimentel and Pimentel, 1996.
[w] Pimentel and Pimentel, 1996.
[x] USDA, 1997b.
[y] USDA, 1997a.
[z] Batty and Keller, 1980.
[aa] Irrigation for 100 cm of water per hectare costs $1,000 (Larsen et al., 2002).
[bb] Larson and Cardwell, 1999.
[cc] USDA, 2002.
[dd] USDA, 1991.
[ee] Input 100,000 kcal per kg of herbicide and insecticide.
[ff] Input 860 kcal per kWh and requires 3 kWh thermal energy to produce 1 kWh electricity.
[gg] Goods transported include machinery, fuels, and seeds that were shipped an estimated 1,000 km.
[hh] Input 0.83 kcal per kg per km transported.
[ii] Average. USDA, 2006; USCB, 2004–2005.

Table 15.2 Inputs per 1000 liters of 99.5% ethanol produced from corn[a]

Inputs	Quantity	kcal × 1000	Dollars $
Corn grain	2,690 kg[b]	2, 355[b]	265.27
Corn transport	2,690 kg[b]	322[c]	21.40[d]
Water	15,000 L [e]	90[f]	21.16[g]
Stainless steel	3 kg[i]	165[o]	10.60[d]
Steel	4 kg[i]	92[o]	10.60[d]
Cement	8 kg[i]	384[o]	10.60[d]
Steam	2,646,000 kcal[j]	2,646[j]	21.16[k]
Electricity	392 kWh[j]	1,011[j]	27.44[l]
95% ethanol to 99.5%	9 kcal/L[m]	9[m]	40.00
Sewage effluent	20 kg BOD[n]	69[h]	6.00
Distribution	331 kcal/L[p]	331	20.00[p]
TOTAL		7,474	$454.23

[a] Output: 1 liter of ethanol = 5,130 kcal (Low heating value). The mean yield of 2.5 gal pure EtOH per bushel has been obtained from the industry-reported ethanol sales minus ethanol imports from Brazil, both multiplied by 0.95 to account for 5% by volume of the #14 gasoline denaturant, and the result was divided by the industry-reported bushels of corn inputs to ethanol plants. (See http://petroleum.berkeley.edu/patzek/BiofuelQA/Materials/TrueCostofEtOH.pdf; Patzek, 2006)
[b] Data from Table 15.1.
[c] Calculated for 144 km roundtrip.
[d] Pimentel, 2003.
[e] 15 liters of water mixed with each kg of grain.
[f] Pimentel et al., 2004.
[g] Pimentel et al., 2004.
[h] 4 kWh of energy required to process 1 kg of BOD (Blais et al., 1995).
[i] Estimated from the industry reported costs of $85 millions per 65 million gallons/yr dry grain plant amortized over 30 years. The total amortized cost is $43.6/1000L EtOH, of which an estimated $32 go to steel and cement.
[j] Illinois Corn, 2004. The current estimate is below the average of 40,000 Btu/gal of denatured ethanol paid to the Public Utilities Commission in South Dakota by ethanol plants in 2005.
[k] Calculated based on coal fuel. Below the 1.95 kWh/gal of denatured EtOH in South Dakota, see j).
[l] $.07 per kWh (USCB, 2004–2005).
[m] 95% ethanol converted to 99.5% ethanol for addition to gasoline (T. Patzek, personal communication, University of California, Berkeley, 2004).
[n] 20 kg of BOD per 1000 liters of ethanol produced (Kuby et al., 1984).
[o] Newton, 2001.
[p] DOE, 2002.

apportioned energy costs of the stainless steel tanks and other industrial materials, is $454.23 per 1000 liters of ethanol produced (Table 15.2).

15.2.2 Net Energy Yield

The largest energy inputs in corn-ethanol production are for producing the corn feedstock, plus the steam energy, and electricity used in the fermentation/distillation process. The total energy input to produce a liter of ethanol is 7,474 kcal (Table 15.2).

However, a liter of ethanol has an energy value of only 5,130 kcal. Based on a net energy loss of 2,344 kcal of ethanol produced, 46% more fossil energy is expended than is produced as ethanol.

15.2.3 Economic Costs

Current ethanol production technology uses more fossil fuel and costs substantially more to produce in dollars than its energy value is worth on the market. Clearly, without the more than $3 billion federal and state government yearly subsidies, U.S. ethanol production would be reduced or cease, confirming the basic fact that ethanol production is uneconomical (National Center for Policy Analysis, 2002).

Federal and state subsidies for ethanol production that total more than $6 billion/year for ethanol are mainly paid to large corporations (Koplow, 2006), while corn farmers are receiving a minimum profit per bushel for their corn (Pimentel and Patzek, 2005). Senator McCain reports that direct subsidies for ethanol, plus the subsidies for corn grain, amount to 79¢ per liter (McCain, 2003).

If the production cost of a liter of ethanol were added to the tax subsidy cost, then the total cost for a liter of ethanol would be $2.47. The mean wholesale price of ethanol was almost $1.00 per liter without subsidies. Because of the relatively low energy content of ethanol, 1.6 liters of ethanol have the energy equivalent of 1 liter of gasoline. Thus, the cost of producing an equivalent amount of ethanol equal a liter of gasoline is $3.00 ($11.34 per gallon of gasoline). This is more than 53¢ per liter, the current cost of producing a liter of gasoline. The subsidy per liter of ethanol is 60 times greater than the subsidy per liter of gasoline! This is the reason why ethanol is so attractive to large corporations.

15.2.4 Cornland Use

Currently, about 18 billion liters of ethanol (5 billion gallons) are being produced in the United States each year (Kansas Ethanol, 2006). The total amount of petroleum fuels used in the U.S. was about 2,500 billion liters (USCB, 2004–2005). Therefore, 18 billion liters of ethanol (energy equivalent to 11.2 billion liters of petroleum fuel) provides only 1% of the petroleum utilized last year. To produce this 18 billion liters of ethanol, about 5.7 million ha or 20% of U.S. corn land is used. Expanding corn-ethanol production to 100% of U.S. corn production would provide just 7% of the petroleum needs of the U.S.

15.2.5 By Products

The energy and dollar costs of producing ethanol can be offset partially by by-products, like the dry distillers grains (DDG) made from dry-milling of corn. From about 10 kg of corn feedstock, about 3.3 kg of DDG with 27% protein content can be harvested (Stanton, 1999). This DDG is suitable for feeding cattle that are ruminants, but has only limited value for feeding hogs and chickens. In practice, this DDG is generally used as a substitute for soybean feed that contains 49% protein (Stanton, 1999). However, soybean production for livestock feed is more energy efficient than corn production, because little or no nitrogen fertilizer is needed for the production of this legume (Pimentel et al., 2002). In practice, only 2.1 kg of soybean protein provides the equivalent nutrient value of 3.3 kg of DDG. Thus, the credit fossil energy per liter of ethanol produced is about 445 kcal (Pimentel et al., 2002). Factoring this credit for a non-fuel source in the production of ethanol reduces the negative energy balance for ethanol production from 46% to 39% (Table 15.2). The high energy credits for DDG given by some are unrealistic because the production of livestock feed from ethanol is uneconomical given the high costs of fossil energy, plus the costs of soil depletion to the farmer (Patzek, 2004).

The resulting overall energy output/input comparison remains negative even with the credits for the DDG by-product.

15.2.6 Environmental Impacts

Some of the economic and energy contributions of the by-products are negated by the widespread environmental pollution problems associated with ethanol production. First, U.S. corn production causes more soil erosion that any other U.S. crop (Pimentel et al., 1995; NAS, 2003). In addition, corn production uses more herbicides and insecticides and nitrogen fertilizer than any other crop produced in the U.S., and these chemicals invade ground and surface water, thereby causing more water pollution than any other crop (NAS, 2003).

As mentioned, the production of 1 liter of ethanol requires 1,700 liters of freshwater both for corn production and for the fermentation/distillation processing of ethanol (Pimentel and Patzek, 2005). In some Western irrigated corn acreage, like some regions of Arizona, ground water is being pumped 10-times faster than the natural recharge of the aquifers (Pimentel et al., 2004).

All these factors confirm that the environmental and agricultural system in which U.S. corn is being produced is experiencing major degradation. Further, it substantiates the conclusion that the U.S. corn production system, and indeed the entire ethanol production system, is not environmentally sustainable now or for the future, unless major changes are made in the cultivation of this major food/feed crop. Because corn is raw material for ethanol production, it cannot be considered a renewable energy source.

Furthermore, pollution problems associated with the production of ethanol at the chemical plant sites are emerging. The EPA (2002) already has issued warnings to ethanol plants to reduce their air pollution emissions or be shut down. Another pollution problem concerns the large amounts of wastewater produced by each ethanol plant. As noted, for each liter of ethanol produced using corn, from 6–12 liters of wastewater are produced. This polluting wastewater has a biological oxygen demand (BOD) of 18,000 to 37,000 mg/liter depending of the type of plant (Kuby et al., 1984). The cost of processing this sewage in terms of energy (4 kWh/kg of BOD) was included in the cost of producing ethanol (Table 15.2).

Basically the major problem with corn and all other biomass crops is that they collect on average only 0.1–0.2% of the solar energy per year. At a fairly typical gross yield of 3,000 liters of ethanol per hectare per year, the power density achieved is only 2.1 kW/ha. That is compared with the gross power density achieved via oil, after delivery for use, on the order of 2,000 kW/ha. (A.R.B. Ferguson, personal communication, Optimum Population Trust, November 6, 2005). If all the current 28 million hectares of corn production were to be devoted only to growing corn for ethanol, then this acreage would supply only 7% of U.S. petroleum needs (USDA, 2003).

15.2.7 Food Security

At present, world agricultural land supplies more than 99% of all world food (calories), while aquatic ecosystems supply less than 1% (FAO, 2006). Worldwide, during the last decade, per capita available cropland decreased 20% and irrigation land 12% (Brown, 1997). Furthermore, per capita grain production has been decreasing, in part due to increases in the world population (FAO, 2006). Worldwide diverse cereal grains, including corn, make up 80% of the human food supply (Pimentel and Pimentel, 1996).

The expanding world population that now numbers 6.5 billion, further complicates and stresses the food security problem now and for the future (PRB, 2006). Almost a quarter million people are added each day to the world population, and each of these human beings requires adequate food. Today, the malnourished people in the world number about 3.7 billion (WHO, 2000). This is the largest number of malnourished people and proportion ever reported in history. Malnourished people are highly susceptible to various serious diseases. The World Health Organization reports a rapid rise in the number of people in the world who are infected with diseases like tuberculosis, malaria, and AIDS (Kim, 2002; Pimentel et al., 2006).

15.2.8 Food Versus the Fuel Issue

Using corn, a basic human food resource, for ethanol production, raises ethical and moral issues (Pimentel and Patzek, 2005). Expanding ethanol production entails diverting valuable cropland from the production of corn needed to nourish people. The energetic and environmental aspects, as well as the moral and ethical issues also deserve serious consideration. With oil and natural gas shortages now facing the United States, ethanol production is forcing the U.S. to import more oil and natural gas to produce ethanol and other biofuels (Pimentel and Patzek, 2005).

Furthermore, increasing oil and natural gas imports drives up the price of oil and gas; this is especially critical for the poor in developing countries of the world. The impact is documented by the fact that worldwide per capita fertilizer use has been declining for the last decade (FAO, 2006).

15.3 Cellulosic Ethanol

15.3.1 Properties of Cellulose

The term "cellulosic ethanol" is imprecise. It is meant to suggest that certain components of wood and green plant materials (cellulose, pectins, and hemicelluloses) can be chemically separated (from mostly lignin in wood) and partially split into hexose and pentose monomers, which are then fermented to produce ethanol. Low energy industrial processes of ethanol production from biomass do *not* exist.

Cellulose is the principal structural component of cell walls in higher plants. It is the most abundant form of living terrestrial biomass (Pimentel, 2001). For hundreds of millions of years, cellulose has protected plants from elements and animals, and from chemical attacks by fungi and bacteria. Cotton is 98% pure cellulose; flax is 80%, and wood is 40–50% cellulose, with the remaining 50–60% made up from other complex polysaccharides (20–35% hemicelluloses and 15–35% lignin).

The special properties of cellulose result from the association of the long, straight polymeric chains to form fibers called micro-fibrils, which are stronger than steel. The micro-fibrils then form larger fibers, which are laid down in a cris-cross pattern, and intermixed with gel-like polysaccharides, hemicelluloses and pectins, that function as biocement (Taiz and Zeiger, 1998). In some ways this structure resembles fiberglass and other composite materials, in which rigid crystalline fibers are used to reinforce a more flexible matrix.

The beta-glycosidic bonds are crucial in determining the structural properties of cellulose, and thus the strength of the cellulose fibers. Because of the beta-bonds, the chain assumes an extended rigid configuration, with each glucose residue turned 180 degrees from its neighbor (Taiz and Zeiger, 1998). Another consequence of alternating top/bottom glucose residues is that OH groups of adjacent chains allow very extensive hydrogen-bonding between chains. This extensive inter-chain hydrogen-bonding and rigid beta-configuration makes cellulose fibers very strong and able to resist strong sodium hydroxide and acid solutions.

In summary, close to one billion years of plant evolution have made cellulose very stable and resistant to biochemical attacks. Cellulose can be quickly decomposed and hydrolyzed only by mechanical grinding or steam exploding and severe chemical attack by hot concentrated sulfuric acid or sodium hydroxide. Biochemical enzymatic attacks take a long time and have low efficiency.

The process of separating cellulose fibers from the rest of woody biomass is well-known, fast, efficient, and very energy intensive. It is called the paper kraft-process. The kraft process is used in production of paper pulp and involves the use of caustic sodium hydroxide and sodium sulfide to extract the lignin from the wood fiber in large pressure vessels called digesters. The process name is derived from German "kraft," meaning strong. Unfortunately, the best energy efficiency of this process is ∼6200 kcal/kg of paper, more than the high heating value of pure ethanol. Therefore a much milder, enzymatic process must be used to obtain simple sugars from cellulose.

15.3.2 Disadvantages of Cellulosic Ethanol

15.3.2.1 Contamination

For corn starch fuel ethanol, normal fermentation times in batch mode (there are no continuous reactors in operation) are 48 hours; up to 72 hours is acceptable. These estimates do not include downtime, cleaning, start up, etc. Over 72 hours the number of failures increases exponentially due to contamination with bacteria: acetogens and others. As described in the literature, typical enzyme processes for lignocellulosic alcohol take 5–7 days, i.e., about 120–170 hours. This spells big problems if lignocellulosic ethanol producers ever go outside the laboratory or pilot scale (sterile fermenters) to a conventional fermentation vessel, which can *not* be sterilized for 120–170 hours.

15.3.2.2 Biomass Availability

Natural *net* productivity of a mature ecosystem (an earth household, e.g., a forest or grassland) is low on human time scale. (Very slow carbon burial occurs on geological time scale.) What is produced by autotrophic plants, rock weathering and floods, is consumed by heterotrophs (bacteria, fungi, and animals that are continuously recycled as nutrients for the plants). Some bacteria and fungi, in return for the food from plant roots, capture nitrogen from the air and convert it to ammonia, thus providing natural fertilization. Therefore, "biowaste" is an engineering classification of plant (and animal) parts unused in an industrial process. This dated human concept is completely alien to natural ecosystems, which must recycle their matter completely in order to survive. Excessive "biowaste" removal robs ecosystems of vital nutrients and species, and degrades them irreversibly. As discussed in (Patzek and Pimentel, 2005), those ecosystems from which we remove biomass at high rate (crop fields, tree plantations) must be heavily subsidized with fossil energy and earth minerals.

15.3.2.3 Enzyme Yield vs. Rate

The rate of lignocellulose hydrolysis and fermentation can be increased by enough pre-treatment (such as ball milling to exceedingly fine dust, at enormous energy costs, or steam exploding with acid pre-treatment), but rates will slow down rather rapidly before high yields are obtained. The main problem is the number of binding sites available; the outside-in rate limitation phenomenon. It simply takes time to chew into the sturdy lignocellulosic particles. Of course, one could run the lignocellulose through the kraft-like process. This cannot be done, however, for lignocellulosic ethanol because energy losses would be severe. One can get rather good yields and rates if one performs energy-intensive and unaffordable pretreatment, or (relatively) high yields with modest pre-treatment if one waits long enough (ideally for weeks). Thus, despite claims to the contrary, a real industrial process for lignocellulosic ethanol does not exist, and may *never* have a sufficiently favorable energy balance.

15.3.2.4 Thermodynamics

Current energy efficiency of producing cellulosic ethanol is so low that all other investigated paths to liquid biofuels are better; see (Patzek and Pimentel, 2005).

15.4 Switchgrass Production of Ethanol

The average energy input per hectare for switchgrass production is only about 3.9 million kcal per year (Table 15.3). With an exceptional average yield of 10 t/ha/yr, this suggests for each kcal invested as fossil energy the return is 11 kcal – an excellent return. This return is *impossible* to realize for more than one year in environments other than an ecologically-balanced prairie. Nonetheless, massive industrial *monocultures* of switchgrass are proposed and studied.[1] If pelletized for use as a fuel in stoves, the return is reported to be about 1:14.6 kcal (Samson et al., 2004). The 14.6 is higher than the 11 kcal in Table 15.3, because here a few more inputs were included than in the Samson et al. (2004) report. If the realistic sustained yield of switchgrass were 1–4 t/ha/yr, the return of 14.6 would drop to 1.5–4.5, similar to corn. The cost per ton of switchgrass pellets ranges from \$94 to \$130 (Samson et al., 2004). This appears to be an excellent price per ton.

However, converting switchgrass into ethanol results in a negative energy return (Table 15.4). The negative energy return is 68% or a slightly more negative energy return than corn ethanol production (Tables 15.2 and 15.4). The cost of producing a liter of ethanol using switchgrass was 93¢ (Table 15.4). The two major energy inputs for switchgrass conversion into ethanol were steam and electricity production (Table 15.4).

---

[1] Emphases added: "The 19 acre switchgrass biomass production field on the Central Grassland Research Station was seeded to 'Sunburst' switchgrass. The field previously had been seeded to oats which were disked and *sprayed* with the herbicide Banvel prior to planting. Volunteer oats were the main weed problem the establishment year and were mowed. After the volunteer oats were *removed*, an adequate stand developed but estimated first year yield was about **1,120** kg per hectare. At this yield level, cost of harvest is equivalent to the value of the biomass so the field was not harvested. In 2002, the field was fertilized with **60** kg N/ha shortly after spring green up and was harvested after the switchgrass had headed. Harvested biomass yield was **4.9** tons/ha on a dry weight basis. Yields will be harvested on all fields for another three years. The production and economic information from this study will be used in economic analyses to determine the potential profitability of switchgrass grown as a biomass crop in the Northern Plains and to plan research to reduce production costs and increase biomass yields." Source: *Field Scale Evaluation of Switchgrass Grown as a Bioenergy Crop in the Northern Plains*, K.P. Vogel, M.R. Schmer, R.K. Perrin, L.E. Moser, and R.B. Mitchell USDA-ARS and the University of Nebraska-Lincoln, Lincoln, NE, 2002.

Table 15.3 Average inputs and energy inputs per hectare per year for switchgrass production

Input	Quantity	10^3 kcal	Dollars
Labor	5 hr [a]	200[b]	$65[c]
Machinery	30 kg[d]	555	50[a]
Diesel	150 L[e]	1,500	75
Nitrogen	80 kg[e]	1,280	45[e]
Seeds	1.6 kg[f]	100[a]	3[f]
Herbicides	3 kg[g]	300[h]	30 [a]
TOTAL	10,000 kg yield[i]	3,935	$268[j]
	40 million kcal yield	input/output ratio 1:02[k]	

[a] Estimated.
[b] Average person works 2,000 hours per year and uses about 8,000 liters of oil equivalents. Prorated this works out to be 200,000 kcal.
[c] The agricultural labor is paid $13 per hour.
[d] The machinery estimate also includes 25% more for repairs.
[e] Calculated based on data from Brummer et al., 2000.
[f] Data from Samson, 1991.
[g] Calculated based on data from Henning, 1993.
[h] 100,000 kcal per kg of herbicide.
[i] Samson et al., 2000.
[j] Brummer et al. 2000 estimated a cost of about $400/ha for switchgrass production. Thus, the $268 total cost is about 49% lower that what Brummer et al. estimates and this includes several inputs not included in Brummer et al.
[k] Samson et al. (2000) estimated an input per output return of 1:14.9, but we have added several inputs not included in Samson et al. Still the input/output return of 1:11 would be excellent if the sustained yield of 10 t/ha/yr were possible.

15.5 Wood Cellulose Conversion into Ethanol

The conversion of 5,000 kg of wood harvested from a sustainable forest into 1,000 liters of ethanol require an input of about 9.2 million kcal (Table 15.5). Therefore, the wood cellulose system requires slightly more energy to produce the 1,000 liters of ethanol than when using switchgrass (Tables 15.4 and 15.5). About 81% more energy is required to produce a liter of ethanol using wood than the energy harvested as ethanol. This includes harvesting 25% of the wood lignin and filtering out the lignin from the residue after the ethanol has been removed by fermentation. The lignin residue was assumed to have 200% moisture and did not require further drying (thus, more energy) before being burned to produce steam.

The ethanol cost per liter for wood-produced ethanol is slightly higher than the ethanol produced using switchgrass, $1.01 versus 93 ¢, respectively (Table 15.4 and 15.5). The two largest fossil energy inputs in the wood cellulose production system were steam and electricity (Table 15.5). Note that 25% lignin was credited in this system, as it was in the switchgrass calculations.

Table 15.4 Inputs per 1000 liters of 99.5% ethanol produced from U.S. switchgrass[a]

Inputs	Quantity	kcal × 1000	Dollars $
Switchgrass	5,000 kg[b]	1, 968[c]	500
S. Grass transport	5,000 kg[b]	600[c]	30[d]
Water	250,000 L[e]	140[f]	40[m]
Stainless steel	3 kg[g]	165[g]	11[g]
Steel	4 kg[g]	92[g]	11[g]
Cement	8 kg[g]	384[g]	11[g]
Grind switchgrass	5,000 kg	200[h]	16[h]
Sulfuric acid	240 kg[i]	0	168[n]
Steam	8.1 tons[i]	4,404	36
Lignin	1,250 kg[j]	minus 1,500	minus 12
Electricity	666 kWh[i]	1,703	46
95% ethanol to 99.5%	9 kcal/L[k]	9	40
Sewage effluent	40 kg BOD[l]	138[o]	12
Distribution	331 kcal/L[p]	331	20
TOTAL		8,634	$929

[a] Output: 1 liter of ethanol = 5,130 kcal. The ethanol yield here is 200 L/t dry biomass (dbm). Iogen suggests 320 L/t dbm of straw that contains 25% of lignin. This yield is equal to the average yield of ethanol from corn, 317 L/t dbm (2.5 gal/bu). In view of the difficulties with breaking up cellulose fibers and digesting them quickly enough, the Iogen yield seems to be exaggerated, unless significantly more grinding, cell exploding with steam, and hot sulfuric acid are used.
[b] Data from Table 15.3.
[c] Calculated for 144 km roundtrip.
[d] Pimentel, 2003.
[e] 15 liters of water mixed with each kg of biomass.
[f] Pimentel et al., 2004b.
[g] Newton, 2001.
[h] Calculated based on grinder information (Wood Tub Grinders, 2004).
[i] Estimated based on cellulose conversion (Arkenol, 2004).
[j] Wood is about 25% lignin and removing most of the water from the lignin by filtering, the moisture level can be reduced to 200% (Crisp, 1999).
[k] 95% ethanol converted to 99.5% ethanol for addition to gasoline (T. Patzek, personal communication, University of California, Berkeley, 2004).
[l] 20 kg of BOD per 1000 liters of ethanol produced (Kuby et al., 1984).
[m] Pimentel, 2003.
[n] Sulfuric acid sells for $7 per kg.
[o] 4 kWh of energy required to process 1 kg of BOD (Blais et al., 1995).
[p] DOE, 2002.

Table 15.5 Inputs per 1000 liters of 99.5% ethanol produced from wood cellulose[a]

Inputs	Quantity	kcal × 1000	Dollars $
Wood	5,000 kg[b]	800	500[p]
Machinery	10 kg[m]	200[m]	20
Replace nitrogen	100 kg[c]	1,600	56
Wood transport	5,000 kg[d]	600	30
Water	250,000 L[e]	140[f]	40[f]
Stainless steel	3 kg	165[g]	11[g]
Steel	4 kg	92[g]	11[g]
Cement	8 kg	384[g]	11[g]
Grind wood	5,000 kg	200[h]	16[h]
Sulfuric acid	240 kg[i]	0	168[n]
Steam	8.1 tons[i]	4,404	36
Lignin	1,250 kg[j]	minus 1,500	minus 12
Electricity	666 kWh[i]	1,703	46
95% ethanol to 99.5%	9 kcal/L[k]	9	40
Sewage effluent	40 kg BOD[l]	138[o]	12
Distribution	331 kcal/L[q]	331	20
TOTAL		9,266	$1,005

[a] Output: 1 liter of ethanol = 5,130 kcal.
[b] 5,000 kg of wood input required for the production of 1,000 liters of ethanol.
[c] 100 kg of nitrogen in 5,000 wood material (Kidd and Pimentel, 1992).
[d] Calculated for 144 km roundtrip.
[e] 15 liters of water mixed with each kg of biomass.
[f] Pimentel et al., 2004.
[g] Newton, 2001.
[h] Calculated based on grinder information (Wood Tub Grinders, 2004).
[i] Estimated based on cellulose conversion (Arkenol, 2004).
[j] Wood is about 25% lignin and removing most of the water from the lignin by filtering, the moisture level was reduced to 200% (Crisp, 1999).
[k] 95% ethanol converted to 99.5% ethanol for addition to gasoline (T. Patzek, personal communication, University of California, Berkeley, 2004).
[l] 20 kg of BOD per 1000 liters of ethanol produced (Kuby et al., 1984).
[m] Mead and Pimentel, 2006.
[n] Sulfuric acid sells for $7 per kg.
[o] 4 kWh of energy required to process 1 kg of BOD (Blais et al., 1995).
[p] Wood material sells for $100 per ton..
[q] DOE, 2002.

15.6 Biodiesel Production

The monoesters commonly known as biodiesel are usually produced through the transesterification of vegetable oils (triglycerides or fatty acids) with the fossil methane-derived methanol. Both liquid oils and solid fats are triglycerides or fatty esters of glycerin. Raw triglyceride vegetable oils have properties similar to those of petroleum-based diesel fuels and can be used as a direct replacement without engine modification for a limited operating duration. Long-term diesel operation on raw vegetable oils causes numerous problems including injector coking, contamination of lubrication oil, engine deposits, and increased emissions. These problems are primarily the result of the high viscosity of the triglyceride oils compared to petroleum-based diesel fuels.

The biodiesel recipe is 100 kg of soybean oil, 15 kg of methanol, 1 kg of catalyst (sodium hydroxide or lye). The products are 100 kg biodiesel, 10 kg glycerol, 5 kg methanol (reusable after separation and purification) and 1 kg of soaps. The lye catalyst is neutralized with hydrochloric or sulfuric acid and becomes a waste product.

15.7 Soybean Conversion into Biodiesel

Various vegetable oils have been converted into biodiesel and they work well in diesel engines. An assessment of producing sunflower oil proved to be energy negative and costly in terms of dollars (Pimentel, 2001). Although soybeans contain less oil than sunflower, about 18% soy oil compared with 26% oil for sunflower, soybeans can be produced without or nearly zero nitrogen (Table 15.6). This makes soybeans advantageous for the production of biodiesel. Nitrogen fertilizer is one of the most energy costly inputs in crop production (Pimentel and Patzek, 2005).

The yield of sunflower is also lower than soybeans, about 1,500 kg/ha for sunflower compared with 2,890 kg/ha for soybeans (USDA, 2004). The production of 2,890 kg/ha of soy requires an input of about 3.0 million kcal per hectare and costs about $473/ha (Table 15.6).

With a yield of oil of 18% then 5,556 kg of soybeans are required to produce 1,000 kg of oil (Table 15.7). The production of the soy feedstock requires an input of 5.7 million kcal. The second largest input is steam that requires an input of 1.4 million kcal (Table 15.7). The total input for the 1,000 kg of soy oil is 13.8 million kcal. In addition, 125 kg of methanol must be added to produce biodiesel fuel. The methanol has an energy value of 587,500 kcal. With soy oil having an energy value of 9 million kcal, then there is a net loss of 53% in energy. A credit should be taken for the soy meal that is produced; this has an energy value of 7.4 million kcal, but it must be emphasized that this soy meal is not liquid fuel but livestock feed. The price per kilogram of soy biodiesel is about $1.12. Note, soy oil has a specific gravity of about 0.92; thus soy biodiesel value per liter is 97 ¢ per liter. This makes soy oil about 1.8 times more expensive than diesel fuel; diesel costs about 53¢ per liter to produce (USCB, 2004–2005).

Table 15.6 Energy inputs and costs in soybean production per hectare in the U.S

Inputs	Quantity	kcal × 1000	Costs $
Labor	7.1 hrs[a]	284[b]	92.30[c]
Machinery	20 kg[d]	360[e]	148.00[f]
Diesel	38.8 L[a]	442[g]	20.18
Gasoline	35.7 L[a]	270[h]	13.36
LP gas	3.3 L[a]	25[i]	1.20
Nitrogen	3.7 kg[j]	59[k]	2.29[l]
Phosphorus	37.8 kg[j]	156[m]	23.44[n]
Potassium	14.8 kg[j]	48[o]	4.59[p]
Limestone	2000 kg[v]	562[d]	46.00[v]
Seeds	69.3 kg[a]	554[q]	48.58[r]
Herbicides	1.3 kg[j]	130[e]	26.00
Electricity	10 kWh[d]	29[s]	0.70
Transport	154 kg[t]	40[u]	46.20
TOTAL		2,959	$472.84
Soybean yield 2,890 kg/ha[w]		10,404	kcal input:output 1:3.52

[a] Ali and McBride, 1990.

[b] It is assumed that a person works 2,000 hrs per year and utilizes an average of 8,000 liters of oil equivalents per year.

[c] It is assumed that labor is paid $13 an hour.

[d] Pimentel and Pimentel, 1996.

[e] Machinery is prorated per hectare and a 10 year life of the machinery. Tractors weigh from 6 to 7 t and harvesters from 8 to 10 tons, plus plows, sprayers, and other equipment.

[f] College of Agri., Consumer & Environ. Sciences, 1997.

[g] Input 11,400 kcal per liter.

[h] Input 10,125 kcal per liter.

[i] Input 7,575 kcal per liter.

[j] Economic Research Statistics, 1997.

[k] Patzek, 2004.

[l] Hinman et al., 1992.

[m] Input 4,154 kcal per kg.

[n] Cost 62¢ per kg.

[o] Input 3,260 kcal per kg.

[p] Costs 31¢ per kg.

[q] Pimentel et al., 2002.

[r] Costs about 70¢ per kg.

[s] Input 860 kcal per kWh and requires 3 kWh thermal energy to produce 1 kWh electricity.

[t] Goods transported include machinery, fuels, and seeds that were shipped an estimated 1,000 km.

[u] Input 0.83 kcal per kg per km transported.

[v] Mississippi State University Extension Service, 1999.

[w] USDA, 2004.

Soybeans are a valuable crop in the United States. The target price reported by the USDA (2003) is 21.2 ¢/kg while the price calculated in Table 15.6 for average inputs per hectare is 16¢/kg. Our calculated price is lower.

Table 15.7 Inputs per 1,000 kg of biodiesel oil from soybeans

Inputs	Quantity	kcal × 1000	Costs $
Soybeans	5,556 kg[a]	5,689[a]	$909.03[a]
Electricity	270 kWh[b]	697[c]	18.90[d]
Methanol	120L[i]	1,248[i]	111.60
Steam	1,350,000 kcal[b]	1,350[b]	11.06[e]
Cleanup water	160,000 kcal[b]	160[b]	1.31[e]
Space heat	152,000 kcal[b]	152[b]	1.24[e]
Direct heat	440,000 kcal[b]	440[b]	3.61[e]
Losses	300,000 kcal[b]	300[b]	2.46[e]
Stainless steel	11 kg[f]	605[g]	18.72[h]
Steel	21 kg[f]	483[g]	18.72[h]
Cement	56 kg[f]	2,688[g]	18.72[h]
TOTAL		13,812	$1,115.37

The 1,000 kg of soy oil plus 125 kg of methanol to produce biodiesel has an energy value of 9 million kcal for the oil plus 125 kg of methanol with an energy value of 587,500 kcal. With an energy input requirement of 13.8 million kcal, there is a net loss of energy of 53%. If a credit of 7.4 million kcal is given for the soy meal produced, then the net loss is less.

The cost per kg of biodiesel is $1.12.

[a] Data from Table 15.6.
[b] Data from Singh, 1986.
[c] An estimated 3 kWh thermal is needed to produce a kWh of electricity.
[d] Cost per kWh is 7¢.
[e] Calculated cost of producing heat energy using coal.
[f] Calculated inputs.
[g] Calculated from Newton, 2001.
[h] Calculated.
[i] Hekkert et al., 2005.

15.8 Canola Conversion into Biodiesel

Another crop that can be converted into biodiesel is canola that produces the most valuable cooking oil. Although soybeans contain less oil than canola, about 18% soy oil compared with 30% oil for canola, soybeans can be produced without or nearly zero nitrogen (Table 15.6). This makes soybeans advantageous for the production of biodiesel. Nitrogen fertilizer is one of the most energy costly inputs in crop production (Pimentel and Patzek, 2005).

The yield of canola is also lower than soybeans, about 1,600 kg/ha for canola compared with 2,890 kg/ha for soybeans (Tables 15.6 and 15.8) (USDA, 2004). The production of 1,568 kg/ha canola requires an input of about 4.4 million kcal per hectare and costs about $573/ha (Table 15.8).

About 3,333 kg of canola oil is required to produce 1,000 kg of biodiesel (Table 15.9). The total energy input to produce the 1,000 of canola oil is 14 million kcal. This suggests a net loss of 58% (Table 15.9). The cost per kg of biodiesel is also high at $1.63.

Table 15.8 Energy inputs and costs in canola production per hectare in the North America

Inputs	Quantity	kcal × 1000	Costs $
Labor	7 hrs[a]	280[b]	91.00[c]
Machinery	20 kg[d]	360[e]	148.00[f]
Diesel	65 L[a]	740[g]	35.00
Nitrogen	120 kg[a]	1,920[h]	75.00[i]
Phosphorus	101 kg[a]	417[j]	71.00[k]
Potassium	14.8 kg[l]	48[m]	4.59[n]
Sulfur	22 kg[a]	10[l]	10.00
Limestone	1000 kg[a]	281[d]	23.00
Seeds	5 kg[o]	40[p]	35.00
Herbicides	1.5 kg[q]	150[p]	30.00
Insecticides	1 kg[q]	100	20.00
Electricity	10 kWh[a]	29[r]	0.70
Transport	100 kg[s]	26[t]	30.00
TOTAL		4,401	$573.29
Canola yield 1,568 kg/ha[u]		5,645	kcal input:output 1:1.06

[a] Smathers, 2005.
[b] It is assumed that a person works 2,000 hrs per year and utilizes an average of 8,000 liters of oil equivalents per year.
[c] It is assumed that labor is paid $13 an hour.
[d] Pimentel and Pimentel, 1996.
[e] Machinery is prorated per hectare with a 10 year life of the machinery. Tractors weigh from 6 to 7 t and harvesters from 8 to 10 tons, plus plows, sprayers, and other equipment.
[f] College of Agri., Consumer & Environ. Sciences, 1997.
[g] Input 11,400 kcal per liter.
[h] Patzek, 2004.
[i] Hinman et al., 1992.
[j] Input 4,154 kcal per kg.
[k] Cost 70¢ per kg.
[l] Pimentel and Pimentel, 2007.
[m] Input 3,260 kcal per kg.
[n] Costs 31¢ per kg.
[o] Molenhuis, 2004.
[p] Pimentel et al., 2002.
[q] Estimated.
[r] Input 860 kcal per kWh and requires 3 kWh thermal energy to produce 1 kWh electricity.
[s] Goods transported include machinery, fuels, and seeds that were shipped an estimated 1,000 km.
[t] Input 0.83 kcal per kg per km transported.
[u] USDA, 2004.

15.9 Conclusion

Several physical and chemical factors limit the production of biofuels such as ethanol and biodiesel from plant biomass. Fossil energy inputs needed in the production of ethanol from corn or cellulosic wood material are several times more than the ethanol energy output. For biodiesel produced from soybeans, fossil energy inputs

Table 15.9 Inputs per 1,000 kg of biodiesel oil from canola

Inputs	Quantity	kcal × 1000	Costs $
Canola	3,333 kg[a]	9,355[a]	$1,419.00[a]
Electricity	270 kWh[b]	697[c]	18.90[d]
Methanol	120L[i]	1,248[i]	111.60
Steam	1,350,000 kcal[b]	1,350[b]	11.06[e]
Cleanup water	160,000 kcal[b]	160[b]	1.31[e]
Space heat	152,000 kcal[b]	152[b]	1.24[e]
Direct heat	440,000 kcal[b]	440[b]	3.61[e]
Losses	300,000 kcal[b]	300[b]	2.46[e]
Stainless steel	11 kg[f]	158[g]	18.72[h]
Steel	21 kg[f]	246[g]	18.72[h]
Cement	56 kg[f]	106[g]	18.72[h]
TOTAL		14,212	$1,625.34

The 1,000 kg of biodiesel produced has an energy value of 9 million kcal. With an energy input requirement of 14.2 million kcal, there is a net loss of energy of 58%. If a credit of 4.6 million kcal is given for the canola meal produced, then the net loss is less.

The cost per kg of biodiesel is $1.63.

[a] Data from Table 15.6.
[b] Data from Singh, 1986.
[c] An estimated 3 kWh thermal is needed to produce a kWh of electricity.
[d] Cost per kWh is 7¢.
[e] Calculated cost of producing heat energy using coal.
[f] Calculated inputs.
[g] Calculated from Newton, 2001.
[h] Calculated.
[i] Hekkert et al., 2005.

are 40% greater than contained in the biodiesel fuel produced. Giving credit for the byproducts produced can reduce the fossil energy inputs only from 10% to 20%.

An extremely low fraction of the sunlight reaching a hectare of cropland is captured by green plant biomass. On average only 0.1% of the sunlight is captured by plants. This value is in sharp contrast to photovoltaics that capture more than 10% of the sunlight, or approximately 100–fold more sunlight than the green plant biomass.

The environmental impacts of producing either ethanol or biodiesel from biomass are enormous. These include: severe soil erosion; heavy use of nitrogen fertilizer; and use of large quantities of pesticides (insecticides and herbicides). In addition to a significant contribution to global warming, there is the use of 1,000–2,000 liters of water required for the production of each liter of either ethanol or biodiesel. Furthermore, for every liter of ethanol produced there are 6–12 liters of sewage effluent produced.

Burning food crops, such as corn and soybeans, to produce biofuels, creates major ethical concerns. More than 3.7 billion humans are now malnourished in the world and the need for food is critical.

Energy conservation strategies combined with active development of renewable energy sources, such as solar cells and solar-based methanol synthesis systems, should be given priority.

References

Ali, M. B. & McBride, W. D. (1990). Soybeans: State level production costs, characteristics, and input use, 1990. Economic Research Service. Stock no. ERS SB873. 48pp.

Arkenol. (2004). Our technology: Concentrated acid hydrolysis. Retrieved August 2, 2004, from www.arkenol.com/Arkenol%20Inc/tech01.html

Batty, J. C. & Keller, J. (1980). Energy requirements for irrigation. (In: D. Pimentel (Ed.), *Handbook of energy utilization in agriculture* (pp. 35–44). Boca Raton, FL: CRC Press).

Blais, J. F., Mamouny, K., Nlombi, K., Sasseville, J. L., & Letourneau, M. (1995). Les mesures deficacite energetique dans le secteur de leau. J.L. Sassville and J.F. Balis (eds). Les Mesures deficacite Energetique pour Lepuration des eaux Usees Municipales. Scientific Report 405. Vol. 3. INRS-Eau, Quebec.

Brees, M. (2004). Corn silage budgets for Northern, Central and Southwest Missouri. Retrieved September 1, 2004, from http://www.agebb.missouri.edu/mgt/budget/fbm-0201.pdf

Brown, L. R. (1997). *The agricultural link: How environmental deterioration could disrupt economic progress.* (Washington, DC: Worldwatch Institute)

Brummer, E. C., Burras, C. L., Duffy, M. D., & Moore, K. J. (2000). *Switchgrass production in Iowa: Economic analysis, soil suitability, and varietal performance.* (Ames, Iowa: Iowa State University).

College of Agricultural, Consumer and Environmental Sciences. (1997). Machinery cost estimates: Summary of operations. University of Illinois at Urbana-Champaign. Retrieved November 8, 2001, from www.aces.uiuc.edu/~vo-ag/custom.htm

Crisp, A. (1999). Wood residue as an energy source for the forest products industry. Australian National University. Retrieved July 10, 2006, from http://sres.anu.edu.au/associated/fpt/nwfp/woodres/woodres.html

DOE. (2002). Review of transport issues and comparison of infrastructure costs for a renewable fuels standard. Washington, DC, U.S. Department of Energy. Retrieved October 8, 2002 from http://tonto.eia.doe.gov/FTPROOT/service/question3.pdf

Economic Research Statistics. (1997). Soybeans: Fertilizer use by state. 1996. Retrieved November 11, 2001, from http://usda.mannlib.cornell.edu/data-sets/inputs/9X171/97171/agch0997.txt

EPA. (2002). More pollution than they said: Ethanol plants said releasing toxins. New York Times. May 3, 2002.

FAO. (2006). *Food balance sheets.* Rome: Food and Agriculture Organization of the United Nations.

Farrell, A. E., Plevin, R. J., Turner, B. T., Jones, A. D., O'Hare, M. O., & Kammen, D. M. (2006). Ethanol can contribute to energy and environmental goals. *Science* 311, 506–508.

Hekkert, M. P., Hendriks, F. H. J. F., Faaij, A. P. C., & Neelis, M. L. (2005). Natural gas as an alternative to crude oil in automotive fuel chains well-to-wheel analysis and transition strategy development. *Energy Policy,* 33(5), 579–594.

Henning, J. C. (1993). Big Bluestem, Indiangrass and Switchgrass. Department of Agronomy, Campus Extension, University of Missouri, Columbia, MO.

Hill, J., Nelson, E., Tilman, D., Polasky, S., & Tiffany, D. (2006). Environmental, economic, and energetic costs and benefits of biodiesel and ethanol biofuels. Retrieved August 31, 2006, from http://www.pnas.org/cgi/content/full/103/30/11206

Hinman, H., Pelter, G., Kulp, E., Sorensen, E., & Ford, W. (1992). Enterprise budgets for Fall Potatoes, Winter Wheat, Dry Beans and Seed Peas under rill irrigation. Farm Business Management Reports, Columbia, Washington State University.

Hoffman, T. R., Warnock, W. D., & Hinman, H. R. (1994). Crop Enterprise Budgets, Timothy-Legume and Alfalfa Hay, Sudan Grass, Sweet Corn and Spring Wheat under rill irrigation, Kittitas County, Washington. Farm Business Reports EB 1173, Pullman, Washington State University.

Illinois Corn. (2004). Ethanol's energy balance. Retrieved August 10, 2004, from http://www.ilcorn.org/Ethanol/Ethan_Studies/Ethan_Energy_Bal/ethan_energy_bal.html

Kansas Ethanol. (2006). Kansas Ethanol: Clean fuel from Kansas farms. Retrieved July 10, 2006 from http://www.ksgrains.com/ethanol/useth.html

Kidd, C. & Pimentel, D. (1992). *Integrated resource management: Agroforestry for development.* (San Diego: Academic Press.)

Kim, Y. (2002). World exotic diseases. (In: D. Pimentel (Ed.), *Biological invasions: Economic and environmental costs of alien plant, animal, and microbe species* (pp. 331–354). Boca Raton, FL: CRC Press)

Koplow, D. (2006). Biofuels – at what cost? Government support for ethanol and biodiesel in the United States. The Global Studies Initiative (GSI) of the International Institute for Sustainable development (IISD). Retrieved February 16, 2007 from http://www.globalsubsidies.org/IMG/pdf/biofuels_subsidies_us.pdf

Kuby, W. R., Markoja, R., & Nackford, S. (1984). Testing and evaluation of on-farm alcohol production facilities. Acures Corporation. Industrial Environmental Research Laboratory. Office of Research and Development. U.S. Environmental Protection Agency: Cincinnati, OH. 100pp.

Larsen, K., Thompson, D., & Harn, A. (2002). Limited and Full Irrigation Comparison for Corn and Grain Sorghum. Retrieved September 2, 2002 from http://www.colostate.edu/Depts/SoilCrop/extension/Newsletters/2003/Drought/sorghum.html

Larson, W. E. & Cardwell, V. B. (1999). History of U.S. corn production. Retrieved September 2, 2004, from http://citv.unl.edu/cornpro/html/history/history.html

Maiorella, B. (1985). Ethanol. In H. W. Blanch, S. Drew & D. I. C. Wang (Eds.), *Comprehensive Biotechnology, Vol. 3.* (Chapter 43). New York: Pergamon Press.)

McCain, J. (2003). Statement of Senator McCain on the Energy Bill. Press Release. Wednesday, November 2003.

Mead, D. & Pimentel, D. (2006). Use of energy analysis in silvicultural decision making. *Biomass and Bioenergy, 30*, 357–362.

Mississippi State University Extension Service. (1999). Agronomy notes. Retrieved July 10, 2006, from http://msucares.com/newsletters/agronomy/1999/199910.html

Molenhuis, J. (2004). Business analysis and cost of production program. Ontario Ministry of Agriculture, Food and Rural Affairs. Retrieved July 10, 2006, from http://www.omafra.gov.on.ca/english/busdev/bear2000/Budgets/Crops/Oilseeds/wcanolahybrid_static.htm

NAS. (2003). *Frontiers in agricultural research: Food, health, environment, and communities.* (Washington, DC: National Academy of Sciences) Retrieved November 5, 2004, from http://dels.nas.edu/rpt_briefs/frontiers_in_ag_final%20for%20print.pdf

NASS. (2003). National Agricultural Statistics Service. Retrieved November 5, 2004, from http://usda.mannlib.cornell.edu

National Center for Policy Analysis. (2002). Ethanol subsidies. Idea House. National Center for Policy Analysis. Retrieved September 9, 2002, from http://www.ncpa.org/pd/ag/ag6.html

Newton, P. W. (2001). Human settlements theme report. Australian State of the Environment Report 2001. Retrieved October 6, 2005, from http://www.deh.gov.au/soe/2001/settlements/acknowledgement.html

Patzek, T. W. (2004). Thermodynamics of the corn-ethanol biofuel cycle. *Critical Reviews in Plant Sciences, 23*(6), 519–567.

Patzek, T. W. and Pimentel, D. (2005). Thermodynamics of energy production from biomass. *Critical Reviews in Plant Sciences* 24(5–6), 327–364.

Patzek, T. W. (2006). Letter to the Editor. *Science, 312,* (23 June 2006), 1747

Pimentel, D. (2001). The limitations of biomass energy. (In R. Meyers (Ed.), *Encyclopedia of physical science and technology. 3rd ed., Vol. 2.* (pp. 159–171). San Diego: Academic Press.)

Pimentel, D. (2003). Ethanol fuels: energy balance, economics, and environmental impacts are negative. *Natural Resources Research, 12*(2), 127–134.

Pimentel, D. & Patzek, T. (2005). Ethanol production using corn, switchgrass, and wood: biodiesel production using soybean and sunflower. *Natural Resources Research, 14*(1), 65–76.

Pimentel, D. & Pimentel, M. (1996). *Food, Energy and Society.* (Boulder, CO: Colorado University Press)

Pimentel, D., Harvey, C., Resosudarmo, P., Sinclair, K., Kurz, D., McNair, M., Crist, S., Sphritz, L., Fitton, L., Saffouri, R., & Blair, R. (1995). Environmental and economic costs of soil erosion and conservation benefits, *Science, 276*, 1117–1123.

Pimentel, D., Doughty, R., Carothers, C., Lamberson, S., Bora, N., & Lee, K. (2002). Energy inputs in crop production: comparison of developed and developing countries. (In R. Lal, D. Hansen, N. Uphoff & S. Slack (Eds.), *Food Security & Environmental Quality in the Developing World.* (pp. 129–151). Boca Raton, FL: CRC Press).

Pimentel, D., Berger, B., Filberto, D., Newton, M., Wolfe, B., Karabinakis, E., Clark, S., Poon, E., Abbett, E., & Nandagopal, S. (2004). Water resources: current and future issues. *BioScience, 54*(10), 909–918.

Pimentel, D., Cooperstein, S., Randell, H., Filiberto, D., Sorrentino, S., Kaye, B., Nicklin, C., Yagi, J., Brian, J., O'Hern, J., Habas, A., & Weinstein, C. (2006). Ecology of increasing diseases: population growth and environmental degradation. *Human Ecology 35*(6), 653–668 DOI 10.1007/s10745-007-9128-3

PRB. 2006. *World population data sheet.* (Washington, DC: Population Reference Bureau).

Samson, R. (1991). Switchgrass: A living solar battery for the praires. Ecological Agriculture Projects, Mcgill University (Macdonald Campus), Ste-Anne-de-Bellevue, QC, H9X 3V9 Canada. Copyright @ 1991 REAP Canada.

Samson, R., Duxbury, P., Drisdale, M., & Lapointe, C. (2000). Assessment of pelletized biofuels. PERD Program, Natural Resources Canada, Contract 23348-8-3145/001/SQ.

Samson, R., Duxbury, P., & Mulkins, L. (2004). Research and development of fibre crops in cool eason regions of Canada. Resource Efficient Agricultural Production-Canada. Box 125, Sainte Anne de Bellevue, Quebec, Canada, H9X 3V9.

Retrieved June 26, 2004, from http://www.reap-canada.com/Reports/italy.html

Shapouri, H., Duffield, J., McAloon, A., & Wang, M. (2004). The 2001 net energy balance of corn-ethanol (Revised). Washington, DC: U.S. Department of Agriculture.

Singh, R. P. (1986). Energy accounting of food processing. (In R. P. Singh (Ed.), *Energy in food processing* (pp. 19–68). Amsterdam: Elsevier.)

Smathers, R. L. (2005). Winter rapeseed after Summer fallow. 2005 Northern Idaho Crop Costs and Returns Estimate. College of Agriculture and Life Sciences. University of Idaho. EBB1-WR-05.

Stanton, T. L. (1999). Feed composition for cattle and sheep. Colorado State University. Cooperative Extension. Report No. 1.615. 7pp.

Taiz, L. & Zeiger, E. (1998). *Plant physiology.* (Sunderland, MA: Sinauer Associates Publishers)

USCB. (2004–2005). *Statistical abstract of the United States 2004–2005.* U.S. Census Bureau. (Washington, DC: U.S. Government Printing Office).

USDA. (1991). Corn-State. Costs of production. U.S. Department of Agriculture, Economic Research Service, Economics and Statistics System, Washington, D.C. Stock #94018.

USDA. (1997a). Farm and ranch irrigation survey (1998). 1997 Census of Agriculture. Volume 3, Special Studies, Part 1. 280pp.

USDA. (1997b). 1997 Census of agriculture. U.S. Department of Agriculture. Retrieved August 28, 2002, from http://www.ncfap.org

USDA. (2002). *Agricultural statistics.* U.S. Department of Agriculture. (Washington, DC: U.S. Government Printing Office).

USDA. (2003) *Agricultural statistics.* U.S. Department of Agriculture. USDA. I – 1 – XV-34p. (Washington, DC: U.S. Government Printing Office).

USDA. (2004). *Agricultural statistics, 2004.* (CD-ROM) U.S. Department of Agriculture/National Agriculture Statistics Service, Washington, DC. A1.47/2:2004.

USDA. (2006). *Agricultural statistics, 2006.* U.S. Department of Agriculture. (Washington, DC: U.S. Government Printing Office)

Wereko-Brobby, C. & Hagan, E. B. (1996). *Biomass conversion and technology.* (Chichester: John Wiley & Sons)

WHO. (2000). Nutrition for health and development: a global agenda for combating malnutrition. Retrieved November 3, 2004, from http://www.who.int/nut/documents/nhd_mip_2000.pdf

Wilcke, B. & Chaplin, J. (2000). Fuel saving ideas for farmers. Minnesota/Wisconsin Engineering Notes. Retrieved September 2, 2004, from http://www.bae.umn.edu/extens/ennotes/enspr00/fuelsaving.htm

Wood Tub Grinders. (2004). Wood Tub Grinders. Tretrieved August 3, 2004, from http://p2library.nfesc.navy.mil/P2_Opportunity_Handbook/7_III_13.html

Chapter 16
Developing Energy Crops for Thermal Applications: Optimizing Fuel Quality, Energy Security and GHG Mitigation

Roger Samson, Claudia Ho Lem, Stephanie Bailey Stamler and Jeroen Dooper

Abstract Unprecedented opportunities for biofuel development are occurring as a result of increasing energy security concerns and the need to reduce greenhouse gas (GHG) emissions. This chapter analyzes the potential of growing energy crops for thermal energy applications, making a case-study comparison of bioheat, biogas and liquid biofuel production from energy crops in Ontario. Switchgrass pellets for bioheat and corn silage biogas were the most efficient strategies found for displacing imported fossil fuels, producing 142 and 123 GJ/ha respectively of net energy gain. Corn ethanol, soybean biodiesel and switchgrass cellulosic ethanol produced net energy gains of 16, 11 and 53 GJ/ha, respectively. Bioheat also proved the most efficient means to reduce GHG emissions. Switchgrass pellets were found to offset 86–91% of emissions compared with using coal, heating oil, natural gas or liquid natural gas (LNG). Each hectare of land used for production of switchgrass pellets could offset 7.6–13.1 tonnes of CO_2 annually. In contrast, soybean biodiesel, corn ethanol and switchgrass cellulosic ethanol could offset 0.9, 1.5 and 5.2 tonnes of CO_2/ha, respectively.

✉ R. Samson
Resource Efficient Agricultural Production (REAP) – Canada, Box 125 Centennial Centre CCB13, Ste. Anne de Bellevue, Quebec, Canada H9X 3V9,
e-mail: rsamson@reap-canada.com

C. Ho Lem
Resource Efficient Agricultural Production (REAP) – Canada, Box 125 Centennial Centre CCB13, Ste. Anne de Bellevue, Quebec, Canada H9X 3V9

S. Bailey Stamler
Resource Efficient Agricultural Production (REAP) – Canada, Box 125 Centennial Centre CCB13, Ste. Anne de Bellevue, Quebec, Canada H9X 3V9

J. Dooper
Resource Efficient Agricultural Production (REAP) – Canada, Box 125 Centennial Centre CCB13, Ste. Anne de Bellevue, Quebec, Canada H9X 3V9

D. Pimentel (ed.), *Biofuels, Solar and Wind as Renewable Energy Systems*,
© Springer Science+Business Media B.V. 2008

The main historic constraint in the development of herbaceous biomass for thermal applications has been clinker formation and corrosion in the boiler during combustion. This problem is being overcome through plant selection and cultural techniques in grass cultivation, combined with advances in combustion technology. In the coming years, growing warm-season grasses for pellet production will emerge as a major new renewable energy technology, largely because it represents the most resource-efficient strategy to use farmland in temperate regions to create energy security and mitigate greenhouse gases.

Keywords Combustion · bioheat · biomass · net energy balance · grass pellets · switchgrass · energy crop · greenhouse gas · thermal energy · energy security · biomass quality · perennial

Acronyms & abbreviations

Bioheat:	biomass use for thermal applications
C_3:	cool season
C_4:	warm season
Cl:	Chlorine
GHG:	greenhouse gas
K:	Potassium
LNG:	liquefied natural gas
N:	nitrogen
RET's:	renewable energy technologies
Si:	Silica
WSG:	warm season grass

16.1 Introduction

In most industrialized countries, thermal energy represents the largest energy need in the economy. Thermal energy is used for space and water heating in the residential, commercial and industrial sectors, low and high temperature process heat for industry, and power applications. Thermal energy can also be used for cooling applications. Rather than supporting biomass for simple thermal applications such as direct heating applications industrialized countries have currently placed emphasis on researching and providing subsidies for more technologically complex innovations such as large industrial bio-refineries. However, governments in industrialized nations who have identified the need to develop biofuels for energy security and greenhouse gas mitigation should look more closely at thermal applications for biomass to fulfill these needs. This review therefore examines energy security in section one, identifying opportunities to grow energy crops on farmland in eastern Canada as a means to collect solar energy and convert it into useful energy products for consumption. The greenhouse gas (GHG) mitigation potential of switching from fossil fuels to various biofuels produced from energy crops is also examined. Section two then overviews recent advances in the emerging agricultural industry growing grasses for bioheat, identifying opportunities and challenges in advancing this technology for commercial applications in temperate regions of the world.

16.2 Energy Crop Production for Energy Security and GHG Mitigation

Since the Arab oil embargo in the 1970s there has been considerable interest in North America in growing both conventional field crops and dedicated energy crops for bioenergy as a means to enhance energy security. The long-term decline in farm commodity prices has also created significant interest in using the surplus production capacity of the farm sector as a means to produce energy while creating demand enhancement for the farm sector. This decline in farm commodity prices, due to innovation in plant breeding and production technology, is accelerating the likelihood that large quantities of biomass energy from farms could penetrate energy markets currently dominated by fossil fuels.

One of the strongest drivers for biofuel development is the GHG mitigation potential of energy crops to produce solid, liquid and gaseous biofuels to replace fossil fuels in our economy. With the increased use of grain crops for liquid biofuels, the past two years have seen a rise in both the demand and price for farm commodities. Also increasing however, are concerns over other important social issues such as the potential for bioenergy to compete with food security, and problems with soil erosion and long-term soil fertility. The production and utilization of crops residues as a global biofuel sources has recently been reviewed (Lal, 2005). The main conclusions were that the most appropriate use of crop residues is to enhance, maintain and sustain soil quality by increasing soil organic matter, enhancing activity and species of soil fauna, minimizing soil erosion and non-source pollution, mitigating climate change by sequestering carbon in the pedosphere, and advancing global food security through enhancement of soil quality. It was recommended that efforts be undertaken to grow biomass on specifically dedicated land with species of high yield potential, suggesting that 250 million hectares (ha) globally could be put into production of perennial energy crops.

The increasing biodiversity loss from agricultural landscapes through crop intensification is also a major environmental concern. The rapid development of liquid biofuels in the tropics in the past decade has also caused significant harm to biodiversity through the conversion of forests into agricultural production. Resource efficient, rather than resource exhausting, bioenergy crop production strategies need to evolve with a priority placed on de-intensification of farm production through the use of perennials and utilization of existing marginal farmlands. This approach would to a much greater extent avoid the biofuel conflicts with food crop production and biodiversity that are now occurring with using annual food crops as biofuels.

To achieve the objective of resource efficient biomass production we must examine some of the basic factors influencing biomass accumulation:

1. There are two main photosynthetic pathways for converting solar energy into plant material: the C_3 and C_4 pathways. The C_4 pathway is approximately 40% more efficient than the C_3 pathway in accumulating carbon (Beadle and Long, 1985).
2. C_4 species use approximately half the water of most C_3 species (Black, 1971).
3. In temperate climates, sunlight interception is often more efficient with perennial plants because annual plants spend much of the spring establishing a canopy and also exhibit poor growth on marginal soils.
4. Some species of warm season grasses are climax community species and have excellent stand longevity (which also results in decreased economic costs for establishing perennial crops through decreased expenditures for seeding, tillage etc.).
5. C_4 species of grasses contain less N than C_3 species and can be more N-use efficient in temperate zones because the N is cycled internally to the root system in the fall for use in the following growing season (Clark, 1977).

It is apparent that the optimal plants for resource-efficient biomass production should be both perennial and C_4 in nature.

16.2.1 Perennial and Annual Energy Crops

In North America, the warm continental climate has produced a diversity of native warm season (C_4) perennial grasses that have a relatively high energy production potential on marginal farmlands. In the more humid zones, these species include switchgrass (*panicum virgatum*), prairie cordgrass (*spartina pectinata*), eastern gamagrass (*tripsacum dactyloides*), big bluestem (*andropogon gerardii vitman*) and coastal panic grass (*panicum amarum A.S. hitchc.*). In semi-arid zones and dry-land farming areas, prairie sandreed (*calamovilfa longifolia*) and sand bluestem (*andropogon hallii*) are amongst the most productive species. All of these species are relatively thin stemmed, winter hardy, highly productive and are established through seed.

Switchgrass was chosen as the model herbaceous energy crop species to concentrate development efforts on in the early 1990s by the U.S. Department of Energy. It had a number of promising features including its moderate to high productivity, adaptation to marginal farmlands, drought resistance, stand longevity, low nitrogen requirements and resistance to pests and diseases (Samson and Omielan, 1994; Parrish and Fike, 2005).

Table 16.1 illustrates that in Ontario, Canada, C_4 species like corn and switchgrass produce considerably higher quantities of energy from farmland than C_3 crops. The perennial crops were also identified to have the lowest fossil energy input requirements. Overall, prior to any conversion process, switchgrass produces 40% more

Table 16.1 Solar energy collection and fossil fuel energy requirements of Ontario Crops per hectare, adapted from Samson et al. (2005)

Crop	Yield (ODT/ha)	Energy content (GJ/ODT)	Fossil energy used (GJ/ODT)	Fossil energy used (GJ/ha)	Solar energy collected (GJ/ha)	Net energy (GJ/ha)
Canola	1.8[a]	25.0	6.3	11.3	45	**33.7**
Soybean	2.2[a]	23.8	3.2	7.0	52.4	**45.3**
Barley	2.8[a]	19.0	3.9	11.0	53.2	**42.3**
Winter Wheat	4.4[a]	18.7	2.9	12.8	82.3	**69.5**
Tame Hay	4.7[a]	17.9	1.0	4.7	84.1	**79.4**
Grain Corn	7.3[a]	18.8	2.9	21.2	137.2	**116.1**
Switchgrass	9	18.8	0.8	7.2	169.2	**162.0**

[a]OMAFRA, (2007)

net-energy gain per hectare than grain corn and five times more net-energy gain per hectare than canola. It also should be noted that corn yields are based on modern hybrid yields in Ontario while switchgrass yields are based on commercial production of the cultivar cave in-rock, an unimproved cultivar that was collected from an Illinois prairie in 1958. Warm season grasses (WSG's) function well as perennial energy crops because they mimic the biological efficiency of the tall-grass prairie ecosystem native to North America. They produce significantly more energy than grain corn while at the same time requiring minimal fossil energy inputs for field operations and less fertilizers and herbicides.

In industrialized countries, the seed portion of annual grain and oilseed crops became the first feedstock for energy applications. However, whole plant annual crops capture much larger quantities of energy per hectare. In Western Europe, whole plant crops such as maize and rye are now commonly harvested for biogas applications. High yielding hybrid forage sorghum, sorghum-sudangrass and millet, also hold promise as new candidates for biogas digestion (Von Felde, 2007; Venuto, 2007). The major advantage of ensiling is that even in relatively unfavourable weather for crop drying, energy crops can be stored and delivered to the digester year round. This is particularly advantageous for thick stemmed species like maize and sorghum which are commonly difficult to dry in areas receiving more than 700 mm of rainfall annually or have harvests late in the year when solar radiation is declining. In combustion applications, thick stemmed herbaceous species have biomass quality constraints which make them difficult to burn (further discussed in Section 16.3). In warm, humid southern production zones in temperate regions, it may also be difficult to dry the feedstock for combustion applications as the material would be more vulnerable to decomposition. In these situations, crop conversion to usable energy would be facilitated by using a biogas conversion system and storing the crop as silage.

Overall, both thick and thin stemmed whole-plant biomass crops can be successfully grown for biogas applications. Highest biogas yields are achieved when a fine chop and highly digestible silage are used. Conversely, thin stemmed, perennial WSG's have been identified as the most viable means to store dry crops for combustion applications and offer the best potential for improved biomass quality for combustion (discussed further in Section 16.3). For liquid fuel production such as cellulosic ethanol, the process is more flexible in terms of the moisture content and chemical composition of the feedstock in the production of energy.

16.2.2 Options for Growing and Using Energy Crops for Energy Security in Industrialized Countries

As greater scarcity of fossil fuels occurs in the next 25–50 years, industrialized countries will undoubtedly seek greater energy security from renewable energy technologies (RET's). Countries will increasingly aim to develop bioenergy production and conversion technologies which are efficient at using energy crops grown on both productive and marginal farmland to displace the use of imported fossil fuels. North America, Europe, and China in particular, urgently need to develop effective bioenergy production systems as these areas will become increasingly dependent on importing fossil fuels due to their large economies and declining fossil energy production. While many industrialized countries have imported petroleum fuels from distant producers for many years, the international trade in natural gas use will expand substantially. For example in North America, domestic natural gas production peaked in the United States in 2001 and has declined by 1.7% per year since that time, while in Canada production has been in decline or reached a plateau since 2001. To compensate for declining North American gas production and rising prices, energy intensive natural gas industries have moved offshore and liquid natural gas (LNG) imports have started to come into the United States (Hughes, 2006). LNG imports currently supply approximately 3% of the United States supply and are expected to increase to 15–20% by 2025. Much of this natural gas demand is presently used in thermal applications. For example, the United States relies on natural gas for 20% of its power requirements and for 60% of its home heating requirements (Darley, 2004).

Identifying sustainable bioenergy technologies with a high net energy gain per hectare is essential to reduce imports of natural gas and other fossil fuels into industrialized countries. In particular, there may be opportunities to cost-effectively produce solid and gaseous biofuels in temperate regions to replace high quality fossil fuels in thermal applications. In the past 5 years, petroleum and natural gas prices have increased substantially while thermal coal prices in the world have remained relatively stable. This likely is a function of the changing awareness around supply and demand of fossil fuels. On a global basis, the lifespan of natural gas and oil reserves are less than half that of coal, however many energy analysts foresee a transition from the current global energy economy dominated by petroleum to one where natural gas plays an equally important role. This widening gap between the prices of high-quality fossil fuels like natural gas and petroleum versus coal will make fuels of higher quality ideal candidates for displacement by renewables. Solid and gaseous biofuels could substitute in thermal applications through both heat generation and combined heat and power operations. This is a fitting association as both biomass production and heat demand are relatively disperse, thus biomass could be produced locally to meet local thermal energy needs sustainably. A key tenet of the concept of the *soft energy path* introduced by Lovins (1977) is that both the scale and quality of energy should be matched appropriately with its end use to create a more sustainable energy supply system.

The growing price difference between coal, natural gas and heating oil suggests that high-quality fossil fuels will be increasingly utilized for high-quality end uses such as transportation fuels and industrial products while lower-quality fuels like coal will be increasingly used for low-end thermal applications. Due to the polluting nature of coal and the increasing emphasis on reducing carbon emissions through taxes and cap and trade systems, there also will be substantial opportunities for biomass to substitute for coal in thermal applications (discussed further in Section 16.2.3). The following section explores the thermodynamics around converting biomass into solid and gaseous products versus their present utilization opportunities as liquid fuels in temperate regions of the world.

16.2.2.1 Opportunities to use Ontario Farmland for Improving Energy Security

This analysis examines present or currently proposed strategies to use biomass derived from farmland in the province of Ontario for generating solid, gaseous and liquid biofuel products. Ontario has a continental climate and cropping patterns that are somewhat similar to other regions in the temperate world including the Great Lake states of Michigan and Wisconsin in the United States, countries in central Europe such as Hungary, and the Northeastern provinces of China. As such, it represents a useful case study for the bioenergy opportunities for continental climates in the temperate world. Ontario produces very limited quantities of fossil fuels. Coal and coal products in Ontario are primarily used for power generation and for large industrial applications, such as the steel and cement industry. Petroleum products are mainly used in the transport sector in Ontario, with some additional use as heating oil. Ontario imports natural gas from western Canada, petroleum from the world market, and coal mainly from the Northeastern United States. Within the next 2–5 years, two LNG terminals on Canada's east coast will begin supplying eastern Canadian energy user's imported liquefied natural gas from either Russia or producers in the Middle East. Declining western Canadian supplies will likely not be sufficient to enable export production to reach Ontario in the coming years. Thus the Ontario economy, which is heavily dependent on natural gas for residential and commercial heating applications and process heat for industry and power generation, will begin to rely on distant foreign natural gas resources.

16.2.2.2 Harvesting Energy from Ontario Farmland for Biofuel Applications: A Case Study

To optimize energy security and GHG mitigation potential from bioenergy, a case study has been developed to compare alternative bioenergy crops and conversion technologies in Ontario. The comparison crops include soybean, corn, corn silage and switchgrass, which are well adapted to Ontario's warm continental climate summer. The main agricultural zones in the province experience a frost free period typically from mid May to mid to late Sept and about 900 mm of annual precipitation. Soybean, corn and corn silage are commonly grown in Ontario while switchgrass and other native warm season grasses such as big bluestem and coastal panic grass are emerging crops that are native to the region. Switchgrass has been selected to represent the WSG's in the analysis as it has undergone the furthest development of all native grasses for energy use in North America. In Ontario, approximately 500 ha of native grasses are presently under bioenergy production in 2007. It is anticipated that a portfolio of warm season species will be developed as future energy crops, with mixed seedings encouraged to reduce production risks and enhance biodiversity.

As can be seen from Table 16.2, the dry matter production potential prior to processing is highest with the whole corn plant harvested as silage. Switchgrass also produces significant quantities of dry matter and has the added advantage of being able to be grown on marginal farmlands. The yields for switchgrass are estimated to be slightly lower for combustion applications as a delayed harvest technique is used (discussed further in Section 16.3). The net energy gain/ha that results from each energy crop and conversion process is generally highest where whole-plant biomass is used for biogas or bioheat, and lowest where the seed portion of annual crops is used for liquid fuels. From a net energy gain perspective, the two most promising systems for Ontario are corn silage biogas and switchgrass pellets. These technologies have the potential to produce 770–890% more net energy gain/ha than growing grain corn for ethanol. Cellulosic ethanol from grasses is much more efficient than other annual grain or oilseed liquid fuel options for producing net energy gain/ha. However it remains substantially less efficient than direct combustion of energy grasses or corn silage biogas as a means to produce energy from farmland. The energy balance and GHG studies cited in the Tables (16.2 and 16.3), largely omit a full accounting of energy use. For example energy inputs associated with plant construction are generally not included and if these energy inputs were included the results would be less favourable especially for the more capital intensive technologies such as corn and cellulosic ethanol. Bioheat from pellets has a much lower capital investment requirement per unit of renewable energy produced (Bradley, 2006; Mani et al., 2006) and as such a full life cycle analysis would have less impact on its energy balance.

The main problem of cellulosic ethanol is that, even with current technology, less than half of the energy in the original feedstock is recovered in the ethanol. This analysis illustrates that upgrading the energy quality of biomass from a solid form to a liquid form appears to be quite expensive thermodynamically. While advances in cellulosic ethanol technology can be expected in the coming years, the prediction of a technology that would be cost-competitive at $1.00/gallon with gasoline by the year 2000 (Lynd et al., 1991), was and remains far from reality. There are currently no commercial cellulosic ethanol plants using agricultural feedstocks in existence despite the generous subsidies for ethanol production available in North America.

Table 16.2 Harvesting Energy from Ontario farmland for biofuel applications: A case study comparing alternative bioenergy crops and conversion technologies in Ontario

Feedstock	Field Yield[a] (tonnes/ha)	Field Yield[b] (ODT/ha)	Losses (%)[c] H = Harvest S = Storage D = Densification	Net Yield (ODT/ha)	Energy Content of feedstock[d] (units)	Total Energy Production (unit/ha)	Conversion[e] (GJ/unit)	Gross Energy (GJ/ha)	Energy Used in Production[f] (GJ/ha)	Net Energy Gain (GJ/ha)
Biogas (Anaerobic Digestion)										
Corn Silage	–	15.6	15% (H/S)	13.3	500 m³/ODT biogas	6625 m³ biogas	0.0232 GJ/m3	153.7	31.0	**122.7**
Perennial Grass Energy Crops	–	10	20% (H/S)	8.0	400 m³/ODT biogas	3200 m³ biogas	0.0232 GJ/m3	74.2	13.0	**61.2**
Bioheat (Direct Combustion)										
Grain Corn	8.6	7.3	–	7.3	18.8 GJ/ODT Heat	137.2 GJ Heat	–	137.2	21.2	**116.0**
Switchgrass Pellet	–	10	18% (H/D)	8.2	18.8 GJ/ODT Heat	154.2 GJ Heat	–	154.2	12.0	**142.2**
Biofuels										
Grain Corn Ethanol	8.6	7.3	–	7.3	473 L/ODT ethanol	3452.9 L ethanol	0.021 GJ/L	72.5	56.6	**15.9**
Switchgrass Cellulosic Ethanol	–	10	5% (H/S)	9.5	340 L/ODT ethanol	3230 L ethanol	0.021 GJ/L	67.8	15.3	**52.5**

Table 16.2 (Continued)

Feedstock	Field Yield[a] (tonnes/ha)	Field Yield[b] (ODT/ha)	Losses (%)[c] H = Harvest S = Storage D = Densification	Net Yield (ODT/ha)	Energy Content of feedstock[d] (units)	Total Energy Production (unit/ha)	Conversion[e] (GJ/unit)	Gross Energy (GJ/ha)	Energy Used in Production[f] (GJ/ha)	Net Energy Gain (GJ/ha)
Soybean Biodiesel	2.6	2.2	–	2.2	224 L/ODT biodiesel	492.3 L biodiesel	0.03524 GJ/L	17.3	6.8	**10.6**

a Corn and soybean yield is 5 year (2002–2007) average in Ontario (OMAFRA, 2007)

b Assuming that corn grain yield is 47% of total plant yield (Zan, 1998), silage corn field yield is equivalent to 7.3 ODT/ha x 2.13 = 15.6 ODT/ha
10 tonne/ha is average of fall and spring field yields in Ontario (Samson, 2007, Samson et al., 2008b)
Ontario's 5 year average soybean yield is 2.6 tonnes per hectare (at 13% moisture content) which results in 2.2 ODT/hectare.

c Harvest and storage losses for corn silage are 15% (Roth and Undersander, 1995)
Harvest and storage losses for energy grass silage production are estimated at 20% (Manitoba Agriculture, Food and Rural Initiatives , MAFRI)
Harvesting, storage and densification losses for switchgrass pellets are estimated to be 18% of field biomass of mature crops (Girouard et al., 1998; Samson et al. 2008b), net yields of 8.2 t/ha can be considered an average of productive and marginal farmlands.
Fall harvesting losses and storage losses for switchgrass used for cellulosic ethanol are estimated to be 5% (Sanderson et al., 1997)

d Corn silage yields 400–600 m³/tonne (dry matter basis) biogas (Braun and Wellinger, 2005; Lopez et al. 2005)
Grass silage yields 350–450 m³/tonne (dry matter basis) biogas (De Baere, 2007; Berglund and Börjesson, 2006; Mähnert et al., 2005)
Grain corn has an energy content of 18.8 GJ/ODT (Schneider and Hartmann, 2005)
Switchgrass has an energy content of 18.8 GJ/ODT (Samson et al., 2005).
Corn ethanol yields 473 L/ODT (Farrell et al., 2006)
Switchgrass ethanol yield is estimated at 340 litres per ODT (Spatari et al., 2005; Iogen Corporation, 2008)
Soybean biodiesel yields 224 L/ODT (Klass, 1998)

e Biogas energy = 0.0232 GJ/m3 (Klass, 1998).
Ethanol energy = 0.021 GJ/litre (Klass, 1998; Smith et al., 2004)
Electrical energy = 0.0036 GJ/kWh (Klass, 1998)
Methyl ester soybean biodiesel = 0.03524 GJ/litre (Klass, 1998)

f Biogas

The energy used in the production of the corn silage biogas equals the gross methane energy consumed to warm the digester, methane leakage, and the energy used in production and conversion. The methane consumed to warm the digester is 3.5% (Gerin et al., 2008) of the original gross methane produced (3.5% of 153.7 GJ/ha = 5.4 GJ/ha). The energy used in corn silage production and biogas conversion is equivalent 25.6 GJ/ha, this assumes energy production to produce corn silage is the same as corn production in Ontario at 20.59 GJ/ha (see grain corn estimate below). 1% methane leakage (1% of 148.3 = 1.5 GJ/ha (Zwart et al., 2007)), plus 2.5% of energy used for biodigester processing (Gerin et al., 2008) (2.5% of 148.3 GJ/ha = 3.7 GJ/ha). Total input is 5.4 + 25.6 = 31.0 GJ/ha.

The energy used in the production of the switchgrass silage biogas equals the gross methane energy consumed to warm the digester plus the energy used in production and conversion. The methane consumed to warm the digester is 3.5% (Gerin et al., 2008) of the original gross methane produced (3.5% of 74.2 GJ/ha = 2.6 GJ/ha). The energy used in switchgrass production and biogas conversion is equivalent 10.4 GJ/ha, comprised of 7.9 GJ/ha for switchgrass production (Samson et al., 2000), 1% methane leakage (1% of 71.6 = 0.7 GJ/ha (Zwart et al., 2007)), plus 2.5% (Gerin et al., 2008) of energy used for biodigester processing (2.5% of 71.6 GJ/ha = 1.8 GJ/ha). Total input is 2.6 + 10.4 = 13 GJ/ha.

Bioheat

The energy input for corn production in Ontario has been estimated to be 2.9 GJ/ODT (Samson et al., 2005) which assuming a field yield of 7.3 ODT/ha equals 21.17 GJ/ha.

The energy input for switchgrass pellets is 12 GJ/ha, based on field energy inputs of 7.9 GJ/ha and 4.1 GJ/ha for pellet processing and marketing (Samson et al., 2000).

Biofuels

The energy output:input ratio for corn ethanol is 1.28:1 (Wang et al., 2007), this results in an energy input of 72.5/1.28 = 47.9 GJ/ha.

The energy output:input ratio switchgrass cellulosic ethanol is 4.44 (average of Lynd, 1996; Sheenan et al., 2004; Lynd and Wang, 2004), this results in an energy input of 67.8/4.44 = 15.3 GJ/ha.

The energy output:input ratio for soybean biodiesel is 2.56:1 (average from Hill et al., 2006, and Sheenan et al., 1998), this results in an energy input of 17.3/2.56 = 6.8 GJ/ha.

Table 16.3 Net GHG offsets from various bioenergy technologies through fuel switching applications for fossil fuels in Ontario, Canada

Fossil Fuel Traditional Use		Renewable Alternative Fuel Use		Net offset emissions including N_2O	
Energy Type	$kgCO_{2e}/GJ$	Energy type	$kgCO_{2e}/GJ$	$(kgCO_{2e}/GJ)$	%[h]
Gasoline Transport	99.56[a]	Corn Ethanol	62.03[c]	21.13[h]	21
		Cellulosic Ethanol	23.40[b]	76.16[b]	77[g]
Diesel Transport	98.54[a]	Soybean Biodiesel	36.36[d]	49.73[h]	50
		Canola Biodiesel	28.77[d]	57.09[h]	58
Coal	93.11[a]	Switchgrass Pellets	8.17[e]	84.94	91
		Wood pellets	13.14[f]	79.97	86
		Straw pellets	9.19[f]	83.92	90
Heating Oil	87.90[a]	Switchgrass Pellets	8.17[e]	79.73	91
		Wood pellets	13.14[f]	74.76	85
		Straw pellets	9.19[f]	78.71	90
Liquefied Natural Gas	73.69[i]	Switchgrass Pellets	8.17[e]	65.52	89
		Wood pellets	13.14[f]	60.55	82
		Straw pellets	9.19[f]	64.5	88
Natural Gas	57.57[a]	Switchgrass Pellets	8.17[e]	49.40	86
		Wood pellets	13.14[f]	44.43	77
		Straw pellets	9.19[f]	48.38	84

[a] Natural Resources Canada, (2007)
[b] Emissions estimated from cited GHG savings
[c] EIA, (2006)
[d] (S&T)^{2}Consultants Inc., (2002)
[e] Samson et al., (2000)
[f] Jungmeier et al., (2000)
[g] Average from Wang et al., (2007), and Spatari et al., (2005)
[h] Samson et al., (2008a)
[i] LNG imported from Russia into North America estimated to have 28% higher GHG emissions then North American NG production due to methane leakage and energy associated with Russian pipelines, LNG liquification, ocean transport and heating during re-gasification (Heede, 2006; Jaramillo et al., 2007; Uherek, 2005)

Biogas production from energy crops represents a more thermodynamically efficient option than converting plant matter into liquid fuels. Considering the case of corn silage, $500 m^3$ of biogas can be produced from one tonne of feedstock (Table 16.2) which is equivalent to 11.6 GJ/ODT or 61.7% conversion efficiency. In contrast with current projected cellulosic ethanol product yields of 340 l of ethanol (Iogen Corporation, 2008), 7.1GJ/tonne of energy is recovered, a 38% conversion efficiency. In Germany, there has been significant scale-up of energy crops grown for biogas applications. In 2006, there were an estimated 3500 biogas digestors in the country that were mainly operating on energy crops such as corn silage, rye silage, and perennial grasses as well as manure and food processing wastes (House et al., 2007).

Some of the main problems facing the cellulosic ethanol industry are: (1) a chronic underestimation of feedstock procurement costs required by farmers in industrialized countries to make the technology viable on a large-scale; and

(2) projected commercial plant construction costs have risen dramatically, especially those for stainless steel and skilled labour costs. The economics now favour larger plants, with most plants foreseen to have a feedstock requirement of one million tonnes per year or more. Considering the increasing depletion of fossil energy resources in industrialized countries projected for the future, it may be difficult for such large amounts of affordable biomass to be procured and transported to bioethanol plants, especially when they are competing with local biogas and bio-heat plants that are more thermodynamically efficient and have significantly lower processing and transport costs. A centralized biogas digester producing 3 MW of thermal energy or a 50,000 tonne per year bioheat pellet plant have much smaller land area footprints than a 700,000 tonne per year cellulosic ethanol plant. The land area to be planted, based on the switchgrass yields in Table 16.2, would be 6000ha and 75,000ha for a switchgrass pellet and cellulosic ethanol plant respectively. If 1 in 4 ha surrounding the cellulosic ethanol plant was planted to switchgrass, the plants feedstock supply would be drawn from a land area covering 300,000 ha and stretch out a radius of 310km from the plant. The economic premium offered to produce liquid biofuels as a substitute for gasoline may not be sufficient to recover the large thermodynamic loss required for production and conversion of solid plant matter into liquid fuel in these large biorefineries. It is, by comparison, more efficient to use whole-plant biomass in pellet or biogas form to substitute for natural gas. As such, in temperate regions of industrialized countries, which are densely inhabited and have high local demands for heat and power, bioheat and biogas are the technologies likely to succeed if there is any level of parity in the government incentives applied to bioheat, biogas and liquid biofuels.

16.2.3 Greenhouse Gas Mitigation from Bioheat and Other Biofuels Options

With increasing concern about global climate change, it is of paramount importance that cost-effective emission reduction strategies evolve from producing bioenergy from farmland in industrialized countries. Efforts to import biofuels from tropical countries to date have resulted in rapid deforestation of native forests for palm oil production, particularly in Malaysia and Indonesia. Sugar cane cultivation for ethanol production is now expanding into traditional grazing lands in countries like Brazil, causing the cattle industry to expand into tropical forests. While certification systems may evolve for sustainable importation of tropical biofuels into industrialized countries, it is essential that effective domestic strategies are developed in industrialized countries to reduce the need for these imports. Developing nations in the tropics will themselves require large volumes of biofuels for their internal needs, further increasing the pressure on industrialized nations to become energy self-sufficient.

An important driver for the development of bioenergy will be the economic competitiveness of various technologies as greenhouse gas mitigation strategies. Thus, it is important that the economics of various solid and liquid biofuels options be compared. There are several factors which are fundamental to the economic competitiveness of various agricultural biomass production and conversion chains to reduce greenhouse gases effectively including:

1. optimizing the amount of energy produced from each hectare of marginal and arable farmland (explored in Section 16.2 above);
2. the net GHG offset provided by displacing a GJ of fossil fuels used for a particular application with a renewable energy for the same application (fuel switching); and
3. the cost of production of the processed bioenergy product relative to the fossil fuel it is displacing.

The net offsets from various bioenergy technologies through fuel switching applications for fossil fuels in Ontario, Canada are summarized in Table 16.3. The net GHG offsets are highest with switchgrass pellets (86–91%), moderate with soybean biodiesel (50% offset) and low with corn ethanol (21%). The low GHG offsets from corn ethanol is confirmed by two recent analyses in the United States which determined the GHG offset potential of corn ethanol to be 15% (Farrell et al., 2006) and 19% (Wang et al., 2007) respectively in the current state of the industry.

The reason for the high offset potential of switchgrass pellets is that they require modest amounts of energy for switchgrass feedstock production and pellet processing. As well there is no change of physical state that occurs, so nearly all of the energy content of the grass is available in a pellet form. Switchgrass production also has no significant landscape emissions as N_2O emissions are low for perennial grasses and the soil carbon sequestered is expected to offset the low amounts of N_2O emissions that occur (Adler et al., 2006). Soybean biodiesel (or canola biodiesel) represents a moderately efficient offset potential because the liquid fuel production process is not energy intensive and the crop has moderate energy inputs and N_2O emissions in North America (Samson et al., 2008a). Each GJ of soybean biodiesel produced displaces approximately half the GHG emissions of diesel fuel. However it still represents a largely ineffective approach to mitigate greenhouse gasses from farmland in temperate regions, as the soybean yield is low and the oil content in the soybean seed is low. With soybean oil being a high quality vegetable oil selling for premium prices in 2007–2008 around $1000/tonne, biodiesel is far from being an economically viable biofuel unless heavily subsidized.

The reasons for the low offset potential of corn are: (1) the technology relies heavily on carbon intensive fuels such as coal and natural gas for processing; (2) corn is an energy intensive annual crop to produce; (3) there are relatively high N_2O losses from each hectare of corn production, which has a strong impact on overall emissions; and (4) comparatively low amounts of energy are captured in the field and converted into a final energy product. In the province of Ontario, Canada, the combined federal and provincial incentives in 2007 amounted to 16.8 cents per litre of ethanol produced (or $8.00 CAN/GJ assuming an energy value of 0.021GJ/litre). With only 21.13 kg CO_{2e} offsets per GJ of fuel, it takes 47.3 GJ of ethanol to offset one tonne of carbon dioxide. This is equivalent to a subsidy of $379 (CAN) (Samson et al., 2008a) (1 \$CAN = \$1 USD in October 2007). Even larger federal and state subsidies are available for promoting corn ethanol production in certain states in the United States.

The main advantage of cellulosic ethanol from switchgrass over corn ethanol is that the heat and power for plant processes are provided by lignin, a by-product in cellulosic ethanol processing. Cellulosic ethanol results in a moderately high offset potential of 76.5% compared to the use of gasoline. Nevertheless this use of lignin causes a parasitic impact on the net GHG mitigation per ha that can be provided in comparison to using the grass for other bioenergy applications. With relatively modest volumes of energy recovered from each tonne of biomass of 340 l ethanol/tonne, the technology on a per hectare basis represents only a moderately efficient approach at using farmland to mitigate greenhouse gases. The technology can be best categorized as having a medium energy output per hectare and a moderate to high GHG offset when displacing fossil fuels. Overall, Table 16.4 illustrates that using Ontario farmland to produce switchgrass ethanol has the potential to offset approximately 5,164 kg CO_{2e}/ha tonnes of GHG emissions. It is significantly superior to corn ethanol and soybean biodiesel if current commercialization problems can be overcome.

From Table 16.4, it can be observed that corn ethanol and soybean biodiesel cannot be considered effective greenhouse gas mitigation policies with less than 1,500 kg CO_{2e}/ha offsets. Per hectare, corn ethanol has modest energy production and poor net GHG offsets, while per hectare soybean biodiesel has a poor liquid fuel output and only a moderate GHG offset. This analysis demonstrates that solid biofuels represent a highly promising means for Ontario to mitigate greenhouse gases, particularly compared with liquid fuel options. Figure 16.1 graphically represents these findings.

The advanced boiler technology currently available to burn pellets offers the same combustion efficiency as natural gas combustion appliances (Fiedler, 2004). When switchgrass pellets are used to displace coal, the highest overall GHG displacement potential can be achieved at 13,098 kg CO_{2e}/ha. The lowest GHG

Table 16.4 Evaluation of different methods of producing GHG offsets from Ontario farmland using biofuels

Feedstock	Gross Energy (GJ/ha)	Fossil Fuel Substitution	Net GHG emission offsets kgCO$_2$e/GJ	Total GHG emission offsets kgCO$_2$e/ha
BioHeat				
Switchgrass Pellets	154.2	Coal	84.94	**13098**
Switchgrass Pellets	154.2	Heating Oil	79.73	**12294**
Switchgrass Pellets	154.2	Liquefied Natural Gas	65.52	**10103**
Switchgrass Pellets	154.2	Natural Gas	49.4	**7617**
Biofuels				
Switchgrass Cellulosic Ethanol	67.8	Transport gasoline	76.16	**5164**
Grain Corn Ethanol	70.6	Transport gasoline	21.13	**1492**
Soybean Biodiesel	18.2	Transport diesel	49.73	**905**

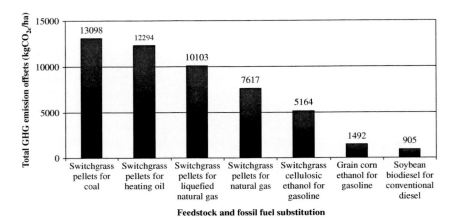

Fig. 16.1 Evaluation of different methods of producing GHG offsets from Ontario farmland using biofuels

emission potential of switchgrass pellets is at 7,617 kg CO_{2e}/ha when they are used to displace natural gas. When switchgrass pellets replace imported LNG from Russia, approximately 10 tonnes of CO_{2e}/ha is abated. From Ontario's perspective, an effective policy strategy for GHG mitigation would clearly be to replace foreign imports of LNG and coal with domestically produced pellets within the province. A \$2(CAN)/GJ incentive for switchgrass pellet producers would cost an average of \$24, \$31 and \$40/tonne of CO_2 offset to displace the use of coal, liquefied natural gas and conventional gas, respectively. In contrast, Ontario has combined federal and provincial wind energy incentives of \$15.28/GJ (6.5 cents/kWh), soybean biodiesel incentives of \$5.68/GJ (20 cents/L) and corn ethanol incentives of \$8.00/GJ (16.8 cents/litre). The corresponding costs of these offsets are \$50, \$98 and \$379/tonne CO_{2e} for wind, biodiesel and corn ethanol, respectively (Samson et al., 2008a). Two other recent studies also found that with carbon taxes under \$100/tonne, bioheat is considerably less expensive GHG offset strategy than producing liquid fuels in temperate regions (Grahn et al., 2007). To create more effective use of taxpayers' money in reducing GHG emissions, policy makers need to understand the offset potential of the various technologies and create mechanisms to allow GHG reduction to happen competitively within the marketplace.

Another problematic example exists with energy crop use for biogas systems, which is currently strongly supported as a RET in Germany. Energy crop biogas systems appear to be facing several challenges in being an efficient GHG mitigation technology. Few detailed studies have been completed but there appears to be some identified limitations. When examining only energy related GHG emissions, power generation from energy crop biogas is a highly effective GHG mitigation technology compared to using fossil fuels for power production (Gerin et al., 2008). However, two GHG emission problems have been identified with energy crop biogas for power generation which are methane leakage from digesters (estimated at 1%) and the high N_2O emissions associated with maize cultivation (Crutzen et al., 2007). In one preliminary study from Western Europe there was no net GHG benefit from maize silage biogas because of these two aforementioned problems (Zwart et al., 2007). It is likely the use of deeper rooted and more nitrogen efficient annual crops such as sorghum or perennial species such as highly digestible warm season grasses may help reduce GHG emissions from feedstock production. As well, energy crop cultivation in less humid regions would reduce the N_2O loss problem. New design features of digesters and larger centralized biogas digesters may help reduce methane losses that are currently occurring. Energy crops used in biogas digesters in the future will likely play an important role in providing GHG friendly thermal energy for combined heat and power applications. However, presently only manure biogas digesters have been found to have positive impacts on GHG mitigation (Zwart et al., 2007). If governments created incentives for RET's based on their actual GHG mitigation efficient approaches to reduce emissions would be likely be stimulated and more efficient progress in mitigating GHG's would be realized through bioenergy technologies.

16.3 Optimization of Energy Grasses for Combustion Applications

From the previous analysis it is evident growing energy grasses for bioheat represents the most outstanding option for using one hectare of farmland to produce renewable energy and mitigate GHG's from an agricultural production system. If energy crop grasses are to evolve as a major new RET for energy security and GHG abatement for the industrialized world, it is imperative that considerable research and development efforts to expand this opportunity be undertaken. Historically, the major limitation to the development of grasses for bioheat applications has been the difficultly associated with burning energy grasses efficiently in conventional biomass boilers. In particular, the relatively high alkali and chlorine contents of herbaceous plants are widely known to lead to clinker formation and corrosion of boilers. These biomass quality problems have resulted in slow commercialization of grass feedstocks as agro-pellets for use in small scale boilers (Elbersen et al., 2002; Obernberger and Thek, 2004). Despite this, the problems with burning grasses have now become reasonably well understood and constraints are being resolved through several strategies. Plant selection and breeding together with delayed harvest management can be used to reduce the chlorine, alkali and silica content in native grasses, reducing clinker formation and corrosion in boilers. Utilizing advanced combustion systems which are specifically designed to burn high-ash; herbaceous fuels can also reduce problems with ash accumulation in burners (Obernberger and Thek, 2004). However, high-ash fuels can still pose major convenience issues, particularly when used in pellet stoves and small scale boilers. Strategies to lower the ash content and the undesirable chemical elements in grasses are essential if commercial markets are to be fully developed.

16.3.1 Improving Biomass Quality for Combustion

The most serious biomass quality problem with herbaceous feedstocks is the alkali and chlorine content in the feedstock material, which has potential for fouling and corroding boilers during combustion (Passalacqua et al., 2004). Particulate emissions are strongly related to fuel type, and specifically, to the content of aerosol-forming compounds such as potassium (K), chlorine (Cl), sodium (Na), sulphur (S) and even lead and zinc in the fuel (Hartmann et al., 2007). Using fuels that are low in the "dust critical" elements K, Cl, Na and S is of particular importance for achieving high-quality biomass fuels and lowering particulate emissions during biomass combustion. The major factors affecting the level of aerosol-forming compounds are fertilization practices, choice of species, stem thickness, time of crop harvest, relative maturity of the cultivar, and the level of precipitation in a region (Samson et al., 2005; Samson, 2007). Chlorine is particularly problematic as it increases the ash-sintering effect of fuels containing potassium and makes these elements migrate from the fuel bed to the boiler walls, forming clinkers (Godoy and Chen, 2004). The nitrogen content of feedstocks has little impact on the efficiency of the combustion process but burning high-N fuels is undesirable from an environmental standpoint as this contributes to NO_x pollution. However, delayed harvest switchgrass has relatively low N contents that are comparable to wood (Samson et al., 2005; Adler et al., 2006). Reducing the moisture content of feedstocks to below 15% is also important as this eases storage problems from decomposition and can reduce or even eliminate the need to dry materials before pelletizing them.

16.3.1.1 Nutrient Management

Both potassium and chlorine are known to be effectively leached out of thin-stemmed grasses in humid climates. As potassium is water soluble, the potassium content in plants can decrease appreciably following senescence of materials during the end of growing season, particularly if significant rainfall occurs during this period. Prairie ecology studies have also demonstrated that potassium in unharvested material is efficiently recycled into the soil over the late fall and winter (Koelling and Kucera, 1965; White, 1973). Kucera and Ehrenreich, (1962) in Missouri found potassium content of native prairie plants to decline from 1.34% K_2O in mid-June, to 0.63% by mid-September, and to 0.05% by the end of November. Koelling and Kucera (1965) found the average potassium content of big bluestem in the Missouri prairies to decrease from 1.28% K_2O in July, to 0.33% in September, and to 0.13% in November. Over-wintering further reduced levels to 0.07% by May the following year. It is also of interest to note that native prairie materials likely have significantly earlier maturity dates (and hence time for fall leaching) than purpose grown energy grasses. In Quebec, Cave-in-Rock switchgrass harvested in early October was found to contain 0.95% potassium, while over-wintered switchgrass harvested in mid-May was found to contain just 0.06% potassium (Goel et al., 2000). In the case of potassium, it appears that harvesting in the fall at least several weeks after materials senesce, or alternately harvesting over-wintered material, provides significant reductions in the potassium content of feedstocks. Chlorine is also highly water soluble in herbaceous biomass feedstocks (Sander, 1997). Like potassium, the chlorine content of perennial grass feedstocks is reduced if a late-season or overwintering harvest management regime is practiced. Burvall (1997) found an 86% reduction in chlorine content of reed canarygrass when it was over-wintered in Sweden.

16.3.1.2 Harvest Management and Cultivar Selection

Despite the benefits that overwintering can provide, letting grasses remain unharvested through the winter can also reduce the eventual biomass yield obtained in the spring. In Southwestern Quebec, spring-harvested switchgrass yields were found to be approximately 24% lower than that of fall-harvested switchgrass (Goel et al., 2000). This loss was due likely to both the late season translocation of materials to the root system in winter (Parrish et al., 2003), and the physical loss of material, mainly from leaves and seed heads during the winter season (Goel et al., 2000). Compared to fall harvested material, spring-harvested switchgrass lost 4% of dry matter from the stem component, 11% from leaf sheaths, 30% from leaves and 80% from seed heads (Goel et al., 2000). Field observations have indicated that when the material is completely dry in late winter and early spring, the majority of breakage losses occur during storm events. As well, some decomposition occurs in the field when material lodges in late summer and early fall and plants come into contact with the soil.

A new delayed harvest technique was assessed in the spring of 2007 in Ontario by REAP-Canada (Samson et al., 2008b) to minimize winter breakage and spring harvest losses from feedstocks, while maintaining the benefits of nutrient leaching that are associated with overwintering. Under this system, the material is mowed into windrows in mid-November and directly baled off the windrow in the spring. Results to date are promising as yields were 21% higher than spring mowed and harvested material. The fall mowing technique also caused faster spring drying of windrowed material, but recovery of material below 10% moisture was achieved in early May in both systems. Finally, the fall mowing technique encouraged earlier soil warming and than spring mowed areas, promoting earlier regrowth of the switchgrass.

Selecting for increased stem and leaf sheath content and developing warm season grass varieties that more efficiently retain their leaves through the winter could help reduce overwintering losses. Another strategy that has proven effective to reduce potassium and chlorine content in feedstocks is to utilize earlier-maturing warm season grass varieties that senesce earlier in the fall (Bakker and Elbersen, 2005). Early maturity enables a more extended period between senescence and late fall harvest for nutrients to be leached from the stem material. Thin stemmed grasses have also been identified to have higher nutrient leaching potential compared than thicker stemmed grasses. Lowland switchgrass cultivars with tall, moderately coarse stems, such as Alamo and Kanlow have been found to be moderately higher in K and Cl than upland switchgrass with short, fine stems at the end of the season (Cassida et al., 2005). The average outer diameter of lowland ecotypes of switchgrass has been found to range from 3.5 mm (Igathinathane et al., 2007) to 5 mm (Das et al., 2004) and to have a stem wall thickness of approximately 0.7 mm (Igathinathane et al., 2007). The problem of biomass quality appears to be even more serious in miscanthus than switchgrass. Thick stemmed miscanthus ecotypes are known to have high potassium and chlorine contents, especially when combined with late maturity (Jørgensen, 1997). Comparatively, the average stem diameter of miscanthus is 8.8–9.2 mm, with a stem wall thickness of 1.3–1.5 mm (Kaack and Schwarz, 2001). No biomass quality reports from Europe could be identified which indicated miscanthus sinensis giganteus was able to reach the minimal biomass quality targets of 0.2% K and 0.1% chlorine for power generation in Denmark outlined by Sander (1997). Thick stems also make it more difficult to dry material. REAP-Canada identified that even in fall harvested upland switchgrass, while most plant components had moisture contents below 15%, the stems still tended to retain significant moisture (Samson et al., 2008b). Spring harvesting of material can enable bales to be collected below 12% moisture. The low moisture content of grasses at spring harvest is a significant advantage that grass energy crops hold over woody energy crops. The moisture content of willows at harvest for willows can be 50%. High moisture woody materials can use 21% of the raw material to provide energy for the drying process if made into pellets (Bradley, 2006). Spring harvested grasses thus have a major biomass quality advantage for pellet processing because of the dryness of the material.

Overall, grass pellets appear to represent the most promising solution to the strong international growth in demand for fuel pellets, a growth that cannot be met with supplies of wood residues forecast for the future. Many combustion issues have now been resolved in replacing wood pellets with grass pellets. Research indicates native warm-season grass pellets grown in North-eastern North America can approach a comparable content of aerosol forming compounds as that found wood residue pellets. However, the overall ash content of grass pellets typically remains considerably higher than wood. Wood residue pellets of highest quality are sold as premium grade when they achieve less than 1% ash. Typically, the European market trades wood pellets with 0.6% ash in this category (Obernberger and Thek, 2004). However, grasses harvested in North-eastern North America are generally in the 3-5% ash range (Samson et al., 2005). Even higher contents of ash are experienced in switchgrass growing regions with less favourable rainfall to evaporation ratios such as western Canada (Jefferson et al., 2004) and the Western United States (Cassida et al., 2005).

16.3.1.3 Impacts and Management of Silica

Silica levels in grasses must also be reduced if grass pellets are to enter into the high-end residential wood pellet market that currently has products trading in Europe at approximately $250/tonne. Producing fuels with lower silica levels has many benefits. Low silica containing fuels have higher energy contents, reduce abrasion on metal parts such as pellet dies during the densification processes, and improve convenience in reducing ash removal requirements. When burned in pellet appliances, high-ash grass pellets with high silica contents can also produce a low-density ash that retains the shape of the former pellet. As an example, consider that the bulk density of reed canary grass ash has been assessed to be half that of wood ash (Paulrud, 2004). Thus the residual ash leftover after burning grass pellets in the 3–5% ash range can take up to 10–20 times the volume of the ash from burning 0.6% ash wood pellets. To burn 3–5% ash grass pellets, ash pans will need to be modified in smaller appliances to create larger ash collecting areas. Combustion units burning high-ash grass pellets will require more frequent cleaning and may experience increased operational problems such as automatic shutdown of the combustion appliance if the ash builds up into the combustion chamber. Conversely, silica is generally not a problematic element for commercial combustion boilers. Paulrud et al., (2001), working with reed canary grass, found that the relative content of K and Ca in the ash was more important for agglomeration and clinker formation than the silica content. High-ash agro-pellets (approximately 5% ash) with low to moderate levels of aerosol forming compounds are readily burned in most coal boiler technologies and greenhouse producers in Canada are now installing multifuel boilers capable of burning both coal and agro-pellets.

A comprehensive strategy will be required to reduce the silica content of grasses to make them more convenient for combustion applications and to improve their energy content. The understanding of silica uptake into the plant is improving amongst agronomists and plant breeders. The main cultural factors which appear to have potential to reduce the silica content are: soil type, production region, photosynthetic cycle of the biomass crop and the choice of grass species and variety. The main breeding strategies to reduce silica content include increasing the stem to leaf ratio of the species and reducing silica transport into the plant. As well, fractionation of plant components can help create lower silica containing feedstocks.

The translocation and deposition of silica in plants is heavily influenced by the soluble levels of silica in the soil, present as monosilicic acid or $Si(OH)_4$ (Jones and Handreck, 1967). Clay soils have higher monosilicic acid levels than sandy soils, and therefore produce feedstocks with higher silica levels. A Scandinavian study found silica levels in reed canarygrass to be highly influenced by soil type; reed canarygrass had silica levels of 1.3%, 1.9% and 4.9% on sandy, organic, and clay soils, respectively (Pahkala et al., 1996). In Denmark, high silica contents in wheat straw were strongly correlated with clay contents of soils (Sander, 1997). A main difference in silica content between perennial grass species can also be the photosynthetic mechanism of the grass and the amount of water being transpired by the plant. Warm season (C_4) grasses on average, use half as much water as C_3 grasses per tonne of biomass produced (Black, 1971). The decreased water usage reduces the uptake of silicic acid and decreases the ash content of the plant.

Within warm season grasses, water use per tonne of biomass produced is highest in regions which have a low rainfall to evaporation ratio, and where biomass crops are grown on marginal soils (Samson et al., 1993; Samson and Chen, 1995). A combination of these conditions may explain some of the higher values obtained by a survey from the United States reporting switchgrass ash contents of 2.8–7.6% (McLaughlin et al., 1996). Regions with a rainfall to evaporation ratio greater than

100% would be expected to have substantially lower ash contents than short grass prairie regions where the rainfall to evaporation ratio is 60%. This is illustrated in analysis from Quebec and Western Europe where silica levels of lower than 3% are commonly obtained in overwintered materials. Plant species have widely differing levels of silica. By comparing the speed of silica uptake with that of water uptake, three modes of silica uptake have been suggested by Takahashi et al., (1990). These modes are active (higher than water uptake), passive (similar with water uptake) and rejective (slower than water uptake). However, Van Der Vorm (1980), found no evidence of passive uptake. A gradual transition was found between metabolic absorption to metabolic exclusion which depended on the silica concentration. In all species examined, including 3 monocots (rice, sugar cane and corn), there was preferential absorption at low concentrations and exclusion at high concentrations (Van Der Vorm, 1980). As silica uptake by rice is significantly higher than other agronomic species, considerable efforts and achievements have been made in understanding and characterizing the process. This now has included molecular mapping studies of the silica transport mechanism (Ma et al., 2004). It may be possible that some reductions in the silica content of warm season grasses could be made in warm season grass breeding programs by reducing silica transport into the plant. It should however be noted that sugar cane and rice plant breeders are currently trying to increase the content of silica in these species because silica plays an important role in reducing plant stresses, increasing resistance to diseases, pests, and lodging, and decreasing transpiration (Ma, 2003).

Silica is mainly deposited in the leaves, leaf sheaths and inflorescences of plants (Lanning and Eleuterius, 1989). Lanning and Eleuterius (1987) working in Kansas prairie stands found switchgrass silica contents to be lowest in stems and higher in leaf sheaths, inflorescences and leaf blades. Silica levels are suggested to have evolved to be high in inflorescence structures to prevent the grazing of seed heads. Due to the low stem silica content, the overall silica concentration of grasses decrease as the stem content increases. Pahkala et al., (1996) examined 9 different varieties of reed canarygrass and found varieties to range from 2.3% to 3.2% silica content, with the lower silica containing varieties having a higher biomass stem fraction. Thus, selection for increased stem content is desirable for improving biomass quality for combustion purposes. This is demonstrated in Table 16.5 where stems had on average 1.03% ash and leaves had 6.94% ash. The impact of ash content on the energy content of the feedstock is evident as the leaves also contained approximately 6% less energy than stems. Stems contained on average 19.55 GJ/ODT which is 98% of the average energy content of high quality wood pellets of 20 GJ/ODT (Obernberger and Thek, 2004).

The differences in silica content between the various components of grasses has been known for more than 20 years. It also appears there are substantial inherent differences between the silica contents of warm season grass species. Two of the 3 main tallgrass prairie species in North America are big bluestem and switchgrass. The overall silica content of big bluestem may be amongst the lowest of the native North American grasses. In studies of plants harvested from a native prairie,

Table 16.5 Energy and ash contents (%) of spring harvested switchgrass (Samson et al., 1999b)

Component	Sandy Loam Soils Spring 1998	Clay Loam Soils Spring 1998	Average
Switchgrass Ash Contents (%)			
Leaves	6.20	7.67	6.94
Leaf sheaths	2.46	3.67	3.04
Stems	1.08	0.98	1.03
Seed heads	2.38	n/a	2.38
Weighted Average:	**2.75**	**3.21**	**2.98**
Switchgrass Energy Contents (GJ/ODT)			
Leaves	18.44	18.38	18.41
Leaf sheaths	19.19	18.27	18.73
Stems	19.41	19.69	19.55
Seed heads	19.49	n/a	19.49
Weighted Average:	**19.11**	**19.07**	**19.09**

relatively low silica contents of 0.29, 1.69, 2.08, and 2.89% were reported for the stems, leaf sheaths, inflorescences and leaves, respectively. In contrast, switchgrass averaged 1.03, 3.89, 3.41 and 5.04% for stems, leaf sheaths, inflorescences and leaves, respectively (Lanning and Eleuterius, 1987). As switchgrass is known to grow in wetter zones in the prairies, the higher levels of silica found may be a result of where the plants were collected within the prairie remnant. Big bluestem is known to have the additional advantage of having a high percentage of its dry matter in the stem fraction and a smaller inflorescence than native ecovars of switchgrass. Typically, the stem fraction of mature native big bluestem ecovars (e.g. cultivars not selected for forage quality) is approximately 60% of the above ground biomass, while in upland switchgrass ecovars the stem typically comprises 45–50% of the biomass in mature plants (Boe et al., 2000; Samson et al., 1999a). Further analysis of species and components of grasses as well as cultivars of grasses is required to more effectively understand how to reduce silica levels.

In the search for low silica herbaceous feedstocks for the pulp and paper industry, there has been considerable research and commercial development in Scandinavia on fractionation technologies to separate the low silica containing stems from the other plant components (Pahkala and Pihala, 2000; Finell et al., 2002; Finell, 2003). Several approaches to dry fractionation have been developed and integrated into commercial straw pulping facilities in Denmark (Finell et al., 2002). The basic process of disc mill fractionation developed by UMS A/S in Denmark is overviewed by Finell (2003) and includes keys steps of bale shredding with a debaler, hammer milling, disc milling, pre-separation (separating leaf meal and internode chips) and then a final sifting to further refine the accepted fraction of internode chips for pulping. In the case of reed canary grass, typically 40–60% of the plant could be recovered for pulping applications with the residual material used as a commercial pellet fuel (Finell, 2003).

This technology can also be applied to the fractionation of warm season grasses to developing fuels for use in the residential and commercial pellet markets. Fractionation of stems from species such as big bluestem would produce pelletized fuels in the range of 1% ash if the feedstock was grown on sandy soils in regions with a favourable rainfall to evaporation ratio. The higher-ash leaf, leaf sheath and inforescence material could then be used as a high-ash commercial pellet fuel for larger-scale thermal applications.

16.4 Outlook

This review supports other recent studies that have found energy crop development for thermal energy applications holds significant potential for industrialized nations as a means to create energy security and clean energy through GHG mitigation. From an energy security standpoint, it appears that the conversion of whole plant biomass from annual C_4 grasses into biogas or bioheat represent the most promising energy production technologies available. With current understanding of the GHG mitigation issue, direct combustion applications of perennial grasses to displace coal, natural gas and heating is the leading strategy to use farmland to mitigate greenhouse gases. The large N_2O emissions associated with the cultivation of corn in humid temperate climates impairs the effectiveness of corn as a feedstock to produce low GHG loading gaseous and liquid biofuels. In this respect, more research on N-efficient annual crops and higher digestibility perennial biogas species could help strengthen the GHG mitigation potential of biogas from energy crops in the future. In the case of bioheat from grasses, the research challenges ahead include the improvement of biomass quality to develop pellet fuels with low contents of silica and aerosol-loading elements.

Some of the largest hurdles to overcome in the emergence of second generation bioenergy technologies are not technological issues, but rather policy barriers. Governments have a major influence on which crops and technologies are scaled up for commercialization through the use of incentives or subsidy programs. It would be highly recommended to encourage policies to avoid picking technology winners in the development of energy security and greenhouse gas mitigation technologies from RET's. Rather, governments should encourage results-based management approaches to address policy issues and examine means to create parity in incentives in the green energy marketplace. This could include the creation of carbon taxes, green carbon incentives, CO_2 trading systems or incentives per GJ of energy produced. Both progressive policy and technology development need to be developed together for renewable energy to work for environmental protection and energy security in industrialized nations.

Acknowledgments The authors gratefully acknowledge financial support from the Biocap Canada Foundation, Natural Resources Canada and the Ontario Ministry of Agriculture, Food and Rural Affairs-Alternative Renewable fuels Fund.

References

Adler, P. A., Sanderson, M. A., Boateng, A. A., Weimer, P. J., & Jung, H. G. (2006). Biomass yield and biofuel quality of switchgrass harvested in fall and spring. *Agron. J., 98*, 1518–1525

De Baere, L. (2007). Dry Continuous Anaerobic Digestion of Energy Crops. (Paper presented at the 2nd International Energy Farming Congress, Papenberg, Germany)

Bakker, R. R., & Elbersen, H. W. (2005). Managing ash content and quality in herbaceous biomass: An analysis from plant to product. (Paper presented at the 14th European Biomass Conference & Exhibition, Paris, France)

Beadle, C. L., & Long, S. P. (1985). Photosynthesis-Is it limiting to biomass production? *Biomass, 8*, 119–168

Berglund, M., & Börjesson, P. (2006). Assessment of energy performance in the life-cycle of biogas production. *Biomass and Bioenergy, 30*, 254–266

Black, C. C. (1971). Ecological implications of dividing plants into groups with distinct photosynthetic production capacities. *Advanced Ecological Resources, 7*, 87–114

Bradley, D. (2006, May). GHG impacts of pellet production from woody biomass sources in BC, Canada. **Retrieved July, 2007, from** www.joanneum.at/iea-bioenergy-task38/projects/task38casestudies/can2-fullreport.pdf

Braun, R., & Wellinger, A. (2005). Potential of Co-digestion. (Prepared under IEA Bioenergy, Task 37, Energy from Biogas and Landfill Gas)

Boe A., Bortnem R., & Kephart, K. D. 2000. Quantitative description of the phytomers of big bluestem. *Crop Science, 40*, 737–741

Burvall, J. (1997). Influence of harvest time and soil type on fuel quality in reed canary grass (*Phalaris Arundinacea* L.). *Biomass and Bioenergy, 12*(3), 149–154

Cassida, K. A., Muir, J. P., Hussey, M. A., Read, J. C., Venuto, B. C., & Ocumpaugh, W. R. (2005). Biofuel component concentrations and yields of Switchgrass in South Central U.S. environments. *Crop Science, 45*, 682–692

Clark, F. E. (1977). Internal cycling of nitrogen in shortgrass prairie. *Ecology, 58*, 1322–1333

Crutzen, P., Mosier, A., Smith, K., and Winiwarter, W. 2007. N_2O release from agro-biofuel production negates global warming reduction by replacing fossil fuels. *Atmos. Chem. Phys.*, (7), 11191–11205.

Darley, J., (2004, August). *High Noon for Natural Gas: The New Energy Crisis*. (Chelsea Green Publishing, ISBN 1-931498-53-9)

Das, M. K., Fuentes, R. G., & Taliaferro, C. M. (2004). Genetic variability and trait relationships in switchgrass. *Crop Science, 44*, 443–448

Elbersen, H. W., Christian, D. G., Bacher, W., Alexopoulou, E., Pignatelli, V., & van den Berg, D. (2002). Switchgrass Variety Choice in Europe. (Final Report FAIR 5-CT97-3701 "Switchgrass")

EIA. (2006). Emissions of Greenhouse Gasses in the United States 2005. Energy Information Administration, official energy statistics from the U.S. Government. **Retrieved July, 2007, from** http://www.eia.doe.gov/oiaf/1605/ ggrpt/index.html

Farrell, A. E., Plevin, R. J., Turner, B., Jones, A. D., O'Hare, M., & Kammen, D. M. 2006. Ethanol can contribute to energy and environmental goals. *Science, 331*, 506–508

Fiedler, F. (2004). The state of the art of small-scale pellet based heating systems and relevant regulations in Sweden, Austria and Germany. *Renewable and Sustainable Energy Reviews, 8*, 201–221

Finell, M. (2003). The use of reed canary-grass (Phalaris arundinacea) as a short fibre raw material for the pulp and paper industry. (Doctoral thesis prepared for the Swedish University of Agricultural Sciences, Grafiska Enheten, SLU, Umea, Sweden)

Finell, M., Nilsson, C., Olsson, R., Agnemo, R., & Svensson, S. (2002). Briquetting of fractionated reed canary-grass for pulp production. *Industrial Crops and Products, 16*(3), 185–192

Gerin, P. A., Vliegen, F., & Jossart, J. M. (2008). Energy and CO_2 balance of maize and grass as energy crops for anaerobic digestion. *Bioresource Technology, 99*(7), 2620–2627

Girouard, P., Samson, R., & Mehdi, B. (1998). Harvest and Delivered Costs of Spring Harvested Switchgrass. (Final report prepared by REAP-Canada final for Natural Resources Canada, Ottawa Ontario)

Godoy, S., & Chen, H. G. (2004). Potassium release during straw devolatilization. (Paper presented at the 2nd World Conference on Biomass for Energy, Industry and Climate Protection, Florence, Florence, Italy, and WIP-Munich, Munich, Germany)

Goel, K., Eisner, R., Sherson, G., Radiotis, T., & Li., J. (2000). Switchgrass: A potential pulp fibre source. *Pulp & Paper-Canada, 101*(6), 51–45

Grahn, M., Azar, C., Lindgren, K., Berndes, G., & Gielen, D. (2007). Biomass for heat or as transportation fuel? A comparison between two model-base studies. *Biomass & Bioenergy 31, 747–758*

Hartmann, H., Turowski, P., Robmann, P., Ellner-Schuberth, F., & Hopf, N. (2007). Grain and straw combustion in domestic furnaces – influences of fuel types and fuel pretreatments. (Paper presented at the 15th European Biomass Conference and Exhibition, Berlin, Germany)

Heede, R. (2006, May). "LNG Supply Chain Greenhouse Gas Emissions for the Cabrillo Deepwater Port: Natural Gas from Australia to California." (Prepared by Climate Mitigation Services). **Retrieved July, 2007, from** http://edcnet.org/ProgramsPages/ LNGrptplusMay06.pdf

Hill, J., Nelson, E., Tilman, D., Polasky, S., & Tiffany, D. (2006). Environmental, economic, and energetic costs and benefits of biodiesel and ethanol biofuels. *PNAS, 30*(3), 11206–11210

House, H. K., DeBruyn, J., & Rodenburg, J. (2007) A Survey of Biogas Production Systems in Europe, and Their Application to North American Dairies. (Paper presented at the Sixth International Dairy Housing Conference, Minneapolis, Minnesota)

Hughes, J. D. (2006). Natural gas at the cross roads. Canadian Embassy report. **Retrieved Aug, 2007, from** http://aspocanada.ca/images/stories/pdfs/hughes_ north_vancouver_nov%2026_ 2006.pdf

Igathinathane, C., Womac, A. R., Sokhansanj, S., & Narayan, S. (2007). Size reduction of wet and dry biomass by linear knife grid device. (Paper number 076045 presented at the American Society of Agricultural and Biological Engineers (ASABE) Annual Meeting, San Antonio, Texas)

Iogen Corporation. (2008). Cellulose ethanol. Retrieved Feb, 2008, from http://iogen.ca/ cellulose_ethanol/what_is_ethanol/process.html

Jaramillo, P., Griffen, M. W., & Matthews, H. S. (2007). Comparative life-cycle air emissions of coal, domestic natural gas, LNG, and SNG for electricity generation. *Environmental Science and Technology, 41*(17), 6290–6296

Jefferson, P. G., McCaughey, W. P., May, K., Woosaree, J., & McFarlane, L. (2004). Potential utilization of native prairie grasses from western Canada as ethanol feedstock. *Canadian Journal of Plant Science, 84*, 1067–1075

Jones, L. H. P., & Handreck, K. A. (1967). Silica in soils, plants and animals. *Advances in Agronomy, 19*, 107–149

Jørgensen, U. (1997). Genotypic variation in dry matter accumulation and content of N, K and Cl in Miscanthus in Denmark. *Biomass & Bioenergy, 12*(3), 155–169

Jungmeier, G., Canella, L., Stiglbrunner, R., & Spitzer, J. (2000). LCA for comparison of GHG emissions of bio energy and fossil energy systems. (Prepared for Joanneum Research, Institut für Energieforschung, Graz. Report No.± IEF/B/06-99)

Kaack, K., & Schwarz, K-U. (2001). Morphological and mechanical properties of Miscanthus in relation to harvesting, lodging, and growth conditions. *Industrial Crops and Products, 14*, 145–154

Koelling, M. R., & Kucera, C. L. (1965). Dry matter losses and mineral leaching in bluestem standing crop and litter. *Ecology, 46*, 529–532

Klass, D. L. (1998). *Biomass for Renewable Energy, Fuels, and Chemicals*. (London, U.K.: Academic Press)

Kucera, C. L., & Ehrenreich, J. H. (1962). Some effects of annual burning on Central Missouri Prairie. *Ecology, 43*(2), 334–336

Lal, R. (2005). World crop residues production and implications of its use as a biofuel. *Environment International, 31*, 575–584

Lanning, F. C., & Eleuterius, L. N. (1987). Silica and ash in native plants of the central and South-eastern regions of the United States. *Annals of Botany, 60*, 361–375

Lanning, F. C., & Eleuterius, L. N. (1989). Silica deposition in some C_3 and C_4 species of grasses, sedges and composites in the USA. *Annals of Botany, 64*, 395–410

López, C. P., Kirchmayr, R., Neureiter, M., Braun, R. (2005). Effect of physical and chemical pre-treatments on methane yield from maize silage and grains. (Poster presented at the 4th International Symposium on Anaerobic Digestion of Solid Waste, Copenhagen, Denmark)

Lovins, A. (1977). *Soft Energy Paths: Towards a Durable Peace.* (San Francisco Friends of the Earth, and Cambridge Massachusetts Ballinger Publishing Co.)

Lynd, L. R., Cushman, J. H., Nichols, R. J., and Wyman, C. E. (1991). Fuel ethanol from cellulosic biomass. *Science, 251*(4999), 1318–1323

Lynd, L. R. (1996). Overview and evaluation of fuel ethanol from cellulosic biomass: Technology, economics, the environment, and policy. *Ann. Rev. Energy Environ., 21*, 403–465

Lynd, L. R., & Wang, M. Q. (2004). A product-nonspecific framework for evaluating the potential of biomass-based products to displace fossil fuels. *J. Ind. Ecol., 7*(3–4), 17–32

Ma, J. F. (2003). Functions of silicon in higher plants. *Prog Mol Subcell Biol., 33*, 127–47

Ma, J. F., Mitani, N., Nagao, S., Konishi, S., Tamai, K., Iwashita, T., & Yano, M. (2004). Characterization of the silicon uptake system and molecular mapping of the silicon transporter gene in rice. *Plant Physiology, 136*, 3284–3289

Mähnert, P., Heiermann, M., & Linke, B. (2005). Batch- and Semi-continuous Biogas Production from Different Grass Species. (Produced by Leibniz-Institute of Agricultural Engineering Potsdam-Bornim, Potsdam, Germany)

Mani, S., Sokhansanj, S., Bi, X., and Turhollow, A. (2006). Economics of producing fuel pellets from biomass. *App Eng Agri.* 22(3), 421–426.

Manitoba Agriculture, Food and Rural Initiatives (MAFRI). (2006). Factsheet: *Harvesting and Storage of Quality Hay and Silage,* **Retrieved Aug, 2007, from** http://www.gov.mb.ca/ agriculture/ crops/forages/bjc01s02.html

McLaughlin, S. B., Samson, R., Bransby, D., & Wiselogel, A. (1996). Evaluating physical, chemical and energetic properties of perennial grasses as biofuels. (Paper presented at Bioenergy 96: The 7th National Bioenergy Conference of the South Eastern Regional Biomass Energy Program, Nashville, TN)

Natural Resources Canada, GHGenius version 3.9. (2007). **Retrieved July, 2007, from** http:// www.ghgenius.ca/

Obernberger, I., & Thek, G. (2004). Physical characterization and chemical composition of densified biomass fuels with regard to their combustion behaviour. *Biomass and Bioenergy, 27*, 653–669

Ontario Ministry of Agriculture Food and Rural Affairs (OMAFRA). (2007) Estimated areas, yield, production, average farm price and total farm value of principal field crops, in metic units, annual. **Retrieved Feb, 2008, from** http://www.omafra.gov.on.ca/english/ stats/crops/estimate_metric.htm

Pahkala, K., Mela, T., Hakkola, H., Jarvi, A., & Virkajari, P. (1996). Production and use of agrofibre in Finland. (Agricultural Research Centre of Finland. Part 1 of the Final report for study: Production of agrofibre crops - Agronomy and varieties)

Pahkala, K., & Pihala, M. 2000. Different plant parts as raw material for fuel and pulp production. *Industrial Crops and Products, 11*, 119–128

Passalacqua, F., Zaetta, C., Janssone, R., Pigaht, M., Grassi, G., Pastre, O., Sandovar, A., Vegas, L., Tsoutsos, T., Karapanagiotis, N., Fjällström, T., Nilsson, S. & Bjerg, J. (2004). Pellets in southern Europe; The state of the art of pellets utilization in southern Europe: New perspectives of pellets from agri-residues. (Paper presented at the 2nd World Conference on Biomass for Energy, Industry and Climate Protection, ETA-Florence, Florence, Italy, and WIP-Munich, Munich, Germany.)

Parrish, D. J., Wolf, D. D., Fike J. H., & Daniels, W. L. (2003). Switchgrass as a biofuel crop for the upper southeast: Variety trials and cultural improvements. (Oak ridge National Laboratory, Oak Ridge, TN. Final Report for 1997 to 2001, ORNL.SUB-03-19SY163C/01)

Parrish, D. J., & Fike, J. H. (2005). The biology and agronomy of switchgrass for biofuels. *Critical Reviews of Plant Sciences, 24,* 423–459

Paulrud, S. 2004. Upgraded biofuels-effects of quality on processing, handling characteristics, combustion and ash melting. (Doctoral dissertation prepared for the Unit of Biomass Technology and Chemistry, SLU) *Acta Universitatis agriculturae Suecia. Agraria, 449*

Paulrud, S., Nilsson, C., & Öhman, M. (2001). Reed canary-grass ash composition and its melting behaviour during combustion. *Fuel, 80,* 1391–1398

Roth, G., and Undersander, D. (1995). Corn silage production, management and feeding. *North Central Regional Publication, 574*

Samson, R., Girouard, P., Omielan, J., & Henning, J. (1993). Integrated production of warm season grasses and agroforestry for biomass production. (Paper presented at Energy, Environment, Agriculture and Industry: The 1st Biomass Conference of the Americas, Golden, CO)

Samson, R. A., & Omielan, J. (1994). Switchgrass: A potential biomass energy crop for ethanol production. (Paper presented at the 13th North American Prairie Conference, Windsor, Ontario, Canada)

Samson, R., & Chen, Y. (1995). Short rotation forestry and the water problem. (Paper presented at the Natural Resources Canada Canadian Energy Plantation Workshop, Ottawa, Ontario)

Samson, R. A., Blais, P-A., Mehdi, B., & Girouard, P. (1999a). Switchgrass Plant Improvement Program for Paper and Agri-Fibre Production in Eastern Canada. (Final report prepared by REAP-Canada for the Agricultural Adaptation Council of Ontario)

Samson, R., Girouard, P., & Mehdi, B. (1999b). Establishment of Commercial Switchgrass Plantations. (Final report prepared by REAP-Canada for Natural Resources Canada)

Samson, R., Drisdelle, M., Mulkins, L., Lapointe, C., & Duxbury, P. (2000). The use of switchgrass as a greenhouse gas offset strategy. (Paper presented at the Fourth Biomass Conference of the Americas, Buffalo, New York)

Samson, R., Mani, S., Boddey, R., Sokhansanj, S., Quesada, D., Urquiaga, S., Reis, V., & Ho Lem, C. (2005). The potential of C_4 perennial grasses for developing a global BIOHEAT industry. *Critical Reviews in Plant Science, 24,* 461–495

Samson, R. (2007). Switchgrass Production in Ontario: A Management Guide. Resource Efficient Agricultural Production (REAP) – Canada. **Retrieved Aug, 2007, from** http://www.reap-canada.com/library/Bioenergy/2007%20SG%20production%20guide-FINAL.pdf

Samson, R., Bailey-Stamler, S., & Ho Lem, C. (2007). The Emerging Agro-Pellet Industry in Canada. (Paper presented at the 15th European Biomass Conference and Exhibition, Berlin, Germany)

Samson, R., Bailey Stamler, S., Dooper, J., Mulder, S., Ingram, V., Clark, K. and Ho Lem, C. (2008a). Analysing Ontario Biofuel Options: Greenhouse Gas Mitigation Efficiency and Costs. (Final report prepared by REAP-Canada to the BIOCAP-Canada Foundation, Kingston, Ontario)

Samson, R., Bailey-Stamler, S., & Ho Lem, C. (2008b). Optimization of Switchgrass Management for Commercial Fuel Pellet Production (Final report prepared by REAP-Canada for the Ontario Ministry of Food, Agriculture and Rural Affairs (OMAFRA) under the Alternative Renewable Fuels Fund)

Sander, B. (1997). Properties of Danish biofuels and the requirements for power production. *Biomass and Bioenergy, 12*(3), 173–183

Sanderson, M. A., Egg, R. P., & Wiselogel, A. E. (1997). Biomass losses during harvest and storage of switchgrass. *Biomass & Bioenergy, 12*(2), 107–114

Schneider C., & Hartmann, H. (2005). Maize as energy crops for combustion-optimization of fuel supply. (Paper presented at the 14th European Biomass Conference & Exhibition, Paris, France)

Sheenan, J., Camobreco, V., Duffield, J., Graboski, M., & Shapouri, H. (1998). Life cycle inventory of biodiesel and petroleum diesel for use in an urban bus. (Prepared for National Renewable Energy Laboratory (NREL) Project SK-580-24089UL).

Sheehan, J., Aden, A., Paustian, K., Killian, K., Brenner, J., Walsh, M., & Nelson, R. (2004). Energy and environmental aspects of using corn stover for fuel ethanol. *J. Ind. Ecol., 7*(3–4), 117–146

Smith S. J., Wise, M. A., Stokes, G. M., & Ermonds, J. (2004). Near-Term US Biomass Potential: Economics, Land-Use, and Research Opportunities. (Prepared by Battelle Memorial Institute, Joint Global Change Research Institute, Maryland)

Spatari, S., Zhang, Y., & Maclean, H. (2005). Life cycle assessment of Switchgrass and corn stover derived ethanol fuelled automobiles. *Environ. Sci. Technol., 39*, 9750–9758

(S&T)[2] Consultants Inc. (2002). Assessment of biodiesel and Ethanol diesel blends, greenhouse gas emissions, exhaust emissions, and policy issues. (Prepared for Natural Resources Canada), **Retrieved July, 2007, from** http://www.greenfuels.org/biodiesel/pdf/res/200209_Assessment_of_Biodiesel_and_EDiesel.pdf

Takahashi, E., Ma, J. F., & Miyake, Y. (1990). The possibility of silicon as an essential element for higher plants. *Comments Agricultural and Food Chemistry, 2*, 99–122

Uherek, E. (2005). Natural gas: Are pipeline leaks warming our planet? Atmospheric Composition Change (ACCENT). **Retrieved Aug, 2007, from** http://www.atmosphere.mpg.de/enid/Nr_3_Sept__2__5_methane/energy/R__Methane_emission_from_pipelines_4pd.html

Van Der Vorm, P. D. J. (1980). Uptake of Si by five plant species, as influenced by variations in Si-supply. *Plant and Soil, 56*, 153–156

Venuto, B. C. (2007). Producing biomass from sorghum and sorghum by sudangrass hybrids. (Paper presented at the 2nd International Energy Farming Congress, Papenberg, Germany)

Von Felde, A. (2007). Advances of energy crops from the viewpoint of the breeder. (Paper presented at the 2nd International Energy Farming Congress, Papenberg, Germany)

Wang, M., Wu, M., & Huo, H. (2007). Life-cycle energy and greenhouse gas emission impacts of different corn ethanol plant types. *Environmental Research Letters, 2*, 1–13

White, E. M. (1973). Overwinter changes in the percent Ca, Mg, K, P and in vegetation and mulch in an eastern South Dakota prairie. *Agronomy Journal, 65*, 680–681

Zan, C. (1998). Carbon Storage in Switchgrass (*Panicum virgatum* L.) and Short-Rotation Willow (*Salix alba x glatfelteri* L.) Plantations in Southwestern Quebec. (Masters Thesis prepared for the Department of Natural Resource Sciences, McGill University, Montreal, Quebec, Canada)

Zwart, K., Oudendag, D. & Kuikman, P. (2007). Sustainability of co-digestion. (Paper presented at the 2nd International Energy Farming Congress, Papenberg, Germany)

Chapter 17
Organic and Sustainable Agriculture and Energy Conservation

Tiziano Gomiero and Maurizio G. Paoletti

Abstract In the last decades biofuels have been regarded as an important source of renewable energy and at the same time as an option to curb greenhouse gas emissions. This is based on a number of assumptions that, on a close look, may be misleading, such as the supposed great energy efficiency of biofuels production. Large scale biofuels production may, on the contrary, have dramatic effects on agriculture sustainability and food security. In this chapter we explore the energy efficiency of organic farming in comparison to conventional agriculture, as well as the possible benefits of organic management in term of Green House Gasses mitigation.

Organic agriculture (along with other low inputs agriculture practices) results in less energy demand compared to intensive agriculture and could represent a mean to improve energy savings and CO_2 abatement if adopted on a large scale. At the same time it can provide a number of important environmental and social services such as: preserving and improving soil quality, increasing carbon sink, minimizing water use, preserving biodiversity, halting the use of harmful chemicals so guaranteeing healthy food to consumers. We claim that more work should be done in term of research and investments to explore the potential of organic farming for reducing environmental impact of agricultural practices. However, the implications for the socio-economic system of a reduced productivity should be considered and suitable agricultural policies analysed.

The chapter is organised as follows: Section (17.1) provides the reader with a definition of organic agriculture (and sustainable agriculture) and a brief history of the organic movement in order to help the reader to better understand what is presented later on; Section (17.2) reviews a number of studies on energy efficiency in organic and conventional agriculture; Section (17.3) compares CO_2 emissions

✉ T. Gomiero
Department of Biology, Padua University, Italy, Laboratory of Agroecology and Ethnobiology, via U. Bassi, 58/b, 35121-Padova, Italy
e-mail: paoletti@bio.unipd.it

M.G. Paoletti
Department of Biology, Padua University, Italy, Laboratory of Agroecology and Ethnobiology, via U. Bassi, 58/b, 35121-Padova, Italy

D. Pimentel (ed.), *Biofuels, Solar and Wind as Renewable Energy Systems*,
© Springer Science+Business Media B.V. 2008

from organic and conventional managed farming systems; Section (17.4) analyses the possible use of agricultural "waste" to produce cellulosic ethanol; Section (17.5) provides some comments concerning the possible production of biofuels from organically grown crops; Section (17.6) concludes the chapter presenting a summary of the review.

Keywords Biofuels · organic agriculture · conventional agriculture · energy use · GHGs emissions · soil ecology · biodiversity

17.1 Organic Agriculture: An Overview

In the last decades the effects of oil crises on world economies along with the environmental impact caused by fossil fuels (e.g. climate change, emission of pollutants) led political leaders and scientists to search for alternative and sustainable energy sources (EC, 2005; EEA, 2006; IPCC, 2007; Goldemberg, 2007). One of these alternatives has been indicated in the use of biomass, in particular to supply biofuels (ethanol, biodiesel). In this chapter we will explore, instead, the possible role of alternative agriculture practices, referring in particular to organic agriculture, in contributing to energy saving and CO_2 sequestration.

If organic agriculture allows for improving energy efficiency and reducing CO_2 and other Green House Gasses (GHGs) emissions it would deserve much attention from policymakers and scientists alike and to be supported world wide. It has to be pointed out that organic agriculture provides many beneficial "byproducts" both for the environment (e.g. eliminating the use of agrochemicals such as synthetic fertilisers and pesticides, increasing organic matter content and conservation of soil fertility, preservation of biodiversity, reduced water consumption) and for human health (e.g. exposure to harmful chemicals, avoiding risks from possible side effects of Genetic Modified Organisms – GMO – use in agriculture).

We wish to underline that, whilst focusing mainly on the energetic performances of organic agriculture and its possible role in CO_2 abatement, we are aware that a much more comprehensive treatment is necessary in order to assess the benefits and/or drawbacks of organic agriculture. Such an analysis is a difficult one, because of the complex nature of agroecosystems.[1]

Agroecosystems interface at different scales with ecosystems (from soil ecology to landscape to global biogeochemical cycles), climate (from local to regional characteristics), economic systems (from local household economy to the global food market), social systems (such as employment opportunities, competition

---

[1] Miguel Altieri, for instance, provides the following definitions "**Agroecosystems** are communities of plants and animals interacting with their physical and chemical environments that have been modified by people to produce food, fibres, fuel and other products for human consumption and processing. **Agroecology** is the holistic study of agroecosystems including all the environmental and human elements. It focuses on the form, dynamics and functions of their interrelationship and the processes in which they are involved." (Altieri, 2002, p. 8, bold is in the original).

for water use, heath risk from agrochemicals use) (Altieri, 1987; Conway, 1987; Giampietro, 2004, Pimentel and Pimentel, 2007a). It has to be stressed that the very same existence of ecosystems depend on biodiversity in the form of: cultivated species, soil and aboveground organisms which help to preserve soil fertility, pests and alley organisms which help to limit pest damages, landscapes and ecosystems.

Agroecosystems play multiple functions that cannot be properly understood by relying only on a single indicator, be it economic (e.g. US\$/ha or US\$/hr of work) or biophysical (e.g. energy efficiency). In order to gain a better perception of agroecosystem performances many aspects have to be considered at the same time, and the whole system has to be viewed as an integrated system (Altieri, 1987; Conway, 1987; Paoletti et al., 1989; Ikerd, 1993; Wolf and Allen, 1995; Bland, 1999; Gliessmann, 2000; Kropff et al., 2001; Giampietro, 2004; Pimentel et al., 2005; Gomiero et al., 2006).

In this section we will provide a brief introduction to the history and principles of organic agriculture. The concept of "sustainable agriculture" is also briefly presented. We will summarise some issues concerning the multifunctional role of agriculture and organic agriculture and will discuss some methodological problems that arise when comparing organic and conventional agriculture farming systems.

17.1.1 Defining Organic Agriculture

Organic agriculture refers to a farming process regulated by international and national institutional bodies which certify organic products from production to handling and processing. Organic agriculture regulations ban the use of agrochemicals such as synthetic fertilisers and pesticides and the use of GMO, as well as many synthetic compounds used as food additives (e.g. preservatives, colouring). Organic farming aims at providing farmers with an income while at the same time protecting soil fertility (e.g. crops rotation, intercropping, polyculture, cover crops, mulching) and preserving biodiversity (even if concern towards local floras and fauna as goals for organic farming are often little understood by consumers and policymarkers), the environment and human health. Pests control is carried out by using appropriate cropping techniques, alley insects and natural pesticides (mainly extracted from plants).

According to The International Federation of Organic Agriculture Movements (IFOAM)[2] organic agriculture should be guided by four principles:

- *Principle of health*: Organic Agriculture should sustain and enhance the health of soil, plant, animal, human and planet as one and indivisible.
- *Principle of ecology*: Organic Agriculture should be based on living ecological systems and cycles, work with them, emulate them and help sustain them.

---

[2] IFOAM is a grassroots international organization born in 1972, today it includes 750 member organizations belonging to108 countries (for details see http://www.ifoam.org/index.html).

- *Principle of fairness*: Organic Agriculture should build on relationships that ensure fairness with regard to the common environment and life opportunities
- *Principle of care*: Organic Agriculture should be managed in a precautionary and responsible manner to protect the health and well-being of current and future generations and the environment.

Eve Balfour (1899–1990), who was one of IFOAM's founders, said that the characteristics of truly sustainable agriculture can be summed up with the word "*permanence*" (Balfour, 1977). According to IFOAM organic agriculture is a holistic production management system which promotes and enhances agroecosystem health, including biodiversity, biological cycles, and soil biological activity. An organic production system is, then, designed to:

- enhance biological diversity within the whole system;
- increase soil biological activity;
- maintain long-term soil fertility;
- recycle plant and animal waste in order to return nutrients to the land, thus minimizing the use of non-renewable resources;
- rely on renewable resources in locally organized agricultural systems;
- promote the healthy use of soil, water and air as well as minimize all forms of pollution that may result from agricultural practices;
- handle agricultural products with emphasis on careful processing methods in order to maintain the organic integrity and vital qualities of the product at all stages;
- become established on any existing farm through a period of conversion, the appropriate length of which is determined by site-specific factors such as the history of the land, and type of crops and livestock to be produced.

In Europe, the first regulation on organic farming was drawn up in 1991 (Regulation EEC N° 2092/91 – EEC, 1991). Since its implementation in 1992, many farms across the EU have applied to get the label "organic" for their products and many others have converted to organic production methods.[3]

Organic standards prohibit the use of synthetic pesticides and artificial fertilizers, the use of growth hormones and antibiotics in livestock production (a minimum usage of antibiotics is admitted in very specific cases and is strictly regulated). Genetically modified organisms (GMOs) and products derived from GMOs are explicitly excluded from organic production methods.[4]

In the USA, congress passed the Organic Foods Production Act (OFPA) of 1990. The OFPA required the U.S. Department of Agriculture (USDA) to develop national

---

[3] Some authors (e.g. Vogl et al., 2005; Courville, 2006) express concern about the excessive bureaucratic control as it poses a burden to organic farmers in form of time and money.

[4] A new regulation entered in to force in 2007 (EC, 2007) in which the two main novelties are that: food will only be able to carry an organic logo (certified as organic) if at least 95% of the ingredients are organic (non-organic products will be entitled to indicate organic ingredients on the ingredients list only), and that although the use of genetically modified organisms will remain prohibited, a limit of 0.9% will be allowed as accidental presence of authorised GMOs.

standards for organically produced agricultural products to assure consumers that agricultural products marketed as organic meet consistent, uniform standards. The OFPA and the National Organic Program (NOP) regulations require that agricultural products labelled as organic originate from farms or handling operations certified by a State or private entity that has been accredited by USDA (USDAa, 2007). According to USDA National Organic Standards Board (2007) definition, April 1995: "*Organic agriculture is an ecological production management system that promotes and enhances biodiversity, biological cycles and soil biological activity. It is based on minimal use of off-farm inputs and on management practices that restore, maintain and enhance ecological harmony.*" (also in USDAb, 2007).

Internationally, organic agriculture has been officially recognised by the Codex Alimentarius Commission,[5] which in the Guidelines for the Production, Processing, Labelling and Marketing of Organically Produced Foods[6] states at point 5 that: "*Organic Agriculture is one among the broad spectrum of methodologies which are supportive of the environment. Organic production systems are based on specific and precise standards of production which aim at achieving optimal agroecosystems which are socially, ecologically and economically sustainable.*" (Codex Alimentarius, 2004, p. 4).

According to the review carried out by Willer and Yussefi (2006)[7] more than 31 million certified hectares (including fully converted land as well as "in conversion" land area) are managed organically by at least 623.174 farms worldwide. Currently, the major part of this area is located in Australia[8] (12.1 million hectares), China (3.5 million hectares), and Argentina (2.8 million hectares). In the USA, in 2005, for the first time, all 50 States had some certified organic farmland. In 2005 U.S. producers dedicated over 4.0 million acres of farmland (1.6 million ha) to organic production systems: 1.7 million acres of cropland (690.000 ha) and 2.3 million acres of rangeland and pasture (910.00 ha). California remains the leading State in certified organic cropland, with over 220,000 acres (89.000 ha), mostly for fruit and vegetable production (USDAc, 2007).

---

[5] The Codex Alimentarius Commission was created in 1963 by FAO and WHO to develop food standards, guidelines and related texts such as codes of practice under the Joint FAO/WHO Food Standards Program. The main purposes of this Program are protecting health of the consumers and ensuring fair trade practices in the food trade, and promoting coordination of all food standards work undertaken by international governmental and non-governmental organizations. (Codex Alimentarius web page at http://www.codexalimentarius.net/web/index_en.jsp)

[6] The Codex Alimentarius Commission began in 1991, with participation of observer organizations such as IFOAM and the EU, to elaborate Guidelines for the production, processing, labelling and marketing of organically produced food. In June 1999 first the plant production and in July 2001 the animal production was approved by the Codex Commission. The requirements in these Codex Guidelines are in line with IFOAM Basic Standards and the EU Regulation for Organic Food (EU Regulations 2092/91 and1804/99). There are, however, some differences with regard to the details and the areas, which are covered by the different standards.

[7] Not all countries supplied the data.

[8] Most of this area is pastoral land for low intensity grazing. Therefore, one organic hectare in Australia is not directly equivalent to one organic hectare in an European country, for example, due to its level of productivity. Comparing countries must be done carefully.

According to the data collected from Willer and Yussefi (2006), the main land uses in organic farming worldwide, as a share (%) of the total global organic area, are as follows: 5% permanent crops,[9] 13% arable land,[10] 30% permanent pasture,[11] 52% certified land which use is not known.

Broadening the scale of organic farming marketing, however, may lead farmers to shift once again into monoculture and industrial agriculture forced by the pressure of agrifood corporations that buy and distribute their organic products, and the market itself (Guthman, 2004). It has to be pointed out that also in the case of organic products, national and international trade results in increasing "food miles" (the distance that food travels from the field to the grocery store) which means increased energy consumption and CO_2 emissions (Pimentel et al., 1973; Steinhart and Steinhart, 1974; DEFRA, 2005; Pretty et al., 2005; Schlich and Fleissner, 2005; Foster et al., 2006, Pimentel and Pimentel, 2007b). To avoid such a problem environmental groups and organic associations are advising consumers to consume locally produced food as part of environmental friendly eating habits.

17.1.2 Sustainable Agriculture

According to Kirschenmann (2004), Wes Jackson was the first to use the term in his publication *New Roots for Agriculture* (1980). The term didn't emerge in popular usage until the late 1980s, however, the notion of land stewardship is a very old one. Sustainable agriculture must, as defined by the U.S. Department of Agriculture in the 1990 Farm Bill: "... *over the long term, satisfy human needs, enhance environmental quality and natural resource base, make the most efficient use of non-renewable resources and integrate natural biological processes, sustain economic viability, and enhance quality of life.*" (USDA, 1990).

Sustainable agriculture does not refer to a prescribed set of practices and it differs from organic agriculture because, in sustainable agriculture, agrochemicals (synthetic fertilizers and pesticides) still play a role. However their use is kept to a minimum, and conservative practices (crop rotation, integrated pest management, natural fertilization methods, minimum tillage, biologic control) are fully integrated in farm management. Sustainable agriculture should aim at: preserving the natural resource base, relying on minimum artificial inputs from outside the farm system,

---

[9] Land cultivated with crops that occupy the land for long periods and need not to be replanted after each harvest, such as cocoa, coffee; this category includes land under flowering shrubs, fruit trees, nut trees and vines, but excludes land under trees grown for wood or timber.

[10] Land under temporary crops, temporary meadows for mowing or pasture, land under market and kitchen gardens and land temporarily fallow (less than five years). Abandoned land resulting from shifting cultivation is not included in this category. Data for "arable land" are not meant to indicate the amount of land that is potentially cultivable.

[11] Land used permanently (five years or more) for herbaceous forage crops, either cultivated or growing wild (wild prairie or grazing land).

recovering from the disturbances caused by cultivation and harvest while at the same time being economically and socially viable (Poincelot, 1986; NRC, 1986; Gliessman, 1990; Dunlap et al., 1992; Feenstra et al., 1997).

Although sustainable agriculture practices are adopted by an increasing number of farmers only organic agriculture is regulated by laws and needs to strictly follow a specific set of norms. A farmer practicing sustainable agriculture can, if in need, spray synthetic pesticides, or add synthetic fertilisers and still claim that he/she is practicing sustainable agriculture. For instance, most no-till agriculture use consistent amounts of broad spectrum herbicides, such as glyphosate, that have severe impact on soil organisms (Paoletti and Pimentel, 2000). In organic agriculture, if trace of agrochemicals are found in the soil or in a product certified as organic the farmer will lose the certification of organic producer and will not be permitted to sell the products as organically grown.

Within the domain of sustainable agriculture fall some other definitions and practices such as integrated agriculture, precision agriculture and permaculture.

Integrated agriculture is a farming method that combines management practices from conventional and organic agriculture. For instance, when possible, animal manure instead of chemical fertilizer is employed. Pest management (integrated pest management) is carried on combining several methods: using crop rotation, the release of parasitoids, cultivating pest-resistant varieties, and using various physical techniques, leaving pesticide as the last resort. Integrated agriculture is not regulated by specific regulations but its goal is still to reduce as much as possible both farm management costs and its environmental impact, aiming at the long term sustainability of farming practices (Edens, 1984; Poincelot, 1986; Mason, 2003; Pretty, 2005). In some cases groups of farmers can subscribe specific protocols that limit the kind and the amount of chemicals in their farming practices in order to improve the marketability of their products as well as saving on management costs (e.g. fruits producers in some areas in Northern Italy).

Mollison and Holmgren, in their book *Permaculture one: A perennial agriculture for human settlements* (Mollison and Holmgren, 1978) coined the term *"permaculture"*, a contraction of "permanent agriculture". Permaculture puts the emphasis on management design and on the integration of the elements in a landscape, considering the evolution of landscape over time. The goal of permaculture is to produce an efficient, low-input integrated culture of plants, animals, people and structure. An integration that is applied at all scales from home garden to large farm (see also http://www.permaculture-info.co.uk/).

Precision agriculture, (also known as "precision farming", "site-specific crop management", "prescription farming", "variable rate technology") developed in the 1990s, refers to agricultural management systems carefully tailoring soil and crop management to fit the different conditions found in each field. Precision agriculture is an information and technology based agricultural management system (e.g. using remote sensing, geographic information systems, global positioning systems and robotics) to identify, analyze, and manage site-soil spatial and temporal variability within fields for optimum profitability, sustainability, and protection of the environment (Lowenberg-DeBoer, 1996; National Research Council, 1998; Srinivasan, 2006).

Precision agriculture is now taught in many universities around the world (see for instance http://precision.agri.umn.edu/links.shtml).

However, it is extremely difficult to determine whether or not certain agricultural practices are sustainable because sustainability cannot be associated with any particular set of farming practices or methods. The sustainability of an agriculture practice will mostly depend on the peculiarities of the context in which it is used and implies a constant process of monitoring and revaluation (Ikerd, 1993; Gliessmann, 2000; Giampietro, 2004; Gomiero et al., 2006; Pimentel and Pimentel, 2007a).

17.1.3 Brief History of Organic Movement

In order to help the reader to better understand the foundation of organic farming, it may be useful to provide a brief sketch of the story of organic agriculture movement. For details on this topic we will refer the reader to the extensive work of Conford (2001) or, for a more concise summary, to Kristiansen (2006), Heckman, (2006), Gold and Gates (2006). Historical information can be found also at the website of the main organic associations such as the British "Soil Association" (http://www.soilassociation.org) or the international representative IFOAM (http://www.ifoam.org).

Until the early 1900s agriculture was necessarily "organic" as there were not yet chemicals available to be used in agriculture. Attempt to coherently organize best management practices for agriculture sustainability can be traced back in the writings of Roman authors such as Marcus Porcius Cato (234–149 B.C.) "De agri coltura", Marcus Terentius Varro (116–27 B.C.) "De re rustica", Lucius Iunius Moderatus Columella (I cen. B.C.) "De re rustica".

It was in North Europe in the late 1930s that a movement perceiving itself as "alternative" to the new agriculture came into existence. Those earlier "alternative farmers" were against the use of synthetic chemicals in agriculture (fertilisers and, later on, pesticides) and wanted to base their agricultural practices on natural principles and processes.

The first organised movement in this sense appeared in Germany at the end of 1930s originally from the lectures given in 1924 by the Austrian philosopher and scientist Rudolf Steiner (who developed also Anthroposophy) to groups of farmers, agronomists, doctors and lay people. The experimental circle of anthroposophical farmers immediately tested Steiner's indications in daily farming practice. Three years later a co-operative was formed to market biodynamic products forming the association Demeter.[12] In 1928 the first standards for Demeter quality control were formulated. Biodynamic agriculture, as this method is named, is well grounded in the practical aspects of organic farming, but it also concerns lunar and astrological scheduling, communication with "nature spirits" and the use of special potencies or preparations, that are derived by what might be called alchemical means

---

[12] For details see Demeter web page at http://www.demeter.net

(Koepf et al., 1996; Conford, 2001; Koepf, 2006). These later relations are not easily "measurable" in scientific terms but performances can be assessed relying on usual agronomic indicators.

While Rudolf Steiner was establishing the roots for the growth of the biodynamic movement, in India Sir Albert Howard (1873–1947), a British agronomist who spent 25 years there, was trying to develop a coherent and scientifically based directive for preserving soil and crop health, and once back to the UK he worked to promote his new view (Howard, 1943; Conford, 2001). Howard believed that reliance on chemical fertilization could not address problems such as loss of soil fertility, pests etc. He was convinced that most of the agricultural problems were dependent on the mismanagement of soil. He maintained that the new agrochemical approach was misguided, and that it was a product of reductionism by "laboratory hermits" who paid no attention to how nature worked. In his milestone book *An Agricultural Testament* (1943), Howard described a concept that was to become central to organic farming: "*the Law of Return*" (a concept expressed also by Steiner). The Law of Return states the importance of recycling all organic waste materials, including sewage sludge, back to farmland to maintain soil fertility and the land humus content (Howard 1943; Conford, 2001).

The first use of the word "*organic*" seems to be traced to Walter Northbourne who in 1940 published in the UK an influential book, *Look to the Land*, in which he elaborated on the idea of the farm as an "organic whole", where farming has to be performed in a biological completeness (Conford, 2001). The term "organic", then, in its original sense, describes a holistic approach to farming: fostering diversity, maintaining optimal plant and animal health, and recycling nutrients through complementary biological interactions.

In 1943 in the UK, Lady Eve Balfour published *The Living Soil* where she described the direct connection between farming practice and plant, animal, human and environmental health. The book exerted an important influence in the public opinion leading in 1946 to the foundation in the UK of "The Soil Association" by a group of farmers, scientists and nutritionists. In the following years, the organisation also developed organic standards and its own certification body.

In the USA the idea of organic agriculture was introduced and promoted by Jerome I. Rodale whose key ideas about farming came mostly from the work of Albert Howard. However, Rodale expanded Howard's ideas in his book *Pay Dirt* (Rodale, 1945) adding a number of other "good farming practices". In 1940, in an article published in *Fact Digest,* Rodale introduced the term "organic agriculture" in the USA and techniques such as crop rotation and mulching, that have, since then, become accepted organic practices in the USA.

17.1.4 The Multifunctional Meaning of Organic Agriculture

In this chapter we will not deal with all those complex issues concerning the management of organic agroecosystems and the multifunctional role of agriculture and organic agriculture in particular. The interested reader can refer to specific literature e.g. Altieri (1987), Gliessmann (1990; 2000), Zimmer (2000), Lampkin (2002), for the previous and to Stölze et al. (2000), FAO (2002), Lotter (2003), Kristiansen et al., (2006); Badgley et al., (2007) for the latter.

Here we just provide a brief note on the issues reporting in Table 17.1 a qualitative assessment, comparing organic vs. conventional farming systems, for a number of environmental indicators as reviewed by Stölze et al. (2000).

From Table 17.1, concerning environmental impact, the overall organic agriculture performances are, in most cases, better or much better than those of conventional

Table 17.1 Overall qualitative assessment of organic farming systems relative to conventional farming. From Stölze et al., (2000) modified. (Organic farming performs: ++ much better, + better, 0 the same, − worse, − much worse). The work of Stölze et al., (2000) review about 300 references

Indicator	Qualitative assessment				
	++	+	**0**	−	−
Ecosystem					
Floral diversity	++	+			
Faunal diversity (invertebrate and vertebrate)	++	+			
Habitat diversity		+	0		
Landscape		+	0		
Soil					
Soil organic matter		+	0		
Biological activity	++				
Structure			0		
Erosion	++	+	0	−	
Ground and Surface Water					
Nitrate leaching		+	0	−	
Pesticides	++				
Greenhouse emissions					
CO_2		+			
N_2O		+	0		
CH_4		+	0		
NH_3		+			
Farm input and output					
Nutrient use		+			
Water use			0		
Energy use		+	0		
Animal welfare and health					
Husbandry		+	0		
Health		+	0		
Quality of product food					
Pesticides residues	++	+			
Nitrate		+	0		
Mycotoxins			0		
Heavy metals			0		
Desirable substances		+	0		
BSE risk		+			
Antibiotics	++				

practices. Such better performances are also reported in other reviews such as FAO (2002), Lotter (2003), and Kasperczyk and Knickel (2006) as well as in long term monitoring trials such as Reganold (1995), Reganold et al. (1987), Paoletti et al. (1993), Matson et al. (1997), Drinkwater et al., (1998), Rigby and Cáceras (2001), Siegrist et al. (1998), Mäder et al. (2002), Pimentel et al. (2005), Badgley et al. (2007). However, it has to be pointed out that in some cases performance can vary according to specific crop species and crop patterns and in relation to the environmental context where agricultural activity is performed.

Considering, for instance, the case of biodiversity. A wide meta-analysis by Bengtsson et al., (2005) indicated that organic farming often has positive effects on species richness and abundance, but that its effects are likely to differ between organism groups and landscapes. Bengtsson et al., (2005) suggest that positive effects of organic farming on species richness can be expected in intensively managed agricultural landscapes, but not in small-scale landscapes comprising many other biotopes as well as agricultural fields. A review of the literature carried out by Hole et al., (2005) confirms the positive effect of organic farming on biodiversity, but they point out that such benefits may be achieved also by conventional agriculture when carefully managed, and indicate the need for long term, system-level studies of the biodiversity response to organic farming. As some authors point out (Thies and Tscharntke, 1999; Hole et al., 2005; Pimentel et al., 2005; Roschewitz et al., 2005), in fact, measures to preserve and enhance biodiversity should be landscape and/or farm specific.

A problem of scale has also to be taken into consideration when it comes to assessing the sustainability of a wide agriculture conversion to organic practices. On this issue a debate in the open and it has reached the major scientific journals. In a recent exchange of points of view in *Science*, Goklany (2002), for instance, stated that if typical cereal yields under organic farming are 60–70% of those of conventional farming, then between 43% and 67% more land would be needed to keep production constant, further diminishing the environmental and biodiversity advantages of organic farming. (Note that some works report comparable yield per ha of organic and conventional crops; see, for instance, apple production in Reganold et al., 2001, and the figures for corn and soy in the 22 years long Rodale experiment in Pimentel et al., 2005). Mäder et al., (2002), on the other hand, argue that in the past three decades, agricultural yields have doubled, but worldwide, one third of arable land has been lost to erosion, there has been a dramatic increase in chemicals usage and an alarming decline in biodiversity of crops, wild flora and fauna (see also the alarm lunched by Krebs et al., 1999). Mäder et al., (2002) point out that the external costs of intensive conventional agriculture have been huge and that although organic farming may need more land to produce the same yield, notwithstanding, it maintains long term soil fertility and biodiversity of the cropped land. Results from long term experiments, such as the Rodale Trial (Pimentel et al., 2005), are quite encouraging reporting comparable yields for corn and soybean grown under organic and conventional farming practices.

Whether organic food is better or equal in terms of quality (e.g. higher content of mineral, vitamins) is also an issue, with some experts stating that organic food is better than conventional, while others claiming that data does not provide significative evidences of differences between the two. For this issues we refer the reader to specific literature (e.g. Adam, 2001; Brandt and Mølgaard, 2001; 2006; Heaton, 2001; Trewavas, 2001; Lu et al., 2006; Winter and Davis, 2006).

We wish to conclude by underlining that there is an urge to develop more ecological agriculture practices (Altieri, 1987; Pimentel et al., 1995; Tilman et al., 2001; 2002). Recently, also the Millennium Ecosystem Assessment (2005) recommended the promotion of agricultural methods that increase food production without harmful tradeoffs from excessive use of water, nutrients, or pesticides. FAO (2002; 2003; 2004) also stressed the need to reduce the environmental impact of agriculture practice as it poses a risk to the sustainability of the agriculture and food security itself. In this sense organic agriculture can represent an interesting option that deserves to be explored as at the same time aims at preserving soil fertility, biodiversity landscape ecological functionality. As stated by FAO (2004, p. iii): *"Evidence suggests that organic agriculture and sustainable forest management not only produce commodities but build self-generating food systems and connectedness between protected areas. The widespread expansion of these approaches, along with their integration in landscape planning, would be a cost efficient policy option for biodiversity."*

17.1.5 Some Methodological Remarks

Comparing organic and conventional farming systems is not a simple task, as they have different functioning and goals. Summarising, we could say that conventional farming aims at achieving maximum yield and profit, while organic farming, other than yield and profit, aims at achieving long term sustainability (e.g. improving soil fertility) and minimizing environmental impact of farming activities (e.g. reducing pollution, minimizing use of water and energy, increasing and preserving biodiversity). The fact that the national accounting system does not consider, as costs or benefits for farmers and the society as a whole, items such as: soil loss, water pollution and depletion, energy use, biodiversity loss, health issues due to chemical contaminants in food and environment, spread of pests resistance, introduces a bias in the comparative analysis.[13]

Concerning such difficulties in comparison we wish to underline three important points:

---

[13] When careful studies are carried out such costs can be huge. For instance, estimates concerning soil erosion indicate that for the USA the cost of unsustainable soil management practices is 44 billion US$ each year (Pimentel et al., 1995). Soil erosion, as other issues mentioned, is not just a matter of short term economic accounting, it concerns the very same long term food security of a country (Howard, 1943; Carter and Dale, 1975; Hillel, 1991; Diamond, 2005; Pointing, 2007). The Dust Bowl that in the 30's hit the USA southern plains, has been the dramatic result of soil mismanagement. According to Donald Worster (2004, p. 63): *"The Dust Bowl rightly become the dominant national symbol of this bankruptcy and ecological decay, fusing into itself all the environmental complexities of the time"*. Climatic changes associated to poor rotation of crops, loss of soil organic matter and natural vegetation, can promote resurgence or incoming for new pests (Paoletti et al., 2008).

1. **Holistic approach**. When comparing organic and conventional farming systems inherent differences should be taken into account. Comparisons based only on economic analysis, or any other single indicator, compromises the understanding of the complex realities of farming systems. However, comparative studies often focus on specific crops (often a single one) and a short period of time. Furthermore, analysis are also focusing on specific indicators such as yield and economic accounting (e.g. Lockeretz et al., 1981), energy efficiency (e.g. Refsgaard et al., 1998), or environmental impact (e.g. Reganold et al., 1987; Paoletti et al., 1993; Drinkwater et al., 1998; Hansen et al., 2001; Mäder et al., 2002). There are a few works which attempt an integrated analysis based upon long term data (e.g. Reganold et al., 2001; Pimentel et al., 2005; Pimentel, 2006a). Longer-term studies (e.g. a minimum of 10 years) should be encouraged to gather information about multiple sustainability of different farming systems in the long run.

2. **Energy accounting**. Results from energy assessments are often difficult to compare because of the variety of methodologies and accounting procedures employed (e.g. Stölze et al., 2000; Hansen et al., 2001; comment on energy use in Germany husbandry in Hass et al., 2001; critics to the energy assessment by Refsgaard et al., 1998 in Dalgaard et al., 2001). Some authors (e.g. Foster et al., 2006) note that few studies cover the whole "farm to fork" life cycle of the agriculture system and this is necessary to get a comprehensive energy analysis of the products in the agrifood system.

3. **Internalisation of externalities**. Further investigations should include also the (energetic) cost of "externalities" such as of soil and water, loss of biodiversity, loss of environmental quality and de-contamination, the whole CO_2 (and GHGs) emissions due to long distance commodities trade compared to locally grown and consumed organic products etc. Indicators able to internalise those "hidden" energetic and economic costs should be employed. Emergy (spelled with an "m") proposed by H.T. Odum (1988; 1996), for instance, is one of the indicators that could help to integrate hidden costs. Emergy is a measure of solar energy used in the past along the way to get the final product or service, and thus different from a measure of the energy content now[14] (e.g. Odum, 1988; 1996; Ulgiati et al., 1994; Ulgiati and Brown, 1998; Haden, 2003). Such an analysis, however, is far from easy to use correctly to provide meaningful results (see for instance the critics moved by debate Castellini et al., (2006) who attempted a comparative emergetic assessment (using Emergy indicator) of two poultry farms in Italy[15] and critics by Maud (2007).

---

[14] If the autolithotrophic food chain is to be better measured, geochemical energy may also have to be accounted for in the future (Stevens and Mckinley, 1995; Stevens, 1997).

[15] They found a Emergy flow for conventional poultry farm of 724.12 10^{14} solar em joule/cycle, while Emergy flow for organic poultry farm was just 92.16 10^{14} solar em joule/cycle. But productive and economic performances are not mentioned.

To avoid, or better to reduce, bias and/or flaws in the analysis, sound comparisons should embrace a multicriterial approach where environmental, social and economic criteria are considered at the same time and at different scales (Giampietro et al., 1994; Wolf and Allen, 1995; Gomiero et al., 1997; Bland, 1999; Dalgaard et al., 2003; Giampietro, 2004; Gomiero et al., 2006).

17.2 Organic Agriculture: An Energy-Saving Alternative?

Energy saving, along with reduction of CO_2 emissions, is an important indicator to assess the sustainability of agricultural practices. In this section we will review a number of studies that compare the energy efficiency of organic and conventional farming systems.

17.2.1 Energy Analysis

Detailed comparisons of energy performance of organic and conventional farming systems were initiated by Pimentel and colleagues in early 1980 (Pimentel et al., 1983). Since then, the interest for such comparison kept growing and a number of works have been produced on the subject, although with different approaches and methodologies that sometimes make results difficult to compare. In Table 17.2 a number of studies are summarised that compare organic and conventional energetic performances.

Because of the typology of accounting or data reporting, some data found in literature are better summarised in term of ratio of energy input/output. Figures are reported in Table 17.3.

The data indicates, for most of the cases, a lower energy consumption for organic farming both for unit of land (GJ/ha), from 10% up to 70%, and per yield (GJ/t), from 15% to 45%. The main reasons for higher efficiency in the case of organic farming are: (1) lack of input of synthetic N-fertilizers (which require a high energy consumption for production and transport and can account for more than 50% of the total energy input), (2) low input of other mineral fertilisers (e.g. P, K), lower use of highly energy-consumptive foodstuffs (concentrates), and (3) the ban on synthetic pesticides and herbicides (Lockeretz et al., 1981; Pimentel et al., 1983; 2005; Refsgaard et al. 1998; Cormack, 2000; Haas et al., 2001; FAO, 2002; Lampkin, 2002; Hoeppner et al., 2006).

It seems that the energetic performances of different farming systems depend on the crops cultured and specific farm characteristics (e.g. soil, climate). For instance, organic potatoes vary from about −20% to + 30% (Table 17.2). Pimentel et al. (1983), who reported lower energy efficiency in organic potatoes, ascribed it to reduced yield due to insect and disease attacks that could not be controlled in the organic system. In the case of apples there is a striking difference between data reported by Pimentel et al. (1983) and Reganold et al. (2001). This can be brought about by different management techniques and their improvement in the last 20 years.

Table 17.2 Fossil energy consumption for different crops: organic vs. conventions (based on Stölze et al., 2000; FAO 2002 and other references (*))

Product and reference	Energy consumption (GJ/ha)			Energy consumption (GJ/t)		
	Conv.	Organic	Org. as % of conv.	Conv.	Organic	Org. as % of conv.
Winter wheat						
Alfoldi et al. (1995)	18.3	10.8	−41	4,21	2.84	−33
Haas & Köpke (1994)	17.2	6.1	−65	2.70	1.52	−43
Reitmayr (1995)	16.5	8.2	−51	2.38	1.89	−21
Potatoes						
Haas & Köpke (1994)	24.0	13.1	−46	0.80	0.07	−18
Alfoldi et al. (1995)	38.2	27.5	−28	0.07	0.08	+7
Reitmayr (1995)	19.7	14.3	−27	0.05	0.07	+29
Mäder et al. (2002)[som]	28.42	40.69	−30	3.70	3.98	−7
Citrus						
Barbera and La Mantia (1995)	43.3	24.9	−43	1.24	0.83	−33
Olive						
Barbera and La Mantia (1995)	23.8	10.4	−56	23.84	13.0	−45
Apple						
Geier et al. (2001)	37.35	33.8	−9.5	1.73	2.13	+23
Milk						
Cederberg and Mattsson (1998)	22.2	17.2	−23	2.85	2.41	−15
Refsgaard et al. (1998)*	–	–	–	3.34	2.16/ 2.88	−35/−13
Cederberg and Mattsson (1998) in Haas et al. (2001)*	–	–	–	2.85	2.4	−8
Haas et al. (1995) in Haas et al. (2001)*	19.4	6.8	−65	–	–	–
Haas et al. (2001)*	19.1	5.9	−69	2.7	1.2	−54

(som): Supporting Online Material (data from)

According to estimates carried out by the Danish government, upon 100% conversion to organic agriculture 9–51% reduction in total energy use would result, depending on the level of import of feeds and the amount of animal production (Hansen et al. 2001).

17.2.1.1 Energy Efficiency Under Extreme Climate

Long-term crop yield stability and the ability to buffer yields through climatic adversity are critical factors in agriculture's ability to support society in the future. A number of studies have shown that under drought conditions, crops in organically

Table 17.3 Comparison of energy efficiency (input/output) per unit of production of organic as % of conventional farming systems (figures from different studies)

Farming system	Reference	Energy Efficiency organic as % of conventional
Analysis for crops under organic and conventional management		
wheat in USA	Pimentel et al. (1983)	+29/+70
wheat in Germany (various studies)	Stölze et al. (2000)	+21/+43
wheat in Italy	FAO (2002)	+25
corn in USA	Pimentel et al. (1983)	+35/+47
apples USA	Pimentel et al. (1983)	−95
potatoes in Germany (3 studies)	Stölze et al. (2000)	+7/+29
potatoes USA	Pimentel et al. (1983)	−13/− 20
rotations of different production systems in Iran	Zarea et al. (2000) (in FAO, 2002)	+81
rotations of different production systems in Poland	Kus and Stalenga (2000) (in FAO, 2002)	+35
Danish organic farming	Jørgensen et al., (2005)	+10
whole system analysis (Midwest – USA) with comparable output	Smolik et al., (1995)	+60/+70
crop rotations (wheat-pea-wheat-flax and wheat-alfalfa-alfalfa-flax) in Canada	Hoeppner et al., (2006)	+20%
Results from Long Term Agroecosystem Experiments		
apples USA	Reganold et al., (2001)	+7
various crop systems	Mäder et al., 2002;	+20/+56%
organic and animals	Pimentel et al., (2005)	+28
organic and legumes	Pimentel et al., (2005)	+32

managed systems produce higher yields than comparable crops managed conventionally. This advantage can result in organic crops outyielding conventional crops by 70–90% under severe drought conditions (Lockeretz et al., 1981; Stanhill, 1990; Smolik et al., 1995; Lotter et al., 2003). Others studies have shown that organically managed crop systems have lower long-term yield variability and higher cropping system stability (Smolik et al., 1995; Lotter et al., 2003).

According to Lotter et al., (2003) the primary mechanism of higher yield in organic crops is due to the higher water-holding capacity of the soils in those treatments. Soils in the organic plots capture more water and retain more of it, up to 100% higher than conventional, in the crop root zone. Such characteristics make organic crop management techniques a valuable resource in this present period of climatic variability, providing soil and crop characteristics that can better buffer environmental extremes, especially in developing countries.

However, it has to be pointed out that local specificity plays an important role in determining the performance of a farming system: what is sustainable for one region may not be for another region or area (Smolik et al., 1995). So, more work has to be done to acquire knowledge about the comparative sustainability of other farming systems.

17.2.1.2 Organic Farming for Developing Countries

Energy and economic savings from organic farming can offer an important opportunity for developing countries to produce crops with limited costs and environmental impact. Some authors claim that organic farming can reduce food shortage by increasing agricultural sustainability in developing countries, contributing quite substantially to the global food supply, while reducing the detrimental environmental impacts of conventional agriculture (Netuzhilin et al., 1999; Paoletti et al., 1999; Pretty and Hine, 2001; FAO, 2002; Pretty et al., 2003; Badgley et al., 2007). Pretty and Hine (2001) surveyed 208 projects in developing tropical countries in which contemporary organic practices were introduced, they found that average yield increased by 5–10% in irrigated crops and 50–100% in rainfed crops. However, those claims have been challenged by different authors (e.g. McDonald et al., 2005; Cassman, 2007; Hudson Institute, 2007; Hendrix, 2007), who dispute the correctness of both the accounting and comparative methods employed. Hudson Institute (2007) refers that in most of the farming cases accounted as organic by Pretty and Hine (2001) chemical fertilisers and/or pesticides have been regularly applied. The latter may be a sound observation. However, we argue that the amount of inputs employed plays a critical role in maintaining the long term sustainability of farming systems. So, although the "organic certification" cannot apply to a farm which uses pesticides, we should recognise the effort to keep the amount at a minimum and the use stack to the real needs. We should aim at is of reducing as much as possible our impact. In this sense organic farming is paving the way to gain knowledge and experience about best practices making them available to all.

17.2.2 A Trade off Perspective

In order to gain an useful insight on the sustainability of a farming system different criteria such as land, time and energy, should be employed at the same time (Smil, 2001; Giampietro, 2004; Pimentel and Pimentel, 2007a). Data on energy efficiency cannot be de-linked from total energy output and from the metabolism of the social system where agriculture is performed. Great energetic efficiency may implie low total energy output that for a large society with limited land may not be a sustainable option menacing food availability.

Models for energy assessment for Danish agriculture developed by Dalgaard et al., (2001), to compare energy efficiency for conventional and organic agriculture, were used to evaluate energy efficiency for eight conventional and organic crop types on loamy, sandy, and irrigated sandy soil. Results from the model indicated that energy use was generally lower in the organic than in the conventional system (about 50%), but yields were also lower (about 40–60%). Consequently, conventional crop production had the highest energy expenditure production, whereas organic crop production had the highest energy efficiency. The same results have been produced also by Cormack (2000) for the UK, modelling a whole-farm system using typical crop yields. (However, it has to be said that in some long term trials yield difference for some crops, in terms of ton/ha, between organic and conventional crops has been minimal or negligible; e.g. Reganold et al., 2001; Delate et al., 2003; Vasilikiotis, 2000; Pimentel et al., 2005).

This inverse relation between total productivity and efficiency seems typical for traditional and intensive agriculture. When comparing corn production in intensive USA farming system and Mexican traditional farming system it resulted that the previous had an efficiency (output/input) of 3.5:1 while the latter of 11:1 (using only manpower). However, when coming to total net energy production, intensive farming system accounted for 17.5 million kcal/ha yr^{-1} (24.5 in output and 7 in input), while traditional just 6.3 million kcal/ha yr^{-1} (7 million in output and 0.6 million in input) (Pimentel, 1989).

In Europe, the yield from arable crops was 20–40% lower in organic systems and the yield from horticultural crops could be as low as 50% of conventional. Grass and forage production was between 0% and 30% lower (Stockdale et al., 2001; Mäder et al., 2002). This led Stockdale et al. (2001) to conclude that when calculating the energy input in terms of unit physical output, the advantage to organic systems was generally reduced, but in most cases that advantage was retained.

The productivity of labour is another key indicator that has to be considered to assess the socio-economic sustainability of the farming enterprise. Although performing better in terms of energy efficiency, organic farms require more labour

Table 17.4 A comparison of the rate of return in calories per fossil fuel invested in production for major crops – average of two organic systems over 20 years in Pennsylvania (based on Pimentel, 2006a, modified)

Crop	Technology	Yield (t/ha)	Labour (hrs/ha)	Energy (kcal x 10^6)	kcal (output/input)
Corn	Organic[1]	7.7	14	3.6	7.7
Corn	Conventional[2]	7.4	12	5.2	5.1
Corn	Conventional[3]	8.7	11.4	8.1	4.0
Soybean	Organic[4]	2.4	14	2.3	3.8
Soybean	Conventional[5]	2.7	12	2.1	4.6
Soybean	Conventional[6]	2.7	7.1	3.7	3.2

[1] Average of two organic systems over 20 years in Pennsylvania
[2] Average of conventional corn system over 20 years in Pennsylvania
[3] Average U.S. corn.
[4] Average of two organic systems over 20 years in Pennsylvania
[5] Average conventional soybean system over 20 years in Pennsylvania
[6] Average of U.S. soybean system

than conventional ones from about 10% up to 90% (in general about 20%), with lower values for organic arable and mixed farms and higher for horticultural farms (Lockeretz et al., 1981; Pimentel et al., 1983, 2005; FAO, 2002; Foster et al., 2006). Case studies in Europe for organic dairy farms report a comparable request of labour (FAO, 2002). Little data exists on pig and poultry farms, but labour per hectare of utilized agricultural area seems to be similar to conventional farms, as livestock density is reduced (FAO, 2002).

Again, is has to be reported that in some long terms trials productivity per ha and hr of work for organic and conventional crops (corn and soybean) were comparable (Pimentel et al., 2005; Pimentel, 2006a), Table 17.4.

Figures from Table 17.4 are very interesting as they compare four key indicators in a 20 years old trials. Data indicates that corns and soybean organic systems perform much better or, at worst, are comparable to conventional systems.

To carry on extensive long term trials for diverse crops in diverse areas is of fundamental importance to understand the potential of organic farming as well as to improve farming techniques moving agriculture towards a more sustainable path.

17.3 CO_2 Emissions and Organic Management

Because of the role played in GHGs emissions by agriculture, it is important to analyse whether there are possibilities to reduce the environmental impact of agriculture activities.

Agriculture accounted for an estimated emissions of 5.1 to 6.1 Gt CO_2-eq/yr in 2005 (10–12 % of total global anthropogenic emissions of GHGs. CH_4 contributes 3.3 Gt CO_2-eq/yr and N_2O 2.8 Gt CO_2-eq/yr. Of global anthropogenic emissions in 2005, agriculture accounts for 10 about 60% of N_2O and about 50% of CH_4 (IPCC, 2007).

CO_2 emissions come mainly from fertilizer industry, the machinery used on the farm and, according to the production system and to the changes in land use, from the carbon present in the soil. Deforestation is also an important contributor to the CO_2 emissions by agriculture. NH_4 emissions come from livestock, mainly from enteric fermentation but also from manure and rice fields. N_2O comes mainly from the soil (denitrification) and to a lesser extent from animal manure (IPCC, 2007).

Biofuels are believed to be able to curb GHGs emissions because plants absorb the CO_2 that is emitted by biofuels combustions, so closing the cycle. However, GHGs other than CO_2 should be accounted for when assessing the impact of agriculture, and in particular of intensive agriculture. Recently, Crutzen et al., (2007, p. 11192) stated that "... *when the extra N_2O emissions from biofuel production is calculated in "CO_2-equivalent" global warming terms, and compared with the quasi-cooling effect of "saving" emissions of fossil fuel derived CO_2, the outcome is that the production of commonly used biofuels, such as biodiesel from rapeseed and bioethanol from corn (maize), can contribute as much or more to global warming by N_2O emissions than cooling by fossil fuel savings*". It has also been argued that microbes convert much more of the nitrogen in fertiliser to N_2O than previously thought, up to 3–5%, more than twice the figure of 2% used by the IPCC. For rapeseed biodiesel, which accounts for about 80% of the biofuel production in Europe, for instance, the relative warming due to N_2O emissions is estimated at 1.0–1.7 times larger than the quasi-cooling effect due to saved fossil CO_2 emissions. For corn bioethanol, dominant in the US, the figure is 0.9 to 1.5 (Crutzen et al., 2007). According to the authors only cane sugar bioethanol – with a relative warming of 0.5–0.9 – looks like a viable alternative to conventional fuels. The recent works by Fargione et al., (2008) and Searchinger et al., (2008) come to the conclusion that when considering the "carbon-debt", that is to say, the release of carbon when converting rainforests, peatlands, savannas, or grasslands to produce food-based biofuels, the overall greenhouse emissions is greatly increased, at least for the next centuries. These results make clear that biofuels are not a viable solution to reduce carbon emissions.

17.3.1 Carbon Sink Under Organic and Conventional Agriculture: The Production Side

The important role of properly managed agriculture as an accumulator of carbon has been addressed by many authors (e.g. Drinkwater et al., 1998; Pretty et al., 2002; Holland, 2004; Janzen, 2004; Lal, 2004; IPCC, 2007; Keeney, 2007). This carbon can be stored in soil by: (1) increasing carbon sinks in soil organic matter and above and below ground biomass (e.g. through adopting rotations with cover crops and green manures to increase biomass, agroforestry, conservation-tillage systems, avoiding soil erosion), (2) reducing direct and indirect carbon emissions, for instance adopting energy saving measures (e.g. reducing use of agrochemicals, pumped irrigation and mechanical power which account for most of the energy input). Besides to that, some authors (e.g. Pretty et al., 2002; Lal, 2004; IPCC, 2007) suggest that CO_2 abatements by agriculture can be achieved by (3) growing annual crops for biofuel production (e.g. ethanol from maize and sugar cane), and annual and perennial crops (e.g. grasses and coppiced trees) for combustion and electricity generation. This latter option has also been suggested for organic farming (Jørgensen et al., 2005). It has also been suggested that organic farms can develop biogas digesters to produce methane for their home use (Pretty et al., 2002; Hansson et al., 2007) or biofuel to become self-sufficient for motor fuels (Hansson et al., 2007). However, for the later case, the assumptions of the model are arguable and from the same model presented by the authors biofuel produced in that way results more expensive than conventional.

Agricultural activities play an important role in CO_2 and other GHGs (in particular NH_4 and N_2O which have a much greater) . Contribution to CO_2 emissions derives from consumption of energy in form of oil and fuel both directly (e.g. field works, machinery) and indirectly (e.g. production and transport of fertilisers and pesticides, changes in soil ecology that releases carbon in the atmosphere).

It is important to evaluate whether under organic management GHGs can be reduced. In the last decades CO_2 emissions assessment from organic and conventional agriculture has been carried out in different countries mainly concerning:

- emissions for different crops and milk production,
- calculations on CO_2 emissions per hectare, based on average farm characteristics (crop management, rotation).

Data on CO_2 emissions for different crops and for milk with respect to organic and conventional farming are reported in Table 17.5.

Figures from Table 17.5 indicate that CO_2 emissions in organic agriculture are lower on a per hectare scale. However, on an per output unit scale, results differ. The lower emissions of CO_2 per ha in organic farming can be explained by the lack of agrochemicals (pesticides and in particular of nitrogen ferlizers which production requires high energy input) and a lower use of high energy consuming feedstuffs for livestock.

Concerning organic agriculture data for the whole Global Warming Potential (GWP) of the different farming systems, such as methane and NO_x emissions are,

Table 17.5 CO_2 emissions (kg) for some productions (based on Stölze et al., 2000 and other references (*))

Study	CO_2 emission (kg CO_2/ha)			CO_2 emission per production unit (kg CO_2 /t)		
	Conv.	Organic	Org. as % of conv.	Conv.	Organic	Org. as % of conv.
Winter wheat						
Rogasik et al. (1996)	826	443	−46	190	230	+21
Haas & Köpke (1994)	928	445	−57	149	110	−21
Reitmayr (1995)	1001[if]	429	−57	145[if]	100	−21
Potatoes						
Rogasik et al. (1996)	1661	1452	−13	46	62	+35
Haas & Köpke 1994	1437	965	−33	46	48	0
Reitmayr (1995)	1153[if]	958	−17	30[if]	45	+50
Milk						
Lundström (1997)	–	–	–	203	212	+4
Haas et al., (2001)*	9400	6300	−67	1280[a]	428[a]	+65%
Haas et al., (2001)*				1300[b]	1300[b]	0
Crop management rotation						
Haas & Köpke, (1994) in Stölze et al., (2000)*	1250	500	-40%	–	–	–
SRU, (1996) in Stölze et al., (2000)*	1750	600	−34%	–	–	
Rogasik et al., (1996) in Stölze et al., (2000)*	730	380	−52%	–	–	–

[if] integrated farming
[a] considering only CO_2 emission
[b] summing up CH_4 and N_2O emissions as CO_2 equivalents, the CH_4 and N_2O emissions are comparably low, but due to the high Global Warming Potential (GWP) of these trace gases their climate relevance is much higher.

in most of the cases, lacking. A comprehensive accounting is important due to the high GWP of those gases.

In Table 17.2, for instance, the study by Hass et al., (2001) for German dairy reports an energy use for organic agriculture less than half per unit of milk of the conventional farming and less than one-third per unit land. But because of slightly higher methane emissions per unit of organic produced milk and the high GWP of methane, authors estimated that the final GWP of the two farming system was equivalent.

We believe that emissions per ton of food produced should be a more relevant indicator to assess the environmental impacts of the farming system for a low per ha emissions can be easily achieved by being content with a minimum yield that from the point of view of food production (as well as economic) can be unsustainable. For instance, production of potatoes in organic farming is associated with lower CO_2 emissions per ha but tends toward higher CO_2 emissions per ton due to a lower productivity. Lower CO_2 emissions per ha in organic farming is reported due to synthetic nitrogen fertilisation used in conventional farming (Stölze et al., 2000). Estimates on the CO_2 emissions per ton gives different results depending on the assumption of yield levels. It is interesting to note the wide range of values of kg CO_2/t, with winter wheat ranging from -21% to $+21\%$ and potatoes from 0% to $+50\%$. In such trials annual climatic variation and assumptions in setting up system analysis can play an important role in determining the final figures.

Stölze et al., (2000) in their review of European farming systems, saw trends towards lower CO_2 emissions in organic agriculture but were not able to conclude that overall CO_2 emissions are lower per unit of product in organic systems compared to the conventional ones. Authors note that the 30% higher yields in conventional intensive farming in Europe can average out the CO_2 emissions per unit of products.

Many authors stressed the importance of energy saving in agriculture and the possible role of organic or sustainable practice in this direction (Pimentel et al., 1973; 2005; Lockeretz, 1983; Poincelot, 1986; Pimentel and Pimentel, 2007a). Smith et al. (2008) estimated a global potential mitigation of 770 $MtCO_2$-eq/yr by 2030 from improved energy efficiency in agriculture (e.g. through reduced fossil fuel use).

17.3.2 Overall Carbon Sink Potential in Organic Farming

Organic agriculture also plays a role in enhancing carbon storage in soil, for instance in the form of soil organic matter (see Section 4). So it is important to evaluate the contribute that organic agriculture has to offer in this sense.

Results from the 15-years study in the USA, where three district maize/soybean agroecosystems, two legume-based and one conventional were compared, led Drinkwater et al., (1998) to estimate that the adoption of organic agriculture practices in the maize/soybean grown region in the USA would increase soil carbon sequestration by 0.13–0.30 10^{14} g yr^{-1}, that equal to 1–2% of the estimated carbon released into the atmosphere from fossil fuel combustion in the USA (referring to 1994 figures of $1.4 \ 10^{15} \ g \ yr^{-1}$).

In the Midwest USA in a 10-year for organic crop systems trial, Robertson et al., (2000) found organic farming system to have about 1/3 of the net GWP of comparable convention crop systems, but 3-fold higher GWP than conventional agriculture under no-till systems, which included embedded energy. They found no difference in nitrous oxide emissions and methane oxidation between the three systems. Average soil carbon accumulation was $0 \ g \ m^{-2} \ yr^{-1}$ in conventional agriculture, $8 \ g \ m^{-2} \ yr^{-1}$ in organic agriculture and $30 \ g \ m^{-2} \ yr^{-1}$ conventional no-till plots.

In any case, because the soil has a limit to carbon sink, also conversion to organic agriculture only represents a temporary solution to the problem of carbon dioxide emissions. Foereid and Høgh-Jensen (2004) developed a scenario for carbon sink under organic agriculture. The simulations showed a relatively fast increase in the first 50 years of $10–40 \ g \ C \ m^{-2} \ y^{-1}$ on average. The increase then levelled off, and after 100 years it had reached an almost stable level.

However, while organic agriculture surely represents an important option to buy time while offering many beneficial services by reducing the agriculture impact on soil and environment, long term solutions concerning CO_2 emissions from global society should be searched in different energy sources or, more probably, on reducing the energy demand.

17.3.3 Improving Soil and Land Management

According to a review carried out by Pretty et al., (2002) carbon accumulated under improved management within a land use and land-use change ranged from 0.3 up to $3.5 \ tC \ ha^{-1} \ yr^{-1}$. Grandy and Robertson (2007) argue that there is high potential in carbon sequestration and offsetting atmospheric CO_2 increases in agriculture land by reducing land use intensity. They estimated that reducing land use intensity (e.g. by no-till systems) enhanced carbon storage to 5 cm relative to conventional agriculture ranged from $8.9 \ gC \ m^{-2} \ y^{-1}$ (0.89 t/ha y^{-1}) in low input row crops to $31.6 \ gC \ m^{-2} \ y^{-1}$ (3.16 t/ha y^{-1}) in the early successional ecosystem. Following reductions in land use intensity soil C accumulates in soil aggregates, mostly in macroaggregates. The potentially rapid destruction of macroaggregates following tillage, however, raises concerns about the long-term persistence of these carbon pools.

Schlesinger (1999) argues that converting large areas of cropland to conservation tillage, including no-till practices, during the next 30 years, could sequester all the CO_2 emitted from agricultural activities and up to 1% of today's fossil fuel emissions in the United States. Similarly, alternative management of agricultural soils in Europe could potentially provide a sink for about 0.8% of the world's current CO_2 release from fossil fuel combustion.

However, such estimates can be somehow optimistic as they do not consider actual changes. For European Union (EU-15), Pete et al., (2005) point out that because cropland area is decreasing and in most European countries there are no incentives in place to encourage soil carbon sequestration, carbon sequestration between 1990 and 2000 was rather small or negative. Based on extrapolated trends, they predicted carbon sequestration to be negligible or even negative by 2010. Authors argue that the only trend in agriculture that may be enhancing carbon stocks on croplands, at present, is organic farming, but the magnitude of this effect, according to them, is highly uncertain. Smith et al., (2005) state that without incentives for carbon sequestration in the future, cropland carbon sequestration under Article 3.4 of the Kyoto Protocol will not be an option in EU.

17.4 Agricultural "Waste " for Cellulosic Ethanol Production or Back to the Field?

A first generation of fuels and chemicals is being produced from high-value sugars and oils products. A second generation is now being researched and is thought to have greater potential as it should be based on cheaper and more abundant lignocellulosic feedstock Cellulosic ethanol, which can be produced from the woody parts of trees and plants, perennial grasses, or crops residues, is considered a promising improvement in transforming crops into energy as it enable to convert all the green plant into ethanol and not just the seeds as it is in the normal fermentation process (Lynd et al., 1991; Badger, 2002; Goldemberg, 2007; Himmel et al., 2007; Lange, 2007; Solomon et al., 2007; Service, 2007; Solomon et al., 2007; Stephanopoulos, 2007).

According to the survey by Service (2007), in the USA the first production plants will come on line beginning in 2009, with an expected cost of cellulosic ethanol doubling that of corn ethanol, but U.S. Department of Energy is expecting production costs to soon become competitive with corn ethanol. Some authors forecast that the full potential of biofuel production from cellulosic biomass will be obtainable in the next 10–15 years (Service, 2007; Stephanopoulos, 2007). However, optimistic claims were already popular about 20 years ago. For instance, in 1991, on *Science* some experts were already stating that: *"In light of past progress and future prospects for research-driven improvements, a cost-competitive process appears possible in a decade"* (Lynd et al., 1991, p. 1318). Subsidies will be essential to market success of this technology (Solomon et al., 2007), indicating that this option suffers from the same drawbacks that affect other biofuels (see the other chapters of this publication).

Some experts argue that cellulosic ethanol, if produced from low-input biomass grown on agriculturally marginal land or from waste biomass, could provide much greater supplies and environmental benefits than food-based biofuels (Hill et al., 2006; Goldemberg, 2007; Koutinas et al., 2007; Lange, 2007). According to Koutinas et al., (2007, p. 25), for instance: *"... maximizing the usage of biomass components would lead to significant improvement of process economics and waste minimization*". Also the works by Fargione et al., (2008) and Searchinger et al., (2008) after stating that biofuels increase the overall greenhouse emissions, at least for the next centuries, suggest that agricultural waste and residues can be use instead. Transforming agriculture waste into energy may seem an interesting option at first sight, but is it a real viable option?

Smil (1999) argues that more than half of the dry matter produced from agriculture is represented by inedible crop residues. Crop residues have been traditionally used for animal feed, bedding, as well as fuels in many rural areas. According to Pimentel et al., (1981), in the USA, agriculture residues remaining after harvest amount to 17% of the total annual biomass produced with an estimate gross heat energy equivalent of 12% of the energy consumed annually in the USA.

Crop residues play a major role to preserve soil fertility by supplying a source of organic matter. Soil organic matter has a fundamental role in soil ecology: it improves soil structure, which in turn facilitates water infiltration and ultimately the overall productivity of the soil, enhance root growth, and stimulate the increase of soil biota diversity and biomass. Wide evidences clearly indicate that the loss of organic matter poses a threat to long term soil fertility and in turn to the very same human life (Howard, 1943; Allison, 1973; Carter and Dale, 1975; Hillel, 1991; Pimentel et al., 1981; 1995; Drinkwater et al., 1998; Rasmussen et al., 1998; Smil, 1999; Lal, 2004; Pimentel, 2007). Soil biodiversity, then, has important ecological functions in agroecosystems influencing, among other things, soil structure, nutrients cycling and water content, and enhancing resistance and resilience against stress and disturbance (Paoletti and Pimentel, 1992; Paoletti and Bressan, 1996; Matson et al., 1997; Coleman et al., 2004; Heemsbergen et al., 2004; Brussaard et al., 2007). It has also to be mentioned that the greater availability of crop residues and weed seeds translate to increasing food supplies for invertebrates, birds and small mammals helping to sustain local biodiversity[16] (Dritschillo and Wanner, 1980; Paoletti et al., 1989; Paoletti and Pimentel, 1992; Paoletti, 2001; Genghini et al., 2006; Holland, 2004; Perrings et al., 2006). Furthermore, as Wardle et al., (2004) argue, aboveground and belowground components of ecosystems have traditionally been considered in isolation from one another, but it is now clear that there is strong interplay between these two systems and they greatly influence one another. This is of key importance, for instance, when coming to biological control of pests. Usefull predators and parasitoids, in fact, in many cases spend underground most of their lifecycle before being active aboveground on the crops, then

---

[16] It has to be mentioned that the impact of intensive agriculture poses a threat to soil ecology in two broad ways (Paoletti and Pimentel, 1992; Pimentel et al., 1995; Matson et al., 1997; Rasmussen et al., 1998; Krebs et al., 1999; Paoletti, 2001): (1) it accelerates soil organic matter oxidation and predisposes soils to increased erosion, (2) heavy application of chemical nitrogen fertilisers increase soil acidity causing numerous detrimental effects on soil quality such as reduction of soil faunal and floral diversity, increase soil-born pathogen activity, retards nutrient cycling, and can restrict water infiltration and plant roots development.

soil quality and management is foremost important in mitigation of most crop pests (Paoletti and Bressan, 1996). Stable litters on topsoil can stimulate some pests such as slugs but can provide feed to detritivores and polyphogous predators and parasitoids that can damage the crops.[17] In this sense, organic agriculture is effective in preserving soil organic matter and preventing soil erosion, as well as an option for carbon sink.

Increasing soil organic matter greatly improves soil quality playing a key role in guaranteeing sustainable crop production and food security. As a side product it provides and effective means for carbon sequestration. Lal (2004) estimated that a strategic management of agricultural soil (e.g. reducing chemical inputs, moving from till to no-till farming[18], contrasting soil erosion, increasing soil organic matter) has the potential to offset fossil-fuels emissions by 0.4 to 1.2 Gt C/yr, that is to say 5 to 15% of the global emissions. Evidences from numerous Long Term Agroecosystem Experiments indicate that returning residue to soil rather than removing them converts many soils from "sources" to "sinks" for atmospheric CO_2(Rasmussen et al., 1998; Lal, 2004).

As Pimentel et al., (1981) early warned, the total net contribution from converting agriculture residues into energy would result relatively small, referring to the overall energy consumption (in the case of the USA 1% of the energy consumed as heat energy), while the effect on soil ecology would be detrimental. As it has been pointed out by Rasmussen et al., (1998): "*If socioeconomic constraints prevent concurrent adoption of residue return to soil, degradation of soil quality and loss of sustainability may result from selective adoption of technology*".

Concerning an extensive use of agricultural waste for energy production, it has to be stressed that when biomass is taken away from, or not returned to the field and burned, this interferes with closing the nutrient cycles and greatly affect soil erosion (Pimentel et al., 1995; Pimentel and Kounang, 1998; Smil, 1999; Pimentel, 2007), leading to a dramatic loss of topsoil being lost from land areas worldwide 10–40 times faster than the rate of soil renewal threatening soil fertility and future human food security (Pimentel et al., 1995; Pimentel, 2006b; 2007). Harvesting crop residues will worsen soil erosion rates from 10-fold to 100-fold (Pimentel, 2007) resulting in a disaster for conventional agriculture and especially for organic agriculture.

It has been suggested that energy from agricultural waste can be obtained also in organic agriculture. Jørgensen et al., (2005), for instance, analysing organic and conventional farming in Denmark, argue that the production of energy in organic farming is very low compared to conventional farming because of the extensive utilisation of straw from conventional that in the organic system is left in the fields (energy content of straw used for energy production was equivalent to 18% of total

---

[17] It has been reported that removing shelterbelts in the rural landscape can cause a loss of litter in topsoil and this can lead to a shift of feeding habits among some detritivores such as the case of the slater *Australiodillo bifrons* , in NSW, Australia, becoming a cereal pest (Paoletti et al., 2008).
[18] No-till farming is also known as *conservation tillage* or *zero tillage*, a way of growing crops from year to year without disturbing the soil through tillage.

energy input in Danish agriculture in 1996). According to Jørgensen et al. (2005), in organic farming energy production can be boosted by utilising farm waste such as: manure and crop residues or adopting short rotation coppice such as Alder[19] (*Alnus* spp.), as energy sources. We argue that this is not a viable option for organic farming (as it is not a viable option for conventional agriculture) and it is actually contrary to the very same principle of organic agriculture that relies on the natural ecological cycles. Under organic agriculture displacing agriculture waste from fields to energy plans will have an even more detrimental effect. This means that the large nutrients void has to be replaced via a massive use of synthetic fertilisers as it is the case in conventional agriculture. Due to the dependence of organic farming from biomass retuning into the fields, bioenergy production based on an extensive use of agricultural waste is not a sustainable option because it will compromise soil health.

17.5 Organically Produced Biofuels?

In this section we examine the position of organic representative concerning biofuels production and the option to produce biofuels according to organic standards.

17.5.1 The Position of the Organic World on Biofuels

National and international organic associations seem to hold different express positions concerning the possible benefits in respect to the benefits of biofuels production for organic agriculture. Some of them are producing positional documents in favour (e.g. IFOAM) and against (e.g. the British Soil Association). Others seem to express contrasting views within themselves (e.g. the Italian Association for Organic Agriculture – Associazione Italiana Agricotura Biologica) or not expressing any opinion on the subject (e.g. the French Fédération Nationale d'Agriculture Biologique).

According to Kotschi and Müller-Sämann (2004), writing for IFOAM, using biomass as a substitute for fossil fuel represents another emissions reduction option. They argue that organic agriculture is well positioned in this sector. It has the advantage that inorganic N-fertilizers are not applied, which cause significant emissions of N_2O and use a lot of energy. IFOAM invites policymakers to consider the potential of organic farming for GHG reduction and develop appropriate programs for using this potential such as: emissions reduction potential, in the sequestration potential, in the possibility for organically grown biomass, or in combinations of all the aspects. This both for developed and developing countries.

The Soil Association, the main certifier and promoter of organic food and farming in Britain, released an official document stating the position of the association

---

[19] Alder is an interesting crop due to its symbiosis with the actinomycete *Frankia*, which has the ability to fix up to 185 kg/ha nitrogen (N_2) from the air (Jørgensen et al., 2005).

concerning biofuels (Soil Association, 2004). The position can be summarised as follows:

- biofuels are highly unlikely to bring the environmental benefits imagined, to assess the impact of biofuels on climate change the effect of the agricultural methods has to be evaluated,
- biofuels produced by conventional agriculture are net user of fossil fuels and then a net CO_2 source. To make biofuel production more sustainable organic methods should be used,
- the use of Genetic Modified crops must be prohibited,
- biofuel production must not displace food production.

Concerning biofuels production, the Soil Association addresses two key issues, (1) a strategic one and a (2) technical one.

(1) it is necessary: (a) to promote energy efficiency by concerning with the impacts of its production and its implication for rural development, and (b) to constrain the need for transport fuel (including food transport that now accounts for a very significant proportion of total transport in the UK, EU road traffic is growing at 2% per year, and this growth would wipe out any contribution from biofuels within just a couple of years),
(2) what can be done: (a) producing biodiesel from oil waste, such as cooking oil reducing the current tax, and (b) developing the anaerobic digestion of slurries and waste to produce biogas to be compressed as vehicle fuel. Residues could be applied to soil and increase soil organic matter, reducing the need for chemical (fossil fuel) based fertiliser.

Roviglioni (2005) writing in *Bioagricoltura*, the journal of the Italian association for organic agriculture (AIAB), states that biofuel can play a role in supplying sustainable energy to farmers and should be developed along with other green energies such as solar, wind. However, the official positions seems have not yet be taken by the AIAB steering committee.

Concerning biofuels Dennis Keeney (the first director at the Leopold Center[20] from 1987 to 1999 and now a Professor Emeritus at Iowa State University) stated that biofuels can represent a way out for farmers from the present crisis: "*This impending social, ecological and economic disaster can be avoided with policies that move us toward perennial biofuels (grasses and trees). These crops, if produced in a sustainable manner, offer large benefits to local economies. The environmental and economic benefits are clear: cellulosic feedstocks from perennials have far higher energy return than corn-based ethanol, and have proven environmental and*

---

[20] The Leopold Center is a research and education center with statewide programs to develop sustainable agricultural practices that are both profitable and conserve natural resources. http://www.leopold.iastate.edu/about/about.htm

biodiversity benefits. Mixed swards of grasses would have more stability and would stretch out the harvest time." (Keeney, 2007).

17.5.2 Organically Produced Energy and Biofuels?

According to the Soil Association (2004) in order to avoid the problems enhanced by conventional agriculture in the production of biofuels two approaches can be adopted: (1) using waste biomass directly (e.g. fuel wood) or indirectly (e.g. biogas), and (2) growing bioenergy crops in a sustainable, or organic, way. According to some authors (e.g. IEA 2002; Jørgensen et al., 2005), some forestry systems could provide an option for sustainably grown biomass for energy production. In such forestry systems nutrient loss can be kept on a low level by on-site foliage and reapplication of wood ash. However, how sustainable these practises are in the long-run needs investigation.

These two options, however, present important differences and have limits to their viability in a context of large scale production. Using waste biomass directly, such as fuel wood may have certainly positive effects, when, for instance, using wood collected in the hedgerows as fuel wood or fruit shell from palm oil, or by capturing methane from anaerobic fermentation of manure. But, in general (as we have seen in Section 5) missing to return agricultural waste to the field is detrimental for the preservation of soil fertility in the long run.

In the second case biogas can be produced while the fermented material can be returned back as fertiliser in the fields.

However, although these energy sources may be relevant at the farm or in rural community level (in particular in developing countries), when coming to discuss these options on a larger-scale perspective it has to be admitted that they cannot cover a significant share of the actual global energy demand.

Recently Ziesemer (2007), reviewing the issue of energy and organic agriculture for FAO, stated that "*Because of its reduced energy inputs, organic agriculture is the ideal production method for biofuels. Unlike the cultivation of staple food crops, in which energy efficiency is just one of many environmental and nutritional aspects of production, biofuels are measured primarily by their energy efficiency. Organic agriculture offers a favourable energy balance because of its lower energy require-ments. As the aim of biofuels is to reduce dependency on non-renewable energy sources and to mitigate environmental damage of fossil fuel emissions, organic pro-duction of biofuels furthers these goals in a way that conventional agriculture does not.*" (Ziesemer, 2007, p. 20).

We argue that although organically grown crops can reach a better energy effi-ciency than conventionally grown, still statements such as that of Ziesemer (2007) miss to consider a number of key points: (1) in most of the cases organic crops have lower productivity per ha (about 20–30%) than conventional crops and generally they require more labour per unit of product. That means that a larger amount of land should be put under cultivation to provide the same quantity of biomass as that produced under conventional agriculture, and that the society has to allocate larger working time in producing its own food, that makes of organic crops a very precious good for the society that cannot be spoiled; (2) agricultural waste and residues is needed to preserve soil fertility and should be returned to the fields, (3) importing organic crops, or "organic biofuels", from developing countries rises important environmental, social and ethical questions that cannot be ignored.

However, if policymakers will continue to promote an extensive production of fuel crops (whatever may be the reason leading to this choice), in order to limit the environmental damages from intensive agricultural practices, a more ecological farming should be employed.

17.6 Conclusion

In the last decades biofuels have been regarded as a promising source of renewable energy while at the same time an option to curb greenhouse gas emissions. This is based on the assumption that biofuels are: (1) *renewable*, crops will store CO_2 while growing absorbing those emitted from combustion closing the cycle, all that without other energy subsidies from fossil fuels, (2) *technological feasible*, we have a sound and effective technology to transform energy stored in the biomass in other forms of energy more useful for us (e.g. liquid fuels), (3) *energetically efficient* to produce, the energy output is substantially higher than energy input, (4) a *viable* option, biofuels production will not interfere with the demand for food from society, is economically affordable and will not threat environment preservation and the nature services.

While point (2) can be hold true (apart from cellulosic ethanol that is still a difficult to produce), the others are disputed. Concerning efficiency, an early warning was launched by David Pimentel (1991). From his comprehensive assessment of energetic, environmental and social issues Pimentel (Pimentel, 1991; 2003; Pimentel et al., 2005) claims that intensive biofuels production: (a) would not improve the USA energy security, (b) is uneconomical, (c) is not a renewable energy source as energy inputs overcome output, (d) it can cause major environmental threats by increasing soil and environmental degradation (in the USA corn production erodes soil some 18 times faster than soil is reformed) and environmental pollution (e.g. by using a large amount of agrochemicals). Furthermore, intensive synthetic fertilization causing the releasing of GHGs with high global warming potential, may contribute to worsen the problem (Crutzen et al., 2007; Fargione et al., 2008; Searchinger et al., 2008).

It has also been pointed out that large scale biofuels production poses major social and ethical issues. Biofuels will compete with food crops for land and water and because most of biofuels will be mostly produced from crops (e.g. corn, sugarcane, wheat, soybean) and this can lead to a boost in the price of staple food and deplete food resources with a dramatic effect on the weaker part of the population to meet their basic food needs (Pimentel, 1991; 1993; 2003; De Oliveira et al., 2005).

Before embracing biofuels, other ways to achieve energy savings and reducing CO_2 emissions should be searched for. In the case of agriculture a possible path can be found in the adoption of less energy intensive agricultural practices such as organic agriculture (and other low inputs agriculture practices). Organic agriculture aims at maintaining the long term sustainability of the agroecosystem as a whole, preserving and improving soil quality, minimizing energy and water use, preserving biodiversity, guaranteeing good quality and safe products to consumers while at the same time proving to generate a proper income for farmers and improving landscape quality.

From the present review we can reach the following conclusions:

Energy efficiency and energy savings: organic agriculture performs much better than conventional concerning energy efficiency (output/input). Generally, however, conventional crop production has the highest total net energy production per unit of cropped land (but in some trials the figures were comparable) and unit of working time. It has to be pointed out that due to the different farming strategies adopted in organic and conventional farms (e.g. integrated cropping systems and rotation adopted in organic farming), and the different response to climate variability of organic and conventional agroecosystems, results obtained from simple, short term comparison of crops productivity may result misleading.

CO_2 abatement: organic agriculture surely represents an important option to supply a carbon sink and so to buy time while searching for more definitive solutions. Soil, in fact, has a limit to carbon sink, so conversion to organic agriculture only represents a temporary solution to the problem of CO_2 offset. Long term solutions concerning CO_2 emissions from global society should be searched for in different energy sources along with the reduction of energy demand.

Use of agriculture waste: due to the dependence of organic agriculture from biomass input to provide nutritional elements to the soil, bioenergy production based on the use of agricultural waste is not a sustainable option in the long run and it will result in the depletion of soil organic matter and nutrients. Using agricultural waste for biofuels production will cause a large nutrients void that should be replaced via a massive use of synthetic fertilisers as it is the case in conventional agriculture. This will result in reducing energy efficiency increasing CO_2 emissions and in a detrimental environmental impact.

Organic biofuels: in the case that policymakers decide to continue to back biofuels production (for whatever reasons), then more sustainable agricultural practices must be adopted so to minimize energy consumption (aiming at improving energy efficiency) and reducing environmental impact (aiming at the long run sustainability of the agroecosystems). Continuing on the path traced by conventional intensive agriculture would threat the food security of nations. It should be reminded, as History teach us, that once the soil fertility, aquifers and biodiversity are gone, there will be no technological fix able to restore them.

Properly managed, organic agriculture could represent an interesting option to reduce energy consumption, CO_2 and other GHGs emissions, as well as to preserve soil health, biodiversity and limiting pollution from chemicals. We believe it is important to carry out large scale experiments with organic or other form of alternative-low impact agriculture practices to monitor and assess their pros and cons at different context and levels. We think this is a strategic investment that can result much more effective (and much less expensive and risky) than, for instance, engineering life.

However, while organic agriculture can offer an option to buy time while securing many beneficial services to the soil and environment sustainability, long term solutions concerning energy consumption and GHGs emissions from global society should be searched in different energy sources and/or, more probably, on reducing the demand side of energy issue, reshaping the structure and functioning of human societies.

References

Adam, D. (2001). Nutritionists question study of organic food. *Nature*, 412, 666.

Allison, F. E. (1973). *Soil organic matter and its role in crop production*. (Elsevier, Amsterdam)

Altieri, M. 2002. Agroecology: the science of natural resource management for poor farmers in marginal environments. *Agriculture, Ecosystems and Environment*, 93, 1–24.

Altieri, M. (1987). *Agroecology: The science of sustainable agriculture*. (Westview Press, Boulder)

Badgley, C., Moghtader, J., Quintero, E., Zakem, E., Chappell, J. M., Avilés-Vázquez, K., Samulon, A., & Perfecto, I. (2007). Organic agriculture and the global food supply. *Renewable Agriculture and Food Systems*, 22, 86–108.

Balfour, E. (1977). Towards a sustainable agriculture – The Living Soil. IFOAM conference in Switzerland in 1977, Retrieved 30 July 30 2007 from http://soilandhealth.org/01aglibrary/010116Balfourspeech.html

Badger, P. C. (2002). Ethanol From Cellulose: A General Review. (In J. Janick, & A. Whipkey (Eds.), *Trends in new crops and new uses* (pp. 17–21). ASHS Press, Alexandria, VA).

Bengtsson, J., Ahnstrom, J., & Weibull, A-C. (2005). The effects of organic agriculture on biodiversity and abundance: a meta-analysis. *Journal of Applied. Ecology,* 42, 261–269.

Bland, W. L. (1999). Toward integrated assessment in agriculture. Agricultural Systems, 60(3), 157–167.

Brandt, K., & Mølgaard, J.P. (2006). Food quality. (In P., Kristiansen, A., Taji, & J., Reganold (Eds), *Organic agriculture. A global perspective*. (pp. 305–328) CSIRO Publishing, Collingwood, Australia)

Brandt, K., & Mølgaard, J. P. (2001). Organic agriculture: does it enhance or reduce the nutritional value of plant foods? *Journal of the Science of Food and Agriculture*, 81, 924–931.

Brussaard, L., de Ruiter, P. C., & Brown, G. G. (2007). Soil biodiversity for agricultural sustainability. *Agriculture, Ecosystems and Environment*, 121, 233–244.

Carter, V. G., & Dale, T. (1975). *Topsoil and civilization*. (Univ. of Oklahoma Press, Revised edition)

Castellini, C., Bastianoni, S., Granai, C., Dal Bosco, A., & Brunetti, M. (2006). Sustainability of poultry production using the emergy approach: Comparison of conventional and organic rearing systems. *Agriculture, Ecosystems and Environment*, 114, 343–350.

Cassman, K. (2007). Editorial response by Kenneth Cassman: can organic agriculture feed the world—science to the rescue? *Renewable Agriculture and Food Systems*, 22(2), 83–84.

Codex Alimentarius. (2004). Guidelines for the production, processing, labelling and marketing of organically produced foods (GL 32 – 1999, Rev. 1 – 2001) Retrieved July 25 2007 from http://www.codexalimentarius.net/web/standard_list.do?lang=en

Conford, P. (2001). *The origins of the organic movement*. (Floris Books, Glasgow, Great Britain).

Coleman, D. C., Crossley, D.A.Jr., & Hendrix, P.F. (2004). *Fundamentals of Soil Ecology*. (Second Edition, Academic Press, Amsterdam)

Cormack, W. F. (2000). *Energy use in organic farming systems.* Final report for project OF0182 for the Department for Environment, Food and Rural Affairs. Retrieved August 12 2007 from http://orgprints.org/8169/01/OF0182_181_FRP.pdf

Conway, G. R. (1987). The properties of agroecosystems. *Agricultural Systems,* 24, 95–117.

Courville, S. (2006). Organic standards and certification. (In P. Kristiansen, A. Taji, J. & J. Reganold (Eds), *Organic agriculture. A global perspective.* (pp. 201–220) CSIRO Publishing, Collingwood, Australia)

Crutzen, P. J., Mosier, A. R., Smith, K. A., & Winiwarter, W. (2007). N_2O release from agro-biofuel production negates global warming reduction by replacing fossil fuels. *Atmospheric Chemistry and Physics.,* 7, 11191–11205. Retrieved September 15 2007 from http://www.atmos-chem-phys-discuss.net/7/11191/2007/acpd-7-11191-2007.pdf

Delate, K, Duffy, M., Chase, C., Holste, A., Friedrich, H., & Wantate, N. (2003). An economic comparison of organic and conventional grain crops in a Long-Term Agroecological Research (LTAR) Site in Iowa. *Journal of Alternative Agriculture,* 18(2), 59–69.

Dalgaard, T., Hutchings, N.J., & Porter, J.R. (2003). Agroecology, scaling and interdisciplinarity. *Agriculture Ecosystems and Environment,* 100, 39–51.

Dalgaard, T., Halberg, N., & Porter, J. R. (2001). A model for fossil energy use in Danish agriculture used to compare organic and conventional farming. *Agriculture, Ecosystems and Environment,* 87, 51–65.

DEFRA (Department for Environment Food and Rural Affairs – UK) (2005). *The validity of food miles as an indicator of sustainable development.* Report number ED50254. Retrieved September 16 2007 from http://statistics.defra.gov.uk/esg/reports/foodmiles/default.asp.

De Oliveira, M. E. D., Vaughan, B. E., & Rykiel, Jr. E. J. (2005). Ethanol as fuel: Energy, carbon dioxide balances, and ecological footprint. *BioScience,* 55(7), 593–602.

Diamond, J. (2005). *Collapse: How societies choose to fail or succeed.* (Penguin, London)

Drinkwater, L. E., Wagoner, P., & Sarrantonio, M. (1998). Legume-based cropping systems have reduced carbon and nitrogen losses. *Nature,* 396, 262–265.

Dritschillo, W., & Wanner, D. (1980). Ground beetle abundance in organic and conventional corn fields. *Environmental Entomology,* 9, 629–631.

Dunlap, R. E., Beus, C. E., Howell, R., &Waud, J. (1992). What is sustainable agriculture? An empirical examination of faculty and farmer definitions. *Journal of Sustainable Agriculture,* 3(1), 5–39.

EC (European Commission) (2007). Council Regulation (EC) No 834/2007, of 28 June 2007 on organic production and labelling of organic products and repealing Regulation (EEC) No 2092. Retrieved July 30 2007 from http://eur-lex.europa.eu/LexUriServ/site/en/oj/2007/l_189/l_18920070720en00010023.pdf

EC (European Commission) (2005). Annex to the communication from the Commission biomass action plan impact assessment. Retrieved 20 July 2007 from http://ec.europa.eu/energy/res/biomass_action_plan/doc/sec_2005_1573_impact_assessment_en.pdf

Edens, T. (1984). *Sustainable agriculture and integrated farming systems.* (Michigan State Univ. Pr.)

EEA (European Environmental Agency) (2006). How much bioenergy can Europe produce without harming the environment? Report No 7/2006 Roland Siemons, Martijn Vis, Douwe van den Berg (BTG) Ian Mc Chesney MBA, Mark Whiteley MSc (ESD). Retrieved June 15 2007 from 1http://reports.eea.europa.eu/eea_report_2006_7/en/eea_report_7_2006.pdf

EEC (European Economic Community) (1991). Council Regulation (EEC) No 2092/91 of 24 June 1991 on organic production of agricultural products and indications referring thereto on agricultural products and foodstuffs (OJ L 198, 22.7.1991, p. 1)

Fargione, J., Hill, J., Tilman, D., Polasky, S., & Hawthorne, P. (2008). Land Clearing and the Bio-fuel Carbon Debt. Science, www.sciencexpress.org/7 February 2008/Page 1/10.1126/science. 1152747

FAO (2004). The scope of organic agriculture, sustainable forest management and eco-forestry in protected area management. (FAO, Rome) Retrieved August 15 2007 from http://www.fao.org/docrep/007/y5558e/y5558e00.htm#toc

FAO (2003).*World agriculture: towards 2015/2030 – An FAO perspective*. (FAO, Rome) Retrieved
    15 August 15 2007 from http://www.fao.org/DOCREP/005/Y4252E/Y4252E00.HTM

FAO (2002). *Organic agriculture, environment and food security*. Environment and Natu-
    ral Resources Service Sustainable Development Department. Retrieved July 20 2007 from
    http://www.fao.org/DOCREP/005/Y4137E/y4137e00.htm#TopOfPage

Feenstra, G., Ingels, C., & Campbell, D. (1997). *What is sustainable agriculture?* Retrieved July
    30, 2007 from University of California Sustainable Agriculture Research and Education Pro-
    gramme Web Site: http://www.sarep.ucdavis.edu/concept.htm

Foereid, B., & Høgh-Jensen, H. (2004). Carbon sequestration potential of organic agriculture in
    northern Europe – a modelling approach. *Nutrient Cycling in Agroecosystems*, 68, 13–24.

Foster, C., Green, K., Bleda, M., Evans, B., Flynn, A., & Myland, J. (2006). *Environ-
    mental impact of food production and consumption*. A report to the Department of
    Environment, Food and Rural Affair (DEFRA). Manchester Business School, DEFRA,
    London. Retrieved July 15 2007 from http://www.defra.gov.uk/science/project_data/Document
    Library/EV02007/EV02007_4601_FRP.pdf

Genghini, M., Gellini, S., & Gustin, M. (2006). Organic and integrated agriculture: the effects
    on bird communities in orchard farms in northern Italy. *Biodiversity and Conservation*, 15,
    3077–3094.

Giampietro, M. (2004). *Multi-scale integrated analysis of agroecosystems*. (CRC Press, Boca
    Raton, London)

Giampietro, M., Bukkens S. G. F., & Pimentel, D. (1994). Models of energy analysis to assess the
    performance of food systems. *Agricultural Systems*, 45(1), 19–41.

Gliessmann, S. R. (2000). *Agroecology: Ecological processes in sustainable agriculture*. (Lewis
    Publisher, Boca Raton, New York)

Gliessmann, S. R. (Ed.) (1990). *Agroecology: Researching the ecological basis for sustainable
    agriculture*. (Springer-Verlag, New York)

Goklany, I. M. (2002). The ins and outs of organic farming. *Science*, 298, 1889.

Goldemberg, J. (2007). Ethanol for a sustainable energy future. *Science*, 315, 808–810.

Gold, M. V., & Gates, J. P. (2007). Tracing the evolution of organic/sustainable agriculture: A se-
    lected and annotated bibliography, Beltsville, Md.: United States Dept. of Agriculture, National
    Agricultural Library, [1988] ; updated and expanded, May 2007 Retrieved July 25 2007 from
    http://www.nal.usda.gov/afsic/pubs/tracing/tracing.shtml

Gomiero, T., Giampietro, M., & Mayumi, K. (2006). Facing complexity on agro-ecosystems: a
    new approach to farming system analysis. *International Journal of Agricultural Resources,
    Governance and Ecology*, 5(2/3), 116–144.

Gomiero, T., Giampietro, M., Bukkens, S. M., & Paoletti, G. M. (1997). Biodiversity use and
    technical performance of freshwater fish culture in different socio-economic context: China
    and Italy. *Agriculture, Ecosystems and Environment*, 62 (2,3), 169–185.

Grandy, A. S., & Robertson, G. P. (2007). Land-use intensity effects on soil organic carbon accu-
    mulation rates and mechanisms. *Ecosystems*, 10, 58–73.

Guthman, J. (2004). *Agrarian dreams: The paradox of organic farming in California*. (University
    of California Press, Los Angeles)

Haas, G., Wetterich, F., & Kopke, U. (2001). Comparing intensive, extensified and organic grass-
    land farming in southern Germany by process life cycle assessment. *Agriculture, Ecosystems
    and Environment*, 83, 43–53.

Haden, A. C. (2003). Emergy evaluations of Denmark and Danish agriculture assessing the limits
    of agricultural systems to power society. Ekologiskt Lantbrunknr, 37 March 2003. Retrieved
    July 24 2007 from http://www.cul.slu.se/information/publik/ekolantbruk37.pdf

Hansen, B., Fjelsted, H., Kristensen, E. S. (2001). Approaches to assess the environmental impact
    of organic farming with particular regard to Denmark. *Agriculture, Ecosystems and Environ-
    ment*, 83, 11–26.

Hansson, P-A., Baky, A., Ahlgren, S., Bernesson, S., Nordberg, A., Nore'n, O., & Pettersson, O.
    (2007). Self-sufficiency of motor fuels on organic farms – Evaluation of systems based on fuels
    produced in industrial-scale plants. *Agricultural Systems*, 94, 704–714.

Heaton, S. (2001). Organic farming, food quality and human health: A review of the evidence, (S(Association, Bristol, UK) Retrievd the 12 June 2007 from http://www.soilassociation.org/We SA/saweb.nsf/9f788a2d1160a9e580256a71002a3d2b/de88ae6e5aa94aed80256abd0037848($FILE/foodqualityreport.pdf

Heemsbergen, D. A., Berg, M. P., Loreau, M., van Hal, J. R., Faber, J. H., & Verhoef, H. 2004. Biodiversity effects on soil processes explained by interspecific functional dissimilari *Science*, 306, 1019–1020.

Heckman, J. (2006). A history of organic farming: Transitions from Sir Albert Howard's v in the soil to the USDA National Organic Program. *Wise Traditions in Food, Farming, a the Healing Arts,* winter 2006. Retrieved July 30 2007 from http://www.westonaprice.o farming/history-organic-farming.html

Hendrix, J. (2007). Editorial response by Jim Hendrix. *Renewable Agriculture and Food Systen* 22(2), 84–85.

Hill, J., Nelson, E., Tilman, D., Polasky, S., & Tiffany, D. (2006). Environmental, economic, a energetic costs and benefits of biodiesel and ethanol biofuels. *PNAS*, 103, 11206–11210.

Hillel, D. (1991). *Out of the earth: Civilization and the life of the soil.* (University of Califor Press).

Himmel, M. E., Ding, S-Y, Johnson, D. K., Adney, W. S., Nimlos, M. R., Brady, J. W., Foust, T. D. (2007). Biomass recalcitrance: Engineering plants and enzymes for biofuels p duction. *Science*, 315, 804–807.

Hoeppner, J., Hentz, M., McConkey, B., Zentner, R., & Nagy, C. (2006). Energy use and efficie in two Canadian organic and conventional crop production systems. *Renewable Agricult and Food Systems*, 21(1), 60–67.

Hole, D. G., Perkings, A. J., Wilson, J. D., Alexander, I. H., Grice, P. V., & Evans, A.D. (20(Does organic farming benefits biodiversity. *Biological Conservation*, 122, 113–130.

Holland, J. M. (2004). The environmental consequences of adopting conservation tillage in Eurc reviewing the evidence. *Agriculture, Ecosystems and Environment*, 103, 1–25.

Howard, A. (1943). *An agricultural testament.* (Oxford University Press, New York)

Hudson Institute (2007). "Organic Abundance" Report: Fatally Flawed. Retrieved Septembei 2007 from http://www.cgfi.org/cgficommentary/organic-abundance-report-fatally-flawed

IEA (International Energy Agency) (2002). Sustainable production of woody biomass for ene International Energy Agency (IEA), Retrieved July 30 2007 from http://www.ieabioene com/library/157_PositionPaper-SustainableProductionofWoodyBiomassforEnergy.pdf

Ikerd, J. E. (1993). The need for a system approach to sustainable agriculture. *Agriculture, Eco tems and Environment*, 46,147–160.

IPCC (Intergovernmental Panel on Climate Change) (2007). *Climate Change 2007: The Pl cal Science Basis. Contribution of Working Group I to the Fourth Assessment Report oj Intergovernmental Panel on Climate Change* [Solomon, S., D. Qin, M. Manning, Z. C M. Marquis, K.B. Averyt, M. Tignor and H.L. Miller (eds.)]. (Cambridge University P Cambridge, United Kingdom and New York, NY, USA) Retrieved June 15 2007 1 http://www.ipcc.ch/

Janzen, H. H. (2004). Carbon cycling in earth systems—a soil science perspective. *Agricul Ecosystems and Environment*, 104, 399–417.

Jørgensen, U., Dalgaar, T., & Kristensen, E. S. (2005). Biomass energy in organic farming – potential role of short rotation Coppice. *Biomass and Bioenergy*, 28(2), 237–248.

Kasperczyk, N., & Knickel, K. (2006). Environmental impact of organic agriculture P., Kristiansen, A., Taji, J., Reganold, J., (Eds), *Organic agriculture. A global perspec* (pp. 259–294) CSIRO Publishing, Collingwood, Australia)

Keeney, D. (2007). Sustainable biofuels: A new challenge for the Leopold Center. Leopold ter for Sustainable Agriculture 2007 Leopld Letters, Spring. Retrieved July 30 2007 http://www.leopold.iastate.edu/pubs/nwl/2007/2007-1-leoletter/anniversary.htm

Kirschenmann, F. (2004). A brief history of sustainable agriculture. *The Networker*, vol. 9, i March 2004. Retrieved July 30 2007. http://www.sehn.org/Volume_9-2.html

Krebs, J. R., Wilson, J. D., Bradbury, R. B., & Siriwardena, G. M. (1999).The second Silent Spring? *Nature*, 400, 611–612.

Kristiansen, P. (2006). Overview of organic agriculture. (In P. Kristiansen, A. Taji, & J. Reganold (Eds.), *Organic agriculture. A global perspective*. (pp. 1–24) CSIRO Publishing, Collingwood, Australia)

Kristiansen, P., Taji, A., & Reganold, J. (Eds) (2006). *Organic agriculture. A global perspective*. (CSIRO Publishing, Collingwood, Australia)

Koepf, H. H. (2006). *The biodynamic farm*. SteinerBooks, Dulles, VA.

Koepf, H. H., Schaumann, W., & Haccius, M. (1996). *Biologisch- Dynamische Landwirtschaft Eine Einführung*. Ulmer (Eugen, Germany) in German. (trad. Biodynamic agriculture)

Kotschi., J., & Müller-Sämann, K. (2004). The Role of Organic Agriculture in Mitigating Climate Change. (IFOAM – Bonn) Retrievend June 15 2007 from http://www.ifoam.org/press/positions/pdfs/Role_of_OA_migitating_climate_change.pdf

Koutinas, A. A., Wang, R.-H., & Webb, C. (2007). The biochemurgist: Bioconversion of agricultural raw materials for chemical production. *Biofuels, Bioprod. Bioref.*, 1, 24–38.

Kropff, M. J., Bouma, J., & Jones, J. W. (2001). Systems approaches for the design of sustainable agro-ecosystems. *Agricultural Systems*, 70(2–3): 369–393

Lange, J.-P. (2007). Lignocellulose conversion: an introduction to chemistry, process and economics. *Biofuels, Bioprod. Bioref.*, 1, 39–48.

Lal, R. (2004). Soil carbon sequestration impact on global climate and food security. *Science*, 304, 1623–1627.

Lampkin, N. (2002). *Organic Farming* (revised edition). (Old Pond Publishing, Suffolk, UK)

Lynd, L. R., Cushman, J. H., Nichols, R. J., & Wyman, C. E. (1991). Fuel ethanol from cellulosic biomass. *Science*, 251, 1318–1323.

Lockeretz, W. (1983). Energy price increases: How strong an incentive for decreasing energy use in agriculture? *Biological Agriculture and Horticulture*, 1, 255–267.

Lockeretz, W., Shearer, G., & Kohl, D. H. (1981). Organic farming in the Corn Belt. *Science*, 211, 540–546.

Lowenberg-DeBoer, J. (1996). Precision farming and the new information technology: implications for farm management, policy, and research: Discussion. *American Journal of Agricultural Economics*, 78(5), 1281–1284.

Lotter, D. W. (2003). Organic agriculture. *Journal of Sustainable Agriculture*, 21(4), 59–128.

Lotter, D. W., Seidel, R., & Liebhart, W. (2003). The performance of organic and conventional cropping systems in an extreme climate year. *American Journal of Alternative Agriculture*, 18(3),146–154.

Lu, C., Toepel, K., Irish, R., Fenske, R.A., Barr, D.B., & Bravo, R. (2006). Organic diets significantly lower children's dietary exposure to organophosphorus pesticides. *Environmental Health Perspectives*, 114(2), 260–263.

Mäder, P., Fließbach, A., Dubois, D., Gunst, L., Fried, P., & Niggli, U. (2002). Soil fertility and biodiversity in organic farming. *Science*, 296, 1694–1697.

Mäder, P., Fließbach, A., Dubois, D., Gunst, L., Fried, P., & Niggli, U. (2002). The ins and outs of organic farming. *Science*, 298, 1889–1890.

Mason, J. (2003). *Sustainable Agriculture*. (CSIRO Publishing; 2nd edition)

Matson, P. A., Parton, W. J., Power, A. G., Swift, M. J. (1997). Agricultural intensification and ecosystem properties. *Science*, 277, 504–509.

Maud, S. (2007). Sustainability of poultry production. *Agriculture, Ecosystems and Environment*, 120, 470–471.

McDonald, A. J., Hobbs, P. R., & Riha, S. J. (2005). Does the system of rice intensification outperform conventional best management? A synopsis of the empirical record. *Field Crops Research*, 96,(1), 31–36.

Millennium Ecosystem Assessment (2005). *Ecosystems and human well-being: Synthesis*. (Island Press, Washington, DC)

Mollison, B., & Holmgren, D. (1978). *Permaculture one: A perennial agriculture for human settlements*. (Trasworld Publishers, London, UK)

National Organic Standards Board (2007). Organic definition passed by the NOSB at its Apr 1995 meeting in Orlando, FL. Retrieved July 30 2007 from the Organic Trade Associatio http://www.ota.com/definition/nosb.html

National Research Council (1998). *Precision agriculture in the 21st century: Geospatial and ir formationtechnologies in crop management.* (National Academies Press)

Netuzhilin, I. Cerda, H., López-Hernández, D., Torres, F., Chacon, P., & Paoletti. M. G. (1999 Biodiversity tools to evaluate sustainability in savanna-forest ecotone in the Amazoni (Venezuela). (In: M.V., Reddy (ed), Management of tropical agroecosystems and the benefici soil biota. (pp. 291–352) Science Publishers Inc., Enfield, New Hampshire)

NRC (National Research Council) (1986). *Alternative agriculture.* (National Academy Pres Washington, D.C.)

Odum, H. T. (1996). *Environmental accounting: Emergy and environmental decision makin* (Wiley, New York)

Odum, H. T. (1988). Self-Organization, tranformity, and information. *Science,* 242, 1132–1139.

Paoletti, M. G. (2001). Biodiversity in agroecosystems and bioindicators of environmental healt (In M. Shiyomi & H. Koizumi) (Eds.), *Structure and function in agroecosystems design ar management.* (pp. 11–44) (CRC press, Boca Raton, FL, USA)

Paoletti, M. G., & Bressan, M. (1996). Soil invertebrates as bioindicators of human disturbanc *Critical reviews in plant sciences,* 15(1), 21–62.

Paoletti, M. G., & Pimentel, D. (2000). Environmental risks of pesticides versus genetic enginer ing for agricultural pest control. *J. Agricultural and Environmental Ethics,* 12(3), 279–303.

Paoletti, G. M., & Pimentel, D. (1992). *Biotic diversity in agroecosystems.* (Elsevier, Amsterdar

Paoletti, M. G., Stinner, B. R., & Lorenzoni, G. G. (Eds.), (1989). *Agriculture, ecology and en ronment.* (Elsevier, Amsterdam)

Paoletti M. G., Tsitsilas A., Thomson L. J., Taiti S., & Umina, P. A. (2008). The flood bug, *Ar traliodillo bifrons* (*Isopoda: Armadillidae*): A potential pest of cereals in Australia? *Appl. Soil Ecology,* 39(1), 76–83.

Paoletti, M. G., Favretto, M. R., Marchiorato, A., Bressan, M., & Babetto, M. (1993). Biodivers in pescheti forlivesi. In: Paoletti M.G. *et al. Biodiversità negli Agroecosistemi.* (Osservato Agroambientale, Centrale Ortofrutticola, Forlì), pp. 20–56. (in Italian)

Paoletti M.G., Giampietro, M., Han, C-R., Pastore, G., Bukkens, S. G.F., & Baudry, J. (Er (1999). Agricultural intensification and sustainability in PR China. *Critical Reviews of Pl Sciences,* 18(3), 257–487.

Perrings, C., Jackson, L., Bawa, K., Brussaard, L., Brush, S., Gavin, T., Papa, R., Pascual, U., & Ruiter, P. (2006). Biodiversity in agricultural landscapes: saving natural capital without los interest. *Conservation Biological,* 20, 263–264.

Pete, S., Olof, A., Thord, K., Paula, P., Kristiina, R., Mark, R., & Bas, W. (2005). Carbon seques tion potential in European croplands has been overestimated. *Global Change Biology,* 11(2153–2163.

Pimentel, D. (2007). Soil erosion. (In D. Pimentel, & M. Pimentel (Eds.) *Food, energy, and soci Third edition.,* (201–214) CRC Press)

Pimentel, D., & Pimentel., M. (2007a). *Food, energy, and society: Third edition.* (CRC Press, F Raton, FL)

Pimentel, D., & Pimentel., M. (2007b). *Transport of agriculture supplies and food.* (In D. Pime & M. Pimentel (Eds.) *Food, energy, and society: Third edition.,* (257–259) CRC Press)

Pimentel, D. (2006a). Impacts of organic farming on the efficiency of energy use in agricu an organic center state of science review. (The Organic Center), Retrieved September 15 ; from http://www.organic-center.org/reportfiles/ENERGY_SSR.pdf

Pimentel., D. (2006b). Soil erosion: A food and environmental threat. *Environment, Develop. and Sustainability,* 8(1), 119–137.

Pimentel, D. (2003). Ethanol fuels: Energy balance, economics, and environmental impact negative. *Natural Resources Research,* 12(2), 127–134.

Pimentel, D. (1993). Economic and energetics of organic and convention farming. *Journal of. cultural and Environmental Ethics,* 6, 53–60.

Pimentel, D. (1991). Ethanol fuels: Energy security, economics, and the environment. *Journal of Agricultural and Environmental Ethics*, 4(1), 1–13.

Pimentel, D. (1989). Energy flow in food system. (In D. Pimentel, & C. W. Hall (Eds.). *Food and natural resources*. (pp. 1–24) Academic press, New York)

Pimentel, D., & Patzek, T. (2005). Ethanol production using corn, switchgrass, and wood: biodiesel production using soybean and sunflower. *Natural Resources Research*, 14(1), 65–76.

Pimentel, D., & Kounang, N. (1998). Ecology of soil erosion in ecosystems. *Ecosystems*, 1, 416–426.

Pimentel, D., Berardi, G., & Fast, S. (1983). Energy efficiency of farming systems: organic and conventional agriculture. *Agriculture Ecosystems and Environment*, 9, 359–337.

Pimentel, D., Hepperly, P., Hanson, J., Douds, D., & Seidel, R. (2005). Environmental, energetic, and economic comparisons of organic and conventional farming systems. *BioScience*, 55(7), 573–582.

Pimentel, D., Hurd, E., Bellotti, A. C., Forster, M. J., Oka, I. N., Sholes, O.D., & Whitman, R. J. (1973). Food production and the energy crisis. *Science*, 182, 443–449.

Pimentel, D., Harvey, C., Resosudarmo, P., Sinclair, K., Kurz, D., McNair, M, Crist, S., Sphpritz, L., Fitton, L., Saffouri, R., & Blair, R. (1995). Environmental and economic costs of soil erosion and conservation benefits. *Science*, 267,1117–23.

Pimentel, D., Moran, M. A., Fast, S., Weber, G., Bukantis, R., Balliett, L., Boveng, P., Cleveland, C., Hindman, S., & Young, M. (1981). Biomass energy from crop and forest residues. *Science*, 212, 1110–1115.

Poincelot, R. P. (1986). *Towards a more sustainable agriculture*. (AVI Publishing co. Co.)

Pointing, C. (2007). *A new green history of the world*. (Vintage books, London)

Pretty, J. (Eds.) (2005). *The earthscan reader in sustainable agriculture*. (Earthscan Publisher, London)

Pretty, J., & Hine, R. (2001). *Reducing food poverty with sustainable agriculture: a summary of new evidence*. Final report from the 'SAFE World' Research Project, University of Essex. Retrieved June 20 2007 from http://www.essex.ac.uk/ces/esu/occasionalpapers/SAFE%20FINAL%20-%20Pages1-22.pdf

Pretty, J. N., Morison, J. I. L., & Hine, R. E. (2003). Reducing food poverty by increasing agricultural sustainability in developing countries. *Agriculture, Ecosystems and Environment*, 95, 217–234.

Pretty, J. N., Ball, A. S., Xiaoyun, L., & Ravindranath, N. H. (2002). The role of sustainable agriculture and renewable-resource management in reducing greenhouse-gas emissions and increasing sinks in China and India. *Philosophical transactions of the Royal Society of London. Series B, Biological sciences A*, 360, 1741–1761.

Pretty, J. N., Ball, A. S., Lang, T., & Morison, J. I. L. (2005). Farm costs and food miles: An assessment of the full cost of the UK weekly food basket. *Food Policy*, 30, 1–19.

Rasmussen, P. E., Goulding, K. W. T., Brown, J. R., Grace, P. R., Janzen, H. H., & Körschens, M. (1998). Long-Term Agroecosystem Experiments: Assessing agricultural sustainability and global change. *Science*, 282, 893–896.

Reganold, J. P. (1995). Soil quality and profitability of biodynamic and conventional farming systems: a review. *American Journal of Alternative Agriculture*, 10(1), 36–46.

Reganold, J., Elliott, L., & Unger, Y. (1987). Long-term effects of organic and conventional farming on soil erosion. *Nature*, 330, 370–372.

Reganold, J., Glover, J., Andrews, P., & Hinman, H. (2001). Sustainability of three apple production systems. *Nature*, 410, 926–929.

Refsgaard, K., Halberg, N., & Kristensen, E. S. (1998). Energy utilization in crop and dairy production in organic and conventional livestock production systems. *Agricultural Systems*, 57(4), 599–630.

Rigby, D., & Cáceras, D. (2001). Organic farming and the sustainability of agricultural systems. *Agricultural Systems*, 68, 21–40.

Robertson, G. P., Paul, E. A., & Harwood, R. R. (2000). Greenhouse gases in intensive agriculture: Contributions of individual gases to the radiative forcing of the atmosphere. *Science*, 289, 1922–1925.

Rodale, J. I. (1945). *Paydirt: Farming & gardening with composts.* (Devin-Adair Co., New York)

Roschewitz, I., Gabriel, D., Tscharntke, T., & Thies, C. (2005). The effects of landscape complexity on arable weed species diversity in organic and conventional farming. *Journal of Applied Ecology*, 42, 873–882.

Roviglioni, R. (2005). Bio consuma meno energia. *BioAgricoltura*, marzo/aprile: 5-7. (in Italian)

Searchinger, T., Heimlich, R., Houghton, A., Dong, F., Elobeid, A., Fabiosa, J., Tokgoz, S., Hayes, D., & Yu, T-H., (2008). Use of U.S. Croplands for Biofuels Increases Greenhouse Gases Through Emissions from Land Use Change www.sciencexpress.org/7 February 2008/Page1/ 10.1126/science.1151861

Service, R. F. (2007). Biofuel researchers prepare to reap a new harvest. *Science*, 315, 1488–1491.

Siegrist, S., Schaub, D., Pfiffner, L., & Mäder, P. (1998). Does organic agriculture reduce soil erodibility? The results of a long-term field study on loess in Switzerland. *Agriculture, Ecosystem and Environment*, 69, 253–264.

Schlesinger, W. H. (1999). Carbon and agriculture: Carbon sequestration in soils. *Science*, 284, 2095.

Schlich, E., & Fleissner, U. (2005). The ecology of scale: Assessment of regional energy turnover and comparison with global food. *The International Journal of Life Cycle Assessment*, 10(3), 213–223.

Smil, V. (2001). *Feeding the world: A challenge for the twenty-first century.* (MIT Press, Cambridge, MC)

Smil, V. (1999). Crop residues: Agriculture's largest harvest. *BioScience*, 49(4):299–308.

Smith, P., Andrén, O., Karlsson,T., Perälä, P., Regina, K., Rounsevell, M., & Van Wesemael, B. (2005). Carbon sequestration potential in European croplands has been overestimated. *Global Change Biology*, 11(12), 2153–2163.

Smith, P., Martino, D., Cai, Z., Gwary, D., Janzen, H. H., Kumar, P., McCarl, B., Ogle, S., O'Mara, F., Rice, C., Scholes, R. J., Sirotenko, O., Howden, M., McAllister, T., Pan, G., Romanenkov, V., Schneider, U., Towprayoon, S., Wattenbach, M., & Smith, J. U. (2008): Greenhouse gas mitigation in agriculture. *Philosophical Transactions of the Royal Society of London, B*, 363, 789–813.

Smolik, J. D., Dobbs, T. L., & Rickerl, D. H. (1995). The relative sustainability of alternative, conventional and reduced-till farming system. *American Journal of Alternative Agriculture*, 10(1), 25, 25–35.

Soil Association (2004). Towards a UK strategy for biofuels – Soil Association response to Department of Transport consultation, July 2004. Retrieved July 30 2007 from http://www.soilassociation.org/web/sa/saweb.nsf/b0062cf005bc02c180256a6b003d987f/ 5401483e80739c88802570cb005919a6?OpenDocument

Solomon, B. D., Barnes, J. R., & Halvorsen, K. E. (2007). Grain and cellulosic ethanol: History, economics, and energy policy. *Biomass and Bioenergy*, 31, 416–425.

Srinivasan, A. (Ed.) (2006). *Handbook of precision agriculture: Principles and applications.* (Food Products Press)

Stanhill, G. (1990). The comparative productivity of organic agriculture. *Agriculture Ecosystems and Environment*, 30(1–2), 1–26.

Steinhart, J. S., & Steinhart, C. E. (1974). Energy Use in the U.S. Food System. *Science*, 184, 307–316.

Stephanopoulos, G. (2007). Challenges in engineering microbes for biofuels production. *Science*, 315, 801–804.

Stevens, T. O. (1997). Lithoautotrophy in the subsurface. *FEMS Microbiology Reviews*, 20, 327–337.

Stevens, T. O., & Mckinley, J. P. (1995). Lithoautotrophic microbia, ecosystems in deep basalt aquifers. *Science*, 270, 450–454.

Stockdale, E. A., Lampkin, N. H., Hovi, M., Keatinge, R., Lennartsson, E. K. M., Macdonald, D. W., Padel, S., Tattersall, F. H., Wolfe, M. S., & Watson, C. A. (2001). Agronomic and environmental implications for organic farming systems. *Advances in Agronomy*, 70, 261–327.

Stölze, M.,. Piorr, A. Häring, & Dabbert, S. (2000). The environmental impact of organic farming in Europe. In: *Organic Farming in Europe: Economics and Policy*. University of Hohenheim: Hohenheim, Germany. Retrieved July 30 2007 from http://orgprints.org/2366/02/Volume6.pdf

Thies, C., & Tscharntke, T. (1999). Landscape structure and biological control in agroecosystems. *Science*, 285, 893–895.

Tilman, D., Cassman, K. G., Matson, P. A., Naylor, R., & Polasky, S. (2002). Agricultural sustainability and intensive production practices. *Nature*, 418, 671–677.

Tilman, D., Fargione, J., Wolff, B., D'Antonio, C., Dobson, A., Howarth, R., Schindler, D., Schlesinger, W.H., Simberloff, D., & Swackhamer, D. (2001). Forecasting agriculturally driven global environmental change. *Science*, 292, 281–284.

Trewavas, A. (2001). Urban myths of organic farming. *Nature*, 410, 409–410.

Ulgiati, S. & Brown, M. T. (1998). Monitoring patterns of sustainability in natural and man-made ecosystems. *Ecological Modelling*, 108, 23–36.

Ulgiati, S., Odum, H.T., & Bastioni, S. (1994). Emergy use, environmental loading and sustainability: an emergy analysis of Italy. *Ecological Modelling*, 73, 215–268.

USDAa (2007). Background information. Retrieved July 30 2007 from http://www.ams.usda.gov/nop/FactSheets/Backgrounder.html

USDAb (2007). Organic production/Organic food: Information access tools. Retrieved July 30 2007 from http://www.nal.usda.gov/afsic/pubs/ofp/ofp.shtml.

USDAc (2007). Organic production. Retrieved July 30 2007 from http://www.ers.usda.gov/data/organic/

USDA (1990). Food, Agriculture, Conservation, and Trade Act of 1990 (FACTA), Public Law 101-624, Title XVI, Subtitle A, Section 1603, Government Printing Office, Washington, DC, 1990 NAL Call # KF1692.A31 1990.

Vasilikiotis, C. (2000). Can organic farming "feed the world"? Retrieved July 12 2007 from http://www.cnr.berkeley.edu/~christos/articles/CV-Organic%20Farming.pdf

Vogl, C. R., Kilcher, L., & Schmidt, H. (2005). Are standards and regulations of organic farming moving a way from small farmers' knowledge? *Journal of Sustainable Agriculture*, 26(1), 5–25.

Wardle, D. A., Bardgett, R. D., Klironomos, J. N., Setälä, H., van der Putten, W. H., & Wall, D. H. (2004). Ecological linkages between aboveground and belowground biota. *Science*, 304, 1629–1633.

Wes, J. (1980). *New roots for agriculture*. (University of Nebraska Press, Lincoln NE)

Willer, H., & Yussefi, M. (Eds.) (2006). The World of organic agriculture: Statistics and emerging trends. International Federation of Organic Agriculture Movements (IFOAM), Bonn Germany & Research Institute of Organic Agriculture FiBL, Frick, SwitzerlandSOEL-Survey 2006 http://www.soel.de/inhalte/publikationen/s/s_74_08.pdf

Winter, C. K., & Davis, S. F. (2006). Organic Foods. *Journal of Food Science*, 71(9), 117–124.

Wolf, S. A., & Allen, T. F. H. (1995). Recasting alternative agriculture as a management model: The value of adept scaling. *Ecological Economics*, 12, 5–12.

Worster, D. (2004). *Dust Bowl: The Southern Plains in the 1930s*. (Oxford Univ. Press, New York)

Ziesemer, J. (2007). *Energy use in organic food systems*. (FAO, Rome) Retrieved October 4 2007 from http://www.fao.org/docs/eims/upload/233069/energy-use-oa.pdf

Zimmer, G. F. (2000). *The biological farmer: A complete guide to the sustainable & profitable biological system of farming*. (Acres USA)

Chapter 18
Biofuel Production in Italy and Europe: Benefits and Costs, in the Light of the Present European Union Biofuel Policy

Sergio Ulgiati, Daniela Russi and Marco Raugei

Abstract We present and critically evaluate in this paper biofuel production options in Italy, in order to provide the reader with the order of magnitudes of the performance indicators involved. Also, we discuss biofuel viability and desirability at the European level, according to the recent EU regulations and energy policy decisions.

Fuels from biomass are most often proposed as substitutes for fossil fuels, in order to meet present and future shortages. Although the scientific literature on biofuel production techniques is abundant, comprehensive evaluations of large-scale biofuel production as a response to fossil energy depletion are few and controversial. The complexity of the assessments involved and the ideological biases in the research of both opponents and proponents of biofuel production make it difficult to weigh the contrasting information found in the literature. Moreover, the dubious validity of extrapolating results obtained at the level of an individual biofuel plant or farm to entire societies or ecosystems has rarely been addressed explicitly. After questioning the feasibility of a large-scale biofuels option based upon yields from case studies, we explore what are the constraints that affect the option even in the case of improved production performance.

Keywords Biomass · biofuels · carbon dioxide emissions · land requirement

✉ S. Ulgiati
Department of Sciences for the Environment, Parthenope University of Napoli,
Centro Direzionale – Isola C4, 80143 Napoli, Italy
e-mail: sergio.ulgiati@uniparthenope.it.

D. Russi
Autonomous University of Barcelona, Department of Economics and Economic History,
Edifici B, Campus de la UAB, 08193 Bellaterra (Cerdanyola del V.), Barcelona, Spain

M. Raugei
Department of Sciences for the Environment, Parthenope University of Napoli,
Centro Direzionale – Isola C4, 80143 Napoli, Italy

D. Pimentel (ed.), *Biofuels, Solar and Wind as Renewable Energy Systems*,
© Springer Science+Business Media B.V. 2008

18.1 Introduction

Two kinds of biofuels are generally considered available and feasible, i.e. bio-ethanol and biodiesel, although some expectations are also being placed on future bio-hydrogen generation. Bio-ethanol is obtained through fermentation and distillation of sucrose-producing plants (sugar cane, sugar beet) or cereals (mostly maize), and is usually mixed with petrol, either directly at the pump (splash blends), or before distribution (tailor blends). New production methods for bio-ethanol are also being developed, which make use of ligno-cellulosic biomass. This is however still at the R&D stage, and is currently referred to as a "second-generation" biofuel.

The second type of biofuel (named biodiesel or Vegetable Oil Methyl Esters – VOME) is produced from vegetable oils, and the crops that are most widely employed in Europe and in the USA are sunflower, rapeseed (canola) and soy. Palm trees are also a very promising raw material in tropical countries. Biodiesel is obtained through a chemical process called trans-esterification, which consists of making the vegetable oil react with methanol, thus yielding biodiesel and glycerine as co-products, and can only be mixed with fossil diesel.

Biofuels raise increasing hopes as substitutes for fossil fuels, and therefore as a contribution towards the reduction of the associated problems of greenhouse effect, high energy expenditures, and energy dependency. Moreover, it is often claimed that biofuels are not only "green" on a global scale (reducing of greenhouse effect) but also on a local scale (reducing urban pollution). Finally, biofuels are seen by many as a motor of rural development.

The European Union transportation sector is responsible for about 20% of total greenhouse gas emissions (AA. VV., 2005). The 2001 European Commission White Paper on Tranport Policy (AA. VV., 2001) estimated that between 1990 and 2010 European CO2 emissions from transportation sector are likely to increase up to 50%, reaching about 1.1 Gt and that road transportation is the main responsible for such a trend with 84% of total emissions (with minor shares from sea, railway and air transportation modalities). The same document claimed that "Reducing dependence on oil from the current level of 98%, by using alternative fuels and improving the energy efficiency of modes of transport, is both an ecological necessity and a technological challenge." Consistently with these estimates, the European Union published "An EU Strategy for Biofuels" (AA. VV., 2006) pointing out the need for a production of about 17.5 Mt of biofuels by the year 2010 and the allocation to energy cropping of an agricultural land between 5 and 10 Mha out of the total 140 Mha globally cropped within the EU Member States. By the year 2020 these values are expected to double.

In the year 2004 the EU biofuel production was 2.4 Mtoe, equal to the 0.8% of total consumption of liquid fuels within the EU. Bioethanol production was 0.5 Mtoe and biodiesel production 1.9 Mt. Total biomass use for energy within EU is about 40 Mtoe/year, out of which 18% in Finland, 17% in Sweden, 13% in Austria, 2% in Italy. In general, biomass use in Europe is still very small, in spite of claimed needs and expectations.

The European Directive 2003/30/EC established that the biofuel share of the energy use in the transport sector should reach 2% by 2005 and 5.75% by 2010

(EU, 2003). As a consequence, in Italy, the national law No. 81 of 11 March 2006 (dedicated to urgent norms for agriculture and agro-industry) required all fuel manufacturers to release to the market biofuels for at least 1% of the total energy content of the diesel and petrol sold in the previous year. Such percentage must be increased by one unit per year until the year 2010, in order to reach the 5.75% required by the European Union.

The latest European energy strategy, agreed in March 2007, increased the target to 10% within 2020.[1] These targets are quite ambitious considering that the actual biofuel share of the energy used for transport was only 0.9% in 2005.[2] Therefore, in order to get closer to the European requirements, an enormous effort is needed to spur a large-scale biofuel production.

In fact, biofuels are not competitive with fossil fuel-derived products if left to the free market. In order to make their price similar to those of petrol and diesel, they need to be subsidized by three means: (1) European agricultural subsidies, granted through the Common Agricultural Policy (CAP); (2) laws requiring a minimum percentage of biofuels in the fuels sold at the pump (biofuel obligations) and (3) de-fiscalization, since energy taxes make up for approximately half of the traditional fuel price.

These three political measures all need financial means, which are provided by the European Commission (agricultural subsidies), the governments (reduction in energy revenues), and car drivers (increase in the final fuel price). For this reason, there is compelling urge for an integrated analysis to discuss whether investing public resources in biofuels (and employing a large extension of agricultural land for that) is at all an advisable strategy. Such analysis should not be limited to energy yield or economic cost considerations, but also include relevant social and environmental factors.

In the following sections we will attempt an integrated assessment of the costs and benefits of a large scale biofuel sector in Europe, from environmental, social and economic points of view, and in the light of the results we will discuss whether promoting biofuels is really an advisable strategy. The starting point for such an assessment is a case study on biofuel production in Italy, given the present state of Italian agriculture and land use, from which larger-scale perspectives for Europe will be extrapolated.

18.2 To What extent Would a Large Scale Biofuel Production Really Replace Fossil Fuels?

18.2.1 Biomass and Biofuels

The terms biomass and biofuels are most often used as synonyms, as if liquid transportation fuels were the only way to extract energy out of photosynthetic substrates.

---

[1] It is to be noted that the European energy strategy places special emphasis on biofuels and indicates a specific target only for them. For the other renewable sources it limits itself to indicating an overall share of 20% on the total energy use.

[2] EUROSTAT data-base.

"Biomass" indicates all kinds of organic materials (mainly compounds of carbon, nitrogen, hydrogen and oxygen) derived from photosynthesis, including the whole metabolic chain through animals and human societies, yielding animal products and all kinds of waste materials from the use and processing of organic matter use. While it is not always true that the main value of biomass relies in its actual energy content, it cannot be disregarded that biomass can be converted to energy via several conversion patterns, including processing to biofuels (Fig. 18.1).

"Biofuels" in general indicates liquid products from biomass processing, to be used for transportation purposes. The same term sometimes also refers to gaseous compounds (biogas). It clearly appears that biomass (including waste materials) is the substrate generated via photosynthetic or metabolic processes, while biofuel is only one of the possible products of biomass processing (together with heat, biogas, electricity, chemicals). Misunderstanding the difference between biomass and bio-fuels leads to erroneous estimates about the potential of energy biomass in support to human activities. Processing biomass into biofuels requires specifically-grown substrates and several conversion steps, each one characterized by its own efficiency and conversion losses. Instead, direct biomass conversion to heat or waste biomass conversion to biogas is most often characterized by better performance, and is there-fore more likely to provide a contribution to at least a small fraction of the energy requirement in sectors other than transportation systems. A correct understanding of the role of biomass would help meeting the EU requirements for increased share of biomass energy, without competing with food production (cropping for energy) and wilderness conservation (energy forest plantations). In the following of the present

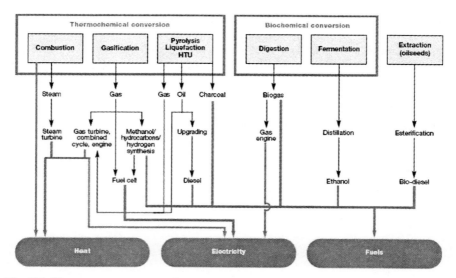

Fig. 18.1 Biomass to energy conversion patterns.
Source: Turkenburg et al., 2000

paper, however, we will limit our focus to biofuels from sugar, cellulose and seed-oil substrates, in order to check their availability, feasibility, and desirability.

18.2.2 An Overview of Results

The systems considered in the following data set are: (i) corn-bioethanol; (ii) sunflower-biodiesel; and (iii) fast-growing wood production for methanol. The productivity of biomass is based on average values found for the Italian agriculture. Conversion of these substrates to biofuel was estimated using data from commercially available technologies from literature.

To ensure that all significant input and output flows have been accounted for, a preliminary mass balance was performed, at the local and global scales. The local scale is the spatial scale within which the process actually occurs. Inputs accounted for at this scale are those that actually cross the local system boundaries. The global scale is the scale of the larger region (or the biosphere as well) within which all the processes that supply inputs to the ethanol system occur. For instance, the electricity input has no associated mass or emissions at the local scale, but the mass of fuel oil burnt and chemicals released for electricity production are accounted for on the global scale. The fuel oil input on the local scale requires an additional crude oil investment (and related emissions) on the global scale, for extraction, processing and transport. Local scale evaluation offers useful information about the investigated process and possible technological improvements. Global scale evaluation offers a better picture of the relationship between the investigated process and the environment (when considered both as a source and a sink), in order to understand sustainability.

Mass evaluation on the global scale was performed according to the Mass Flow Accounting method (Schmidt-Bleek, 1993; Fischer-Kowalski 1998; Bargigli et al., 2004). It provides indicators of the indirect demand for abiotic and biotic material input as well as water (the so-called material intensities) and quantify the contribution of the process to the withdrawal and depletion of material resources on the large scale. The amount of matter that is processed and diverted from its natural pattern was also assumed as a measure of potential environmental disturbance by some authors (Hinterberger and Stiller, 1998). A similar procedure for the calculation of direct and indirect energy flows has also been performed (Embodied Energy Analysis, Herendeen, 1998; 2004) in order to assess the energy cost of one unit of output (either substrate or biofuel) and the overall efficiency of biofuel production processes. From the embodied energy data and fuel used directly we also calculated the local- and global-scale airborne emissions. Finally, the Emergy Synthesis method (Odum, 1996; Brown and Ulgiati, 2004) was used to assess the ecological metabolism of each investigated pattern, based on the quantification of the environmental support needed for the process to occur.

Table 18.1a lists the main input flows to typical corn and sunflower productions in Italy, while the main input flows to industrial bioethanol and biodiesel production processes are shown in Table 18.1b. Table 18.2 compares the mass- and energy-based

Table 18.1a Input flows to corn and sunflower production (average estimates per hectare per year, local scale, Italy 2004) – Section 18.2.2

Description of flow	Units	Corn	Sunflower
Loss of topsoil (due to erosion)	t/ha/yr	20.0	17.2
Nitrogen fertilizer (N)	kg/ha/yr	169.4	103.2
Phosphate fertilizer (P2O5)	kg/ha/yr	82.0	86.0
Potash fertilizer (K2O)	kg/ha/yr	//	129.0
Insecticides, pesticides and herbicides	kg/ha/yr	5.4	4.3
Diesel	kg/ha/yr	150.0	117.0
Lubricants	kg/ha/yr	3.7	4.1
Petrol	kg/ha/yr	3.0	//
Water for irrigation	t/ha/yr	400.0	1283.0
Electricity for irrigation pumps	GJ/ha/yr	2.0	//
Diesel for irrigation pumps	kg/ha/yr	//	90.3
Steel for agricultural machinery (annual share)	kg/ha/yr	13.6	5.2
Seeds	kg/ha/yr	16.2	5.0
Human labor	hrs/ha/yr	25.0	32.7
Annual services (cost of input flows)	$/ha/yr	890.0	292.9
Additional input flows due to the harvest of 70% of residues (increased soil erosion and water use are not accounted for)			
Nitrogen harv. in residues	kg/ha/yr	78.8	50.0
Phosphorus harv. in resid.	kg/ha/yr	18.2	25.0
Potash harvested in residues	kg/ha/yr	//	55.6
Diesel for residues	kg/ha/yr	9.0	41.3
Machinery for residues (annual share)	kg/ha/yr	2.6	0.6
Labor	hrs/ha/yr	2.7	1.0
Main output flows			
Seeds, dry matter	t/ha/yr	6.1	1.8
Residues in field as such, dry matter	t/ha/yr	4.6	2.6

indicators calculated for bioethanol, biodiesel and biomethanol, under the following assumptions:

a. Use of 70% of residues as process energy source (the remaining 30% being left in field) and credit to DDGS and seed oil cakes equal to their replacement value, i.e. the energy value of the substitute product replaced in animal nutrition.
b. Use of 70% of residues as process energy source (the remaining 30% being left in field), but with no energy credit for animal feed replacement.
c. No residues as process energy source, but energy credit for animal feed replacement.
d. No residues as process energy source and no energy credit for animal feed replacement.

Overall indicators of material demand may appear larger than expected. This is an outcome of the adopted large-scale approach. For example, 1 g of processed iron requires about 4 to 5 g of iron ore plus other biotic and abiotic materials (including large amounts of water) that are directly and indirectly involved in the process.

Table 18.1b Input flows to industrial bioethanol and biodiesel production (average estimates per hectare per year, local scale, Italy 2004)–Section 18.2.2

Description of flow	Units	Bioethanol	Biodiesel
Dry grains to be converted	t/ha/yr	6.1	1.8
Residues in field as such, dry matter	t/ha/yr	4.6	2.6
Steel for transp. machinery (annual share)	kg/ha/yr	2.4	0.3
Diesel for transport of seeds to plant	kg/ha/yr	3.0	0.9
Steel for plant machinery (annual share)	kg/ha/yr	44.1	4.1
Cement in plant construction (annual share)	kg/ha/yr	78.4	35.2
Energy for hot water/steam generation (assuming partial use of agricultural residues)	GJ/ha/yr	0.1	2.3
Process electricity	GJ/ha/yr	2.4	0.3
Process and cooling water	t/ha/yr	16.2	//
Yeast	kg/ha/yr	5.1	//
Petrol (denaturant)	kg/ha/yr	11.1	//
Ammonia (from natural gas)	g/ha/yr	35.6	//
Exane for oil extraction	kg/ha/yr	//	1.2
Methanol for blending with seed oil	kg/ha/yr	//	87.1
Lime (calcium oxide)	g/ha/yr	9.3	//
NaCl	kg/ha/yr	4.6	//
Enzymes (alpha-amylase)	kg/ha/yr	9.1	//
Sludge polymer	g/ha/yr	93.7	//
BFW Chemicals	g/ha/yr	234.2	//
Labor for plant construction and operation	hrs/ha/yr	3.2	0.8
Annual capital cost and services	$/ha/yr	222.4	238.6
Main output flows			
Fuel product (Ethanol /biodiesel)	t/ha/yr	2.0	0.9
Feedstock product (DDGS/seed cake)	t/ha/yr	2.2	1.3
Glicerin	kg/ha/yr	//	87.1

The same holds for electricity, fuels, and fertilizers. Furthermore, since the mass of biofuels is always much lower than the mass relative to the processed substrate, the large scale assessment increases the value of all indicators per unit of net product, as clearly shown in Table 18.2. Water appears to be the dominant (and maybe limiting) factor, as will be discussed later on, although abiotic inputs as well as disaggregated data about fertilizers and pesticides are also sources of concern.

The overall energy advantage, on a purely thermodynamic level, is indicated by the output/input energy ratio, also expressed in Table 18.2 as a crude oil equivalent cost per unit of output. First of all, the increase of the unit energy cost (in terms of oil equivalent per gram of product) from the production of substrate to the production of the fuel is remarkable for all the crops considered. This indicates an energy bottleneck (and a significant energy loss) in the conversion step from substrate to fuel. Producing the substrate provides a concentration of net (photosynthetic) energy, while converting it to biofuel erodes most of the initial energy availability. The energy "gain" of agricultural substrate production ranges approximately from 2 to 4 (Table 18.2), whereas it drops down to about 1 (and less) after the conversion to biofuel. Finally, the best net-to-gross ratio is obtained by: ethanol in the option (a); methanol in option (b); and biodiesel in option (c). Anyway, all these values are in the range 1.1–1.5, which is not enough to ensure a self-sufficient production

Table 18.2 Global matter and energy flows and ratios in selected substrate and biofuel production in Italy (average values, 2004) – Section 18.2.2

Substrate production (wet matter)		Corn	Sunflower	Wood
Oil equivalent demand per unit of substrate	g/g	0.09	0.24	0.05
Fertilizers and pesticides demand per unit of substrate	g/g	0.04	0.15	0.03
Material intensity, abiotic factor	g/g	1.73	5.33	n.a.
Material intensity, biotic factor	g/g	0.09	0.31	n.a.
Material intensity, water factor	g/g	1238.20	1128.74	n.a.
Soil erosion	g/g	2.26	7.82	n.a.
Labor and services demand per unit of substrate	hrs/kg	0.003	0.015	0.002
Land demand per unit of substrate	m²/kg	1.32	4.55	0.003
Economic cost per unit of substrate	$/kg	0.16	0.13	n.a.
Biofuel production		**Ethanol**	**Biodiesel**	**Methanol**
Oil equivalent demand per unit of biofuel	g/g	0.60	0.82	0.108
Fertilizers and pesticides demand per unit of biofuel	g/g	0.15	0.37	0.114
Material intensity, abiotic factor	g/g	7.45	13.97	n.a.
Material intensity, biotic factor	g/g	0.35	0.79	n.a.
Material intensity, water factor	g/g	4811.21	2852.61	n.a.
Soil erosion	g/g	8.78	19.74	n.a.
Labor demand per unit of biofuel	hrs/kg	0.02	0.04	0.01
Land demand per unit of biofuel	m²/kg	5.10	11.48	12.6
Net energy yield	MJ/Ha	1.89E+04	4.88E+03	1.40E+03
Net energy return per hour of applied labor	MJ/hr	613.55	145.77	133.08
Economic cost per unit of biofuel	$/kg	0.50	0.61	n.a.
Waste and releases				
CO_2 released per unit of substrate	g/g	0.32	0.98	0.38
CO_2 released per unit of biofuel	g/g	2.02	3.21	1.54
Industrial wastewater released per unit of biofuel	g/g	9.08	n.a.	n.a.
Energy efficiency		**Corn**	**Sunflower**	**Wood**
Energy output/(direct and indirect) energy input for substrate		3.82	2.59	4.24
Energy output/(direct and indirect) energy input for biofuel		**Ethanol**	**Biodiesel**	**Methanol**
(a) Use of residues as energy source, credit for feedstock		1.50	1.21	(*)
Net-to-gross energy ratio		**0.33**	**0.17**	(*)
(b) Use of residues as energy source, no credit for feedstock		1.15	0.98	1.10
Net-to-gross energy ratio		**0.13**	**<0**	**0.09**
(c) No residues as energy source, credit for feedstock use		0.65	1.51	(*)
Net-to-gross energy ratio		**<0**	**0.34**	(*)
(d) No residues as energy source, no feedstock credit		0.58	1.16	(*)
Net-to-gross energy ratio		**<0**	**0.14**	(*)

of biofuel, due to the feedback loop discussed above. Much to our surprise, the biodiesel option performs even worse than the bioethanol option, in spite of the often claimed performance of oilseed crops.

18.2.3 The Energy Return on Investment (EROI)

For an energy process to be feasible, the energy it provides must be higher than the energy it requires. When the energy cost of recovering a barrel of oil becomes greater than the energy content of the oil extracted, production will be discontinued, no matter what the monetary price may be. This requires the definition of the "energy cost" of energy, and the introduction of the so-called EROI (Energy Return on Investment, Cleveland et al., 1984; Cleveland, 2005). (Fig. 18.2)

In short, the EROI is defined as the ratio of the energy that is obtained as output of a given energy extraction process to the total energy that is invested for its extraction, processing, and delivery, including the energy embodied in the goods and machinery used. The lower the EROI, the smaller the net advantage provided by a given energy source. Investing one joule in a source with high EROI, provides a net return of many joules in support of the investor's economy. Fossil sources provided high EROI's in the past, up to 100:1, but values have been declining down to the present 20:1, as shown by Cleveland (2005), due to the exploitation of the most favourable and higher quality fossil reservoirs, and are expected to decrease further. Figure 18.2 also defines the net energy of a source and shows the relation of EROI to the net-to-gross ratio, the latter being the fraction that the net energy is of the total energy delivered by a process to the investor. A net-to-gross ratio lower than one means that a source does not deliver any net energy. Such a ratio can be used as a measure of the ability of a source (or a fuel) to support societal activities. Society needs energy to run economic (agriculture, industry) and service (transportation, education, health sectors, etc) activities. A high EROI allows society to run more activities out of a small investment in the energy sector. When EROIs of energy sources decline, the same gross energy expenditure translates into a smaller net, after subtracting conversion losses and energy investment. Figure 18.3 describes four scenarios of different EROI values. The higher EROI (20:1) characterizes the present situation of fossil fuels, the lower (1.2:1) characterizes the present situation of most biofuels.

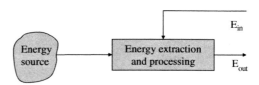

Fig. 18.2 Definition of EROI – Energy Return on Investment

Net Energy$= E_{out} - E_{in}$

EROI $= E_{out}/E_{in}$

Net-to-Gross Ratio $= (E_{out} - E_{in})/ E_{in} = 1 - 1/EROI$

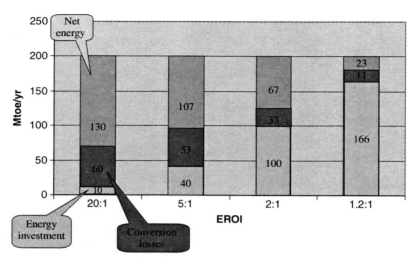

Fig. 18.3 Comparison of the energy investment needed and net energy available for Italy 2004
Note: total energy expenditure of Italy 2004 (200 Mtoe/yr) dealt with according to the assumed use
of four energy sources with different EROIs (Energy Return on Investment). The higher EROI
(20:1) characterizes the present situation of fossil fuels, the lower (1.2:1) characterizes the present
situation of most biofuels

It clearly appears that the net energy available to a society running on biofuels
would be much smaller (23 Mtoe/yr out of 200 Mtoe/yr of gross energy expenditure) and therefore not much would be left to support development and growth. Of
course, it is possible to decrease conversion losses, use resources more effectively,
increase recycling patterns, decrease luxury consumption, reverse population trends,
and still keep a life style at an acceptable level (Odum and Odum, 2001; 2006) even
running on lower EROI sources. However, Fig. 18.3 together with a careful look
at the breakdown of societal energy consumption in the different sectors (health
and education, primary production, transportation) indicates that EROI values lower
than 4:1 are unlikely to support a developed society. Such a threshold value for the
EROI is typical of average renewable energies (solar and wind), but is not typical of
the present biofuel sector.

18.2.3.1 EROI and Biofuels

A biofuel option should therefore provide more energy than is invested, to be
energetically and economically viable, i.e. should have a high EROI and a high
net-to-gross ratio. This is almost never the case with the processes investigated in
this chapter. For example, the output/input energy ratio of bioethanol production
from corn is 0.58, with no positive return in terms of net-to-gross ratio (option d,
Table 18.2). If so, there is no reason for investing in the form of crude oil more
energy than is recovered in the form of ethanol. Improvement of the global efficiency of the process may come from a better use of agricultural and distillation

by-products. Higher EROIs are calculated for alternatives where DDGS and residues are used (respectively 0.65 and 1.15 in Table 18.2). However, only when the two by-product use options, residues and DDGS, are used together as in alternative (a), we get a significant improvement of the EROI up to a value of 1.50. Similar considerations apply to biodiesel, for which the best performing option is option (c), with no residues as energy source, credit for feedstock use, yielding an EROI equal to 1.51. A very low EROI equal to 1.10 is shown by methanol from wood, also by using all available residues as process heat.

Comparison with previous studies confirms our results by providing even worse performances. CCPCS (1991) reported an output/input energy ratio of 1.02 for ethanol from corn in France (country average), without residue use. Marland and Turhollow (1991) calculate an EROI = 1.13 for average USA. Their figure increases up to about 1.27 when an energy credit is assigned for use of coproducts. Shapouri et al. (1995) calculated a value of 1.01 as an average of nine states in the U.S., without any use of co-products. When these Authors assigned an energy credit for DDGS, their average energy ratio increased to 1.24. For comparison, it is worth noting that Giampietro et al. (1997) calculate EROIs in the range 2.5–3.5 (net-to-gross ratio= 0.6/0.7) for Brazilian sugarcane, with bagasse used to supply process heat. This last result is likely to be among the best performances for ethanol production from any crops that have been published.

For a more complete and more up-to-date comparison, it is worth mentioning a study about the production of soybean in Brazil and export to Europe for fuel and feedstock purpose, as a consequences of the recent European directives in matter of biofuels (Cavalet, 2007; Cavalet and Ortega, 2007). The Authors calculated firstly an EROI of 2.30 by allocating a large amount of input energy to soy cakes to be used as animal feedstock, and then a more realistic 1.23 without such an allocation. In fact, when a large production of biofuels is performed in order to meet the required replacement of fossil fuels, the related production of animal feedstock largely exceeds the demand of the livestock sector, so the produced DDGS and oilseed cakes are rather to be considered a waste to be disposed of than an additional useful product.

It is worth noting that there is still large uncertainty about data, conversion coefficients and results with bioenergy production worldwide. Hoogwijk et al. (2003) and Berndes et al. (2003) evaluated the results of 17 earlier studies on the subject and extrapolated a final evaluation of biomass potential up to the year 2050. These authors, who are not in principle negative to bioenergy use, point out that "the main conclusion of the study is that the range of the global potential of primary biomass (in about 50 years) is very broad quantifed at 33-1135 EJy^{-1}." (Hoogwijk et al., 2003). Such a large range indicates how uncertain a biomass based development is. The same authors identify the reasons for the uncertainty by underlining that "crucial factors determining biomass availability for energy are: (1) the future demand for food, determined by population growth and diet; (2) the type of food production systems that can be adopted world-wide over the next 50 years; (3) productivity of forest and energy crops; (4) the (increased) use of bio-materials; (5) availability of degraded land; (6) competing land use types, e.g. surplus agricultural land used for reforestation. It is therefore not "a given" that biomass for energy can become available at a large-scale " (Hoogwijk et al., 2003) and conclude that "the question how an expanding bioenergy sector would interact with other land uses, such as food production, biodiversity, soil and nature conservation, and carbon sequestration has been insufficiently analyzed in the studies. It is therefore difficult to establish to what extent bioenergy is an attractive option for climate change mitigation in the energy sector" (Berndes et al. 2003).

18.2.4 The Claim for Renewability

Table 18.3, based on the approach of eMergy synthesis (Odum, 1996; Brown and Ulgiati, 2004), looks at biofuels from another point of view, their global renewability. EMergy measures the direct and indirect environmental support to the process generating a given output. That is, it assesses solar and solar-equivalent flows of available energy invested over the whole chain of transformations leading to the final product. The eMergy intensity of a product (so-called transformity, or specific eMergy) is therefore a measure of the ecological renewability of that product, i.e. how much it takes in terms of embodied time and space to make the product

Table 18.3 Solar transformities of selected fuels and biofuels. (figures also include the eMergy associated to labor and services) – Section 18.2.4

Fuel	Transformity (sej/J)	Reference
Coal	6.70E+04	(Odum et al., 2000)
Natural Gas	8.04E+04	(Odum et al., 2000)
Crude oil	9.05E+04	(Odum et al., 2000)
Refined fuels (petrol, diesel, etc)	1.11E+05	(Odum et al., 2000)
Hydrogen from water electrolysis ($^\circ$)	1.39E+05	(Brown and Ulgiati, 2004)
Hydrogen from steam reforming of natural gas	1.93E+05	(after Raugei et al, 2005)
Hydrogen from water electrolysis (*)	4.04E+05	(Brown and Ulgiati, 2004)
Methanol from wood	2.66E+05	This work
Bioethanol from corn	1.89E+05	This work
Ethanol from sugarcane	1.86E+05–3.15E+05	Ulgiati, 1997
Biodiesel	2.31E+05	This work
Electricity from renewables (§)	1.10E+05–1.12E+05	(Brown and Ulgiati, 2004)
Electricity from fuel cells natural gas powered	2.18E+05–2.68E+05	(after Raugei et al, 2005)
Electricity from thermal plants (#)	3.35E+05–3.54E+05	(Brown and Ulgiati, 2004)

($^\circ$) using wind- and hydro-electricity
(§) wind and hydro
(*) Using coal and oil powered thermoelectricity
(#) coal and oil powered thermal plants

Note: Transformities have been recently revised, based on a recalculation of energy contributions done in the year 2000 by Odum et al. (2000). Prior to 2000, the total emergy contribution to the geobiosphere that was used in calculating emergy intensities was 9.44×10^{24} seJ/yr. Adopting a higher global emergy reference base – 15.83×10^{24} seJ/yr – changes all the emergy intensities which directly and indirectly were derived from it. This explains a slight difference with values previously published.

available. A careful look at Table 18.3 shows that the transformities calculated for biofuels are never lower than those for fossil fuels. Biofuel transformity values are in the same range as electricity and hydrogen from fossil fuel powered plants. This simply indicates that, since biofuels are produced via multi-step processes all characterized by conversion losses and supported by non-negligible amounts of fossil fuels, they share the same non-renewable characteristics as other fossil fuel powered processes. Actually according to this index they are even performing worse than fossil fuels themselves.

18.3 Physical Constraints Other than Energy

18.3.1 Land and Water Constraints

The available amount of arable land and water are usually neglected in most analyses. To feed people adequately about 0.5 ha of arable land per capita is needed (Lal, 1989), yet only 0.27 ha per capita worldwide (WRI, 1994) and 0.25 ha per capita in Italy are available (ISTAT, 2007). The world population increase and the parallel increase of land erosion and degradation are not likely to help solve food shortages and malnutrition. Intensive agriculture is undoubtedly increasing soil erosion worldwide (Pimentel et al., 1995). Crop yields on severely eroded soil are lower than those on protected soils because erosion reduces soil fertility and water availability, infiltration rates, water-holding capacity, nutrients, organic matter, soil biota, and soil depth (OTA, 1982, 1993; El-Swaify et al., 1985; Troeh et al., 1991). Cropping for energy will compete with arable land use for food production. Available arable land is already a scarce resource. Worldwide, only Canada, USA, Argentina and France are able to export significant amounts of cereals (Giampietro et al., 1997). Wackernagel and Rees (1996), after introducing their "ecological footprint" concept, calculated that only Canada and Australia have footprints that exceed their endowment of ecologically productive land. Cropping marginal or set aside lands for fuel would negatively affect wildlife (one of the main reasons set aside policies have been introduced) and would provide lower yields due to lower productivity of marginal lands and higher energy demand for cultural practices. However, even if competition with food were not taken into account, in the hope that better yields or genetic improvements could help solve this problem, the need of high biomass yields for efficient biofuels production would cause an additional pressure on land and accelerate the process of soil erosion and depletion. Topsoil formation by natural processes is a very slow process, and organic matter in soil should be considered a nonrenewable resource.

Table 18.2 shows that $5.1 \, m^2$ of land are needed in Italy to yield 1 kg of ethanol and $11.5 \, m^2$ per kg of biodiesel. Total energy use in the transport sector in Italy is about 44.4 million tons of oil equivalent per year (ISTAT, 2007), i.e. about 31% of the overall energy use in the country. How much land is actually available in Italy for biofuel production? A careful look at Table 18.4 offers a clear picture of the

Table 18.4 Land used for agriculture in Italy, 2004 – Section 18.3.1

	Land (Thousand ha)	%	Production (thousand tonnes)	Yield (t per ha)
Cereals	4,276	30	23,596	5.5
−Wheat	2,354	17	8,777	3.7
Leguminous	71	0	140	2.0
Tubers	74	1	1,885	25.5
Open air vegetables	473	3	14,101	29.8
Greenhouse vegetables	34	0.2	1,585	46.1
Fruit trees	445	3	6,200	13.9
Olive trees	1,135	8	4,678	4.1
Vineyard	787	6	8,973	11.4
Citruses	168	1	3,531	21.0
Temporary fodder plants	2,019	14	59,654	29.6
Perennial fodder plants	4,205	30	19,321	4.6
Industrial cultivations	498	4	23,166	46.5
- Rapeseed	3	0	5	1.8
- Sunflower	124	1	278	2.2
TOTAL	14,185	100	166,829	

Source: ISTAT, 2007

agricultural land allocation in Italy. Cereals account for 30% of total arable land, while temporary and perennial fodder plantations globally account for 44% thereof. Stable agricultural uses for citruses, fruit and olive trees as well as vineyard account for 18%. These plantations, which require decades and large investments for establishment and full production, are unlikely to change even under the pressure of higher income promises from cropping for biofuels. All other crops only account for an additional 8%. In order to meet the required 5.75% biofuel replacement required by the EU by the year 2010, not less than 2.5 million ha are needed, i.e. about the 17.6% of total arable land. Such a figure can only be understood in its full meaning if we consider that:

a. our calculations are based on best available agricultural yields, while instead Table 18.4 shows average nationwide yields for cereals and oilseed crops smaller than the ones we used;
b. Italy imports food and meat from outside, including large amounts of cereals and oilseed crops.

It is impossible to think of changing the present land use in favour of fuel crops. Actually, this is happening in some parts of Italy and Europe and is already generating an increase of the price of food crops, as we will see in more details later on in this paper. Marginal lands, often claimed to be available for energy cropping, do not provide any significant return in terms of yield, income and energy. They are very often abandoned due to lack of water, small fertility, high erosion, distance from markets, etc., and they are unlikely to be returned to full production, in spite of claims of bio-industry supporters. However, the best estimates of marginal land in Italy indicate about 3 million hectares of available land (Nebbia, 1990). Such optimistic assumption, if validated, would only cover the requirement for a 5.75% replacement and would never be enough to cover any significant fraction of the country's energy demand.

Finally, direct water demand (most of which is used in the agricultural phase and only a small fraction thereof in the industrial conversion phase) is 4.8 kg per g of ethanol produced and 2.8 kg water per g of biodiesel produced (Table 18.2). Under the same assumptions used for land demand (i.e. 5.75% of the total energy used for transportation in Italy replaced by biofuels), we would have an additional direct water demand of about 14.5 billion m^3 of water. This additional water demand would be about 5% of total annual rainfall water, to be diverted from other uses towards cropping for fuel. Water issue is already a strategic issue in Italy, due to competing uses for agriculture and industry, and it is projected to become even more crucial at both national and European levels in the near future. It is therefore not easy to think of increased water demand for energy cropping.

18.3.2 Carbon Dioxide Emissions and the Global Warming Constraint

The climate debate has been particularly rich in the last three decades. Although it is not the goal of this paper to go into the details of such topic, we can at least provide a short evaluation of the possible advantages in this regard. Fossil fuels release carbon dioxide when they are processed and when they are used. Instead, release of carbon dioxide from biofuel production is claimed to be zero, due to the photosynthetic activity of plants. Therefore, we did not include CO_2 emission from biofuel use in our evaluation. Of course in order to meet global warming concerns, release of carbon dioxide due to biofuel production should at least be lower than that from an equivalent amount of fossil energy used in the transportation system.

Carbon dioxide emissions associated to biofuel production occur during biomass production and during its conversion to fuel. A fraction of the emissions in the agricultural phase is due to soil oxidation. Fertile topsoil (i.e. the upper 0.2–0.4 m of soil layer) typically contains about 100 tonnes of organic matter (or 3–4% of total soil weight) per hectare (Medici and Martinelli, 1963; Follet et al., 1987; Triolo et al., 1984; Triolo, 1988). Organic matter is mostly stored in soil layers close to the surface. Excess soil tilling and soil erosion by rain and wind bring organic matter in contact with atmospheric oxygen. Subsequent oxidation of organic matter (like fuel combustion) will cause a release of CO_2 into the atmosphere. Assuming that: (a) 3% of topsoil is organic matter, (b) 70% of organic matter is water, and that (c) dry organic matter oxidation releases roughly 3 grams CO_2/g of oxidized organic matter; the mass of CO_2 that is released per gram of soil eroded is therefore:mass of CO_2 (grams) = 0.03 * (1-0.70) * 3 = 0.027 g of CO_2 per gram of soil.

Typical soil erosions for industrialized corn production are in the range of 13–17 tonnes per hectare per year (17 tonnes/ha in Tuscany, Italy, according to Magaldi et al., 1981; 13 tonnes/ha in the US, according to USDA, 1993; 1994), equivalent to a CO_2 emission of 0.35–0.46 tonnes per ha per year, the same amount that would be released by the combustion of 110–145 kg of petrol. Soil erosion is affected by many different factors (characteristics of crop, soil slope, soil quality, cultural techniques, wind and rain, etc.) and wide ranges in soil erosion data are reported in the literature. Therefore, our estimate only aims at providing a reliable order of magnitude.

Table 18.2 shows that 1.5–3.2 g CO_2 would be released per g of net biofuel produced, including process energy and CO_2 from soil oxidation. Making the system fully independent of fossil fuels by reinvesting a fraction of bioenergy in the process would decrease the related emissions of carbon dioxide but would increase the carbon dioxide released from topsoil oxidation. In fact, the larger area needed to make the system fossil-fuel independent would partially offset the decreased emissions from less fossil fuels use. We estimated that CO_2 emissions from topsoil oxidation is about 10–20% of total CO_2 emissions from a fossil fuelled biofuel making process. This would translate into a release of up to about 50% of the present CO_2 emissions for a fossil-free process, and would still provide a net global warming advantage. No advantage, however, would result for all other less favourable assumptions. The benefit of a lower carbon dioxide release decreases accordingly and may become unimportant, unless a significant improvement of the net-to-gross ratio is achieved.

18.4 The Large-Scale Picture. An Overview of Substitution Scenarios

Net-to-gross ratios previously calculated are crucial for the construction of Table 18.5a, where two options for bioenergy supply to the Italian transportation sector are discussed. Table 18.5b is a list of selected parameters used in the calculation of the scenarios shown in Table 18.5a.

A low net-to-gross ratio would amplify the demand for arable land, irrigation and process water, among other factors, due to the internal loop required to make the system self-sufficient. The best EROIs calculated in our study (Table 18.2) are 1.50 for bioethanol, with residues used as process energy source and energy credit assigned to DDGS, and 1.51 for biodiesel, with energy credit given to oilseed cakes. This corresponds to a net-to-gross ratio equal to 0.33–0.34. It would have the consequence that three liters of biofuel must be produced per litre delivered to society, if we foresee a production process that is independent of fossil fuels input. This would make the demand for land, water and all other factors three times larger, i.e. put additional strain on resources that are already scarce and insufficient to achieve food security and ensure environmental protection worldwide.

A comparison of the environmental consequences of replacing 5.75% of the total petrol and diesel used in Italy respectively by means of bioethanol and biodiesel (either used alone or in blends with fossil fuels) is provided. Columns A, C, and E show the additional   amounts of seeds, land, water, labour, and chemicals, which would be needed to replace respectively 5.75% of petrol, diesel and total transportation fuels, as well as their percent of total present use in Italy. The amount of animal feed generated as by-product for covering 5.75% of Italian petrol is also shown. Fractions of total use calculated in Columns A, C and E are already non-negligible

Table 18.5a Scenarios of substitution of fossil fuels with biofuels –Section 18.4

Replacing-->		A 5.75% of gasoline used nationwide with bioethanol	B Amplification due to the net-to-gross factor	C 5.75% of diesel used nationwide with biodiesel	D Amplification due to the net-to-gross factor	A+C Total both fuels	B+D Total both fuels
Amount of energy replaced	J	6.38E+16	=	4.25E+16	=	1.06E+17	=
Amount of substitute fuel needed	Kg	2.38E+09	=	1.06E+09	=	3.44E+09	=
Amount of seeds needed (w.m.)	Kg	9.24E+09	2.77E+10	2.68E+09	7.91E+09	1.19E+10	3.56E+10
% of 2004 production in Italy		**81.3%**	**243.8%**	**243.8%**	**719.0%**	**95.6%**	**285.8%**
Land demand	Ha	1.22E+06	3.65E+06	1.22E+06	3.59E+06	2.43E+06	7.23E+06
% of arable land available		**9.3%**	**27.8%**	**9.3%**	**27.4%**	**18.5%**	**55.2%**
Water demand	m³	1.15E+10	3.44E+10	3.02E+09	8.91E+09	1.45E+10	4.33E+10
% of 2004 water use in Italy		**131.7%**	**395.1%**	**34.7%**	**102.4%**	**166.4%**	**497.5%**
% of 2004 rainfall in Italy		**4.2%**	**12.7%**	**1.1%**	**3.3%**	**5.3%**	**16.0%**
Labor demand	hours	4.76E+07	1.43E+08	4.24E+07	1.25E+08	9.00E+07	2.68E+08
% of 2004 agric. work force		**2.3%**	**6.8%**	**2.0%**	**5.9%**	**4.3%**	**12.7%**
Release of chemicals	kg	3.57E+08	1.07E+09	3.92E+08	1.16E+09	7.49E+08	2.23E+09
% of 2004 agric. chemicals		**6.5%**	**19.4%**	**7.1%**	**21.0%**	**13.6%**	**40.4%**
Coproducts as livestock feed	kg	5.17E+09	1.55E+10	1.41E+09	4.16E+09	6.58E+09	1.97E+10
% of total 2004 livestock feed		**40.4%**	**121.1%**	**67.1%**	**198.0%**	**44.0%**	**131.5%**

Two options considered:
(a) 5.75% energy replacement.
(b) amplification of inputs due to the need of a sustainable production decoupled from fossil fuel inputs.

Table 18.5b Parameters used for scenarios drawn in Table 18.5a. – Section 18.4

Arable land in Italy, beginning 2004	ha	1.31E+07	ISTAT, 2007
Annual rainfall in Italy, nationwide, 2004	m3	2.71E+11	ISTAT, 2007
Chemicals used in Italian agriculture, 2004	Kg	5.52E+09	ISTAT, 2007
Population of Italy, 2004	#	5.85E+07	ISTAT, 2007
Working hours invested in agriculture, 2004	hours	2.10E+09	ISTAT, 2007
Production of corn in Italy, 2004	Kg	1.14E+10	ISTAT, 2007
Production of oilseeds in Italy, 2004	Kg	1.10E+09	estrapolated from ISTAT, 2007
Present annual use of water in Italy	m3	8.70E+09	ISTAT, 2007
Annual gross energy use in Italy, 2004	J	8.18E+18	(BP Amoco, 2005)
Annual final energy uses in Italy, 2004	J	6.00E+18	ISTAT, 2007
Annual energy used for transport	J	1.85E+18	ISTAT, 2007
Fraction of transport energy that is gasoline	J	1.11E+18	(assumed 60% of total transport)
Fraction of transport energy that is diesel	J	7.40E+17	(assumed 40% of total transport)
H.H.V. of gasoline	J/g	4.40E+04	Boustead and Hancock, 1979
H.H.V. of diesel	J/g	4.48E+04	Boustead and Hancock, 1979
H.H.V. of bioethanol	J/g	2.68E+04	Wyman et al., 1993
H.H.V. of biodiesel	J/g	4.01E+04	Stazione Sperimentale per i combustibili, Milano, 1992
Yield per hectare, bioethanol	Kg/ha	1.96E+03	This work
Yield per hectare, biodiesel	Kg/ha	8.71E+02	This work
Water demand per unit of bioethanol	m^3/Kg	4.81	This work
Water demand per unit of biodiesel	m^3/Kg	2.85	This work
Labor demand per unit of bioethanol	hours/Kg	0.02	This work
Labor demand per unit of biodiesel	hours/Kg	0.04	This work
Demand of chemicals, bioethanol	Kg/Kg	0.15	This work
Demand of chemicals, biodiesel	Kg/Kg	0.37	This work
Coproducts for livestock, bioethanol	Kg/Kg	2.17	DDGS, this study
Coproducts for livestock, biodiesel	Kg/Kg	1.33	Seed oil cakes, this study
Present use of oil seed cakes in Italy	Kg	2.10E+09	estrapolated from ISTAT, 2007
Present use of cereal feed in Italy	Kg	1.28E+10	estrapolated from ISTAT, 2007
Feedstock for animals, used nationwide	Kg/yr	1.49E+10	ISTAT, 2007

and can be source of major concern. Moreover, the non-linear increase of required input associated with the assumption of a self-sufficient production pattern (with no fossil fuel support) is impressive. Although the net-to-gross ratios used in the calculation (columns B, D, and F) are based on a very optimistic process performance (0.33 for bioethanol and 0.34 for biodiesel), a significant increase of input flows and associated emissions compared to the present availability would still be required just to meet a comparatively small 5.75% of present demand of petrol and diesel.

The present Italian energy consumption per person is about 145 GJ/(person*yr). If only 10% of total energy needs should be replaced by biofuels, then four times the actually available arable land would be needed (assuming no internal food production and disregarding the fact that Italy at present imports already cereals and meat which would require more than half its arable land). Similar very worrying considerations apply to soil erosion, water, land and labour demand, etc., as well as to the larger European scale.

In the case of methanol from short-rotation wood, the situation is even worse, due to a net-to-gross ratio equal to 0.09. This represents less than one third of the value found for biofuels from corn and sunflower. Trying to achieve a higher wood productivities per hectare would require a larger input of fertilizers and pesticides, something that again would further decrease both the energy ratio and the net-to-gross ratio. The land demand of SRWC (Short Rotation Wood Crops, monoculture of trees) is expected to be about 0.03 ha/net GJ of methanol in the near future (although it is still 0.06 ha/MJ in the case study considered in this paper). To cover 10% of the 140 GJ consumed per capita in Italy, 0.5 ha per capita of non-arable land (forests and marginal land) should be converted to SRWC. This would translate into a demand for 29 million ha to be converted into monocultures of trees, i.e. a little less than the whole surface of Italy.

18.5 Discussion

18.5.1 The Potential Contribution of Biofuels to the Reduction of Urban Pollution

Biofuels are most often presented as a solution for the problem of urban pollution. In fact, many literature studies have shown that automotive engines do produce less polluting emissions when running on bio-ethanol and biodiesel blends vs. regular oil-derived fuels. However, if the aim is to reduce urban pollution, it is important to put these emission reductions into perspective, and compare the results obtainable through the use of biofuel blends vs. other readily-available fuels.

The internal combustion engine, in its two most widespread variants (i.e. "Otto cycle" running on petrol, LPG or NG and "Diesel cycle" running on diesel oil), is responsible for several classes of airborne emission, i.e. carbon monoxide (CO), nitrogen oxides (NOx), sulphur oxides (SOx), volatile organic compounds (VOC), and particulate matter (PM).

Among these, SOx emissions have been reduced dramatically thanks to the introduction of low-sulphur diesel oil, to the point of having become virtually irrelevant in most cases. However, diesel-fuelled vehicles still emit far larger amounts of NOx and PM per km travelled than similarly powerful petrol, LPG or CNG-fuelled vehicles (respectively around 10 and 20 times as much [Beer et al., 2004; Morris et al., 2003]). These two classes of emissions are arguably the two worst offenders in terms of secondary smog formation, and carcinogenic and respiratory disease potential, respectively. This is a fact which ought to always be kept in mind while evaluating the possible effective strategies to try and curb urban pollution.

Based on the available literature, it can be estimated that extensively employing a 10% biodiesel/diesel blend (which would meet the European target for 2020) in diesel-cycle engines would lead to a reduction in urban PM emissions of around 5%, while NOx emissions would remain virtually unchanged (after EPA, 2002). VOC emissions would be reduced by about 10%.

A 10% bio-ethanol/unleaded petrol blend would not significantly change either NOx or PM emissions with respect to a regular petrol-fuelled vehicle (Vitale et al., 2002). The only emission which would be considerably reduced is benzene (−25%); however, this latter gain would be partly counterbalanced by a rather steep increase in acetaldehyde emission (+133%), deriving from the incomplete oxidation of the bio-ethanol. Acetaldehyde is irritating for the eyes and lungs, and, even more importantly, acts as a precursor to secondary-smog pollutants such as the toxic and strongly irritating peroxy-acetyl nitrates (PAN).

From these emission reduction figures, two incontrovertible conclusions can be drawn:

(1) the results in terms of reduction of the most relevant urban polluting emissions which could be obtained by reaching the European target of 10% market penetration for biofuels would be rather modest;
(2) to simply disincentive the use of diesel- (and biodiesel-) fuelled vehicles in urban areas, in favour of Otto-cycle engines running on regular unleaded petrol, or better still LPG or CNG, would be a far more effective political strategy.

18.5.2 Environmental and Social Impacts of a Large Scale Biofuel Production

As opposed to the modest advantages listed in Sections 18.2 and 18.3, the negative impacts of a large scale biofuel production would be very worrying.

In fact, due to their low energy yield, the land requirement of biofuels is very high. In the European *Biomass Action Plan* (Annex 11)[3] it is calculated that in order to achieve the 5.75% energy target (corresponding to around 1.7% of the

---

[3] COM/2005/628 final.

final energy use, since the transport sector accounts for one third of the total energy demand) about 17 million hectares would be needed, i.e. one fifth of the European tillable land.

The consequence would be an enormous increase in the import (and price) of food and feedstock, and therefore a further reduction of the European alimentary sovereignty. Moreover, importing such a large amount of matter would entail a large energy expenses, especially if it is sourced from across the oceans.

For this reason, the most likely scenario is that Europe will be importing most of the biofuels required to reach the objective stated by the European Commission. As a matter of fact, both in the *Biomass Action Plan* and in the *EU Strategy for Biofuels*[4] it is stressed that Europe will promote the production of raw material for biofuels in extra-European countries:

"Biomass productivity is highest in tropical environments and the production costs of biofuels, notably ethanol, are comparatively low in a number of developing countries. [. . .] Developing countries such as Malaysia, Indonesia and the Philippines, that currently produce biodiesel for their domestic markets, could well develop export potential"[5]

It is easily foreseeable that if the world demand for biofuels increased because of agricultural subsidies and other supporting policies, Southern countries would be stimulated to establish large scale monocultures of sugar cane, palm trees and soy for energy production. This means that at least part of the impacts of energy farming would be exported to Southern countries.

In fact, biofuels are not so green as they may appear at first sight. Due to their low yield, intensive agricultural techniques are normally employed, because otherwise the yield would be even lower and consequently the land requirement higher. For this reason, energy farming is mostly carried out in large monocultures, with heavy use of fertilizers, pesticides, and machinery. The consequences are in many cases soil erosion, reduction of wild and agricultural biodiversity, reduction of water availability and quality. Also, a large-scale biofuel production may lead to an increased use of genetically modified organisms (GMOs). In fact, soy, corn and rapeseed are respectively the first, second and fourth most important GMO crops.[6]

18.5.2.1 Alarming Signs

The European Directive, and in general all biodiesel promoting policies, may favour competition for tillable land and an increasing dependency of Southern countries on the international markets for food supply. The resulting reduction in world food availability could be a particularly serious problem in a context of increasing population and energy demand. A recent example is the doubling of corn price that is taking place in Mexico, which left Mexicans without cheap "tortillas" (the basis

---

[4] COM/2006/34 final

[5] COM/2006/34 final.

[6] Clive J., 2005, http://www.isaaa.org.

of their diet). The phenomenon was mainly caused by the growing demand for corn-derived bio-ethanol in the USA (Mexico is a net importer of corn from the USA). The 2007 FAO Food Outlook (FAO, 2007) confirms an alarming trend for food markets. The increased demand of cereals for biofuel programmes is already competing with food use in international markets making price of cereals to rise. Increased cereal prices translate into increased price of all cereal based products (milk, meat, corn based drinks and hundreds of other goods). Farmers also prefer to shift from food to non-food crops, looking for better and easier income sources. As a consequence of increased competition and decreased offer of cereals, corn price in China has increased by 40% and pig meat by 43% in the first 8 months of 2007 (Rampini, 2007). Beer price in Germany is expected to grow by 5–10% as a consequence of decreased offer and increased price of barley (Calabresi, 2007); pasta in Italy is expected to cost about 20% more in autumn 2007 as a consequence of decreased imports of durum wheat from Canada, diverted to bioethanol production (BBC, 2007).

Also, an increase in the world biofuel demand may encourage tropical countries to replace native forests. The European Directive, and in general all biodiesel promoting policies, may incentive plantations of palm trees, whose oil is cheaper than any other source. Palm plantations are responsible for most deforestation in South-Eastern Asia and represent a real threat to the remaining native forests. For example, between 1985 and 2000 in Malaysia palm plantations caused 87% of the total deforestation and a further 6 million hectares will be deforested to make room for palm trees (Monbiot, 2005). Barta and Spencer (2006) pointed out the economic and environmental consequences of the on-going oil palm plantations business in Indonesia and Borneo, providing alarming signals of deforestation, increased carbon emissions to atmosphere, alteration of water-collection areas, destruction of animal habitats and biodiversity. The same might apply to sugarcane plantations in Brazil. Moreover, taking into account the CO_2 emissions due to inter-continental transport and the increase of CO_2 in the atmosphere due to deforestation (forests are CO_2 sinks), the final result might be an overall increase of the greenhouse emissions instead of the desired reduction. In fact, even though the European Union has declared its intention to track the origin of the imported biofuels in order to ensure that they do not derive from unsustainable practices such as deforestation of native forests, it must be realized that such controls are very hard, if not impossible, to put into practice, and are often rather easy to circumvent. Unfortunately, recurrent failure in similar control systems is already happening in the tropical wood sector, where larger and larger extensions of theoretically protected land are being clear-cut to supply the lucrative western markets.

18.5.3 Biofuels and Rural Development

As shown in the above sections, a large scale biofuel production would not contribute much to the reduction of the greenhouse effect, energy dependency and urban pollution. The only remaining sound argument to promote biofuels may then be to support rural development.

This is an even more attractive target now that the European agriculture is becoming a less and less profitable activity from a strictly economic point of view. Market liberalization and globalization is progressively eroding its added value, because the international food markets deliver much cheaper food products than the European farmers could ever do.

However, society considers that the agricultural sector generates more values than the pure economic ones, and for this reason it must be "artificially" kept alive through public subsidies. In fact, agriculture is multifunctional in nature: besides producing food, it protects the landscape, can maintain biodiversity (but only if properly implemented), the rural architectural patrimony and local knowledge. Also, it creates employment, thereby preventing rural depopulation. For these reasons, agriculture needs to be protected from the fluctuations of the global market. The European Union considers the survival of agriculture so important that it assigns approximately 46% of its budget to the Common Agricultural Policy (CAP) (55 billion Euros in commitment appropriations in 2006).

Nevertheless, the CAP is being increasingly criticized because the agricultural subsidies causes unfair competition with Southern countries, besides being too expensive. Biofuels are often presented as a way out of this impasse: subsidizing energy farming for biofuel production would allow supporting European agriculture, without interfering with the international food market and avoiding food over-production.

However, if the objective of biofuel policies is to promote rural development, other options such as for instance organic agriculture may be a better strategy, instead. Like energy farming, organic agriculture is not yet economically competitive with its conventional alternative (oil products in the case of biofuels and intensive agriculture in the case of organic farming), and would probably not survive without a subsidizing scheme. However, organic agriculture provides many much more valuable services to society than biofuels: maintenance of soil fertility, reduction of water pollution, biodiversity protection, landscape improvement, healthier, safer and tastier food. Also, by reducing the use of fertilizers and pesticides, organic agriculture contributes to reducing the energy demand of the agricultural sector.

18.6 Conclusions

The results of a specific case study for Italy as well as a review of other analyses show that biofuels are, essentially, not yet a viable alternative based on economic, energy and environmental aspects. The constraints are not simply technological, but also based on the large scale consequences of biofuel programmes, although improved efficiency in the conversion process and reduced use of fossil fuels in agricultural production might slightly improve the present figures. In particular, when crop production and conversion to fuel are supported by fossil fuels in the form of chemicals, goods and process energy, the fraction of the fuel energy that is actually renewable (i.e. the net energy available) is negligible. On the other side, if a fraction of the biofuel is fed back to the process, in order to make it independent of fossil fuel inputs, the demand for land, water, fertilizers and labour is amplified accordingly, thus increasing the competition with other uses for the same resources. In fact, the growing population of the planet, coupled with the demand for better nutritional quality in developing countries is likely to increase the demand for water and high quality land, even without cropping for energy. Similarly, the decrease of carbon dioxide emissions per unit of fuel delivered is negligible when the process is supported by biofuels in alternative to fossil inputs.

For these reasons, biofuels should not be regarded as a contribution to the solution of the problems related to Europe's strong dependency on fossil fuels. In fact, fossil fuels are used in all phases of the biofuel production chain, with the consequence that the energy yield is very low. Therefore, the real fossil fuel savings of a large scale biofuel production, the reduction of the anthropogenic greenhouse emissions and the increase of energy security would be very modest. Also, urban air quality would not show significant improvements.

As opposed to these small advantages, the disadvantages of a large scale biofuel production in terms of land requirement, environmental impact (deforestation, loss of wild and agricultural diversity, over use and contamination of water, etc.) and economic impact (increase in the price of cereals) would be relevant. Obviously, these considerations do not apply to the recycling of spent oils or agricultural residues, nor to small-scale niche productions, all of which may be good strategies instead. However, it must be realized that the latter will never play a really significant role on a large-scale energy policy.

Pessimistic though the present situation may sound, a margin of hope remains in the advent of second-generation biofuels derived from ligno-cellulosic biomass. In fact, these are expected to raise the energy yield by almost one order of magnitude, thereby increasing the energy and economic revenues and at the same time reducing the requirement for large extensions of land. However, some of the issues discussed above would still apply even to second-generation biofuels. In particular, the risks associated to uncontrolled deforestation of native forests, large water demand and reduction in biodiversity (especially if GMOs are employed) should not be underestimated.

All in all, it appears to be evident that the energy and economic profit of the process is so low as to be unfeasible in nearly all cases. The future acceptance and feasibility of biofuels is very likely to be linked to the ability of clustering biofuel production with other agro-industrial activities at an appropriate scale, in order to take advantage of the potential supply of valuable by-products.

As these strategies are strongly linked to the existence of special conditions (large amounts of available land, high productivity of crops, water availability, etc), biofuels are unlikely to become a generalized solution to the foreseen energy shortages, even if their contribution might become environmentally sound and economically profitable at the local scale, where optimization plays a significant role. If optimization strategies are not carefully designed, intensive exploitation of land is more likely to produce "more uniform green deserts" (Taschner, 1991) rather than to become a sustainable energy source for human societies.

References

AA. VV., 2001. White Paper "European transport policy for 2010: time to decide". http://ec.europa.eu/transport/white_paper/index_en.htm, European Commission, Bruxelles. Last web contact 24 August 2007.

AA.VV., 2005. European Environment Agency. Annual European Community greenhouse gas inventory 1990–2003 and inventory report 2005. http://reports.eea.europa.eu/technical_report_2005_4/en. Last web contact 24 August 2007.

AA.VV., 2006. An EU Strategy for Biofuels. European Commission, Bruxelles. http://ec.europa.eu/comm/agriculture/biomass/biofuel/index_it.htm, Last web contact 24 August 2007.

Bargigli S., Raugei M., and Ulgiati S., 2004. Mass flow analysis and mass-based indicators. In: Handbook of Ecological Indicators for Assessment of Ecosystem Health. CRC Press, 439p.

Barta, P., and Spencer, J., 2006. Crude Awakening. As Alternative Energy Heats Up, Environmental Concerns Grow. Wall Street Journal Online, 5 Dicember 2006. http://online.wsj.com/article_email/SB116501541088338547-lMyQjAxMDE2NjA1NTAwMTU1Wj.html. Last web contact 24 August 2007.

BBC, 2007. Italians facing pasta price rise. By David Willey, BBC News, Rome http://news.bbc.co.uk/1/hi/world/europe/6287850.stm; Last Updated: Tuesday, 10 July 2007, 11:52 GMT 12:52 UK.

Beer T., Grant T., Watson H., and Olaru D., 2004. Life-Cycle Emissions Analysis of Fuels for Light Vehicles. Report HA93A-C837/1/F5.2E to the Australian Greenhouse Office.

Berndes, G., Hoogwijk, M., van den Broek, R., 2003. The contribution of biomass in the future global energy supply: a review of 17 studies. Biomass and Bioenergy 25 (2003) 1–28.

Brown, M.T., and Ulgiati, S., 2004. Emergy Analysis and Environmental Accounting. In: Encyclopedia of Energy, C. Cleveland Editor, Academic Press, Elsevier, Oxford, UK pp. 329–354.

Calabresi, M., 2007. How the rush to biofuels boosts corn prices up (in Italian: Così la corsa al biocarburante impenna le quotazioni del mais). La Repubblica, 19 August 2007, p. 19.

Cavalet, O., 2007. PhD thesis at the University of Campinas, preliminary results. Personal communication to one of the Authors of this paper (Ulgiati).

Cavalet, O. and Ortega, E., 2007. Emergy Analysis of Soybean Production and Processing in Brazil. Paper presented to the 5th International Biennial Workshop "Advances in Energy Studies", Porto Venere, Italy, September 2006. Book of Procedings in press.

CCPCS, Commission Consultative pour la Production de Carburant de Substitution, 1991. Rapport des Travaux du Groupe Numero 1. Paris.

Cleveland, C.J., 2005. Net energy from the extraction of oil and gas in the United States. Energy 30:769–782.

Cleveland, C.J., Costanza, R., Hall, C.A.S., and Kaufmann, R., 1984. Energy and the U.S. economy: a biophysical perspective. Science 255:890–897.

El-Swaify S.A., Moldenhauer W.C., Lo A., 1985. Soil Erosion and Conservation (Soil Conservation Society of America, Ankeny, IA).

EPA, 2002. A Comprehensive Analysis of Biodiesel Impacts on Exhaust Emissions, Draft Technical Report EPA420-P-02-001.

EU, 2003. Directive 2003/30/EC of the European Parliament and of the Council of 8 May 2003 on the promotion of the use of biofuels or other renewable fuels for transport. Official Journal of the European Union. L 123/42, 17.5.2003.

FAO, 2007. Food Outlook. http://www.fao.org/giews/english/fo/. Last contact, 24 August 2007.

Fischer-Kowalski M. 1998 Metabolism: The Intellectual History of Material Flow Analysis Part I, 1860–1970 Journal of Industrial Ecology 2(1):61–78.

Follet R.F., Gupta S.C., and Hunt P.G., 1987. Soil Fertility and Organic Matter as Critical Components of Production Systems. Soil Science Society of America and American Society of Agronomy, Madison, WI.

Giampietro, M., Ulgiati, S., and Pimentel, D., 1997. Feasibility of Large-Scale Biofuel Production. Does an Enlargement of Scale Change the Picture? BioScience, 47(9):587–600, 1997.

Herendeen, R.A., 1998. Embodied Energy, embodied everything ... now what? In: Advances in Energy Studies Energy Flows in Ecology and Economy. Ulgiati S., Brown M.T., Giampietro M., Herendeen R.A., and Mayumi K. (Eds). Musis Publisher, Roma, Italy; pp. 13–48.

Herendeen, R.A., 2004. Energy analysis and EMERGY analysis—a comparison, Ecological Modelling 178 (2004) 227–237.

Hinterberger F. and Stiller H. (1998). Energy and Material Flows. In: Advances in Energy Flows in Ecology and Economy. Ulgiati S., Brown M.T., Giampiero M., Herendeen R.A., and Mayumi K. (Eds). Musis Publisher, Roma, Italy; pp. 275–286.

Hoogwijk, M., Faaij, A., van den Broek, R., Berndes, G., Gielen, D., and Turkenburg, W., 2003. Exploration of the ranges of the global potential of biomass for energy. Biomass and Bioenergy 25 (2003) 119–133

ISTAT-ASI, 2007. Istituto Nazionale di Statistica (National Institute of Statistics), Roma. Annuario Statistico Italiano 2007. (Statistical Yearbook, 2007).

Lal R., 1989. In Food and Natural Resources, D. Pimentel and C.W. Hall, Eds. (Academic Press, San Diego, 1989), pp. 85–140.

Magaldi D., Bazzoffi P., Bidini D., Frascati F., Gregori E, Lorenzoni P., Miclaus N. and Zanchi C., 1981. Studio interdisciplinare sulla classificazione e la valutazione del territorio: un esempio nel Comune di Pescia (Pistoia). Istituto Sperimentale Studio e Difesa del Suolo, Firenze, Italy, Annali vol. XII, pp. 31–114 (in Italian).

Marland G., and Turhollow A.F., 1991. CO2 emissions from the production and combustion of fuel ethanol from corn. Energy, 16 (11/12):1307–1316.

Medici L. and Martinelli E., 1963. Chimica Agraria. Società Editrice Dante Alighieri, Milano. pp. 339.

Monbiot G. (2005). Worse than fossil fuel. The Guardian 12 dicembre 2005. http://www.monbiot.com/archives/2005/12/06/worse-than-fossil-fuel.

Morris R.E., Pollack A.K., Mansell G.E., Lindhjem C., Jia Y., and Wilson G., 2003. Impact of Biodiesel Fuels on Air Quality and Human Health. Report NREL/SR-540-33793 to National Renewable Energy Laboratory, USA.

Nebbia, G., 1990. Alcool carburante, Politica e Economia, III, 21(5):9–10.

Odum H.T., and Odum E.C., 2006. The prosperous way down. Energy 31 (2006) 21–32.

Odum H.T. and Odum E.C., 2001. A Prosperous Way Down: Principles and Policies. 326 pp., University Press of Colorado.

Odum H.T., 1996. Environmental Accounting. Emergy and Environmental Decision Making. John Wiley & Sons, N.Y.

OTA, 1982. U.S. Congress, Office of Technology Assessment. Impacts of Technology on U.S. Cropland and Rangeland Productivity. (Washington, DC: US Government printing Office).

OTA, 1993. U.S. Congress, Office of Technology Assessment. Potential Environmental Impacts of Bioenergy Crop Production-Background Paper. OTA-BP-E-118 (Washington, DC: US Government printing Office, September 1993). 71pp.

Pimentel D., Harvey C., Resosudarmo P., Sinclair K., Kurz D., McNair M., Crist S., Shpritz L., Fitton L., Saffouri R., and Blair R., 1995. Environmental and Economic Costs of Soil Erosion and Conservation Benefits. Science, 267:1117–1123.

Rampini, F., 2007. Beijing. The "pig meat" crisis (in Italian: Pechino, scoppia la crisi del maiale). La Repubblica, 1 June 2007, p. 20.

Schmidt-Bleek F., 1993. MIPS re-visited. Fresenius Environmental Bulletin 2:407–412.

Shapouri H., Duffield J.A., and Graboski M.S., 1995. Estimating the Net Energy Balance of Corn Ethanol. U.S. Department of Agriculture, Economic Research Service, Office of Energy and New Uses. Agricultural Economic Report No. 721, pp. 16.

Taschner K., 1991. Bioethanol: a solution or a new problem. Proceedings of the IX International Symposium on Alcohol Fuels (ISAF). Firenze (Italy), 12–15 November 1991. Vol.3: 922–926.

Triolo L., 1988. Agricoltura Energia Ambiente. Tecnologie meccaniche e chimiche. Consumi e inquinamento. Editori Riuniti. pp. 152.

Triolo L., Mariani A., and Tomarchio L., 1984. L'uso dell'energia nella produzione agricola vegetale in Italia: bilanci energetici e considerazioni metodologiche. ENEA, Italy, RT/FARE/84/12.

Troeh F.R., Hobbs J.A., and Donahue R.L., 1991. Soil and Water Conservation (Prentice-Hall, Englewood Cliffs, NJ).

Turkenburg, W.C. (Convening Lead Author), Faaij, A. (Lead Author), et al., 2000. Renewable Energy Technologies. Chapter 7 in World Energy Assessment of the United Nations, UNDP, UNDESA/WEC. UNDP, New York.

USDA. 1993. Agricultural Statistics. United States Department of Agriculture, Washington D.C.

USDA. 1994, United States Department of Agriculture. Summary Report 1992 National Resources Inventory. Soil Conservation Service, U.S. Department of Agriculture, Washington, DC.

Vitale, R., Boulton, J. W., Lepage, M., Gauthier, M., Qiu, X., and Lamy, S., 2002. "Modelling the Effects of E10 Fuels in Canada". Emission Inventory Conference Emission Inventory Conference, Florida, USA.

Wackernagel M. and Rees W., 1996. Our Ecological Footprint. New Society Publishers.

World Resources Institute (WRI) 1994. World Resources 1994–95. New York: Oxford University Press.

Chapter 19
The Power Density of Ethanol from Brazilian Sugarcane

Andrew R.B. Ferguson

Abstract The power density of ethanol produced from sugarcane in Brazil is about 2.9 kW/ha. That is equivalent to capturing a little more than a thousandth part of solar radiation, and is also a little more than a thousandth part of the power density we are used to from oil and gas. So ineffective is 2.9 kW/ha, that about 5 million ha of land would have to be put down to sugarcane *every year* just to satisfy the increase in transportation energy demand that results from the annual expansion of population in the U.S.A.

Keywords Brazil · sugarcane · ethanol · power density

19.1 Introduction

In an eleven page paper, *Sugarcane and Energy*, the relationship between sugarcane and energy has been covered in considerable detail (Ferguson, 1999); however it may be useful to make available a more concise summary of this essential question: what is the power density of ethanol from sugarcane? The question needs to be asked since one great problem with biofuels is their low power density.

The lack of agricultural potential in the USA to achieve anything significant from biofuels has been superbly demonstrated by Donald F. Anthrop, professor emeritus of environmental studies at San Jose State University, in the *Oil and Gas Journal*, Feb.5, 2007. For instance, he brought up the fact that if the whole of the US corn crop were to be devoted to producing ethanol from corn, this would satisfy only 11.5% of gasoline demand in the US. Note, too, that the reference is to gasoline, and since gasoline represents about half of transportation fuels, it could also be said that the ethanol produced would satisfy only about 6% of transport fuel. My thanks go to Walter Youngquist for sending me this important paper.

Donald Anthrop did not cover sugarcane, and since the 'energy fantasists' are not easily brought to see reality, some will doubtless hold on to the hope that the

✉ A.R.B. Ferguson
11 Harcourt Close, Henley-on-Thames, RG9 1UZ, England
e-mail: andrewrbferguson@hotmail.com

D. Pimentel (ed.), *Biofuels, Solar and Wind as Renewable Energy Systems*,
© Springer Science+Business Media B.V. 2008

supposedly huge unused acres of Brazil can come to the rescue. Thus a look at the power density of ethanol from sugarcane would appear to be timely.

As with all liquid biofuels, there are various power densities which could be assessed:

a) The calorific value of the ethanol produced each year per hectare of land.
b) The calorific value of the 'useful' ethanol produced each year per hectare of land, that is *after* subtracting the portion of ethanol that is needed for input into the agricultural and production processes.
c) The calorific value of the ethanol and by-products produced each year after subtracting the calorific value of *all* the inputs. This is the *net* energy capture (or *net* power density).

Choice (c) might seem to be the most revealing analysis, but there are both practical and almost philosophical questions about how to assess the inputs, particularly: (1) to what extent it is misleading to subtract the calorific value of non-liquid inputs from the calorific value of liquid outputs; and (2) what value should be assigned to by-products, especially when some of the by-products could be used to improve soil fertility and prevent erosion.

Albeit at the cost of being potentially misleading, the type (b) analysis gets around that, and so is a useful starting point, but it requires an assessment of the liquid inputs needed, for which data are not always available.

Although using corn (maize) as feedstock to produce ethanol differs in several important respects from using sugarcane, there is bound to be a degree of similarity in the amount of *liquid* inputs needed as a fraction of the *total* inputs. So as a guide, let us look at a statement in Shapouri et al., 2002:

> As discussed earlier, some researchers prefer addressing the energy security issue by looking at the net energy gain of ethanol from a liquid fuels standpoint. In this case, only the liquid fossil fuels used to grow corn and produce ethanol are considered in the analysis. On a weighted average basis, about 83% of the total energy requirements come from non-liquid fuels, such as coal and natural gas.

That is clearly a statement of method (b) above, and it implies that 17% of the inputs need to be in liquid form. However, we should not take corn as being too accurately aligned with sugarcane in this respect, so I build in a 3% error margin, and assume that only 14% of the total inputs needs to be in liquid form.

To establish the power density of sugarcane I have, with the kind permission of David Pimentel, reworked the tables on pages 238–239 of *Food, Energy, and Society* (Pimentel and Pimentel, 1996), which refer to sugarcane production in Brazil, updating the yield to the latest average yield which is being achieved over 5.2 million hectares of sugarcane. From Table 19.2 we have the answer to our question. It is that the power density achieved in producing ethanol from sugarcane in Brazil is about 2.9 kW/ha—but that is on the very lenient measure of accounting only for the liquid inputs.

Table 19.1 Average energy inputs and output per hectare for sugarcane in Brazil

	Quantity/ha	10^3 kcal/ha
Inputs		
Labor	210 hr	157[a]
Machinery	72 kg	1,944
Fuel	262 liters	2,635
Nitrogen (ammonia)	65 kg	1,364
Phosphorus (triple)	52 kg	336
Potassium (muriate)	100 kg	250
Lime	616 kg	192
Seed	215 kg	271
Insecticide	0.5 kg	50
Herbicide	3 kg	300
Total		**7,499**
Output		
Sugarcane (fresh)	71,400 kg[b]	

One thing to note is that sugarcane is usually grown in sunny areas, so the insolation would be around 2200 kW/ha, so the energy capture is only a little more than 0.1% of insolation, that is a bit more than 1 part in a thousand. This is very relevant in the

Table 19.2 Inputs to transform 71,400 kg of Brazilian sugarcane (fresh) to ethanol

	Quantity/ha	10^3 kcal/ha	
Inputs			
Sugarcane (fresh) as per Table 19.1	71,400 kg	7,499	
Transport	71,400 kg	994	
Water	482,140 kg	270	
Stainless steel[c]	12 kg	174	
Concrete[c]	31 kg	58	
Bagasse (fresh) [d]	21,340 kg	38,760	
Pollution	–	–	
Total		**47,755**	
Gross output of ethanol = 5,525 liters =		28,343	
Liquid inputs = 47,755 × 0.14 =		6,686	
So output of 'useful' ethanol		$\overline{21,657}$	= 4,222 liters ethanol/ha/yr.
So power density = 21,657,000 kcal/ha/yr = 90.7 GJ/ha = **2.9 kW/ha**			

[a] There is some debate as to whether the energy associated with the labor input should reflect the lifestyle of the laborers, but that is not germane to this analysis.

[b] The original tables were associated with 54,000 kg of sugarcane. No increase in inputs have been introduced into Table 19.1, and the only items that have been proportionately increased in Table 19.2, to allow for the 71,400 kg of sugarcane, are transport and the heat provided by the bagasse.

[c] The embodied energy associated with these raw materials are amortized over their lifetime.

[d] The calorific value of fresh bagasse is 1816 kcal/kg (see Ferguson, 1999), which is used to calculate the weight. Bagasse is a by-product and is used to produce the heat needed for the transformation process, thus arguably its energy content need not be included in an input/output analysis. It is relevant here anyway because it helps in the assessment of the required liquid inputs.

context of the fact that 'energy fantasists' like to dwell at length on the amount of solar power that is available, as though we are likely to capture much of it.

It is not easy to conceive of the paucity of 2.9 kW/ha. Another useful way to look at the matter is to consider that while it is hard to measure the power density of oil and gas, it is clear that the figures are numerically in the region of solar insolation in the United States, that is about 2000 kW/ha. So capture of sunlight in the form of ethanol achieves a power density that is once again only a bit more than a thousandth part of what we are used to enjoying while oil and gas are available.

A further point of reference is to consider how much land would be needed to provide the burgeoning U.S. population with liquid fuel using ethanol from sugarcane. Dividing transportation fuels by the number of citizens, each American uses, on average, about 3 kW of fuel for transportation (out of a total energy use of about 10.5 kW). Virginia Abernethy (2006) has pointed out that the Census Bureau greatly undercounts the extent of illegal immigration, and that the correct figure for the growth of the U.S. population is between 4.7 and 5.7 million per year. Taking a central figure of 5.2 million, since each American would need $3/2.9 = \mathbf{1.03}$ ha to provide transport fuel from ethanol, there would be a need for an additional 1.03×5.2 million, say 5 million hectares to be put down to sugarcane *every year*, just so as to keep pace with the expansion in population. It is clear that even borrowing land freely from Brazil this becomes impossible within a decade.

There is also this moral question: will conscience allow us to satisfy the motoring public this way when the WHO assesses that 3700 million are suffering from malnutrition and over 800 million from hunger? Not everyone will be as unconcerned about that as President George Bush, who in his State of the Union address called for a 20% cut in gasoline consumption by 2017 and indicated that biofuels would provide a substantial part of the solution. Yet surely his advisers told him that the power density of ethanol from corn, assessed on the same basis as above, is lower than for sugarcane, being about 2776 liters of ethanol/ha/yr = 59.0 GJ/yr = $\mathbf{1.9}$ kW/ha (see OPTJ 3/1, p. 12 for more detail), and other biofuels have even lower power densities (excepting sugarcane). Biofuels can hardly be regarded as even part of the answer when, as we have seen, the growth of biofuels could not match the growth in U.S. population. Insofar as that attempt is made, it will continue to increase the cost of food. Donald Anthrop showed that to be happening, with figures that illustrated a 94% increase in the contract price for corn, between March 2006 and March 2007.

19.2 Errors and the Potential for More Relating to Sugarcane

The subject of sugarcane seems to abound in substantial errors, and perhaps the 'energy fantasists' cling on to them. It may be the very high moisture content of sugarcane (about 70%) which causes confusion. Anyway information sources which are otherwise reliable contain gross errors both about ethanol from sugarcane and sugarcane itself.

The most egregious must surely be that in an old book *Biological Energy Resources*, 1979, by Malcolm Slesser and Chris Lewis. Several times it is repeated therein that the yield of ethanol from sugarcane is about 17 tonnes per hectare per year. That would be 457,300 MJ = **21,520** liters of ethanol. Because Brazil is the place where the 'energy fantasists' assume there are boundless hectares of potential sugarcane land, we have taken Brazil as an example, but even with a high yield of 88 tonnes of sugarcane per hectare, as might be obtained in Louisiana, the ethanol yield would only be about **6290** liters.

Regarding sugarcane itself, Howard Hayden, in the revised edition of his book *The Solar Fraud*, page 242, states that the power density of "Sugar cane (whole plant, tropical conditions, plenty of fertilizer and pesticides)" is **37** kW/ha. That is far too high. Once again taking the high yield of 88,000 kg of fresh sugarcane, the calorific value would be about 88,000 × 1212 kcal/kg = **107** million kcal/ha/yr = 446 GJ/ha/yr = **14** kW/ha. The figure is easy to cross-check, as 88,000 kg at 70% moisture content would contain 26,400 kg of dry matter, and as dry matter has an energy content in the region of 4180 kcal/kg, the calorific value must be in the region of 110 million kcal.

A hope which lingers around (so far only a potential error) is that the by-product bagasse is so plentiful that it can not only provide the heat needed to carry out the distillation processes but also contribute large amounts ('energy fantasists' steer clear of giving actual figures!) of heat for providing electricity. That too has now been quantified, and amounts to only 0.1 kW(e)/ha. Clearly that is hardly significant, and anyhow it is doubtful that the bagasse should be put to that purpose, as the next section makes clear.

19.3 Soil Erosion Problems

It will be noted from Table 19.2 that the heat value of the bagasse used to effect the transformation of the sugarcane to ethanol amounts to about 1.8 times the amount of useful ethanol produced. So it is true to say that the only reason that producing ethanol from sugarcane is not a very substantial energy loser is that the heat can be provided by the bagasse instead of from fossil fuels. However it is doubtful that much of the bagasse should be so used if the sugarcane production is to be truly sustainable, for one dire problem with sugarcane is its tendency to cause soil erosion (Pimentel, 1993). That is a matter of considerable importance to which we will now turn.

Corn has a total yield of around 15 dry tonnes, half being grain and half stover (Pimentel and Pimentel, 1996, p. 36). With reference to corn, David Pimentel has continually stressed the problems arising from soil erosion, and the need to keep all the stover on the ground to maintain the fertility of the soil. Thus in the case of corn about the maximum biomass that should be removed permanently is 7.5 dry t/ha/yr. The Brazilian sugarcane we are considering has an average yield of 71.4 t/ha/yr fresh which is 21 t/ha/yr dry. To remove no more dry matter than recommended for corn, 14 dry t/ha/yr (47 tonnes fresh) of sugarcane biomass should be either left on the soil or returned to it. Also common sense dictates that it is not sustainable to remove 21 dry tonnes of biomass from the land each year without sooner or later causing soil impoverishment and erosion.

We can conclude that while it is possible to deliver a 'useful' 2.9 kW/ha as liquid fuel from Brazilian sugarcane, there would need to be considerable 'external' inputs to replace the heat provided by the bagasse if the process is to be made sustainable by maintaining soil quality and preventing soil erosion. While that is not relevant to the uncontentious power density calculations of this paper, it does remind us that the simplified calculation of power density made here—so as to escape the more philosophical points of *net* energy—does not paint the full dismal picture of the great difficulty of producing liquid fuels sustainably.

References

Abernethy, D.V. 2006. *Census Bureau Distortions Hide Immigration Crisis: Real Numbers Much Higher*. Population-Environment Balance.

Anthrop, D.F. 2007. Limits on energy promise of biofuels. *Oil and Gas Journal*, Feb.5, 2007 (pp. 25–28).

Ferguson, A.R.B. 1999. *Sugarcane and Energy*. Manchester: Optimum Population Trust. 12pp. Archived at www.members.aol.com/optjournal/sugar.doc

Hayden, H.C.. 2004. *The Solar Fraud: Why Solar Energy Won't Run the World* (2nd edition). Vales Lake Publishing LLC. P.O. Box 7595, Pueblo West, CO 81007-0595. 280pp.

OPTJ 3/1. 2003. *Optimum Population Trust Journal*, Vol. 3, No 1, April 2003. Manchester (U.K.): Optimum Population Trust. 32 pp. Archived on the web at www.members.aol.com/ optjournal2/optj31.doc

Pimentel, D. (Ed.) 1993. *World Soil Erosion and Conservation*. Cambridge (UK): Cambridge Uni. Press.

Pimentel, D. and Pimentel, M. 1996. *Food, Energy, and Society*. Niwot Co., University Press of Colorado. 363 pp. This is a revised edition; the first edition was published by John Wiley and Sons in 1979.

Shapouri, H., Duffield, J.A., and Wang, M. 2002. *The Energy Balance of Corn Ethanol: An Update*. United States Department of Agriculture (USDA), Agricultural Economic Report Number 813.

Slesser, M. and C. Lewis. 1979. *Biological Energy Resources*. London: E. & F.N. Spon Ltd.

Chapter 20
A Brief Discussion on Algae for Oil Production: Energy Issues

David Pimentel

Abstract Further laboratory and field research is needed for the algae and oil theoretical system. Claims based on research dating over three decades have been made, yet none of the projected algae and oil yields have been achieved. Harvesting the algae from tanks and separating the oil from the algae, are difficult and energy intensive processes.

Keywords Algae · biomass · energy · harvesting algae

The culture of algae can yield 30–50% oil (Dimitrov, 2007). Thus, the interest in the use of algae to increase U.S. oil supply is based on the theoretical claims that 47,000–308,000 liters/hectare/year (5,000–33,000 gallons/acre) of oil could be produced using algae (Briggs, 2004; Vincent Inc., 2007). The calculated cost per barrel would be only $20 (Global Green Solutions, 2007). Currently, a barrel of oil in the U.S. market is selling for over $100 per barrel. If the production and price of oil produced from algae were true, U.S. annual oil needs could theoretically be met, but only if 100% of all U.S. land were in algal culture!

Despite all the claims and research dating from the early 1970's to date, none of the projected algae and oil yields have been achieved (Dimitrov, 2007). To the contrary, one calculated estimate based on all the included costs using algae would be $800 per barrel, not $20 per barrel previously mentioned. Algae, like all plants, require large quantities of nitrogen fertilizer and water, plus significant fossil energy inputs for the functioning system (Goldman and Ryther, 1977).

One difficulty in culturing algae is that the algae shade one another and thus there are different levels of light saturation in the cultures, even under Florida conditions (Biopact, 2007). This influences the rate of growth of the algae. In addition, wild strains of algae invade and dominate the algae culture strains and oil production by the algae is reduced (Biopact, 2007).

✉ D. Pimentel

College of Agriculture and Life Sciences, Cornell University, 5126 Comstock Hall, Ithaca, NY 15850

e-mail: Dp18@cornell.edu

Another major problem with the culture of algae in ponds or tanks is the harvesting of the algae. Because algae are mostly water, harvesting the algae from the cultural tanks and separating the oil from the algae, is a difficult and energy intensive process. This problem was observed at the University of Florida (Gainesville) when algae were being cultured in managed ponds for the production of nutrients for hogs (Pimentel, unpublished 1976). After two years with a lack of success, the algal-nutrient culture was abandoned.

The rice total yield is nearly 50 tons/ha/yr of continuous culture and this includes both the rice and rice straw (CIIFAD, 2007). The best algal biomass yields under tropical conditions is about 50 t/ha/yr (Biopact, 2007). However, the highest yield of alga biomass produced per hectare based on theoretical calculations is 681 tons/ha/yr (Vincent Inc., 2007). Rice production in the tropics can produce 3 crops on the same hectare of land per year requiring about 400 kg/ha of nitrogen fertilizer and 240 million liters of water (Pimentel et al., 2004).

Obviously, a great deal of laboratory and field research is needed for the algae and oil theoretical system.

References

Biopact. (2007). An in-depth look at biofuels from algae. Retrieved January 7, 2008, from http://biopact.com/2007/01/in-depth-look-at-biofuels-from-algae-html

Briggs, M. (2004). Widescale biodiesel production from algae. Retrieved January 7, 2008, from http://unh.edu/p2/biodielsel/article_alghae.html

CIIFAD. (2007). More rice with less water through SRI – the System of Rice Intensification. Cornell International Institute for Food, Agriculture, and Development Retrieved January 7, 2008, from http://ciifad.cornell.edu/SRI/extmats/philmanual.pdf

Dimitrov, K. (2007). GreenFuel technologies: a case study for industrial photosythetic energy capture. Brisbane, Australia. Retrieved January 7, 2008, from http://www.nanostring.net/Algae/CaseStudy.pdf

Global Green Solutions. (2007). Renewable energy. Retrieved January 7, 2008, from http://www.stockupticks.com/ profiles/7-26-07.html

Goldman, J.C. and Ryther, J.H. (1977). Mass production of algae: bio-engineering aspects. (In A. Mitsui et al. (Eds.), *Biological Solar Energy Conversion*. (pp. 367–378). New York: Academic Press.)

Pimentel, D., Berger, B., Filiberto, D., Newton, M., Wolfe, B., Karabinakis, B., Clark, S., Poon, E., Abbett, E., and Nandagopal, S. 2004. Water resources: Agricultural and environmental issues. *Bioscience* 54(10): 909–918

Vincent Inc. 2007. Valcent Products. Initial data from the Vertigro Field Test Bed Plant reports average production of 276 tons of algae bio mass on a per acre/per year basis. Retrieved January 7, 2008, from http://money.cnn.com/news/newsfeeds/articles/marketwire/0339181.htm

Index